PLANTS OF THE
SAN FRANCISCO BAY REGION

MENDOCINO to MONTEREY

EUGENE N. KOZLOFF

AND

LINDA H. BEIDLEMAN

Sagen Press
Pacific Grove, California

Library of Congress Catalog Card Number 94-68921

ISBN 0-9643756-0-5

First printing, November, 1994

Printed by Alan Lithograph, Inc., Inglewood, CA 90302

Published by Sagen Press, P. O. Box 51042, Pacific Grove, CA 93950
Copies of this book (with tax and postage paid) are available from Sagen Press for $35.00.

CONTENTS

PREFACE

This book has been written for amateur naturalists, students, professional biologists, and others who need an easy-to-use guide to the flora of the San Francisco Bay Region. It covers approximately 2027 species, subspecies, and varieties of higher plants. About 470 of these are aliens that have become firmly established since their accidental or willful introduction; the rest are native.

Nearly all of the plants dealt with are included in *A California Flora*, by Munz and Keck (1959), or in the Supplement to this work (1968). *The Jepson Manual: Vascular Plants of California* (1993), which provides an up-to-date treatment of the flora, lists additional native and non-native species that have been reported from the region. We have included most of these.

In addition to a systematic coverage of higher plants, presented in the conventional order--ferns and fern allies, conifers and their relatives, and families of flowering plants--we provide keys for identification of shrubs and broad-leaved trees. Although all species in these last two categories are also included in keys to the families to which they belong, some users of this book will find it convenient to have shrubs and broad-leaved trees made more accessible by this treatment.

We expect that many persons will peruse the color photographs and line drawings of individual species at least as often as they refer to the text. Both sets of pictures are arranged in the same sequence as the sections that cover the major natural groups of plants and their families. Botanical manuals generally employ a rather extensive vocabulary. By using plain English whenever possible, we have eliminated much unnecessary terminology.

We are grateful for the encouragement we received while this book was developing. Peggy Grier and Myrtle Wolf were especially supportive. Together with Naomi Giddings and Shirley McPheeters, they also checked some keys with actual specimens, and Peggy proofread some of the manuscript. Richard Beidleman prepared the key to grasses, a difficult and time-consuming task, aided in the construction of several other keys and assisted in proofreading. Josh Price provided computer expertise which certainly speeded up this part of the project and greatly reduced frustration. Scott Ranney, at Alan Lithograph, expertly guided us through the printing process.

Most of our photography was accomplished in habitats where native or introduced species were growing wild. A substantial portion of this work was carried out, however, in the University of California Botanical Garden, Berkeley, where Roger Raiche was generous with information, and in the Botanic Garden of the East Bay Regional Park, where Steve Edwards gave valued help. A few pictures were taken in the collection of native plants at Merritt Community College and in other gardens. Our photographic coverage of the genus *Clarkia* would be far less complete than it is if it were not for the courtesy of Leslie Gottlieb, who provided seeds of several species. Ruth and Robert Athearn and Diane and John Wilkinson generously allowed us to use portions of their sunny gardens for growing wildflowers native to our area.

In constructing and testing keys, extensive use was made of pressed herbarium specimens. Barbara Ertter, John Strother, and the late Lawrence Heckard at the University of California, Berkeley, and Sarah Gage, at the University of Washington, facilitated our use of the collections in their care. Early versions of some of our keys were improved by comments from Barbara and John, as well as from Lincoln Constance and Travis Columbus. Members of the team working on *The Jepson Manual*--especially the late James Hickman, Dieter Wilkin, and Susan D'Alcamo--generously allowed us to examine manuscript material and gave special advice when we asked for it.

To the University of Washington Press we are grateful for permission to use numerous illustrations in the excellent five-volume *Vascular Plants of the Pacific Northwest*, prepared by Hitchcock, Cronquist, Ownbey, and Thompson (1955-1969). Nearly all of the drawings taken from this publication were the work of Jeanne R. Janish. The University of California Press allowed us to use some illustrations from Jepson's *Manual of the Flowering Plants of California*. Other illustrations are from sources that are now in the public domain.

Users of this book will discover that there are many details packed into the space between its covers. While it is tempting to blame errors, omissions, and ambiguities on those who helped us, or on publications from which we drew information, we must say right now that we are personally responsible for all shortcomings that you may encounter. If you will be so kind as to tell us about problems you have discovered, you will have our thanks.

And now Linda and Gene thank each other. Our friendship and collegiality easily survived a few episodes of writer's block and some minor disagreements over construction of keys, general style, and organization. The ways in which we resolved our differences led to many improvements in the book. We hope that you enjoy using it, and that it will intensify your appreciation of the flora of this beautiful region of California.

Eugene N. Kozloff
Linda H. Beidleman

INTRODUCTION

Although much of the San Francisco Bay Region is densely populated and industrialized, many thousands of acres within its confines have been set aside as parks and preserves. Most of these tracts were not rescued until after they had been altered. Construction of roads, modification of drainage patterns, grazing by livestock, and introduction of aggressive weeds are just a few of the factors that initiated irreversible changes in the plant and animal life. Yet on the slopes of Mount Diablo and Mount Tamalpais, in the redwood groves at Muir Woods, and in some of the regional parks, one can find habitats that probably resemble those that were present two hundred years ago. Even tracts that are far from pristine have much that will bring pleasure to persons who enjoy the study of nature.

Visitors to our region soon discover that the area is diverse in topography, geology, climate, and vegetation. Hills, valleys, wetlands, and the seacoast are just some of the situations that will have one or more well defined assemblages of plants.

The counties that constitute the San Francisco Bay Region include those that touch San Francisco Bay. Reading a map clockwise, they are Marin, Sonoma, Napa, Solano, Contra Costa, Alameda, Santa Clara, San Mateo, and San Francisco. This book will also be useful in bordering counties, such as Mendocino, Santa Cruz, and Monterey, because many of the plants dealt with occur farther north, east, and south. Some in fact, are more common outside our region than within it. Nevertheless, it is important to keep in mind that the plants of the bordering counties are not comprehensively covered in this manual.

Scientific Names, Common Names, and Geographic Ranges of Plants

The Jepson Manual will be, for many years to come, the definitive reference on the flora of California. We have therefore followed its scheme of classification and its names for species, subspecies, and varieties.

Latin names in brackets are those used by Munz and Keck in *A California Flora* (1959; Supplement 1968; combined edition 1973). Although we provide common names for most species, these may not be the names used by other authors. Some plants have several perfectly acceptable common names, but we give only one.

Additional information that may be provided in connection with the Latin and common names of a plant consists of the region of origin (if the plant is not a native of California), the geographic range, and the extent to which it is rare or endangered. (*See* Abbreviations for explanation of the symbols used.) In the keys to ferns, trees, and shrubs, the family to which each species belongs is also given.

A few words should be said about geographic range. If the range is not mentioned, it may be assumed that the plant has a wide distribution, at least in California, although it is not necessarily plentiful everywhere. When the range is given, it is primarily concerned with the north-south distribution. For example, Ma-s, means that Marin County is the northern limit of the species, and that the range extends well into southern California, perhaps to the state border or beyond it. The species could also occur, however, in the Sierra Nevada or the Sacramento Valley, but this book is not concerned with this portion of its geographic distribution. Similarly, although SFBR means that the species is found through much of the San Francisco Bay Region, its range may include other areas of the state.

Persons already familiar with plants of the region will note that our scientific names reflect many changes that have been published since 1968.

The changes are based on re-examination of each species in the light of names that have been applied to it, and there are some striking departures from names that have been in use for a long time.

Some of the scientific names include the names of subspecies or varieties. The definitions for these entities have not been universally agreed upon by botanists. In general, the term subspecies should be applied to plants that differ rather clearly from those fitting the characteristics of the species, but not to the extent of warranting a separate species name. The term variety should perhaps be restricted to plants that differ from a species or subspecies in relatively unimportant ways, such as the extent to which the leaves and stems are hairy.

Groups of Higher Plants

What botanists call higher plants, or vascular plants, are those that have tissues specialized for transport of water and the subtances dissolved in it. Furthermore, except for ferns and fern allies, they have flowers, cones, or certain other structures for producing seeds.

In the scheme below, the several major categories of higher plants for which keys are provided are given in boldface type. After you become familiar with the criteria on which the groupings are based, you will be able to decide which key to use for identifying a new plant.

A Plants that do not have flowers, cones, or any other structures for producing seeds

Ferns: leaves compound, with many leaflets, or at least much divided, usually in a pinnate pattern

Fern Allies: without the appearance of a fern, the leaves being scalelike, hairlike, quill-like, or with 4 palmately arranged lobes

B Plants with flowers, cones, or other structures in which seeds are produced

Needle-leaved, Cone-bearing Trees and their Relatives: pines, firs redwoods, cypresses, yews, and a few others, usually producing seeds in cones

Broad-leaved Trees: willows, alders, maples, oaks, and other trees with flowers, mostly more than 2 m tall and with a single main trunk. (The flowers are sometimes in catkins that may superficially resemble cones, but the presence of broad leaves will eliminate these trees from the preceding category.)

Shrubs and Woody Vines: usually with several main stems, and woody, except for new growth; shrubs generally upright and at least 1 m tall, but not often more than 2 m tall; vines twining or clambering over shrubs or trees

Families of Herbaceous Flowering Plants: regardless of their height or form, nonwoody, except perhaps at the base; includes plants conventionally called wildflowers and weeds, and also grasses, sedges, rushes, and nonwoody vines such as morning-glories

Monocotyledonous Flowering Plants: lilies, irises, orchids, grasses, sedges, rushes, and their relatives; flower parts commonly in cycles of 3 or 6; main veins of leaves usually nearly parallel

Dicotyledonous Flowering Plants: flower parts usually in cycles of 4 or 5; leaves usually with a single main vein that originates at the base of the blade, then branches, or with 3 or more main veins that diverge from the base of the blade

It must be mentioned that the simple criteria just given for distinguishing monocotyledonous and dicotyledonous plants are not the only ones used by botanists. Some very basic features that separate the two categories are found in the microscopic anatomy of the stems, and also in the fact that when the seed of a monocotyledonous plant germinates, a single leaflike structure (cotyledon) emerges, whereas in a dicotyledonous plant there are two of these structures.

For nonwoody plants, the key to Families of Herbaceous Flowering Plants will lead you to the correct family. In this key, all families other than those designated by an (M) are dicotyledonous. The woody plants covered by the keys to broad-leaved trees, shrubs, and vines are included in the keys for individual families.

Measurements

In the keys, the height of plants and size of plant parts, such as petals and leaves, are given in metric units: meters (m), centimeters (cm), and millimeters (mm). These are routinely used in modern scientific work. Thus one can directly relate measurements in this book to those in *The Jepson Manual* and most other treatises. If you are not already familiar the linear units of the metric system, a few comparisons with conventional units may be helpful. A meter, consisting of 100 centimeters, is slightly more than 39 inches. A centimeter, consisting of 10 millimeters, is about two-fifths of an inch, so 2.5 centimeters equal 1 inch, 5 centimeters equal 2 inches, and 25 centimeters equal 10 inches.

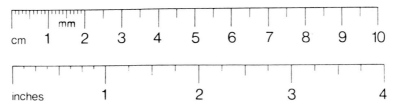

Temperatures are given on the Celsius scale, in which 0^0 equals 32^0 Fahrenheit and 100^0 equals 212^0 Fahrenheit. For altitude, long distances, and volumes, the conventional units--feet, miles, and gallons--are used.

How to Use a Key to Identify a Plant

A key for plant identification takes advantage of contrasting charac-ters, such as pink petals as opposed to blue petals, leaf blades with toothed margins as opposed to leaf blades with smooth margins, and so on. These characteristics are presented in a series of couplets. The two choices in the first couplet are A and AA. If A is the better choice for the plant you are looking at, go to the next couplet--B and BB--under A. Don't wander into the territory under AA! Similarly, if AA happens to be the better choice, stick to the sequence of couplets under AA.

In certain choices, one or more characteristics, enclosed by paren-theses, follow the basic features being contrasted. These characteristics may be helpful in confirming the correctness of a particular choice, but they are not necessarily limited to the species to which that choice leads. Some of the species under choice AA, for instance, may have only basal leaves or be found in wet meadows, but they do not fit the primary features given for choice A.

The more you use keys of this type, the more quickly you will become familiar with the terms commonly used in plant classification. Experience, furthermore, will enable you to make judgments more quickly and to glide over choices that do not pertain to the specimen you are looking at.

If you have not used a key before, the little exercise provided here wiil help you. An "unknown" is shown in the illustration. This plant belongs to the Saxifrage Family, and the pertinent portions of the key to this family are reproduced, with successive correct choices underlined.

If you are not familiar with some of the terms you encounter, look these
up in the Glossary.

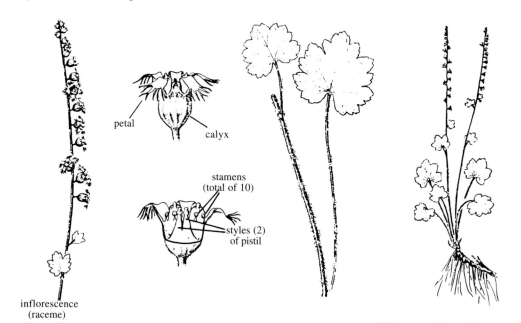

petal

calyx

stamens
(total of 10)

styles (2)
of pistil

inflorescence
(raceme)

Tellima grandiflora, Fringecups

A Flowers solitary at the top of the stem, with white petals 1-1.5 cm
   long, these not lobed (all leaves basal, the blades broad, with smooth
   margins; in wet meadows)
                          Parnassia californica [*P. palustris* var. *californica*]
                                       Grass-of-Parnassus; SB-n
AA Flowers mostly in inflorescences, and if with white petals, these either
   less than 1 cm long or divided into lobes
   B Petals 4 (these purplish brown, almost threadlike); stamens 3 (some
      leaves sprouting plants that may eventually take root)
                                       Tolmiea menziesii (*Plate* 51)
                                       Piggy-back-plant; SFBR-n
   BB Petals 5; stamens 5 or 10
      C Stamens 5 (usually in moist, shaded areas)
         D Leaves scattered along the flowering stems, as well as basal
            (leaf blades up to 8 cm wide, divided into toothed lobes;
            inflorescence usually more than 30 cm long; petals white, 3-4
            mm long, widest above the middle; restricted to shady, wet
            banks and cliffsides)   *Boykinia occidentalis* [*B. elata*]
                                                            Brookfoam
         DD Nearly all leaves basal (not more than 2 leaves along each
            flower stem)
            E Petals yellow-green, divided into 4-7 slender lobes
               (collectively, the petals form what looks like a
               snowflake)   *Mitella ovalis*
                                                     Mitrewort; Ma
            EE Petals white or pink, not divided into slender lobes
               F Style of pistil 2-4 mm long, decidedly extended out of
                  the flower; inflorescence usually open, the flowers
                  not crowded; hairs on calyx not obvious without a hand
                  lens; widespread
                              Heuchera micrantha [includes *H. micrantha*
                                          var. *pacifica*] (*Plate* 95)
                                          Smallflower Alumroot; SLO-n
               FF Style of pistil less than 2 mm long, not extended out
                  of the flower; inflorescence usually dense, the
                  flowers crowded; hairs on calyx dense and obvious
                  without a hand lens; coastal   *Heuchera pilosissima*
                                          Seaside Alumroot; SLO-n
   CC Stamens 10

D Pistil with 2 styles
  E Petals either less than 1.5 mm wide, or broader and
     divided into several lobes
      F Petals white, slender, not divided; flowers in a
         panicle
               Tiarella trifoliata var. *unifoliata* [*T. unifoliata*]
                            Sugarscoop; SFBR-n
      FF Petals greenish-white to nearly crimson, at least 2 mm
         wide, divided into 5-7 lobes (petals sometimes falling
         early); flowers in a slender raceme
                     Tellima grandiflora (*Plate* 51)
                            Fringecups; SLO-n
  EE Petals at least 2 mm wide, not divided into lobes (white,
     sometimes with darker spots)
      F Leaf blades not rounded, longer than wide, not more
         than 3 cm wide (hairy on the upper surface, with
         prominent teeth) (filaments wider below than above;
         inflorescence with not more than 20 flowers, these
         often on one side; widespread) *Saxifraga californica*
                          California Saxifrage
      FF Leaf blades rounded, sometimes wider than long, up to
         10 cm wide
        G Two petals longer than the other 3; leaves up to 10
           cm wide, with white veins on the upper surface and
           red veins on the underside; new plants produced
           from prostrate stems that root at the nodes
                 Saxifraga stolonifera [*S. sarmentosa*]
                       Strawberry-geranium; as; Sn
        GG Petals of more or less equal length; leaves up to 7
          cm wide, with veins of the same color on both
          surfaces; new plants not produced from prostrate
          stems (filaments wider above than below)
                      Saxifraga mertensiana
                      Wood Saxifrage; Sn-n
  DD Pistil with 3 styles

The key to Families of Herbaceous Plants (p. 53), which you would have
had to use to get your unknown to the Saxifrage Family, is more extensive
than most keys for individual families. The plants in a particular family,
being related to one another, share many characteristics. Thus there is not
so much diversity in structure between the genera and species as there is
between the numerous families. The plant you have just practiced
identifying would nevertheless have been rather easy to place in the right
family. The early choices (AA, BB, CC, DD, EE, FF, GG, HH, II, JJ, and KK)
would lead you to Group C and subsequently (choosing AA, B, CC, DD, EE, and
FF) to Subkey 1 (terrestrial plants with alternate or entirely basal
leaves). Choices in this key (AA, BB, CC, DD, EE, F, GG, H, and I) would
take you to Saxifragaceae. That is when you would turn to the key for this
family.

Conservation

The best way to save our native plants and animals is to set aside
large, undisturbed tracts where the flora and fauna are typical. The need
for protection is especially urgent in areas to which rare or endangered
species are restricted. Among private nonprofit organizations, The Nature
Conservancy has been particularly effective in acquiring and managing lands
with a view to preserving natural communities of plants and animals. We
also owe much to federal, state, and local agencies--including those that
control the extensive acreages of watersheds--for protecting the fauna and
flora in parks and other natural areas.
Botanical gardens, some of which are concerned entirely with the
indigenous flora, are doing a good job of displaying native species, as
well as propagating them, thereby reducing the risk that some will become
extinct or excessively rare.

California, like many other states, has laws to prevent the digging or cutting of native plants on public lands. To prevent damage to the native flora, the keys in this book do not deal with features that can be seen only if a plant is removed from the soil. Although the characters of roots, bulbs, and other underground parts are often used to distinguish species belonging to certain genera, the keys in this book concentrate on the habit of growth, stems, leaves, flowers, and fruit. If one has a hand lens and a ruler, one should be able to identify many plants without having to remove even a leaf or flower.

Growing Native Plants

By 1850, numerous species of plants found in western North America, especially California, had been grown in Europe, had been illustrated in botanical journals, and were well known to many gardeners. Appreciation of California plants as subjects for cultivation in areas where they were already growing wild developed slowly, in spite of the efforts of botanical gardens, some influential professional botanists, a few commercial growers, and many enthusiastic amaturs who did so much to encourage the use of natives. In 1965, the California Native Plant Society was founded, and although this organization was from the beginning concerned with protecting the flora, it has also been extremely effective in promoting cultivation of indigenous species. Furthermore, extended periods of relatively low rainfall in parts of California have raised the consciousness of residents with respect to possibilities for using attractive native species on landscapes where water-guzzling exotics would seem to be inappropriate.

Where does one obtain plants for a wild garden? A few commercial nurseries specialize in natives, or at least have extensive stocks. Seasonal sales conducted by regional chapters of the California Native Plant Society and by botanical gardens are excellent sources of plants, bulbs, and seeds, as well as information and advice. Members of the California Native Plant Society receive notices of sales as well as *Fremontia*, a journal that often has interesting articles about growing natives and that also publishes advertisements of commercial growers.

If your property already has native plants, you may wish to leave it just as it is. Most persons will, however, cut back or selectively remove untidy vegetation to make room for other interesting species that will succeed under the same conditions. Furthermore, by growing certain annuals from seed, one can add color in open areas in which there are no attractive wildflowers, or extend the flowering season well beyond that of plants already present.

References

The works listed here deal either with the flora of California as a whole or with that of our region and other portions of central and northern California. Some of the references are out of print, and some are not likely to be readily available except in libraries of larger academic institutions. Furthermore, the nomenclature will deviate, to varying extents, from that in the *Jepson Manual*.

References dealing with the plant communities, topography, climate, geology, and other physical attributes of our region, are listed on p. 22.

Comprehensive Floras of California

Abrams, L. R., and R. S. Ferris. 1923-60. Illustrated Flora of the Pacific States. 4 volumes. Stanford University Press, Stanford, California.
Grillos, S. J. 1966. Ferns and Fern Allies of California. University of California Press, Berkeley, Los Angeles, London.

Hickman, J. C. (editor). 1993. The Jepson Manual. Higher Plants of California. University of California Press, Berkeley, Los Angeles, London.

Jepson, W. L. 1923, 1925. A Manual of the Flowering Plants of California. Associated Students Store, University of California, Berkeley.

Metcalf, W. 1969. Native Trees of the San Francisco Bay Region. University of California Press, Berkeley, Los Angeles, London.

McMinn, H. E. 1939. An Illustrated Manual of California Shrubs. University of California Press, Berkeley, Los Angeles, London.

Munz, P. H., and D. D. Keck. 1968. A California Flora, with Supplement. University of California Press, Berkeley, Los Angeles, London.

Floras of Specific Regions of Central and Northern California

Bowerman, M. L. 1944. The Flowering Plants and Ferns of Mount Diablo, California. Gillick Press, Berkeley, California.

Ferris, R. S. 1968. Native Shrubs of the San Francisco Bay Region. University of California Press, Berkeley, Los Angeles, London.

Ferris, R. S. 1970. Flowers of the Point Reyes National Seashore. University of California Press, Berkeley, Los Angeles, London.

Howell, J. T. 1970. Marin Flora. Manual of the Flowering Plants and Ferns of Marin County, California. Second edition, with supplement. University of California Press, Berkeley, Los Angeles, London.

Howell, J. T., P. H. Raven and P. Rubtzoff. 1958. A flora of San Francisco, California. Wasmann Journal of Biology, 16: 1-155. (Reprinted 1990 by California Native Plant Society, Yerba Buena Chapter, San Francisco.)

Howitt, B. F., and J. T. Howell. 1964. The vascular plants of Monterey County, California. Wasmann Journal of Biology, 22: i-ii, 1-184. (Supplement published 1973 by Pacific Grove Museum, Pacific Grove, California.)

McClintock, E., and W. Knight, with N. Fahy. 1968. A flora of the San Bruno Mountains, San Mateo County, California. Proceedings of the California Academy of Sciences, series 4, 32: 587-677. (Reprinted 1990, as Special Publication no. 8, by California Native Plant Society, Sacramento.)

Peñalosa, J. 1963. A flora of the Tiburon Peninsula, Marin County, California. Wasmann Journal of Biology, 21: 1-74

Sharsmith, H. K. 1945. Flora of the Mount Hamilton Range of California. American Midland Naturalist, 34: 289-382. (Reprinted 1982, as Special Publication no. 6, by California Native Plant Society, Sacramento.)

Sharsmith, H. K. 1965. Spring Wildflowers of the San Francisco Bay Region. University of California Press, Berkeley, Los Angeles, London.

Smith, G. L., and C. L. Wheeler. 1990-1991. A flora of the vascular plants of Mendocino County, California. Wasmann Journal of Biology, 48-49: i-iv, 1-387. (Reprinted 1992 by University of San Frnacisco.)

Thomas, J. H. 1961. Flora of the Santa Cruz Mountains of California. Stanford University Press, Stanford, California.

TOPOGRAPHY, CLIMATE, SOILS, AND PLANT COMMUNITIES OF THE SAN FRANCISCO BAY REGION

Topography and Climate

Long before there was a San Francisco Bay, the westward flow from the large river system of the Central Valley, fed mostly by streams draining the Sierra Nevada, cut through two low points in the Coast Ranges, creating what are now Carquinez Strait and the Golden Gate. Smaller rivers from the Coast Ranges contributed to the scouring process. Later, at the close of the last Ice Age, when glaciers melted, the sea level began to rise. Over a period lasting several thousand years, water from the ocean gradually flooded much of the low-lying area behind the Golden Gate. Thus San Francisco Bay was formed.

The Coast Ranges of the region consist of hills and substantial mountains whose orientation is from northwest to southeast. Close to the ocean, there are the Santa Cruz Mountains, Marin Hills (in which Mount Tamalpais is the highest peak), and the Sonoma Mountains. Inland, there are the Mount Hamilton Range, Diablo Range, and Berkeley Hills, separated by Carquinez Strait and Suisun Bay from the Vaca Mountains and Mayacamas Range. Western portions of the latter are scarcely distinct from the Sonoma Mountains.

The Coast Ranges of the region, as mountains go, are not high--the top of Mount Hamilton is 4261 feet above sea level, Mount Diablo has an altitude of 3849 feet, Mount Tamalpais reaches only to 2610 feet--but they are nevertheless barriers that impede the mixing of the inland air mass with the one that lies over the ocean. In general, therefore, the climate at the coast, influenced by the relatively stable temperature of the ocean, is more moist than the climate east of the mountains. Furthermore, it is cooler in summer and warmer in winter. But the topography of the Coast Ranges is complex. The valleys and ridges run in all directions, and there are numerous low points besides the huge gap at the Golden Gate. The large estuarine water mass of San Francisco Bay is another factor that conspires to create an extremely variable pattern of weather.

The climate of San Francisco Bay Region is complex indeed. For within the nine counties of this area, there are perhaps more different "microclimates" than will be found in almost any other area of comparable size. For instance, on a particular summer day, the city of San Francisco and Marin Hills may be enshrouded by fog, the residents of San Rafael may be enjoying perfect weather, and people who live in Walnut Creek may be starting to complain of the heat. On an annual basis, differences may be even more dramatic. San Rafael has a total rainfall of about 35 inches, whereas San Mateo, about the same distance from the coast, has only 21 inches. Martinez' yearly total is about 14 inches, but Orinda, not far away, gets nearly twice that amount, and even more than Berkeley, which faces the Golden Gate. The disparity between the amount of rain falling on the western slopes of the mountains and hills near the coast and on some areas inland is even more striking. There are places in the Santa Cruz Mountains where the annual rainfall reaches 60 inches, yet in the Santa Clara Valley, just behind the mountains, the total rarely surpasses 15 inches.

An inescapable attribute of some parts of the region is fog. Most of our fogs originate when humid and relatively warm air from the Pacific approaches the coast. The coldness of the ocean water offshore causes the moisture in the air to condense into tiny airborne droplets. Some fogs over San Francisco Bay are due partly to the daily tidal exchanges of water.

9

After water has run out through the Golden Gate at low tide, the incoming rush of colder water at high tide may induce condensation of moisture in the overlying air. Shadows of clouds, which abruptly lower the temperature of the air, may bring about the same effect. Occasionally, fog develops when humid air from the ocean reaches the hills and is cooled as it is deflected upward. The result may be an extensive fog bank lying atop or against the hills, or perhaps just small patches of fog that float not far overhead.

The higher ridges of the Coast Ranges may prevent low-lying fog from reaching the valleys farther east. At the Golden Gate, however, nothing stands in the way of a fog bank being pushed inland. The amount of water that moves through the gap in the form of fog is sometimes enormous. It could be as much as a million gallons an hour. The fog is essentially a cloud lying close to the surface of the water of the Bay or of the ocean. If it rises, then it becomes a cloud of the conventional type.

In summer, when it is hot in the Central Valley, the rising of warm air causes cooler air and fog to be drawn in from the coast. Much of this eastward flow passes through the area, accounting for many of our summer fogs. After reaching the Central Valley, the fog is usually quickly evaporated by the daytime warmth and relative dryness. Nevertheless, the weather in the valley cools down a bit, and because there is less warm air to rise, the influx of air from the coast slows down or stops. Soon, however, the weather in the Central Valley heats up again, and the cycle is repeated. The result: more fog in the region.

So far, this discourse has dealt mostly with fog. What about rain? Rain falls when the fine droplets of water in a cloud coalesce into larger drops. If this type of condensation takes place at a freezing temperature, there will be snow. If, on the other hand, an updraft causes raindrops to be carried upwards until they are frozen, the result will be hail. Rain that freezes as it falls becomes sleet.

In most parts of the region, about 90% of the total annual rainfall occurs from October to April. Within this period, however, there may be dry spells that last several weeks, and in some years the amount of rainfall is much less than normal. The droughts may adversely affect the germination of seeds, the survival of seedlings, and even the growth, flowering, and fruiting of well established plants. As a rule, heavy rainfall in the winter leads to a good display of spring flowers, especially annuals.

Where fogs are prevalent, these also contribute to the moisture requirements of the vegetation. One reason for this is that high humidity and the presence of mist reduce the extent to which plants lose water to the atmosphere by evaporation. Furthermore, as mist condenses on foliage, some of the water may drip down onto the soil. A study in one area a few miles south of San Francisco showed that moisture added to the soil by drip from foliage was considerably more than the amount deposited by rain. That at least some plants can absorb moisture condensing on their leaves is indicated by results of studies of seedlings of Pacific Ponderosa Pine. The foliage of one group of seedlings was exposed to frequent applications of mist, but a plastic shield prevented water from dripping down onto the soil. These seedlings survived better than those of another group, growing in equally dry soil and not receiving the benefit of mist.

Soils

In the region under consideration, there are rocks of several distinctly different categories, including granite and basalt, which are of igneous origin, and shales, sandstones, and conglomerates, which are of sedimentary origin. Thus, because soil is derived from rock, one may expect a wide variety of soil types to exist in the region.

As every gardener and farmer knows, not all plants grow well on all soils. Some of the attributes of a particular soil are shown by its general texture, coarseness or fineness, degree of acidity or alkalinity, the amount of organic matter it contains, its ability to retain moisture, the extent to which it allows air to penetrate, its supply of mineral nutrients, and the possible presence of substances that inhibit the growth of plants.

The majority of soils are mixtures of sand (particles from 0.02 to 2 mm); silt (particles 0.002 to 0.02 mm), and clay (particles smaller than 0.002 mm). Clay is important in deciding how much water a soil will hold, because fine particles have more surface area, in proportion to size, than large particles. In clay, therefore, there is more surface area to which a film of water can adhere. Clay also helps to stabilize the supply of mineral nutrients. The particles, consisting largely of compounds of aluminum, silicon, and oxygen, have a negative charge. Calcium, potassium, and some other ions needed by plants have a positive charge, and are therefore attracted to the clay particles and held so tightly that rainwater draining through the soil does not remove them.

One particular situation common in the area deserves special mention. This is the presence of a type of rock called serpentine, which consists to a large extent of magnesium silicate. It is typically somewhat shiny and slightly greenish, although after being weathered for a long time it becomes reddish brown, and the soil into which it eventually breaks down is usually reddish. Serpentine is abundant in California, and an act of the legislature named it the State Rock. Extensive outcrops can be found on Mount Tamalpais, Mount Diablo, and Mount Hamilton, in the hills directly east of Oakland and Berkeley, and in Edgewood Park in San Mateo County.

Soils in which serpentine is an important component are, as might be expected, rich in magnesium. Plants need this metal in order to synthesize chlorophyll, as well as for other processes. When magnesium is present in large quantities, however, it interferes with the uptake of calcium, which plants also require, and which is scarce in serpentine soils. Inadequate supplies of other nutrients, including nitrogen, make life difficult for plants, too, and the toxic effect of nickel and chromium, sometimes present in unusually high concentrations, is still another problem. These features prevent many plants from growing on serpentine soils. There are, nevertheless, species that tolerate serpentine, and even some that are found only where it is present. Notable serpentine lovers include the following.

Trees
  Cupressus sargentii, Sargent
  Cypress
Shrubs
  Quercus durata, Leather Oak
Herbs
  Allium fimbriatum, Fringed
  Onion
  Allium lacunosum, Wild Onion
  (*Plate 58*)
  Aquilegia eximia, Serpentine
  Columbine
  Aspidotis densa, Indian's-
  dream (*Plate 65*)
  Calystegia collina, Woolly
  Morning-glory
  Carex serratodens, Bifid Sedge
  Eriophyllum jepsonii, Jepson
  Woolly Sunflower
  Hesperolinon micranthum,
  Smallflower Western Flax

Monardella villosa ssp.
  franciscana, Serpentine
  Coyotemint
Muilla maritima, Common Muilla
  (*Plate 60*)
Parvisedum pentandrum,
  Parvisedum
Polygonum douglasii ssp.
  spergulariiforme, Fall
  Knotweed
Streptanthus breweri var.
  breweri, Brewer Streptanthus
Streptanthus glandulosus ssp.
  secundus, Oneside
  Jewelflower
Viola ocellata, Western
  Heart's-ease (*Plate 56*)
Zigadenus micranthus var.
  fontanus, Serpentine Star
  Lily

Principal Plant Communities

A "plant community," in the minds of most botanists, ecologists, and naturalists, is a relatively constant association of several to many species, certain of which predominate. A forest of Coast Redwood is a good example, for the Tanbark Oak, the Western Sword Fern, Redwood Sorrel, and some other plants are commonly found with this coniferous tree over much of its north-south range. Thus the entire assemblage may be called a community.

In classifying vegetational assemblages, even in a rather restricted geographic area, it is best to maintain a flexible attitude. One community may intergrade with another, and an association of plants that seems to fit a particular type of community in a general way may deviate from that type in certain respects. For instance, a Douglas-fir Forest in Humboldt County, or farther north, will have several species of plants that are not present in a Douglas-fir Forest of Marin County. Nevertheless, it is convenient to be able to deal with the Douglas-fir Forest as a entity, so long as we are prepared to accept some variation in its composition.

In subsequent sections of this chapter, an asterisk (*) is used to indicate species that have been introduced.

Valley and Foothill Woodland

In the area, Valley and Foothill Woodland occurs at elevations of about 300 to 3500 feet. It is characteristic of inland habitats that receive about 20 inches of rain each year, and that are rarely touched by fog except in the winter. The summer weather is warm, with daytime temperatures often reaching $90°$ F. Much of the area occupied by the Mount Diablo and Mount Hamilton ranges, and by hills in Napa and Solano counties, has woodland vegetation.

The trees are generally somewhat scattered, so there is a well lighted understory, with a few shrubs and a wide variety of grasses and other herbaceous plants. Most of the more common and conspicuous plants of woodland are listed below.

Trees
  Aesculus californica, California
    Buckeye (*Plate* 30)
  Pinus coulteri, Coulter Pine
    (*Plate* 68)
  Pinus sabiniana, Foothill Pine
    (*Plate* 68)
  Quercus agrifolia, Coast Live
    Oak (*Plate* 28)
  Quercus chrysolepis, Cañyon
    Oak (*Plate* 82)
  Quercus douglasii, Blue Oak
    (*Plate* 81)
  Quercus lobata, Valley Oak
    (*Plate* 81)
  Quercus wislizenii var.
    wislizenii, Interior Live
    Oak (*Plate* 82)
  Umbellularia californica,
    California Bay (*Plate* 34)
Shrubs
  Ceanothus cuneatus, Buckbrush
    (*Plate* 47)
  Eriodictyon californicum,
    Yerba-santa (*Plate* 31)
  Heteromeles arbutifolia, Toyon
    (*Plate* 48)
  Rhamnus californica,
    California Coffeeberry
    (*Plate* 47)

  Symphoricarpos albus var.
    laevigatus, Common Snowberry
    (*Plate* 18)
  Toxicodendron diversilobum,
    Western Poison-oak (*Plate* 4)
Herbs
  Cardamine californica var.
    californica, Common
    Milkmaids (*Plate* 17)
  Collinsia heterophylla,
    Chinesehouses (*Plate* 53)
  Dodecatheon hendersonii,
    Mosquito-bills (*Plate* 45)
  Lithophragma heterophyllum,
    Hill Starflower
  Nemophila menziesii var.
    menziesii, Baby-blue-eyes
    (*Plate* 31)
  Pentagramma triangularis,
    Goldback Fern (*Plate* 2)
  Phoradendron villosum, Oak
    Mistletoe (*Plate* 56)
  Ranunculus occidentalis,
    Western Buttercup
  Saxifraga californica,
    California Saxifrage

Riparian Woodland

The complex of trees, shrubs, and herbaceous plants that usually grow along streams and rivers form the Riparian Woodland community. The vegetation consists mostly of species that require more soil moisture than do the oaks and other constituents of Valley and Foothill Woodland. A few of the especially common plants of Riparian Woodland in the region are listed below.

Trees
Acer macrophyllum, Bigleaf Maple (*Plate* 69)
Alnus rhombifolia, White Alder (*Plate* 74)
Platanus racemosa, Western Sycamore (*Plate* 87)
Populus balsamifera ssp. *trichocarpa*, Black Cottonwood (*Plate* 94)
Salix lasiolepis, Arroyo Willow
Salix lucida ssp. *lasiandra*, Shining Willow (*Plate* 94)
Umbellularia californica, California Bay (*Plate* 34)
Shrubs
Cornus sericea ssp. *occidentalis*, Western Creek Dogwood (*Plate* 21)
Lonicera involucrata var. *ledebourii*, Twinberry (*Plate* 18)
Oemleria cerasiformis, Osoberry (*Plate* 92)
Rhododendron occidentale, Western Azalea (*Plate* 23)

Rhus trilobata, Skunkbush (*Plate* 4)
Ribes sanguineum var. *glutinosum*, Pinkflower Currant (*Plate* 30)
Rosa californica, California Rose (*Plate* 50)
Toxicodendron diversilobum, Western Poison-oak (*Plate* 4)
Vitis californica, California Wild Grape (*Plate* 56)
Herbs
Aralia californica, Elk-clover (*Plate* 6)
Claytonia perfoliata, Miner's-lettuce (*Plate* 90)
Cyperus eragrostis, Tall Cyperus (*Plate* 57)
Mimulus cardinalis, Scarlet Monkeyflower (*Plate* 53)
Thalictrum fendleri var. *polycarpum*, Meadowrue (*Plate* 92)
Woodwardia fimbriata, Giant Chain Fern (*Plate* 2)

Redwood Forest

Natural forests of the Coast Redwood, *Sequoia sempervirens* (*Plate* 3), are found from Monterey County to southern Oregon. The best known and perhaps most nearly pristine grove in our area is the one in Muir Woods, Marin County, but there are a few other good examples. Furthermore, some of the groves that had been cut down for lumber have made a fairly good recovery, and will be found to have many of the plants that are characteristically associated with the Coast Redwood. As often happens when an area is disturbed, however, the balance of species changes. Some become rare or disappear entirely, others become proportionately more common. And opportunists not previously present may add themselves to the mix.

Natural groves are found in areas that receive considerable annual rainfall--at least 35 inches--and that have frequent heavy fogs during the dry season. There are substantial plantations of redwoods in some places where they were not native, but these are not likely to have the usual associates. Many of the plants typically found in Redwood Forests are listed below.

Trees (both of these are more abundant in second-growth forests)
Lithocarpus densiflorus, Tanbark Oak (*Plate* 81)
Umbellularia californica, California Bay (*Plate* 34)
Shrubs
Ceanothus thyrsiflorus, Blue-blossom (*Plate* 47)
Corylus cornuta var. *californica*, Hazelnut (*Plate* 74)

Gaultheria shallon, Salal (*Plate* 23)
Rhododendron macrophyllum, Rosebay (*Plate* 23)
Rosa gymnocarpa, Wood Rose (*Plate* 50)
Rubus parviflorus, Thimbleberry (*Plate* 93)
Vaccinium ovatum, Evergreen Huckleberry (*Plate* 23)

Herbs
  Adenocaulon bicolor,
    Trailplant (*Plate* 72)
  Anemone oregana, Wood Anemone
    (*Plate* 91)
  Asarum caudatum, Wild-ginger
    (*Plate* 6)
  Athyrium filix-femina var.
    cyclosorum, Lady Fern (*Plate* 1)
  Clintonia andrewsiana, Red
    Bead Lily (*Plate* 59)
  Disporum smithii, Largeflower
    Fairybell (*Plate* 102)
  Oxalis oregana, Redwood Sorrel
    (*Plate* 39)
  Polystichum munitum, Western
    Sword Fern (*Plate* 66)
  Scoliopus bigelovii, Fetid
    Adder's-tongue
  Trillium ovatum, Western
    Trillium (*Plate* 103)
  Vancouveria planipetala,
    Inside-out-flower (*Plate* 15)
  Viola ocellata, Western
    Heart's-ease (*Plate* 56)
  Viola sempervirens, Evergreen
    Violet (*Plate* 97)

Closed-cone Pine Forest

A closed-cone pine is one whose cones do not simply open up as soon as the seeds are mature. The scales of the cones are held tightly together by pitch, and they are not likely to separate for several years unless the pitch is melted by fire or exceptionally warm sunshine.

Within our area, the only species forming significant natural stands is the Bishop Pine, *Pinus muricata* (*Plate* 3). Its range, extending from Humboldt County to Baja California, overlaps that of the Beach Pine, *Pinus contorta* (Mendocino County northward) and Monterey Pine, *Pinus radiata* (San Mateo, Monterey, and San Luis Obispo counties). Like these last two species, it is not found far from the coast. In Marin County, an extensive grove of Bishop Pine can be seen at Inverness Ridge, and there are some trees mixed with chaparral shrubs in the Carson Ridge area, northwest of Fairfax and San Anselmo. Other representatives not too far away will be found in Sonoma, San Mateo, and Santa Cruz counties.

The cones of Bishop Pine, often clustered, are prickly and lopsided. The branches may be nearly limited to the upper portions of the trees, which is usually the case in dense groves, or they may be distributed rather evenly from the top down to nearly ground level. Some common associates of the Bishop Pine include the following.

Trees
  Quercus agrifolia, Coast Live
    Oak (*Plate* 28)
Shrubs
  Arctostaphylos, manzanitas;
    various species (*Plates* 22, 23)
  Baccharis pilularis,
    Coyotebrush (*Plate* 8)
  Rhamnus californica,
    California Coffeeberry
    (*Plate* 47)
  Toxicodendron diversilobum,
    Western Poison-oak (*Plate* 4)
  Vaccinium ovatum, Evergreen
    Huckleberry (*Plate* 23)

Douglas-fir Forest

From southern Alaska to northern Sonoma County, there are dense forests in which Douglas-fir, Grand Fir, and Western Hemlock are the characteristic coniferous trees. Farther south--on Mount Tamalpais, on the Inverness Ridge, and in some other portions of our region--forests in which Douglas-fir is conspicuous do not closely resemble those of more northern areas. This is due partly to the absence of many trees, shrubs, and herbaceous plants found in a typical Douglas-fir forest, and partly to the presence of some that are not typical--the Tanbark Oak and California Bay, for instance. In less dense growths there may be Pacific Madrone, California Coffeeberry, Blue-blossom, and Western Poison-oak. Characteristic members of the Douglas-fir Forest are listed below.

Trees
  Abies grandis, Grand Fir (*Plate* 67)
  Arbutus menziesii, Pacific Madrone (*Plate* 22)
  Lithocarpus densiflorus, Tanbark Oak (*Plate* 81)

Trees continued
  Pseudotsuga menziesii,
    Douglas-fir (*Plate* 3)
  Tsuga heterophylla, Western
    Hemlock (*Plate* 3)
  Umbellularia californica,
    California Bay (*Plate* 34)
Shrubs
  Ceanothus thyrsiflorus, Blue-
    blossom (*Plate* 47)

Rhamnus californica,
  California Coffeeberry
  (*Plate* 47)
Toxicodendron diversilobum,
  Western Poison-oak (*Plate* 4)
Herbs
  Polystichum munitum, Western
    Sword Fern (*Plate* 66)

Chaparral

Chaparral is a type of vegetation in which most of the obvious
components are tough-leaved evergreen shrubs that are adapted for life in a
relatively dry habitat. The word comes from Spain, where it has been used
to refer to brushy places dominated by the chaparro, a kind of scrub oak.
Although our chaparral includes a similar species, most of the conspicuous
shrubs belong to other families. The following are particularly prevalent
constituents of the area's chaparral.

Shrubs
  Adenostoma fasciculatum,
    Chamise (*Plate* 48)
  Arctostaphylos glandulosa,
    Eastwood Manzanita (*Plate* 22)
  Arctostaphylos glauca,
    Bigberry Manzanita
  Arctostaphylos manzanita ssp.
    manzanita, Parry Manzanita
  Arctostaphylos stanfordiana,
    Stanford Manzanita (*Plate* 23)
  Arctostaphylos tomentosa ssp.
    crustacea, Brittleleaf
    Manzanita (*Plate* 23)
  Ceanothus cuneatus, Buckbrush
    (*Plate* 47)
  Ceanothus foliosus var.
    foliosus, Wavyleaf Ceanothus
  Ceanothus oliganthus var.
    sorediatus, Jimbrush
  Cercocarpus betuloides,
    Mountain-mahogany (*Plate* 48)

Eriodictyon californicum,
  Yerba-santa (*Plate* 31)
Heteromeles arbutifolia, Toyon
  (*Plate* 48)
Pickeringia montana, Chaparral
  Pea (*Plate* 26)
Quercus berberidifolia, Scrub
  Oak (*Plate* 81)
Ribes malvaceum, Chaparral
  Currant
Toxicodendron diversilobum,
  Western Poison-oak (*Plate* 4)
Herbs
  Castilleja affinis ssp.
    affinis, Common Indian-
    paintbrush (*Plate* 52)
  Pedicularis densiflora,
    Indian-warrior (*Plate* 54)
  Salvia columbariae, Chia
    (*Plate* 33)
  Salvia mellifera, Black Sage
    (*Plate* 33)

Chaparral vegetation is usually dense and often difficult to slog
through. The shrubs are not often more than 2.5 m tall, but the foliage of
most of them rubs us rather harshly. One of the few soft-leaved species in
the list above is Poison Oak, and contact with it is not recommended,
either.

Chaparral is characteristic of hilly areas in which the soil, being
gravelly or sandy, is well drained and does not hold water effectively.
Furthermore, this type of vegetation is typical of areas in which the dry
season lasts for several months. The total annual rainfall of chaparral-
covered habitats may reach 25 inches, but most of this rain will fall
between November and April.

During the dry season, chaparral is extremely vulnerable to fire, partly
because of the density of the vegetation, partly because the foliage is not
"juicy." A chaparral fire can be a very hot one; temperatures of over 600^0 C
have been recorded at ground level, and even 3 cm below the surface the
temperature may exceed 150^0 C. Nevertheless, chaparral is capable of rather
quick recovery. One reason for this is that certain species, after being
burned to the ground, sprout new shoots from the crown. The large burls of
some manzanitas, just above ground level, are testimonials to repeated fires.
It is from these burls that new shoots sprout when recovery begins. Some
species of shrubs, it is true, are destroyed completely, and come back only

if seeds have survived the fire, or if seeds have been carried in from another area.

If there is ample rainfall during the growing season that follows a chaparral fire, there is likely to be a fine show of wildflowers, some of which may have been uncommon for several years preceding the fire. A few of the species that grow luxuriantly in fire-ravaged areas are the following.

Adenostoma fasciculatum, Chamise (*Plate* 48)
Dicentra chrysantha, Golden Eardrops (*Plate* 39)

Emmenanthe penduliflora var. *penduliflora*, Whispering-bells (*Plate* 31)
Papaver californicum, Fire Poppy

An interesting feature of chaparral in our region is that some of the shrubs begin to bloom in December. In the case of certain manzanitas, the opening of flowers is soon followed by production of new foliage; by late spring or early summer, the plants already have the buds of flowers that must wait several months until the arrival of the rainy season. Other shrubs of the chaparral, such as Chamise, flower late in the spring, after they have produced new foliage.

The evergreen members of chaparral generally shed some old leaves in early summer, about the time that formation of new foliage has been completed. Although the shrubs become more or less dormant during the summer months, they retain the functional leaves, so when rains finally come, they are fully equipped for the photosynthetic activity that prepares them for renewed growth.

Hill and Valley Grassland

Grassland is a treeless, shrubless vegetational assemblage that once occupied large areas of California. By the middle of the nineteenth century, three factors had begun to bring about an irreversible change in this kind of habitat. One of these was the conversion of grasslands into cultivated fields. Another was the sharp increase in grazing by cattle, sheep, goats, and horses. Still another was the introduction of non-native plants, especially annual grasses. Most of the native grasses were clump-forming perennials of the type called bunchgrasses. They were literally grazed to death, and their demise was hastened by lack of sufficient rain in certain years, which prevented them from making enough new growth to compensate for the grazing they suffered. The aggressive introduced species--their seeds arriving on the hair of domestic animals, on clothing, and by other means--joined a few native grasses in filling up the space that was opened up for them. For many years, California's grasslands have been golden during the summer months, but it is likely that before the flora was significantly altered, there would have been a substantial green or gray-green component during the dry season.

It is now nearly impossible to find examples of pristine grassland. Nevertheless, there are many localities where bunchgrasses and their natural associates have persisted, even though they have been diluted by introduced species. And it is our grasslands that provide the most spectacular displays of native wildflowers, most of which have survived change better than the native bunchgrasses.

The following list of grassland plants is far from complete, but many of the common species are included.

Achyrachaena mollis, Blow-wives (*Plate* 6)
Amsinckia menziesii var. *intermedia*, Intermediate Fiddleneck (*Plate* 16)
Avena fatua, Wild Oat (*Plate* 104)*

Briza minor, Little Quaking Grass*
Brodiaea elegans, Elegant Brodiaea (*Plate* 58)
Bromus diandrus, Ripgut Grass*
Bromus hordeaceus, Soft Cheat Grass*

Calandrinia ciliata, Redmaids
(*Plate* 45)
Castilleja densiflora, Common
Owl's-clover
Cirsium quercetorum, Brownie
Thistle (*Plate* 9)
Cynosurus echinatus, Hedgehog
Dogtail (*Plate* 63)*
Dichelostemma capitatum,
Bluedicks (*Plate* 102)
Elymus multisetus, Big
Squirreltail (*Plate* 64)
Erodium botrys, Broadleaf
Filaree (*Plate* 29)*
Eschscholzia californica,
California Poppy (*Plate* 40)
Filago californica, California
Fluffweed (*Plate* 11)
Hemizonia congesta ssp.
congesta, Yellow Hayfield
Tarweed (*Plate* 12)
Hordeum jubatum, Foxtail
Barley
Lasthenia californica, California
Goldfields (*Plate* 13)
Lepidium nitidum, Common
Peppergrass (*Plate* 76)
Linanthus parviflorus, Common
Linanthus (*Plate* 42)

Lupinus bicolor, Minature
Lupine
Micropus californicus, Slender
Cottonweed
Microseris douglasii ssp.
douglasii, Douglas
Microseris (*Plate* 73)
Nemophila menziesii var.
menziesii, Baby-blue-eyes
(*Plate* 31)
Plagiobothrys nothofulvus, Common
Popcornflower (*Plate* 16)
Plantago erecta, California
Plantain (*Plate* 41)
Platystemon californicus,
Creamcups (*Plate* 40)
Ranunculus californicus,
California Buttercup (*Plate* 46)
Sanicula bipinnatifida, Purple
Sanicle (*Plate* 5)
Sidalcea malvaeflora ssp.
malvaeflora, Common
Checkerbloom (*Plate* 35)
Sisyrinchium bellum, Blue-
eyed-grass (*Plate* 57)
Viola pedunculata, Johnny-
jump-up (*Plate* 56)

Coastal Prairie

Above cliffs along the coast, landward from the backshores of sandy
beaches, and sometimes in other situations, there are areas in which
perennial grasses, Bracken, and several other types of herbaceous plants
predominate. Typically there are neither trees nor large shrubs. Such
coastal prairies nearly fit our definition of grassland, and the distinction between the two types of plant communities is not sharp. Common
members of this community are listed below.

Calamagrostis nutkaensis,
Pacific Reed Grass
Calochortus luteus, Yellow
Mariposa Lily (*Plate* 58)
Danthonia californica,
California Oat Grass
Deschampsia cespitosa ssp.
holciformis, Pacific Hair
Grass
Dichelostemma capitatum,
Bluedicks (*Plate* 102)
Dichelostemma congestum, Ookow
(*Plates* 60, 102)
Festuca californica,
California Fescue
Festuca idahoensis, Idaho
Fescue
Grindelia hirsutula var.
hirsutula, Hairy Gumplant

Heterotheca sessiliflora ssp.
bolanderi, Bolander Golden
Aster
Iris douglasiana, Douglas Iris
(*Plate* 57)
Lupinus formosus, Summer
Lupine
Lupinus variicolor, Varied
Lupine
Pteridium aquilinum var.
pubescens, Bracken (*Plate* 2)
Ranunculus californicus,
California Buttercup (*Plate* 46)
Sanicula arctopoides, Footsteps-
of-spring (*Plate* 5)
Sisyrinchium bellum, Blue-
eyed-grass (*Plate* 57)

As in the case of other plant communities, not all of the plants in the
list will necessarily be found at a particular site that qualifies as
Coastal Prairie. Furthermore, some habitats that may seem to fit into this
category are not genuine. They have developed in places that once were
brushy or forested, and that were cleared to promote the growth of native
and introduced grasses which would provide favorable food for livestock.

Coastal Scrub and Coastal Forest

Close to the coast, from Monterey County to the Oregon border, there is a type of vegetation called Coastal Scrub. It consists of a mixture of grassland plants and low shrubs. Coastal Scrub does not form a continuous north-south band; it is interrupted by Closed-cone Pine Forest, Coastal Prairie, and other plant communities.

A few good examples of Coastal Scrub are found in San Mateo, Marin, and Sonoma counties. The following plants are common at these localities.

Shrubs
  Artemisia californica, California Sagebrush (*Plate* 8)
  Baccharis pilularis, Coyotebrush (*Plate* 8)
  Cytisus scoparius, Scotch Broom (*Plate* 24)*
  Heteromeles arbutifolia, Toyon (*Plate* 48)
  Mimulus aurantiacus, Bush Monkeyflower (*Plate* 53)
Herbs
  Anaphalis margaritacea, Pearly Everlasting (*Plate* 7)
  Erigeron glaucus, Seaside Daisy (*Plate* 10)
  Eriophyllum lanatum var. *achillaeoides*, Common Woolly Sunflower (*Plate* 11)
  Scrophularia californica, Beeplant (*Plate* 54)
  Sisyrinchium bellum, Blue-eyed-grass (*Plate* 57)
  Triteleia laxa, Ithuriel's-spear (*Plate* 61)
  Zigadenus fremontii, Common Star Lily (*Plate* 61)

Close to the shores of San Francisco Bay (in Oakland and Berkeley, for example) an oak woodland interfingers with the open Coastal Scrub described above. In both of these areas the climate is much more temperate compared with Valley and Foothill Woodland. Some of the components of Coastal Forest are listed below.

Trees
  Corylus cornuta var. *californica*, Hazelnut (*Plate* 74)
  Quercus agrifolia, Coast Live Oak (*Plate* 28)
  Sambucus mexicana, Blue Elderberry (*Plate* 18)
Shrubs
  Rhamnus californica, California Coffeeberry (*Plate* 47)
  Rubus ursinus, California Blackberry (*Plate* 93)
  Toxicodendron diversilobum, Western Poison-oak (*Plate* 4)
Herbs
  Claytonia perfoliata, Miner's-lettuce (*Plate* 90)
  Stachys ajugoides var. *rigida*, Rigid Hedgenettle

Vernal Pools

A vernal pool is a low spot that fills with water during the rainy season, then dries out by the beginning of summer. This type of habitat almost always has an interesting community of herbaceous plants, some of which are not found anywhere else. A few of the more common species found in vernal pools are listed.

  Downingia concolor, Maroonspot Downingia (*Plate* 77)
  Downingia pulchella, Flatface Downingia
  Lasthenia glabrata, Yellowray Goldfields
  Limnanthes douglasii ssp. *douglasii*, Douglas Meadowfoam (*Plate* 34)

Freshwater Marsh

The vegetation of marshy areas around ponds, lakes, and slow-moving, shallow streams consists primarily of cattails, bulrushes, and sedges. There are, however, other types of aquatic or semi-aquatic herbaceous plants in this habitat, and the more commonly encountered species are listed here.

Carex obnupta, Slough Sedge
Cyperus eragrostis, Tall
  Cyperus (*Plate* 57)
Juncus balticus, Baltic Rush
Juncus lesueurii, Salt Rush
  (*Plate* 101)
Mimulus guttatus, Common
  Monkeyflower (*Plate* 53)
Oenanthe sarmentosa, Pacific
  Oenanthe
Potentilla anserina ssp.
  pacifica, Pacific Cinquefoil
  (*Plate* 49)

Scirpus acutus var.
  occidentalis, Hardstem
  Bulrush
Scirpus microcarpus,
  Smallfruit Bulrush
Sparganium eurycarpum, Giant
  Burreed (*Plate* 110)
Typha latifolia, Broadleaf
  Cattail (*Plate* 64)
Veronica americana, American
  Brooklime (*Plate* 96)

The water and muck in a freshwater marsh are typically slightly acid, and the availability of certain mineral nutrients that plants need-- including phosphorus, nitrogen, and molybdenum--is often low. When this is the case, plant productivity is also low. The density of the vegetation is deceptive; it has been achieved slowly. Most of the common plants, incidentally, are perennials that are adept at vegetative propagation.

Our freshwater marshes are not as prevalent or as large as they once were. To a considerable extent, the decrease has been caused by the draining of marshes in order to make more land available for agricultural use, by diverting water needed for irrigation, and by building dams to create lakes.

Backshores of Sandy Beaches

Landward from the line reached by high tides on wave-swept sandy beaches, and also around certain bays, is an area on which loose sand is deposited by wind. This strip, ranging in width from a few meters to hundreds of meters, is called the backshore. The stability of the sand depends on plants that are rooted in it and on the amount of organic matter and clay that it contains. Much of the organic matter will consist of the remains of previous generations of plants. Strong winds, especially when the tides are high, carry salt-laden droplets landward. Furthermore, the sand does not hold water very well and is generally poor in mineral nutrients that plants require. Its surface layer, on warm days, may absorb considerable heat from the sun. Being of a light color, the sand also reflects heat back at the plants. The grayish or whitish hairiness of the foliage of some species is probably protective, since it should help to reflect some of the heat away from them. Some of the backshore plants are moderately to very succulent, storing up water that enables them to withstand extended periods during which there may be neither rain nor fog. As sand is blown about, it may accumulate in dunes, whose stability is constantly challenged. Even if a dune has been fairly thoroughly colonized by plants, a strong wind may uproot enough of the sand-binding plants to cause breakdown of the whole dune. The vegetation of backshores includes certain grasses, many other herbaceous species, and a few shrubs. There are no trees, for the substratum is not conducive to the success of tall plants that are frequently exposed to strong winds.

The most successful colonizer of dunes is *Ammophila arenaria*, called European Beach Grass (*Plate* 62). It was introduced from Europe because of its capability as a sand-binder, and it is more effective in this respect than the native Dune Grass, *Leymus mollis* (*Plates* 63, 108). Where *Leymus* alone grows, the dunes nearest the shore, which are likely to be the least stable, tend to be lower than those in areas where *Ammophila* occurs. Other efficient sand-binders are *Juncus lesueurii*, Salt Rush (*Plate* 101), *Leymus pacificus*, Pacific Wild Rye, and *Poa douglasii*, Sand-dune Blue Grass.

Other plants found on the dunes often form tight low mats, and thus stabilize the substratum. Some of these species are listed below.

Abronia latifolia, Yellow
  Sand-verbena (*Plate* 36)
Abronia umbellata ssp.
  umbellata, Coast Sand-
  verbena (*Plate* 36)
Agoseris apargioides var.
  eastwoodiae, Coast Dandelion
  (*Plate* 6)
Ambrosia chamissonis, Silvery
  Beachweed (*Plate* 7)
Atriplex californica,
  California Saltbush
Atriplex leucophylla, Beach
  Saltbush (*Plate* 20)
Cakile edentula, Searocket*
Cakile maritima, Horned
  Searocket (*Plate* 17)*
Calystegia soldanella, Beach
  Morning-glory (*Plate* 20)
Camissonia cheiranthifolia,
  Beach Primrose (*Plate* 36)

Carpobrotus chilensis, Seafig
  (*Plate* 4)*?
Erigeron glaucus, Seaside
  Daisy (*Plate* 10)
Eschscholzia californica,
  California Poppy (*Plate* 40)
Fragaria chiloensis, Beach
  Strawberry (*Plate* 48)
Heliotropium curassavicum, Seaside
  Heliotrope (*Plate* 16)
Lathyrus littoralis, Beach Pea
  (*Plate* 25)
Lotus salsuginosus, Bird's-
  foot Trefoil
Plantago maritima, Sea
  Plantain (*Plate* 41)
Polygonum paronychia, Beach
  Knotweed (*Plate* 44)
Senecio elegans, Purple
  Ragwort (*Plate* 13)*
Tanacetum camphoratum, Dune
  Tansy (*Plate* 14)

The most conspicuous shrubby plant of the backshore habitat is *Lupinus arboreus*, the Yellow Bush Lupine (Fabaceae) (*Plate* 26). Its flowers are usually yellow, but there are forms with pale blue or pale lilac flowers. A little farther inland, where sand gives way to dirt, one may expect to see *Lupinus chamissonis*, the Chamisso Bush Lupine, whose flowers are always bluish. Some other common shrubs are *Artemisia pycnocephala*, Coastal Sagewort (*Plate* 8), *Baccharis pilularis*, Coyotebrush (*Plate* 8), and *Ericameria ericoides*, Mock-heather (*Plate* 73).

Coastal Salt Marsh

Salt marshes are typically found around the edges of bays and river mouths, where there is little wave action but where fine sediment--silt and clay--can gradually accumulate. The deposition of sediment is followed by invasion of plants, arriving as seeds or fragments that can take root. Successful colonization leads to expansion of the plant community. The plants stabilize the substratum and trap more fine sediment.

The salt marsh is at a level that is submerged only during highest tides. The salinity varies according to the extent of tidal flooding, rainfall, runoff of fresh water, and evaporation. Unless the rate of evaporation cancels out rainfall, and unless there is very little fresh water draining toward the salt marsh, the salinity of the marsh will probably be greater on the seaward side than on the landward side. This accounts, in part, for the differences in vegetation.

The slope of an extensive salt marsh is usually so gradual that the surface of the marsh is nearly level. It is likely, however, to have shallow pools, as well as a system of tidal creeks formed by the erosive action of tidal flow into and out of the marsh.

The pressures of civilization have led to destruction of salt marshes in many parts of the world. In San Francisco Bay, large areas of cheap acreage were needed for airports, industrial sites, housing developments, farmland, race tracks, and dumps, and today only about 25% of the area once occupied by salt marshes remains in a nearly natural state. A cursory look at the shoreline in populated areas will reveal the presence of industries whose wastes affect surviving marshes, as well as ecological aspects of the rest of the Bay.

Most of the flowering plants characteristic of salt marshes are found nowhere else. In the salt marshes of San Francisco Bay, and those of Tomales Bay, Bodega Bay, and some other estuarine habitats within the region for which this book is written, the following species are especially common and widespread.

Atriplex patula var. *obtusa*,
   Common Orache
Cuscuta salina var. *major*,
   Saltmarsh Dodder (*Plate* 21)
Distichlis spicata, Salt Grass
   (*Plate* 106)
Frankenia salina, Alkali-heath
   (*Plate* 28)
Jaumea carnosa, Fleshy Jaumea
   (*Plate* 12)

Limonium californicum,
   California Sea-lavender
   (*Plate* 41)
Plantago maritima, Sea
   Plantain (*Plate* 41)
Puccinellia nutkaensis, Alaska
   Alkali Grass
Salicornia virginica, Virginia
   Pickleweed (*Plate* 20)
Triglochin maritima, Seaside
   Arrow-grass (*Plate* 102)

Castilleja ambigua, Johnny-nip, *Glaux maritima*, Sea-milkwort, and *Salicornia europaea*, Slender Glasswort are less common, or at least not widespread in salt marshes.

On the seaward side of the salt marsh, there will often be *Spartina foliosa*, California Cord Grass (Poaceae). It is submerged by tides lower than those required to inundate plants of the salt marsh proper. *Spartina alterniflora*, Salt-water Cord Grass, recently introduced from the Atlantic coast, is spreading aggressively in some places in San Francisco Bay.

On slightly drier and less saline soil landward of the salt marsh, members of the following characteristic assemblage of plants may be encountered.

Atriplex patula var. *patula*,
   Spear Orache
Baccharis pilularis,
   Coyotebrush (*Plate* 8)
Grindelia stricta var.
   platyphylla, Pacific
   Gumplant

Rumex maritimus, Golden Dock
   (*Plate* 89)
Spergularia macrotheca var.
   macrotheca, Perennial Sand-
   spurry (*Plate* 19)
Tetragonia tetragonioides, New
   Zealand Spinach (*Plate* 4)*

In mucky areas rarely if ever touched by high tides, the water is fresh or at most very slightly brackish. Here one will find *Scirpus acutus* var. *occidentalis*, Hardstem Bulrush (*Plate* 99), *Scirpus robustus*, Robust Bulrush, and *Typha latifolia*, Broadleaf Cattail (*Plate* 64).

References

All of the works listed below deal to a considerable extent with topography, climate, geology, and other physical attributes of our region, and some discuss plant communities. References concerned primarily with the flora of California, the San Francisco Bay Region, and some other portions of the state are listed on p. 6.

Baker, E. S. 1984. An Island Called California, second edition. University of California Press, Berkeley, Los Angeles, London.
Barbour, M. G., R. B. Craig, F. R. Drysdale, and M. T. Ghiselin. 1973. Coastal Ecology: Bodega Head. University of California Press, Berkeley, Los Angeles, London.
Barbour, M. G., and J. Major. (editors). 1988. Terrestrial Vegetation of California, second edition. Special Publication no. 9, California Native Plant Society, Sacramento.
Barbour, M., B. Pavlik, F. Drysdale, and S. Lindstrom. 1993. California's Changing Landscapes. California Native Plant Society, Sacramento.
Bowerman, M. L. 1944. The Flowering Plants and Ferns of Mount Diablo. Gillick Press, Berkeley.
Cooper, W. S. 1967. Coastal Dunes of California. Memoir 104, Geological Society of America.
Dale, R. F. 1959. Climates of the States. California. Weather Bureau, United States Bureau of Commerce.
Evens, J. G. 1988. The Natural History of the Point Reyes Peninsula, Point Reyes National Seashore Association, Point Reyes, California.

22 References

Gilliam, H. 1962. Weather of the San Francisco Bay Region. University of
    California Press, Berkeley, Los Angeles, London.
Howell, J. T. 1970. Marin Flora, second edition, with supplement.
    University of California Press, Berkeley, Los Angeles, London.
McClintock, E., and W. Knight, with N. Fahy. 1968. A Flora of the San Bruno
    Mountains, San Mateo County, California. Proceedings of the California
    Academy of Sciences, series 4, 32: 587-677. (Reprinted 1990, as Special
    Publication no. 8, by California Native Plant Society, Sacramento.)
Munz, P. A., and D. D. Keck. 1949. California Plant Communities. El Aliso,
    2: 87-105, 199-202.
Ornduff, R. 1974. An Introduction to California Plant Life. University of
    California Press, Berkeley, Los Angeles, London.
Schoenherr, A. A. 1992. A Natural History of California. University of
    California Press, Berkeley, Los Angeles, London.
Sharsmith, H. K. 1945. Flora of the Mount Hamilton Range of California.
    American Midland Naturalist, 34: 289-367. (Reprinted 1982, as Special
    Publication no. 6, California Native Plant Society, Sacramento.)
Smith, A. C. 1959. Introduction to the Natural History of the San Francisco
    Bay Region. University of California Press, Berkeley, Los Angeles,
    London.
Sweeney, J. R. 1956. Responses of vegetation to fire. University of
    California Publications in Botany, 28: 143-250.

FERNS AND FERN ALLIES

Ferns

Ferns do not have flowers or seeds. Some or all of their leaves, or a specialized portion of a single leaf, bear structures called sporangia. These are often clustered, forming sori, and the sori may be protected, at least for a time, by flaps or disks called indusia. The function of the sporangia is to produce microscopic spores, which are eventually released and scattered by wind. Those that reach a suitable situation may develop into very small, short-lived plants called prothallia. These represent the sexual generation within the life cycle. Each prothallium produces a few eggs or many sperm, sometimes both. A film of water, from rain or dew, must be present to enable the sperm to swim to the eggs, which are located in little flask-shaped structures. An egg that has been fertilized develops into a new plant of the conspicuous spore-producing generation. At first there is only one tiny leaf, one stem, one root, but soon more leaves and roots are formed, and the young fern becomes independent of the prothallium, which then dies.

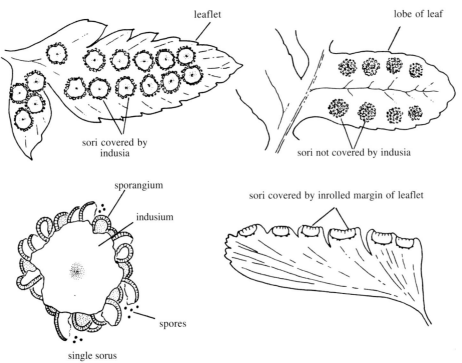

The distinctions between families of ferns are based to some extent on the arrangements of sporangia and sori and on the form of the indusia, when these are present. Key 1 separates the species of ferns; Key 2 will enable one to appreciate the more obvious features that define the 6 families found in the area. Note that 3 families--Blechnaceae, Dryopteridaceae, and Pteridaceae--appear more than once in this key.

Ferns, Key 1: Fern species
A   Each mature plant annually producing a single leaf, this further divided into a vegetative portion and a branching structure that bears numerous globular sporangia (in this species, the vegetative portion of the leaf

persists for another year, so the fern is essentially evergreen; up to
about 50 cm tall, but usually much smaller; generally in sphagnum bogs
and wet meadows)   *Botrychium multifidum*
                     [includes *B. multifidum* ssp. *silaifolium*] (*Plate* 65)
                              Leather Grape-fern; Mo-n; Ophioglossaceae
AA Plants not producing a single leaf that is divided into 2 very different
   portions (in some species, however, there are 2 types of leaves, one
   type being vegetative, the other bearing sporangia)
   B Leaves pinnately lobed or compound, the lobes or leaflets broad
      (except in temporary spore-producing leaves of *Blechnum spicant*), at
      most merely toothed, not deeply divided
      C Leaves compound, the leaflets attached to the rachis by a narrow,
         stalklike portion of the base (large ferns; leaves all similar,
         forming a crown)
         D Leaflets 3-8 cm long; indusium fringed with hairlike
            outgrowths (widespread)   *Polystichum munitum* (*Plate* 66)
                              Western Sword Fern; Dryopteridaceae
         DD Leaflets 2-3 cm long; indusium usually toothed, but not
            fringed with hairlike outgrowths
                         Polystichum imbricans [*P. munitum* var. *imbricans*]
                              Imbricate Sword Fern; Mo-n; Dryopteridaceae
      CC Leaves not compound, but deeply lobed, the lobes attached to the
         rachis by the full width of the base
         D Leaves forming a crown, and with dark, reddish brown or
            purplish brown petioles; leaves of 2 types: persistent
            vegetative leaves, generally more than 40 cm long with lobes
            5-8 mm wide, and temporary (spring-summer) spore-producing
            leaves with lobes about 2 mm wide
                                              Blechnum spicant (*Plate* 65)
                              Deer Fern; SCr-n; Blechnaceae
         DD Leaves arising from a creeping rhizome, not forming a crown,
            and not with dark brown petioles; leaves all of one type, and
            all potentially spore-producing, the sori being on the
            underside
            E Lobes of leaflets dark green, leathery, the unpaired
               terminal one often conspicuous (coastal, typically
               growing on cliffs and trunks of trees)
                                              Polypodium scouleri
                              Leatherleaf Fern; SCr-n; Polypodiaceae
            EE Lobes of leaflets light green, not leathery, and the
               terminal one not especially conspicuous (often growing on
               mossy tree trunks or logs, or in rocky habitats)
               F Veins of lobes translucent; lobes mostly tapering to a
                  point   *Polypodium glycyrrhiza*
                              Licorice Fern; Mo-n; Polypodiaceae
               FF Veins of lobes opaque; lobes mostly rounded at the tip
                  (widespread)   *Polypodium californicum* (*Plate* 1)
                              California Polypody; Polypodiaceae
   BB Leaves usually at least bipinnately compound (if not fully bipinnate,
      the leaflets are at least deeply toothed or lobed), and often
      tripinnate or quadripinnate
      C Leaflets somewhat fan-shaped (lopsidedly so in *Adiantum
         aleuticum*) (petioles nearly black, polished; sori under inrolled
         margin of the leaflets; deciduous)
         D Petiole continued as a single rachis, this branching
            alternately; most leaflets less than twice as long as wide,
            and not obviously lopsided (the stalk positioned in the
            middle) (widespread)   *Adiantum jordanii* (*Plate* 1)
                              California Maidenhair; Pteridaceae
         DD Petiole branching into two rachises, each of these branching
            dichotomously again; most leaflets about twice as long as
            wide, appearing lopsided with respect to the position of the
            slender stalk
                         Adiantum aleuticum [*A. pedatum* var. *aleuticum*] (*Plate* 1)
                              Fivefinger Fern; Pteridaceae
      CC Leaflets not at all fan-shaped
         D Petioles (at least their basal portions) brown or reddish
            brown, shiny
            E Leaves powdery on the underside, the sori occupying much
               of the underside of the leaflets
               F Leaves golden on the underside (widespread)
                                              Pentagramma triangularis
                         [*Pityrogramma triangularis* var. *triangularis*] (*Plate* 2)
                              Goldback Fern; Pteridaceae

FF Leaves whitish on the underside   *Pentagramma pallida*
                    [*Pityrogramma triangularis* var. *pallida*]
                              Silverback Fern; Pteridaceae
EE Leaves not powdery on the underside, the sori at, or close
   to, the margin of the leaflets
   F Leaflets with scales on the underside (and sometimes
     also on the upper surface) (not often found at
     elevations below 2000')
     G   Scales on the underside of leaflets divided into
         narrow lobes (scales not completely covering
         leaflets)   *Cheilanthes gracillima*
                              Lace Fern; SCl-n; Pteridaceae
     GG Scales on the underside of leaflets usually not
         divided (they are also widest at their bases)
         H   Scales usually completely covering underside of
             leaflets   *Cheilanthes covillei* (*Plate* 1)
                         Coville Lace Fern; Me-Mo; Pteridaceae
         HH Scales usually confined to the central portion
             of the leaflets   *Cheilanthes intertexta*
                         Coastal Lip Fern; Me-Mo; Pteridaceae
   FF Leaflets without scales on the underside
     G   Leaflets, or their divisions, rounded at the tip
                                   Pellaea andromedifolia
                         [*P. andromedaefolia*] (*Plate* 2)
                              Coffee Fern; Pteridaceae
     GG Leaflets, or their divisions, pointed at the tip
         H   Most leaflets arranged in groups of 3 on the
             branches of the rachis; outline of leaf blades
             elongated, not triangular
                         Pellaea mucronata (*Plate* 2)
                              Bird's-foot Fern; Pteridaceae
         HH Leaflets not arranged in groups of 3; outline
             of leaf blades not much longer than wide, and
             roughly triangular
             I   Leaves tripinnate; sori in longitudinal
                 rows that run the length of the ultimate
                 leaflets (often on serpentine)
                 Aspidotis densa [*Onychium densum*] (*Plate* 65)
                         Indian's-dream; SLO-n; Pteridaceae
             II Leaves nearly quadripinnate, the ultimate
                 leaflets often being deeply lobed; sori
                 crescent-shaped, near the margin of the
                 leaflets (Coast Ranges)
                                   Aspidotis californica
                         California Lace Fern; Pteridaceae
DD Petioles not brown or reddish brown, and not polished
   E   Leaves (up to more than 1 m long) arising singly from a
       creeping underground rhizome; sori under the inrolled
       margin of the leaflets (outline of leaf blade triangular;
       petiole and blade slightly hairy; deciduous; widespread)
                 Pteridium aquilinum var. *pubescens* (*Plate* 2)
                              Bracken; Dennstaedtiaceae
   EE Leaves (except in *Cystopteris fragilis*) forming a crown,
      rather than arising singly from an underground rhizome;
      sori not under the inrolled margin of the leaflets
      F   Leaves usually less than 30 cm long and arising singly
          from an underground rhizome (the leaves may be
          crowded, but they do not form a crown); delicate;
          mostly in crevices between rocks or at the edge of
          rocks (deciduous)   *Cystopteris fragilis* (*Plate* 65)
                              Brittle Fern; Dryopteridaceae
      FF Leaves usually at least 30 cm long, and forming a
         crown; not especially delicate; not limited to rocky
         habitats
         G   Leaf blades broadest near the middle, becoming
             obviously narrower toward the base (deciduous)
             H   Sori elongated, and arranged end-to-end in 2
                 rows; each ultimate division of the leaf blade
                 narrowing gradually to an acute tip; large,
                 stout species, with leaves commonly more than
                 1 m long   *Woodwardia fimbriata* (*Plate* 2)
                              Giant Chain Fern; Blechnaceae

> HH Sori semicircular or crescent-shaped, not
> arranged in 2 definite rows; ultimate
> divisions of the leaf blade rounded; leaves
> not often more than 1 m long (widespread)
>> *Athyrium filix-femina* var. *cyclosorum*
>> [*A. filix-femina* var. *sitchense*] (*Plate* 1)
>>> Lady Fern; SLO-n; Dryopteridaceae
> GG Leaf blades widest at the base or very close to the
> base, not near the middle
>> H Leaves partly bipinnate, but fully tripinnate
>> at the base
>>> I Leaf blades broadly triangular, the width
>>> at the base being more than half the
>>> length; on decaying logs or in soil that
>>> contains much decaying wood
>>>> *Dryopteris expansa*
>>>> Wood Fern; SM-n; Dryopteridaceae
>>> II Leaf blades narrowly triangular, the width
>>> at the base being less than half the
>>> length; in rocky areas *Polystichum dudleyi*
>>>> Dudley Shield Fern; Ma-SLO; Dryopteridaceae
>> HH Leaves pinnate to nearly or fully bipinnate,
>> but nowhere tripinnate
>>> I Each primary leaflet, next to where it
>>> originates from the rachis, with an
>>> enlarged lobe whose tip is directed toward
>>> the tip of the blade; blade bipinnate only
>>> in the lower half (upper primary leaflets
>>> merely toothed)
>>>> *Polystichum californicum* (*Plate* 1)
>>>> California Shield Fern; Me-SCr
>>>> Dryopteridaceae
>>> II Primary leaflets without one enlarged lobe
>>> that is directed toward the tip of the
>>> blade; blade bipinnate throughout
>>>> *Dryopteris arguta* (*Plate* 1)
>>>> Coastal Wood Fern; Dryopteridaceae

Ferns, Key 2: Fern families
A Individual sporangia about 1 mm wide, without an annulus (a strip of
thick-walled cells that straightens out in dry weather, tearing open the
ripe sporangium); sporangia restricted to one of the two divisions of
the single leaf produced during the growing season (the vegetative
portion of the leaf of the preceding season may have persisted,
however); young leaves not coiled
> Ophioglossaceae, Adder's-tongue Family
>> *Botrychium*
AA Individual sporangia less than 1 mm wide, with an annulus; sporangia
either on vegetative leaves or on specialized leaves, but not on a
division of the only leaf produced during the growing season; young
leaves at first coiled, unrolling as they enlarge (nearly all ferns
found in habitats other than sphagnum bogs fit under this choice)
> B Leaves of 2 different forms, the ultimate lobes of those producing
> sporangia much narrower than those of strictly vegetative leaves
>> C Vegetative and reproductive leaves pinnately lobed
>>> Blechnaceae, Deer Fern Family
>>>> *Blechnum*
>> CC Vegetative and reproductive leaves at least bipinnately compound
>>> Pteridaceae, Brake Fern Family
>>>> *Aspidotis*
> BB Leaves all alike, the sporangia produced on the vegetative leaves
>> C Sori located along the margin of the leaflets or lobes (sometimes
>> covered by the inrolled margin, but not by indusia)
>>> D Leaves produced in clusters; petioles brown to black, shiny,
>>> not grooved; rarely more than 50 cm tall
>>>> Pteridaceae, Brake Fern Family
>>>>> *Adiantum*, *Pellaea*
>>> DD Leaves widely scattered, arising singly from a creeping
>>> underground rhizome; petioles greenish, not shiny, hairy at
>>> the base, and with a prominent groove; up to more than 100 cm
>>> tall
>>>> Dennstaedtiaceae (formerly in Pteridaceae), Bracken Family
>>>>> *Pteridium*
>> CC Sori (or scattered sporangia) not located along the margin of the
>> ultimate leaflets or lobes

D Sporangia scattered along the veins of the leaflets, not
   concentrated in sori; indusia absent
                          Pteridaceae, Brake Fern Family
                            Cheilanthes, Pentagramma
DD Sporangia concentrated in sori; indusia sometimes present
   E Indusia absent
      F Leaves pinnately lobed, but not fully compound
                          Polypodiaceae, Polypody Family
                                        Polypodium
      FF Leaves bipinnately compound, the ultimate leaflets
         again deeply divided
                   Dryopteridaceae (Aspidiaceae), Wood Fern Family
                                          Athyrium
   EE Indusia present
      F Veins of leaves forming a network
                          Blechnaceae, Deer Fern Family
                                        Woodwardia
      FF Veins of leaves ending freely, not forming a network
         G Sori circular, each with an umbrellalike indusium
            attached in the center
                   Dryopteridaceae (Aspidiaceae), Wood Fern Family
                                        Polystichum
         GG Sori somewhat elongated, each with an indusium
            attached at one side (leaves bipinnately compound,
            or bipinnately lobed)
                   Dryopteridaceae (Aspidiaceae), Wood Fern Family
                            Cystopteris, Dryopteris

Fern Allies

The plants called fern allies have life cycles comparable to those of
ferns, and they are similar to ferns in lacking flowers, cones comparable
to those of gymnosperms, and seeds. They are, however, very diverse. The
following key will enable one to distinguish the few families represented
in our region.

A Plants rarely more than 3 cm long, floating on fresh water (sometimes
   stranded on mud above the water line) (leaves very small, scalelike,
   divided into 2 unequal lobes, the lower lobes serving as floats for the
   upper ones, which contain a symbiotic blue-green alga; stems obscure,
   due to the crowded, overlapping leaves, but nevertheless branching above
   every third leaf; roots short, inconspicuous, and unbranched)
                          Salviniaceae, Salvinia Family
                   Azolla filiculoides, Waterfern (*Plate* 66)
AA Plants not especially small and not floating
   B Plants rooted in mud or sand in shallow fresh water (or slightly
      above the water line); leaves either nearly cylindrical, quill-like,
      and clustered on a short, slightly bulbous stem, or arising singly
      or in groups from a creeping stem, in which case they are either
      almost hairlike and not more than 4 cm long or have a terminal blade
      that is divided palmately into 4 wedge-shaped leaflets
      C Leaves clustered on a short, slightly bulbous stem, nearly
         cylindrical, quill-like, with swollen bases (where the sporangia
         are borne)   Isoetaceae, Quillwort Family
                                               Isoetes
      CC Leaves arising singly or in groups along creeping stems, either
         only 2-4 cm long and almost hairlike, or much longer, with hairy
         petioles and 4 palmately arranged, wedge-shaped leaflets (spores
         produced in short-stalked, egg-shaped or spherical structures
         attached to the stems)   Marsileaceae, Marsilea Family
                                        Marsilea, Pilularia
   BB Plants generally not aquatic (although some are found in very wet
      places); leaves scalelike, either united to form sheaths that
      encircle the upright stems at the joints, or crowded along branching
      stems that form mosslike growths
      C Plants not mosslike, the stems hollow, jointed, sometimes
         branched at the joints, and with scalelike leaves that are
         attached to one another in such a way that they form a sheath
         above each joint; spores produced in structures that resemble
         cones of coniferous trees   Equisetaceae, Horsetail Family
                                               Equisetum

CC Plants mosslike, the stems not jointed, branching extensively and mostly completely covered by small, separate, scalelike leaves; spores produced in nearly quadrangular conelike structures (the sporangia, when ripe, are usually orange and clearly evident amongst the specialized leaves of each conelike structure)
Selaginellaceae, Spikemoss Family
Selaginella

Equisetaceae--Horsetail Family

Plants called horsetails and scouring-rushes have underground stems from which the above-ground stems develop. The cell walls of the epidermis contain considerable silica, and may also have silicified projections that make the surface gritty. In times past, before commercial scouring powders had been developed for cleaning pots and pans, pulverized stems of equisetums were used for this purpose.

A   Above-ground stems of 2 types: those that are green, never have conelike structures at the tip, and persist for several months; and those that are not green, have conelike structures at the tip, and are short-lived (they disappear after spores have been produced in the conelike structures) (green stems at least 6 mm thick and up to about 1 m tall; sheaths, consisting of united leaves encircling the stem, with about 20-30 teeth)   *Equisetum telmateia* ssp. *braunii* (*Plate* 2)
Giant Horsetail
AA All above-ground stems green, not short-lived, some bearing conelike spore-producing structures at the tip
   B   Sheaths about as long as wide, with 2 black bands, one just below the teeth, the other near the base; above-ground stems persisting from one year to the next   *Equisetum hyemale* ssp. *affine* (*Plate* 66)
Common Scouring-rush
   BB Sheaths decidedly longer than wide, with a single black band, this just below the teeth; above-ground stems generally annual (but sometimes persisting from one year to the next) *Equisetum laevigatum*
Smooth Scouring-rush

Isoetaceae--Quillwort Family

Quillworts, which grow in shallow water (often partly exposed) or in wet mud, have a crown of slender, cylindrical or 3-angled leaves that arise from a short, 2-or 3-lobed stem. The swollen base of most leaves contains a large sporangium within which spores are produced.

A   Stem 2-lobed at the base; sporangium at the base of each fertile leaf only partly covered by the membrane that grows down over it (leaves up to about 25 cm long)   *Isoetes howellii*
Howell Quillwort
AA Stem 3-lobed at the base; sporangium at the base of each leaf completely covered by the membrane that grows down over it
   B   Plants rarely with more than 20 leaves, these usually less than 7 cm long; leaves without distinct lengthwise strands of supporting tissue; plants maturing while still partly submerged
Isoetes orcuttii
Orcutt Quillwort; CC-s
   BB Plants with up to 75 leaves (sometimes with only a few, however), these usually more than 7 cm long (up to 20 cm long); leaves with 3 distinct lengthwise strands of supporting tissue; plants generally maturing while out of water   *Isoetes nuttallii* (*Plate* 66)
Nuttall Quillwort

Marsileaceae--Marsilea Family

Plants of the Marsilea Family grow partly submerged in shallow fresh water, or in mud that has been exposed by a drop in the water level. Their creeping stems produce single leaves, or clusters of leaves, at close intervals, and also short-stalked globular or ovoid structures within which spores are produced.

A   Leaves up to about 20 cm long, each with a petiole and a blade composed
    of 4 palmately arranged, wedge-shaped leaflets, thus resembling leaves
    of a "four-leaf" clover; at the margin of lakes, ponds, slow streams,
    and in some marshlands   *Marsilea vestita* (*Plate* 66)
                                                          Hairy Pepperwort
AA Leaves rarely more than 4 cm long, the blade not distinctly different
    from the petiole; mostly restricted to vernal pools *Pilularia americana*
                                                                 Pillwort

Salviniaceae--Salvinia Family

   Our only representative of the Salviniaceae is *Azolla filiculoides*,
called Waterfern (*Plate* 66). The upper lobes of the leaves, which are the
ones readily visible, are about 1 mm long, and nearly obscure the stems.
The pale bluish green color of these lobes is due, in part, to the threads
of a symbiotic blue-green alga (*Anabaena azollae*) that grows within them.
The plants usually become reddish in summer. This interesting relative of
the ferns is often abundant in ponds and other freshwater situations where
the water is quiet.

Selaginellaceae--Spikemoss Family

   The genus *Selaginella* is represented in the area by 2 species. These
plants, which superficially resemble mosses, grow mostly in crevices
between rocks on exposed hillsides. They are extremely drought-resistant,
appearing dead during dry weather, becoming green after they have been well
soaked by rains.

A   Stems mostly prostrate, rooting at intervals, sometimes even near the
    tip   *Selaginella wallacei* (*Plate* 66)
                                                        Little Clubmoss; Ma-n
AA Stems at least partly upright, rooting only near the base, where leaves
    do not persist   *Selaginella bigelovii*
                                                           Spikemoss; Sn-s

TREES

Needle-leaved, Cone-bearing Trees and their Relatives (Gymnosperms)

Gymnosperms produce seeds, but they do not have flowers. The name of the group, meaning "naked seed," alludes to the fact that the seeds are not enclosed by a structure fully comparable to the fruit of a flowering plant. They are, in fact, generally exposed, at least by the time they are mature. Although there are exceptions with respect to this point, it is rarely difficult to decide whether a plant is or is not a gymnosperm. Most gymnosperms--and all of our native species--have needlelike, scalelike, or awl-like leaves of types that are not often encountered among flowering plants.

In all of our native gymnosperms other than the Pacific Yew, California Nutmeg, and California Juniper, the seeds are produced in cones that become woody when they mature. Each cone consists of several to many scales, and it is on the surface of these scales that the seeds develop. In junipers, the scales are fleshy and so tightly fused together that the conelike structure is not immediately apparent. In yews, a single seed develops within a shallow cup; in the California Nutmeg, which also belongs to the Yew Family (Taxaceae), the seed becomes completely enclosed, and the fleshy covering superficially resembles the fruit of an olive.

The species of gymnosperms are separated in Key 1, which follows. For those who wish to know the distinctions between the four families of gymnosperms represented in the region, these are summarized in Key 2.

Gymnosperms, Key 1: Gymnosperm species
A Leaves scalelike, except on very small seedlings, on which they may be
    needlelike
    B Leaves in whorls of 3; seed-bearing structures (on trees separate
        from those that produce pollen) somewhat fleshy (1-1.5 cm wide, and
        covered with a whitish deposit)   *Juniperus californica* (*Plate* 3)
                                            California Juniper; Cupressaceae
    BB Leaves in pairs or whorls of 4; seed-bearing structures, when mature,
        woody cones (although the scales may remain fitted closely together)
        C Branchlets flattened; leaves in whorls of 4; cones about twice as
            long as wide (until they open and the 2 outer scales diverge);
            each cone scale with 2 seeds   *Calocedrus decurrens* (*Plate* 67)
                                            Incense Cedar; Cupressaceae
        CC Branchlets cordlike, not flattened; leaves in pairs; cones nearly
            globular, the scales with prominent bumps; each cone scale with
            numerous seeds
            D Upper surface of older leaves without a resin pit
                E Top of tree decidedly conical; new shoots not often more
                    than 1.5 mm wide; cones usually 2-2.5 cm long; up to 7 m
                    tall; not commonly cultivated   *Cupressus abramsiana*
                                        Santa Cruz Cypress; SM-SCr; 1b; Cupressaceae
                EE Top of tree usually flattened or rounded; new shoots more
                    than 1.5 mm wide; cones usually at least 2.5 cm long; up
                    to more than 20 m tall; native to Monterey County, but
                    often planted elsewhere   *Cupressus macrocarpa* (*Plate* 3)
                                            Monterey Cypress; Cupressaceae
            DD Upper surface of older leaves with a resin pit (but the pit
                may be closed)
                E Bumps on cone scales low-conical, not elongated or curved;
                    limited to mountains near the coast (often found on
                    serpentine soils)   *Cupressus sargentii*
                                            Sargent Cypress; Me-s; Cupressaceae
                EE Bumps on cone scales (especially those of the terminal
                    pair) elongated and curved; inland   *Cupressus macnabiana*
                                            McNab Cypress; Sn-n; Cupressaceae
AA Leaves needlelike
    B Needles in bundles of 2, 3, or 5
        C Needles in bundles of 2 (cones lopsided; coastal)

D Cones up to about 9 cm long, usually closed at maturity;
    needles usually 10-12 cm long   *Pinus muricata* (*Plate* 3)
                                                    Bishop Pine; Pinaceae
DD Cones less than 5 cm long, open at maturity; needles mostly
    3-6 cm long   *Pinus contorta*
                                                 Shore Pine; Pinaceae
CC Needles in bundles of 3 or 5
    D Needles in bundles of 5 (cones up to 40 cm long, open at
        maturity)   *Pinus lambertiana* (*Plate* 68)
                                                  Sugar Pine; Pinaceae
    DD Needles in bundles of 3
        E Cones generally at least 15 cm long (open at maturity)
            F Cones mostly 25-30 cm long; attachment of cones off-
                center at the base; needles green, not obviously
                grayish; branches erect   *Pinus coulteri* (*Plate* 68)
                                          Coulter Pine; CC-s; Pinaceae
            FF Cones mostly 15 to 20 cm long; attachment of cones
                centered at the base; needles gray-green; branches
                drooping (widespread)   *Pinus sabiniana* (*Plate* 68)
                                             Foothill Pine; Pinaceae
        EE Cones rarely so much as 15 cm long
            F Needles commonly 12-20 cm long; cones open at maturity
                (Coast Ranges)   *Pinus ponderosa* (*Plates* 3 and 68)
                                         Pacific Ponderosa Pine; Pinaceae
            FF Needles generally less than 12 cm long; cones opening
                if exposed to fire or prolonged drying
                G Cones about twice as long as wide (in mountains
                    away from the coast)   *Pinus attenuata* (*Plate* 67)
                                          Knobcone Pine; SCL-n; Pinaceae
                GG Cones less than twice as long as wide (native to
                    San Mateo County and south, but widely planted
                    elsewhere in the region)   *Pinus radiata* (*Plate* 68)
                                          Monterey Pine; SM-s; 1b; Pinaceae
BB Needles arising individually, not in bundles
    C Needles notched at the tip (the only needle-leaved tree in our
        region that has this characteristic) (seed-bearing cones
        upright, 8-10 cm long, restricted to upper branches, and falling
        apart at maturity)   *Abies grandis* (*Plate* 67)
                                            Grand Fir; Sn-n; Pinaceae
    CC Needles not notched at the tip
        D Needles with sharply pointed tips
            E Needles dark green on both surfaces; seed-bearing
                structures woody cones about 2.5 cm long; trees commonly
                taller than 15 m, and sometimes attaining 100 m (native
                to areas close to the coast, but widely planted
                elsewhere)   *Sequoia sempervirens* (*Plate* 3)
                                         Coast Redwood; Taxodiaceae
            EE Needles dark green on the upper surface, either yellow-
                green or with 3 green lines and 2 white lines on the
                underside; seed-bearing structures (on trees separate
                from those that produce pollen) fleshy and 1-seeded;
                rarely attaining a height of 15 m
                F Needles mostly 3-5 cm long, with 3 green lines and 2
                    white lines on the underside; seed completely enclosed
                    within a green or purple-streaked, fruitlike covering
                    about 2.5 cm long   *Torreya californica* (*Plate* 4)
                                       California Nutmeg; Me-Mo; Taxaceae
                FF Needles mostly 2 cm long, yellow-green on the
                    underside; seed partly enclosed within a red cup about
                    1 cm long   *Taxus brevifolia* (*Plate* 4)
                                          Pacific Yew; SCr-n; Taxaceae
        DD Needles rounded at the tip
            E Needles mostly of 2 sizes, some about 1 cm long, others
                about 1.5 cm long; cones about 2 cm long
                                              Tsuga heterophylla (*Plate* 3)
                                         Western Hemlock; Sn-n; Pinaceae
            EE Needles not obviously of 2 sizes, and generally at least 2
                cm long; cones 6-7 cm long
                F Cones hanging down, not limited to upper branches, not
                    falling apart at maturity; seed-bearing scales
                    alternating with 3-pronged sterile bracts; needles 2-3
                    cm long; at elevations from near sea level to 5000'
                    (widespread)   *Pseudotsuga menziesii* (*Plate* 3)
                                           Douglas-fir; Mo-n; Pinaceae

FF Cones upright, only on upper branches, falling apart
   at maturity; seed-bearing scales not alternating with
   sterile bracts; needles 3-4 cm long; at elevations of
   3000' and higher   *Abies concolor* (*Plate* 67)
                                   White Fir; Pinaceae

Gymnosperms, Key 2: Gymnosperm families
A Seeds produced singly, either incompletely surrounded by a fleshy
   cuplike structure or completely enclosed within a fleshy, fruitlike
   covering that shows no evidence of consisting of coalesced scales;
   leaves always needlelike   Taxaceae, Yew Family
                                            Taxus, *Torreya*
AA Seeds produced on the surface of scales of a woody cone, or within a
   berrylike structure that shows some external evidence of consisting of a
   few coalesced scales; leaves needlelike or scalelike
   B Leaves opposite or whorled, scalelike or awl-like (but needlelike on
      young seedlings); cones woody or fleshy Cupressaceae, Cypress Family
                                  Calocedrus, *Cupressus*, *Juniperus*
   BB Leaves alternate (or in bundles that are alternate), needlelike;
      cones woody
      C Buds of cones enclosed by obvious bracts; leaves normally falling
         separately after aging; leaves sometimes in bundles, sometimes
         separate   Pinaceae, Pine Family
                                  Abies, *Pinus*, *Pseudotsuga*, *Tsuga*
      CC Buds of cones not enclosed by obvious bracts; leaves not falling
         separately after aging, entire twigs being deciduous; leaves
         always separate   Taxodiaceae, Redwood Family
                     Sequoia sempervirens, Coast Redwood (*Plate* 3)

Broad-leaved Trees

A Leaves opposite (or sometimes in whorls)
   B Leaves compound, with 3 or more completely separate leaflets
      C Leaves palmately compound (or, if technically pinnate, with only
         3 leaflets)
         D Leaflets 5-9 (flowers white to pink, 1.5 cm long, in large
            upright clusters; sometimes more than 8 m tall; widespread)
                                   Aesculus californica (*Plate* 30)
                           California Buckeye; Hippocastanaceae
         DD Leaflets 3 (leaves technically pinnate, but look palmate)
            E Some leaflets lobed as well as toothed, the terminal one
               often wider than the other 2; fruit dry, consisting of 2
               winged halves; up to about 20 m tall
                           Acer negundo var. *californicum* (*Plate* 69)
                                           Boxelder; Aceraceae
            EE Leaflets toothed but not lobed, of approximately equal
               size; fruit a fleshy berry (this nearly black, but
               appearing blue because of a whitish coating); up to about
               8 m tall (widespread)
                     Sambucus mexicana [includes *S. caerulea*] (*Plate* 18)
                              Blue Elderberry; La-s; Caprifoliaceae
      CC Leaves pinnately or bipinnately compound
         D Margin of leaflets smooth; terminal leaflet often larger than
            the others (fruit dry, winged; in moist areas)
                                          Fraxinus latifolia
                              Oregon Ash; SCl-n; Oleaceae
         DD Margin of leaflets toothed; all leaflets about the same size
            E Leaflets 2-4 cm long, without an elongated tip; fruit dry,
               winged   *Fraxinus dipetala*
                              California Ash; Oleaceae
            EE Leaflets 5-15 cm long, with an elongated tip; fruit a
               fleshy berry
               F Leaflets usually 7; fruit red; inflorescence conical
                  (usually in damp places)
                           Sambucus racemosa [*S. callicarpa*] (*Plate* 18)
                              Red Elderberry; SM-n; Caprifoliaceae
               FF Leaflets 3-5; fruit black, but appearing blue because
                  of a whitish coating; inflorescence nearly flat-topped
                  (widespread)
                     Sambucus mexicana [includes *S. caerulea*] (*Plate* 18)
                              Blue Elderberry; La-s; Caprifoliaceae
   BB Leaves not compound

    C   Leaves palmately lobed (blades usually more than 10 cm wide;
       flowers greenish, in racemes; fruit dry, consisting of 2 winged
       halves; widespread)   *Acer macrophyllum* (*Plate* 69)
                                      Bigleaf Maple; Aceraceae
   CC Leaves not palmately lobed
      D Flowers in dense, almost spherical heads up to about 3 cm
         wide (leaves up to 15 cm long, with short petioles; corolla
         consisting of a long tube and 4 lobes, white or yellowish;
         fruit dry, widest near the top; in moist habitats)
              Cephalanthus occidentalis var. *californicus* (*Plate* 93)
                  California Button-willow; Na-n; Rubiaceae
    DD Flowers not in almost spherical heads
        E Leaf blades toothed (often inconspicuously in *Viburnum*
           ellipticum)
           F Leaf blades up to more than 10 cm long, often more
              than twice as long as wide, with only one main vein
              (the midrib) and usually with a slender terminal
              portion; flowers solitary or a few on long peduncles
              in the leaf axils, about 1 cm wide, with 5 petals,
              these brownish purple, with small dots; fruit 3-lobed,
              dry at maturity   *Euonymus occidentalis* (*Plate* 19)
                        Western Burningbush; Mo-n; Celastraceae
          FF Leaf blades rarely more than 5 cm long, less than
              twice as long as wide, usually with 3 main veins
              arising at the base, more or less rounded at the tip;
              flowers in umbel-like inflorescences, less than 1 cm
              wide, the corolla white; fruit fleshy, red
                                Viburnum ellipticum
                Oval-leaf Viburnum; CC, Sn-n; Caprifoliaceae
      EE Leaf blades not toothed
        F Leaf blades (these oval or elliptical) leathery;
           flowers in drooping pistillate or staminate catkins
                           Garrya (*see* Garryaceae)
        FF Leaf blades not leathery; flowers not in drooping
           catkins
           G Large tree, up to more than 20 m tall; flowers in
              dense, almost buttonlike clusters, each cluster
              above 4-7 large white or yellowish bracts; fruit
              orange-red; mostly in woods or at the edge of
              woods   *Cornus nuttallii* (*Plate* 21)
                    Mountain Dogwood; SCl, Na; Cornaceae
           GG Small trees or shrubs, rarely more than 5 m tall;
             flowers in umbel-like inflorescences, the bracts
             of these green or grayish green; fruit white,
             cream-colored, or bluish; in moist habitats
                         Cornus (*see* Cornaceae)
AA Leaves alternate
  B Leaves compound
    C Leaves palmately compound (with 3 leaflets) (leaves aromatic;
      flowers small, greenish white, in clusters; rarely more than 4 m
      tall)   *Ptelea crenulata* (*Plate* 51)
                          Hoptree; SCl-n; Rutaceae
   CC Leaves pinnately or bipinnately compound (all except *Juglans*
     californica var. *hindsii*, are escapes from cultivation)
      D Leaves bipinnately compound, the leaflets less than 1 cm long
        E Each of the primary divisions of the leaf with 30-40 pairs
          of leaflets, these usually 3-4 mm long (sometimes
          longer); flowers small, yellow, in several rounded
          clusters on each raceme; fruit with conspicuous
          constrictions   *Acacia decurrens*
                        Green Wattle; au; Fabaceae
        EE Each of the primary divisions of the leaf with 7-10 pairs
          of leaflets, these usually about 5 mm long; flowers few
          in each inflorescence (petals 2-3 cm long, yellow,
          stamens 8-10 cm long); fruit without obvious
          constrictions   *Caesalpinia gilliesii*
                        Bird-of-paradise; sa; Fabaceae
    DD Leaves only pinnately compound, the leaflets at least 1.5 cm
      long
      E Leaves and fruit peppery-smelling (leaves about 20-30 cm
        long, with 30-40 leaflets, these 3-5 cm long; fruit
        round, about 5 mm wide, pinkish red, numerous in each
        much-branched cluster   *Schinus molle*
                Peruvian Peppertree; sa; Anacardiaceae

EE Leaves and fruit not peppery-smelling
    F Evergreen; leaves with an even number of leaflets,
       there being no unpaired terminal leaflet (leaflets 8-
       16, up to 4 cm long; corolla 2.5 cm wide, noticeably
       irregular, yellow)
                  Senna multiglandulosa [*Cassia tomentosa*]
                              Senna; mx; Ma; Fabaceae
    FF Deciduous; leaves with an odd number of leaflets, one
       of these being terminal
       G Leaflets (usually 11-17) up to 2.5 cm long, oval,
          widest near the middle, more or less rounded at
          the tip; base of leaf petiole flanked by a pair of
          spines   *Robinia pseudoacacia*
                       Black Locust; na; Fabaceae
       GG Leaflets generally more than 5 cm long, broadest
          below the middle, narrowing gradually to a pointed
          tip; base of leaf petiole not flanked by spines
          H Leaflets (usually 15-23) evenly finely toothed;
             fruit a walnut, up to about 4 cm wide in the
             husk
               Juglans californica var. *hindsii* [*J. hindsii*]
               Northern California Black Walnut; Na, CC; 1b
                                Juglandaceae
          HH Leaflets (usually 11-31) with smooth margins,
             except near the base, where there are a few
             coarse teeth; fruit 3-5 cm long, flattened,
             the single seed forming a bulge near the
             middle   *Ailanthus altissima*
                Tree-of-heaven; as; Simaroubaceae
BB Leaves not compound (but they may be deeply lobed)
  C Leaves palmately lobed
    D Leaf blades usually more than 10 cm wide, the lobes deeply
      separated, pointed at the tip; deciduous (bark smooth, pale,
      with a pattern; flowers and fruit in hanging globular heads;
      leaves with stipules (these eventually falling off; usually
      near water)   *Platanus racemosa* (*Plate* 87)
                       Western Sycamore; Platanaceae
    DD Leaf blades rarely more than 5 cm wide, the lobes shallowly
      separated, rounded; evergreen (but often losing many leaves)
      (leaf surfaces rough to the touch; flowers showy, the 5-lobed
      calyx orange-yellow, up to 5 cm wide
      Fremontodendron californicum [includes *F. californicum* sspp.
                  crassifolium and *napense*] (*Plate* 55)
                     Flannelbush; Sterculiaceae
  CC Leaves either pinnately lobed or not lobed
    D Leaves (dull green) less than 5 mm long, scalelike and
      overlapping (flowers small, pink, in dense racemes up to 5 cm
      long; naturalized, especially near streams and in
      bottomlands)   *Tamarix ramosissima*
                  Tamarisk; as; Tamaricaceae
    DD Leaves usually more than 5 mm long, not scalelike
      E Branches thorny or ending in sharp spines. (If only the
        leaves are prickly, take choice EE.)
        F Some leaf blades with 1 or 2 especially large teeth
          (almost lobes) below the middle (deciduous; flowers
          few in each terminal cluster, about 1.5 cm wide; fruit
          resembling an apple in appearance and texture,
          yellowish or reddish, up to 1.5 cm long; pedicels of
          fruit at least 2 cm long)   *Malus fusca* (*Plate* 92)
                  Oregon Crabapple; Sn, Na-n; Rosaceae
        FF None of the leaf blades with 1 or 2 especially large
          teeth below the middle
          G Deciduous; leaf blades up to about 8 cm long,
             broadly oval (leaf blades coarsely toothed,
             sometimes almost lobed; flowers about 1 cm wide;
             petals 5, white; stamens usually about 20; fruit
             blackish, about 8 mm long)   *Crataegus suksdorfii*
                   Black Hawthorn; Ma-n; Rosaceae
          GG Evergreen; leaf blades not often more than 3 cm
             long (flowers numerous in each panicle; petals and
             sepals white to blue; fruit 3-lobed, dry)
                          Ceanothus spinosus
              Greenbark Ceanothus; SLO-n; Rhamnaceae
EE Branches neither thorny nor ending in stout spines

F   Leaf blades (or apparent leaf blades) with 2 or more
    main veins originating at the base
    G   "Leaf blades" (they are in reality flattened
        petioles) several times as long as wide; flowers
        small, cream or yellow (escapes from cultivation)
        H   Flowers cream-colored, in clusters on each
            raceme; "leaf blades" up to 10 cm long, with
            3-5 main veins   *Acacia melanoxylon*
                            Blackwood Acacia; au; Fabaceae
        HH Flowers yellow, solitary along the raceme;
            "leaf blades" up to 15 cm long, with 2 or 4
            main veins   *Acacia longifolia*
                            Golden Wattle; au; Fabaceae
    GG Leaf blades not several times as long as wide
        H   Leaf blades (these with 3-5 main veins, and
            mostly about 5 cm long) rounded, but with a
            notch at the base (and often with a slight
            indentation at the tip); flowers 8-12 mm long,
            few in each cluster, resembling those of a
            pea, reddish purple; fruit a dry pod up to 8
            cm long (deciduous)
                            Cercis occidentalis (*Plate* 24)
                            Western Redbud; Sl-n; Fabaceae
        HH Leaf blades (these evergreen, with 3 main
            veins), generally oval, without a notch at the
            base; flowers small, numerous in each
            inflorescence, not resembling those of a pea,
            pale to deep blue; fruit dry, less than 1 cm
            long, 3-lobed
            I   Margin of leaf blades turned under (flowers
                deep blue)   *Ceanothus parryi*
                            Parry Ceanothus; Na, Sn; Rhamnaceae
            II Leaf blades flat, the margin not noticeably
                turned under (flowers pale to deep blue)
                J   Small branches cylindrical, usually
                    gray-green, sometimes with short hairs
                            Ceanothus oliganthus var. *sorediatus*
                                            [*C. sorediatus*]
                                    Jimbrush; Rhamnaceae
                JJ Small branches angled and green at
                    first, then becoming dark and streaked,
                    not hairy (widespread)
                            Ceanothus thyrsiflorus [includes
                            C. thyrsiflorus var. *repens*] (*Plate* 47)
                                    Blue-blossom; Rhamnaceae
FF Leaf blades with only 1 main vein (the midrib)
    originating at the base
    G   Leaves (these evergreen) decidedly aromatic or with
        a strong odor when bruised
        H   Leaf blades (these bluish green, usually about
            6-10 cm long) less than twice as long as wide
            (corolla about 3 cm long, 1 cm wide, mostly
            tubular, but with 5 lobes, pale yellow; well
            established)   *Nicotiana glauca* (*Plate* 54)
                            Tree-tobacco; sa; Solanaceae
        HH Leaf blades at least 3 times as long as wide
            I   Leaf blades (these up to 12 cm long, about
                4 times as long as wide) distinctly toothed
                (large and small teeth often alternating)
                (staminate and pistillate flowers [these
                without petals] in catkinlike structures;
                fruit nearly spherical, rough, covered with
                a whitish wax; primarily near the coast)
                            Myrica californica (*Plate* 85)
                                    Wax-myrtle; Myricaceae
            II Leaf blades not toothed
                J   Leaves olive-green, 3-10 cm long,
                    abruptly pointed; flowers yellow, less
                    than 1 cm wide, in clusters at the ends
                    of branches; stamens 9; fruit ovoid,
                    about 2 cm long, fleshy, with a single
                    large seed (widespread)
                            Umbellularia californica (*Plate* 34)
                                    California Bay; Lauraceae

JJ Leaves pale green, 15-20 cm long,
gradually narrowed to a point; flowers
white, 2 cm wide, mostly solitary in
the leaf axils; stamens more than 25;
fruit a hemispherical, flat-topped
woody capsule about 2 cm high (so
abundant in some places as to be
mistaken for a native)
Eucalyptus globulus
Blue Gum; au; Myrtaceae
GG Leaves neither aromatic nor with a strong odor when
bruised
H Outer bark of much of trunk and main stems
smooth, reddish brown (and peeling away);
corolla urn-shaped, usually less than 1 cm
long, with small lobes
I   Flower panicles showy, up to 15 cm long;
leaves mostly 6-12 cm long, the upper
surface dark green; up to more than 30 m
tall; fruit (bright red, about 1 cm long)
with many small bumps
Arbutus menziesii (*Plate* 22)
Pacific Madrone; Ericaceae
II Flower panicles rarely more than 3 cm long;
leaves 2-7 cm long, often grayish green; up
to 3 m tall; fruit fleshy but usually firm,
smooth   *Arctostaphylos* (*see* Ericaceae)
HH Bark of trunk and main stems, if smooth, not
reddish brown and not peeling away; corolla
not urn-shaped
I   Leaf blades with a prominent notch at the
base (blades coarsely toothed, commonly 5-6
cm long; staminate flowers in drooping
catkins; pistillate flowers [with bright
red stigmas] usually solitary or in pairs;
fruit a nut up to 1.5 cm long, enclosed by
a pair of united, somewhat papery bracts)
Corylus cornuta var. *californica* (*Plate* 74)
Hazelnut; SCr-n; Betulaceae
II Leaf blades without a prominent notch at
the base   Broad-leaved Trees, Subkey
Broad-leaved Trees, Subkey: Leaves alternate, not compound, the blades with
1 main vein (the midrib) originating at the base; branches neither thorny
nor ending in stout spines
A Fruit a nut enclosed within a spiny bur (at least a few burs can usually
be found at any season); leaves dark green on the upper surface, golden
brown on the underside, especially when young; flowers in branched,
catkinlike inflorescences
B Leaves not folded, mostly 7-9 cm long; at elevations up to 1500'; up
to 45 m tall   *Chrysolepis chrysophylla* var. *chrysophylla*
Giant Chinquapin; Ma-n; Fagaceae
BB Leaves often folded along the midrib or wavy along the margin, mostly
3-4 cm long; at elevations up to 6000'; up to 10 m tall (often
somewhat shrubby)   *Chrysolepis chrysophylla* var. *minor* (*Plate* 28)
Golden Chinquapin; Fagaceae
AA Fruit not a nut enclosed within a spiny bur, and plants not conforming
to other features described in choice A
B Trees recognizable as oaks: fruit an acorn (look on the ground under
the tree for acorns or the cuplike involucres that partly enclose
the acorns); staminate flowers in catkins (leaves tough, sometimes
convex on the upper surface or with slightly inrolled margins, often
pinnately lobed, and often with spine-tipped teeth)
Lithocarpus, *Quercus* (*see* Fagaceae)
BB Trees other than oaks: definitely without acorns; flowers either not
in catkins, or both pistillate and staminate flowers in catkins
C Flowers small, deep blue, in crowded inflorescences (margin of
leaves turned under)   *Ceanothus parryi*
Parry Ceanothus; Na, Sn; Rhamnaceae
CC Flowers not deep blue
D Leaves not toothed (there may be glands on the margins,
however)
E   Leaves deciduous, rarely more than 3 cm wide; flowers in
staminate and pistillate catkins, these usually on
separate trees   *Salix* (*see* Salicaceae)

EE Leaves evergreen, usually at least 3 cm wide; flowers in
   showy clusters, the corolla funnel-shaped and lobed, up
   to about 4 cm wide, mostly pink or rose
                          Rhododendron macrophyllum (*Plate* 23)
                              Rosebay; Mo-n; Ericaceae
DD Leaves toothed (but the teeth may be small, blunt, or
   otherwise inconspicuous)
   E   Teeth limited to the upper half or two-thirds of the leaf
       blades (which are not so much as twice as long as wide)
       F   Evergreen; leaf blades usually wedge-shaped at the
           base, and the teeth often blunt; flowers (without
           petals) 1-3 in each axillary cluster; fruit dry, with
           a hairy, persistent style usually at least 5 cm long
           (widespread)   *Cercocarpus betuloides* (*Plate* 48)
                                   Mountain-mahogany; Rosaceae
       FF Deciduous; leaf blades usually rounded at the base,
           and the teeth commonly pointed; flowers (these about 2
           cm wide, with 5 white petals) several in each short
           raceme; fruit about 1 cm long, blue-black, fleshy but
           almost juiceless when ripe
                                Amelanchier alnifolia (*Plate* 48)
                                   Serviceberry; Rosaceae
   EE Most leaf blades with teeth around all of the margin,
       except perhaps at the very base
       F   Most leaf blades somewhat triangular in outline
           (although rounded at the lower corners) (staminate and
           pistillate catkins on separate trees; seeds with long
           white hairs that aid in dispersal by wind; usually
           close to streams, lakes, and ponds)
                                   Populus (*see* Salicaceae)
       FF Leaf blades not at all triangular in outline
           G   Teeth either prickly, or at least stiff and spine-
               tipped (evergreen; leaf blades more than 4 cm
               long, almost leathery)
               H   Teeth prickly; leaf blades less than 3 times as
                   long as wide; flowers in racemes; fruit up to
                   about 2 cm long, red to blue-black, juicy
                   (resembling that of a cultivated cherry)
                                   Prunus ilicifolia (*Plate* 49)
                                      Hollyleaf Cherry; Na-s; Rosaceae
               HH Teeth stiff, with obvious spines; leaf blades
                   at least 3 times as long as wide; flowers in
                   panicles; fruit rarely so much as 1 cm long,
                   red, fleshy (but not especially juicy)
                   (widespread)
                                   Heteromeles arbutifolia (*Plate* 48)
                                      Toyon; Rosaceae
           GG Teeth not at all prickly and not spine-tipped (but
               they may be sharply pointed)
               H   Neither pistillate not staminate flowers in
                   catkins or catkinlike structures
                   I   Some leaf blades with 1 or 2 especially
                       large teeth (almost lobes) below the middle
                       (flowers few in each terminal cluster,
                       about 2 cm wide; fruit yellowish or
                       reddish, up to 1.5 cm long, on pedicels at
                       least 2 cm long; branches sometimes
                       thorny)   *Malus fusca* (*Plate* 92)
                             Oregon Crab-apple; Sn, Na-n; Rosaceae
                   II None of the leaf blades with 1 or 2
                       especially large teeth below the middle
                       J   Leaves up to more than 8 cm long
                           K   Flowers (up to 3 cm wide) solitary
                               or few in each group; fruit 4-5 cm
                               long, brown, tough and eventually
                               splitting (escape from cultivation)
                                   Prunus dulcis [*P. amygdalus*]
                                      Almond; as; Rosaceae
                           KK Flowers numerous (these 5-10 cm
                               long); fruit about 1 cm long, red to
                               nearly black, fleshy
                                             Prunus virginiana
                                      var. *demissa* (*Plate* 93)
                             Western Choke Cherry; Rosaceae

JJ Leaves not more than 5 cm long
  K Flowers up to 7 in each cluster;
     fruit 1.5-2.5 cm long, usually dark
     red, sometimes yellowish
               Prunus subcordata (*Plate* 93)
                 Sierra Plum; Mo-n; Rosaceae
  KK Flowers up to about 15 in each
     raceme; fruit slightly less than 1
     cm long, usually dark red, sometimes
     purple *Prunus emarginata* (*Plate* 49)
                Bitter Cherry; Rosaceae
HH Both pistillate and staminate flowers in
catkins or catkinlike structures (the
pistillate catkins persist at least into the
late spring and summer, and in *Alnus* they
become woody, resembling the cones of
coniferous trees)
  I Most leaves less than 3 cm wide; pistillate
  catkins not becoming woody structures
    J Evergreen; pistillate catkinlike
     structures producing nearly spherical
     fruits covered with a whitish wax;
     leaves (these about 4 times as long as
     wide) usually slightly aromatic and
     conspicuously toothed
            Myrica californica (*Plate* 85)
               Wax-myrtle; Myricaceae
    JJ Deciduous; pistillate catkins producing
     fruits that release seeds covered with
     cottony hairs; leaves not aromatic and
     either not toothed or only with obscure
     teeth   *Salix* (*see* Salicaceae)
  II Most leaves (these deciduous) at least 4 cm
  wide (and about twice as long as wide);
  pistillate catkins becoming woody
  structures that resemble cones of
  coniferous trees (old pistillate catkins
  are likely to be found on or under the tree
  during the winter)
    J Margin of leaf blades slightly inrolled
     (look with hand lens), the large teeth
     usually toothed again; fruit (attached
     to the axis of the catkins) with
     prominent membranous margins that are
     at least half as wide as the thickened
     portion; mostly along the coast, at
     elevations of up to 500'
        Alnus rubra [*A. oregona*] (*Plate* 74)
           Red Alder; SCr-n; Betulaceae
    JJ Margin of leaf blades not inrolled, the
     teeth all about the same size; fruit
     with thin-angled margins, but without
     membranous margins; mostly away from
     the coast, at elevations up to 5000'
     (widespread)
          Alnus rhombifolia (*Plate* 74)
           White Alder; Betulaceae

SHRUBS AND WOODY VINES

A Woody vines
  B Stems spiny except in *Rubus ulmifolius* var. *inermis*, but this is
    recognizable as a blackberry; see choice C, below)
    C Leaves palmately lobed or compound, none modified as tendrils;
      petals and sepals 5; fruit an aggregate of fleshy achenes
                                        Rubus (*see* Rosaceae)
    CC Leaves not lobed or compound, some oval or heart-shaped, others
       modified as tendrils; flowers (staminate and pistillate
       separate) with 6 equal, greenish perianth segments; fruit
       fleshy, but not an aggregate of achenes   *Smilax californica*
                                      Greenbrier; Liliaceae
  BB Stems not spiny; plants not as described in choices C and CC, above
    C Leaves compound (and opposite); flowers with numerous pistils
      (these becoming dry, 1-seeded fruits, each with a persistent,
      feathery style more than 2 cm long) (flowers with white sepals,
      but without a corolla)
      D Leaves with 5-7 leaflets; flowers up to 2 cm wide, numerous
        in each panicle (widespread)   *Clematis ligusticifolia*
                              Virgin's-bower; Ranunculaceae
      DD Leaves usually with 3 (occasionally 5) leaflets; flowers up
        to 5 cm wide, usually single (but sometimes 3) on each
        peduncle   *Clematis lasiantha* (*Plate* 46)
                              Pipestems; Ranunculaceae
    CC Leaves not compound; flowers with a single pistil
      D Leaves alternate, the blades more or less heart-shaped (but
        the notch at the base may be shallow, and in *Vitis*
        californica the blades are also lobed)
        E Leaves 3- or 5-lobed and coarsely toothed; inflorescence a
          panicle of many small flowers (fruit fleshy)
                           Vitis californica (*Plate* 56)
                             California Wild Grape; Vitaceae
        EE Leaves smooth-margined; flowers solitary (without a
          corolla, but with a large, tubular calyx [greenish, with
          purple veins], this bent in such a way that it resembles
          the bowl of a smoker's pipe)
                     Aristolochia californica (*Plate* 6)
                        Dutchman's-pipe; Aristolochiaceae
      DD Leaves opposite, not heart-shaped; flowers with a corolla
        (this cream-colored, yellowish, pink, or purplish), the lower
        portion tubular, the lobes equal or unequal (fruit red)
        E Corolla pink or purplish; leaves of one or more upper
          pairs united (widespread)
                Lonicera hispidula var. *vacillans* (*Plate* 18)
                        Hairy Honeysuckle; Caprifoliaceae
        EE Corolla cream-colored or yellowish; upper pairs of leaves
          not united (or only rarely united)
                  Lonicera subspicata var. *denudata*
                [includes *L. subspicata* var. *johnstonii*]
        Southern Honeysuckle; CC(MD), SCl(MH)-s; Caprifoliaceae
AA Shrubs
  B Stems nearly or completely leafless throughout the year, or leaves
    modified as spines
    C Plants succulent; branchlets jointed, green (leaves represented
      by small scales; in alkali areas)   *Allenrolfea occidentalis*
                        Iodinebush; Al, CC; Chenopodiaceae
    CC Plants not succulent; branchlets not jointed
      D Leaves modified as spines, 5-15 mm long, on spiny branchlets
        (flowers yellow, resembling those of a pea; widely
        established)   *Ulex europaea* [*U. europaeus*]
                                 Gorse; eu; Fabaceae
      DD Plants not spiny
        E Flowers yellow, in composite heads about 8 mm high (rays
          absent)   *Lepidospartum squamatum*
                     Scalebroom; SCl-s; Asteraceae
        EE Flowers not in composite heads
          F Flowers pinkish, regular and not at all like those of
            a pea, 1-2 mm long (naturalized, especially near
            streams and in bottomlands)   *Tamarix ramosissima*
                     Tamarisk; as; Tamaricaceae

G Young branches with raised ridges; some leaves
divided into 3 leaflets
Cytisus scoparius (*Plate* 24)
Scotch Broom; eu; SCr-n; Fabaceae
GG Young branches smooth, without raised ridges;
leaves not divided into leaflets
Spartium junceum (*Plate* 27)
Spanish Broom; me; Fabaceae
BB Plants with well developed leaves (but they may lack leaves during a
few months of each year)
C Some or all leaves compound
D Leaves opposite (leaflets 3-9) *Sambucus* (*see* Caprifoliaceae)
DD Leaves alternate
E Branches either ending in spines or with numerous small
spines along much of their length
F Leaflets 5-9; corolla pink   *Rosa* (*see* Rosaceae)
FF Leaflets usually 3; corolla red-purple
G Flowers irregular, resembling those of a pea;
corolla not more than 1 cm wide; leaflets usually
about 1 cm long; branches ending in stout spines;
in chaparral   *Pickeringia montana* (*Plate* 26)
Chaparral Pea; Me-s; Fabaceae
GG Flowers regular, not resembling those of a pea;
corolla about 2.5 cm wide; leaflets usually more
than 3 cm long; branches with numerous small
spines along much of their length; in moist places
Rubus spectabilis
[*R. spectabilis* var. *franciscanus*] (*Plate* 49)
Salmonberry; SCl-Sn; Rosaceae
EE Branches not spiny
F Leaves with 3 leaflets
G Flowers irregular, resembling those of a pea
(leaflets not toothed; widespread)
H Flowers in few-flowered racemes at the end of
short side branches or at the end of main
stems; leaflets 8-20 mm long (corolla yellow;
escape from cultivation) *Genista monspessulana*
[*Cytisus monspessulanus*] (*Plate* 24)
French Broom; me; Fabaceae
HH Flowers in the leaf axils; leaflets 4-12 mm
long
I Flowers 7-10 mm long, almost sessile; stems
not obviously angled, usually sparingly
leafy; rarely more than 1 m tall
Lotus scoparius (*Plate* 25)
Deerweed; Fabaceae
II Flowers up to 20 mm long, not sessile;
stems obviously angled, usually densely
leafy; often more than 2 m tall (extremely
invasive)   *Cytisus scoparius* (*Plate* 24)
Scotch Broom; eu; Fabaceae
GG Flowers regular, not resembling those of a pea.
(Caution: *Toxicodendron diversilobum* [below] may
cause a skin rash! Do not touch!)
H Corolla purplish red, about 2.5 cm wide
Rubus spectabilis
[*R. spectabilis* var. *franciscanus*] (*Plate* 49)
Salmonberry; SCl-Sn; Rosaceae
HH Corolla greenish or white, less than 2 cm wide
I Leaves with a strong odor similar to that
of citrus leaves; leaflets not toothed
Ptelea crenulata (*Plate* 51)
Hoptree; SCl-n; Rutaceae
II Leaves without a strong odor; leaflets
toothed or lobed
J Leaflets rounded, more or less equal;
flowers and fruit whitish (widespread)
(Caution: touching this plant--even
when it is without leaves--may cause a
serious skin rash.)
Toxicodendron diversilobum
[*Rhus diversiloba*] (*Plate* 4)
Western Poison-oak; Anacardiaceae

JJ Leaflets triangular, the terminal one
larger than the 2 lateral ones; flowers
greenish, fruit reddish
Rhus trilobata [includes *R. trilobata*
vars. *malacophylla* and *quinata*] (*Plate* 4)
Skunkbush; Anacardiaceae

FF Leaves with at least 5 leaflets
  G Leaflets toothed (the teeth prickly in *Berberis*)
    H Leaves generally more than 75 cm long, with 3
primary divisions; leaflets 3-5 on each
division, often more than 8 cm long (sometimes
more than 20 cm), without prickly teeth;
flowers whitish; fruit blackish; often more
than 2 m tall (growing along streams in shaded
canyons) *Aralia californica* (*Plate* 6)
Elk-clover; Araliaceae
    HH Leaves not more than 45 cm long (usually
shorter), pinnately compound; leaflets 5-21,
not more than 8 cm long, with prickly teeth;
flowers yellow; fruit bluish; not more than
1.5 m tall *Berberis* (*see* Berberidaceae)
  GG Leaflets not toothed
    H Flowers (corolla about 3 cm wide, yellow) only
slightly irregular, not closely resembling
those of a pea; leaves pinnately compound,
with 12 to 16 leaflets and no terminal leaflet
(escape from cultivation)
Senna multiglandulosa [*Cassia tomentosa*]
Senna; mx; Ma; Fabaceae
    HH Flowers distinctly irregular, either resembling
those of a pea or with only 1 petal; leaves
pinnately or palmately compound, but if
pinnately compound, with a terminal leaflet
      I Leaves pinnately compound, with 11-27
leaflets; flowers with 1 petal (leaf
rachises with pricklelike glands about 1 mm
long) *Amorpha californica* var. *napensis*
(*Plate* 80)
False Indigo; Ma; Fabaceae
      II Leaves palmately compound, with 5-12
leaflets; flowers with 5 petals (but the 2
lower ones are united to form a keel)
        J Keel of corolla without hairs (use a
lens and look at the edges of the keel
after pushing back the wings)
Lupinus chamissonis
Chamisso Bush Lupine; Ma-s; Fabaceae
        JJ Keel of corolla with hairs (but
sometimes only on the lower or upper
margin)
          K Banner of corolla hairy on the back;
corolla blue, red-purple, or
lavender (widespread)
Lupinus albifrons var. *albifrons*
(*Plate* 26)
Silver Lupine; Fabaceae
          KK Banner of corolla not hairy on the
back; corolla yellow, occasionally
nearly white, lilac, or of mixed
colors (coastal) *Lupinus arboreus*
[includes *L. arboreus* var. *eximius*]
(*Plate* 26)
Yellow Bush Lupine; Fabaceae

CC Leaves not compound
  D Leaves opposite Shrubs, Subkey 1 (p. 47)
  DD Leaves alternate or in clusters at the nodes, but not clearly
opposite (the new leaves of deciduous species, such as those
belonging to the genus *Ribes*, will often be in clusters at
first, and later will be clearly alternate)
    E Plants with spiny branches (the branches may end in stout
spines, or the spines may be scattered along the
branches)
      F Leaves with more than 3 main veins, these arranged
palmately *Ribes* (*see* Grossulariaceae)

FF Leaf blades either with only 1 main vein (the midrib),
  or with 3 main veins originating at the base
  G Leaf blades with 3 main veins originating at the
     base (evergreen)   *Ceanothus* (*see* Rhamnaceae)
  GG Leaf blades with 1 main vein (the midrib)
     H Leaves (2.5-5 cm long) hairy on the underside,
        without teeth or with only inconspicuous
        teeth; fruit orange or yellow-orange
        (evergreen; occasional escape from
        cultivation)   *Pyracantha angustifolia*
                          Firethorn; as; Ma; Rosaceae
     HH Leaves not hairy on the underside, with obvious
        teeth; fruit red or blackish
        I Leaves deciduous, up to about 7 cm long;
           fruit blackish   *Crataegus suksdorfii*
                          Black Hawthorn; Ma-n; Rosaceae
        II Leaves evergreen, not more than 4 cm long;
           fruit red (leaves up to 1.5 cm long; teeth,
           if present, just visible without a hand
           lens; petioles 1-4 mm long)
                          Rhamnus crocea (*Plate* 47)
                   Spiny Redberry; La-s; Rhamnaceae
EE Plants without spiny branches (*Prunus subcordata*, in
   Subkey 2, rarely has spiny branchlets)
   F Flowers in composite heads (these may be small,
      however)
      G Flower heads with both disk flowers and ray flowers
         (these yellow)
         H Rays not more than 5 mm long
            I Leaves (these sometimes lobed) not
               clustered, flattened, densely woolly on the
               underside, up to more than 5 cm long
                          Eriophyllum staechadifolium [includes
                          E. staechadifolium var. *artemisiaefolium*]
                   Seaside Woolly Sunflower; Me-SCr; Asteraceae
            II Leaves (these not lobed) often clustered,
               narrow and nearly cylindrical, up to about
               1.2 cm long   *Ericameria ericoides*
                          [*Haplopappus ericoides*] (*Plate* 73)
                          Mock-heather; Asteraceae
         HH Most rays at least 10 mm long
            I Rays rarely more than 9; phyllaries in 2
               series, those of the upper series much
               larger than those of the lower series;
               leaves less than 2 mm wide, sometimes
               divided into very slender lobes
                          Senecio flaccidus var. *douglasii*
                                   [*S. douglasii*]
                          Bush Groundsel; Me-s; Asteraceae
            II Rays 13-18; phyllaries all about the same
               size, in 2 or 3 overlapping series; leaves
               up to about 3 mm wide, not divided into
               lobes   *Ericameria linearifolia*
                          [*Haplopappus linearifolius*] (*Plate* 10)
                          Interior Goldenbush; La-s; Asteraceae
      GG Flower heads consisting entirely of disk flowers
         H Leaves divided into lobes less than 2 mm wide
            (leaves and stems grayish, with a sagebrush
            odor; common) *Artemisia californica* (*Plate* 8)
                   California Sagebrush; Ma, Na-s; Asteraceae
         HH Leaves sometimes toothed, but not divided into
            slender lobes
            I Flowers white (sometimes tinged with pink
               or red) or greenish
               J Flower heads 12-14 mm high
                          Brickellia californica
                   California Brickellia; Asteraceae
               JJ Flower heads about 5 mm high
                  K Leaves about twice as long as wide;
                     common in dry, chaparral habitats
                          Baccharis pilularis [includes
                          B. pilularis var. *consanguinea*]
                                   (*Plate* 8)
                          Coyotebrush; Asteraceae

KK Leaves about 10 times as long as
wide; near stream banks and ditches
Baccharis salicifolia [*B. viminea*]
(*Plate* 8)
Mulefat; Asteraceae
II Flowers yellow (often becoming reddish)
J Leaves mostly basal, the stem leaves few
and reduced (phyllaries often dark-
tipped)   *Lepidospartum squamatum*
Scalebroom; SCl-s; Asteraceae
JJ Stem leaves well developed
K Flower heads about 1.5 cm high, and
about 3 times as high as wide;
phyllaries distinctly ridged on the
back   *Chrysothamnus nauseosus*
ssp. *mohavensis*
Rubber Rabbitbrush; SCl(MH)-s
Asteraceae
KK Flower heads up to about 1 cm high,
and not appreciably higher than
wide; phyllaries not distinctly
ridged on the back
L Pappus of achenes consisting of
5-8 scales (bark of older stems
grayish or whitish, shredding;
leaves mostly 2-3 cm long, about
10 times as long as wide)
Eastwoodia elegans
Yellow Mock Aster; Al-s, e
Asteraceae
LL Pappus of achenes consisting of
slender bristles
M Leaves 1-4 cm long, usually
less than 5 times as long as
wide, toothed; involucre 5-7
mm high; plants woody only
at the base; in sandy
coastal areas
Isocoma menziesii var.
vernonioides [*Haplopappus*
venetus ssp. *vernonioides*]
Coast Goldenbush; SF-s
Asteraceae
MM Leaves 3-6 cm long, usually
more than 10 times as long
as wide, not toothed;
involucre 4-5 mm high;
plants woody almost
throughout; inland
(widespread)
Ericameria arborescens
[*Haplopappus arborescens*]
Goldenfleece; Asteraceae
FF Flowers not in composite heads (but they may be in
dense inflorescences)
G Leaf blades not palmately lobed (but they may be so
coarsely toothed as to be almost pinnately lobed
H Leaves up to 10 mm long, not more than 1 mm
wide, in clusters (flowers less than 5 mm
wide, with 5 white petals, in panicles 4-12 cm
long; widespread)
Adenostoma fasciculatum (*Plate* 48)
Chamise; Rosaceae
HH Leaves mostly more than 10 mm long,
considerably more than 1 mm wide
I Leaf blades about as wide as long,
palmately veined, with a prominent notch
where the petiole is attached, and often
with a slight indentation at the tip
(deciduous; flowers resembling those of a
pea, the corolla reddish purple; fruit a
dry pod 5-8 cm long)
Cercis occidentalis (*Plate* 24)
Western Redbud; Sl-n; Fabaceae

II Leaf blades, if as wide as long, not
palmately veined, and plants not conforming
to all other features described in choice I
  J Leaves (the blades usually almost
triangular) covered with grayish scales
(similar scales present on young
branches, plants appear grayish or
whitish); restricted to saline and
alkaline soil   *Atriplex lentiformis*
[includes *A. lentiformis* ssp. *breweri*]
Big Saltbush; SF-s, e; Chenopodiaceae
  JJ Leaves not covered with scales; not
primarily in saline and alkaline soil
    K Corolla (1-3 cm wide) 5-lobed, but
opening so flat that there is almost
no tube, predominately white, blue,
lavender, or purple, sometimes with
conspicuous green spots; pistil and
stamens protruding well beyond the
corolla; fruit a fleshy berry, 5-15
mm long, green, yellow, red, or
black when ripe; leaves and stems
usually hairy
                    Solanum (*see* Solanaceae)
    KK Plants not as described in choice K
      L Dwarf, evergreen oaks: fruit an
acorn; staminate flowers in
drooping catkins; leaves tough,
often convex on the upper
surface and usually with a few
sharp-tipped (or at least
pointed) teeth
                    Quercus (*see* Fagaceae)
      LL Shrubs other than dwarf,
evergreen oaks
        M Willows: young stems usually
very flexible; flowers very
small, the staminate flowers
and pistillate flowers in
catkins on separate plants
(the remains of pistillate
catkins and their seeds can
often be found on the ground
long after they have
fallen); fruit releasing
seeds covered with cottony
hairs; leaf shape variable,
but the blades of some
species more than 3 times as
long as wide
                    Salix (*see* Salicaceae)
        MM Shrubs other than willows
                    Shrubs, Subkey 2 (p. 49)
GG At least some of the leaf blades (and usually most
of them) divided palmately into lobes
  H Largest leaves usually more than 10 cm wide
(escapes from cultivation)
    I Leaf lobes about as wide as long; flowers
at least 3 cm wide, with 5 petals, these
pink or purple, with darker veins; fruit
without spiny outgrowths
                    Lavatera (*see* Malvaceae)
    II Leaf lobes much longer than wide; flowers
about 1 cm wide, without petals; fruit
usually with spinelike outgrowths
                    Ricinus communis (*Plate* 24)
                    Castor-bean; eu; Euphorbiaceae
  HH Largest leaves rarely more than 6 cm wide (but
sometimes wider in species of *Malacothamnus*)
    I Leaves with star-shaped hairs on both
surfaces, rather rough to the touch;
evergreen; petals or sepals at least 10 mm
wide

J   Flowers mostly 3-5 cm wide, with orange-
    yellow, often red-tinged sepals that
    resemble petals (petals absent)
            Fremontodendron californicum [includes
            F. californicum sspp. *crassifolium* and
                                    napense] (*Plate* 55)
            Flannelbush; Mo-n; Sterculiaceace
JJ Flowers less than 2 cm wide, with pink
    or rose petals
            Malacothamnus (*see* Malvaceae)
II Leaves without star-shaped hairs on both
    surfaces (but there may be other hairs, and
    sometimes there are star-shaped hairs on
    the underside of leaves of *Physocarpus
    capitatus*); deciduous; petals less than 4
    mm long
    J   Leaf lobes not toothed (the blades not
        hairy); sepals and petals yellow (often
        turning red as they age)
            Ribes aureum var. *gracillimum* (*Plate* 30)
            Golden Currant; Al-s; Grossulariaceae
    JJ Leaf lobes toothed; sepals and petals
        not yellow
        K   Inflorescences umbel-like, mostly
            terminal; petals white, sepals
            green, all nearly equal; leaf blades
            scarcely hairy; petioles not
            glandular
                    Physocarpus capitatus (*Plate* 49)
                            Ninebark; Rosaceae
        KK Inflorescences racemes originating
            in the axils of the leaves; petals
            and sepals both some shade of white
            to rose, the sepals longer than the
            petals; leaf blades hairy, sometimes
            densely so; petioles glandular
            L   Leaves dull green, the upper
                surface glandular and hairy;
                sepals pale pink to bright rose
                        Ribes malvaceum
                Chaparral Currant; CC, Ma-s
                        Grossulariaceae
            LL Leaves bright green, the upper
                surface slightly hairy but not
                glandular; sepals usually pale
                pink, occasionally white
                (widespread)
                Ribes sanguineum var. *glutinosum*
                                    (*Plate* 30)
                        Pinkflower Currant
                        Grossulariaceae

Shrubs, Subkey 1: Shrubs with opposite leaves, these not compound
A   Leaves with spiny teeth   *Ceanothus* (*see* Rhamnaceae)
AA Leaves not spiny, although they may have small teeth or lobes
    B   Branches ending in stout spines (leaves either not toothed or only
        with some irregular teeth, on petioles 2-3 mm long; branches
        grayish; fruit 5-8 mm long, purplish black, with a grayish coating)
                    Forestiera pubescens [*F. neomexicana*] (*Plate* 86)
                            Desert-olive; CC-s; Oleaceae
    BB Branches not ending in spines
        C   At least some of the stem leaves sessile
            D   Flowers regular (fragrant), not more than 5 mm long, in
                crowded racemes 15-20 cm long (corolla lilac to purple, often
                with an orange throat; escape from cultivation)
                                        Buddleja davidii
                            Summer Lilac; as; Buddlejaceae
            DD Flowers irregular, more than 10 mm long, in the axils of the
                leaves (upper leaves often reduced)
                E   Leaves sticky; leaf margins sometimes rolled under
                    (corolla dull orange; widespread)
                                    Mimulus aurantiacus (*Plate* 53)
                        Bush Monkeyflower; Scrophulariaceae
                EE Leaves not sticky; leaf margins not rolled under

      F Leaves with a strong odor resembling that of sage;
         stems 4-angled (inflorescence up to 10 cm long;
         corolla 2.5-3 cm long, white or pink, with darker
         veins and spots, solitary; fertile stamens 4;
         widespread)   *Lepechinia calycina* (*Plate* 32)
                         Pitcher Sage; Lamiaceae
     FF Leaves without a strong odor resembling that of sage;
        stems rounded
        G Uppermost stem leaves (excluding those at the base
           of flower peduncles) 6-10 mm wide
           H Corolla mostly yellow, with some brown and
              purple, 1-1.5 cm long; leaves often toothed
                          Keckiella lemmonii
                Lemmon Penstemon; Sl-n; Scrophulariaceae
           HH Corolla scarlet, 2.5-3.5 cm long; leaves not
              toothed   *Penstemon centranthifolius*
                Scarlet-bugler; La-s; Scrophulariaceae
        GG Uppermost stem leaves (excluding those at the base
           of flower peduncles) 2-3 mm wide (corolla white
           with some rose, about 1.5 cm long; leaves
           sometimes toothed)
           H Calyx and pedicels glandular
                  Keckiella breviflora var. *breviflora*
                Gaping Penstemon; Al-s; Scrophulariaceae
           HH Calyx and pedicels not glandular
                  Keckiella breviflora var. *glabrisepala*
                         Me-Na; Scrophulariaceae
CC All stem leaves with at least a short petiole
  D Leaves toothed (but the teeth are often inconspicuous in
     Paxistima myrsinites)
     E Leaves with 3 main veins (sometimes 5) originating at the
        base of the blade (deciduous; leaf blades mostly 3-5 cm
        long, elliptical, often with a slight indentation where
        the petiole is attached, toothed in the upper half;
        flowers about 7 mm wide, in a rather dense inflorescence;
        corolla white; fruit about 1 cm long, red)
                          Viburnum ellipticum
          Oval-leaf Viburnum; CC, Sn-n; Caprifoliaceae
     EE Leaves with only 1 main vein (the midrib)
        F Leaves with a strong odor of sage; corolla 2-lipped
           (evergreen; leaves mostly 3-6 cm long, hairy on the
           underside; corolla usually lavender or pale blue,
           sometimes white or pinkish, about 1.5 cm, long;
           widespread)   *Salvia mellifera* (*Plate* 33)
                      Black Sage; CC-s; Lamiaceae
        FF Leaves without an odor of sage; corolla not 2-lipped
          G Leaves not stiff; normally deciduous, but sometimes
             persisting through the winter (leaves usually at
             least 7 cm long, more than twice as long as wide,
             light green, evenly toothed, and tapering
             gradually to a slender tip; petals [5] about 5 mm
             long and wide, brownish purple, with small dots;
             fruit 3-lobed, leathery)
                      Euonymus occidentalis (*Plate* 19)
             Western Burningbush; Mo-n; Celastraceae
         GG Leaves rather stiff (and usually less than 3 cm
            long); evergreen
           H Flowers (about 3 mm wide) in small axillary
               clusters; petals 4, brownish red (leaves with
               inconspicuous teeth)
                    Paxistima myrsinites (*Plate* 19)
                Oregon Boxwood; Ma-n; Celastraceae
           HH Flowers in conspicuous many-flowered panicles
             or umbels; petals 5, white or some shade of
             blue or lavender   *Ceanothus* (*see* Rhamnaceae)
  DD Leaves not toothed (but they may be wavy-margined or divided
     into lobes)
     E Staminate and pistillate flowers on separate plants, and
        in drooping catkins 5-20 cm long (evergreen)
                         Garrya (*see* Garryaceae)
     EE Stamens and pistils present in all flowers, these not in
        catkins
        F Largest leaf blades not more than 5 cm long (usually
          at least slightly hairy)

G Leaves evergreen; fruit dark when mature
    H Leaves often rather stiff; flowers (these many
      in each panicle or corymb) with 5 sepals and 5
      petals, the petals blue, lavender, or white;
      fruit dry at maturity
                      Ceanothus (*see* Rhamnaceae)
    HH Leaves not stiff; flowers solitary, with 4
      calyx lobes and 4 petals, the petals cream-
      white and pink; fruit fleshy (escape from
      cultivation)
             Luma apiculata [*Eugenia apiculata*]
                      Temu; sa; Ma; Myrtaceae
GG Leaves deciduous; fruit white (flowers pink or
  white, in clusters)
    H Leaves grayish green on both surfaces (with 3-4
      pairs of lateral veins); inflorescence a
      spreading panicle (in moist areas)
                    Cornus glabrata
                Brown Dogwood; Cornaceae
    HH Leaves not grayish green; inflorescence
    racemelike, usually few-flowered
      I   Plants often more than 100 cm tall,
        upright; leaves not hairy on the underside;
        corolla with a nectar gland below only 1
        lobe; fruit usually about 10 mm long
        (widespread)
            Symphoricarpos albus var. *laevigatus*
                      [*S. rivularis*] (*Plate* 18)
              Common Snowberry; Caprifoliaceae
      II Plants rarely more than 50 cm tall, usually
        sprawling; leaves usually hairy on the
        underside; corolla with nectar glands below
        all 5 lobes; fruit about 8 mm long
            Symphoricarpos mollis (*Plate* 18)
          Creeping Snowberry; Me-s; Caprifoliaceae
FF Largest leaf blades more than 5 cm long (up to 12 cm
  long) (deciduous; in moist places)
  G Leaves hairy, especially on the underside (leaves
    with 4-7 pairs of lateral veins; flowers in small
    corymbs; petioles 5-8 mm long)
            Cornus sericea ssp. *occidentalis*
                [*C. occidentalis*] (*Plate* 21)
            Western Creek Dogwood; Cornaceae
  GG Leaves not hairy
    H Flowers in dense, spherical heads (these
      usually 2-2.5 cm wide) at the end of stems
      (corolla white or faintly yellowish,
      consisting of a tube and 4 lobes)
         Cephalanthus occidentalis var. *californicus*
                      (*Plate* 93)
         California Button-willow; Na-n; Rubiaceae
    HH Flowers not in dense, spherical heads (usually
    along streams or in other moist habitats)
      I   Leaves with a pleasantly spicy fragrance
        when bruised; flowers solitary, with many
        perianth segments, these up to 2.5 cm long,
        reddish brown; fruit leathery, becoming
        hard   *Calycanthus occidentalis* (*Plate* 17)
                 Spicebush; Na-n,e; Calycanthaceae
      II Leaves without a spicy odor; flowers in
        pairs, consisting of a tube and 5 lobes,
        the corolla about 1.5 cm long, yellow;
        fruit fleshy, black (peduncles of flowers
        arising within an involucre consisting of 2
        pairs of bracts)
            Lonicera involucrata var. *ledebourii*
                      (*Plate* 18)
              Twinberry; Caprifoliaceae

Shrubs, Subkey 2: Shrubs with alternate leaves, these neither compound nor
palmately lobed, and without spiny branches (except rarely in *Prunus
subcordata*)
A Blades of most leaves at least 3 times (sometimes 4-5 times) as long as
  wide

B   Leaves not toothed (except perhaps on young seedlings or on juvenile
    foliage that sometimes appears on a few branches of a mature plant)
    C   Fruit a spiny bur containing acornlike nuts (at least some burs
        usually present on or around each plant); leaves (5-8 cm long)
        dark green on the upper surface, golden brown on the underside,
        at least when young; inflorescence catkinlike; evergreen
                            Chrysolepis chrysophylla var. *minor* (*Plate* 28)
                                        Golden Chinquapin; Fagaceae
    CC Fruit not a spiny bur; leaves not golden brown on the underside;
        inflorescence not catkinlike
        D   Evergreen; leaves commonly more than 12 cm long, stiff;
            flowers in umbel-like clusters; corolla 3-4 cm wide, with a
            funnel-shaped lower portion and 5 lobes, mostly pink or rose
                            Rhododendron macrophyllum (*Plate* 23)
                                        Rosebay; Mo-n; Ericaceae
        DD Deciduous; leaves rarely so much as 12 cm long, not stiff;
            flowers in racemes (the staminate and pistillate flowers on
            separate plants); corolla consisting of 5 petals about 1 cm
            long, white or pale pink (fruit slightly longer than wide,
            orange when partly ripe, becoming blue-black; widespread)
                Oemleria cerasiformis [*Osmaronia cerasiformis*] (*Plate* 92)
                                        Osoberry; Rosaceae
BB Most leaves distinctly toothed (look carefully, because the teeth may
    be very small and closely spaced)
    C   Young branches and upper surfaces of leaf blades sticky; corolla
        1-1.5 cm long, with a tubelike lower portion and 5 short lobes,
        white to purplish (evergreen; leaf blades 7-10 cm long, often
        blackened by a fungus; widespread)
                            Eriodictyon californicum (*Plate* 31)
                                        Yerba-santa; SB-n; Hydrophyllaceae
    CC Young branches and upper surfaces of leaf blades not sticky;
        corolla not as described in choice C
        D   Marginal teeth unequal (larger ones, in general, alternating
            with 1 or 2 smaller ones) (evergreen; leaf blades mostly 8-12
            cm long, bright green, not hairy, often aromatic when
            bruised; staminate and pistillate flowers in separate
            catkinlike inflorescences; fruit purplish, covered with a
            whitish wax)   *Myrica californica* (*Plate* 85)
                                        Wax-myrtle; Myricaceae
        DD Marginal teeth almost equal
            E   Deciduous; leaves hairy; each marginal tooth tipped by a
                tapering hair; flowers about 3 cm wide, with a funnel-
                shaped lower portion and 5 lobes, white or cream-white,
                often pink-tinged and with considerable yellow on the
                uppermost lobe; mostly along streams and in other moist
                habitats   *Rhododendron occidentale* (*Plate* 23)
                                        Western Azalea; SCr-n; Ericaceae
            EE Evergreen; leaves not hairy; marginal teeth without hairs;
                flowers not as described in choice E; in dry habitats
                F   Leaves gray-green, not stiff, the teeth small; flowers
                    with 4 yellow petals 2-3 cm long; fruit slender, up to
                    10 cm long   *Dendromecon rigida* (*Plate* 39)
                                        Bush Poppy; Sn-s; Papaveraceae
                FF Leaves dark green, stiff, the teeth prominent; flowers
                    with 5 white petals about 3 mm long; fruit rounded, up
                    to 1 cm long (bright red; widespread)
                            Heteromeles arbutifolia (*Plate* 48)
                                        Toyon; Rosaceae
AA Blades of leaves rarely so much as 3 times as long as wide
    B   Evergreen (except *Ceanothus integerrimus*); leaves often stiff, with a
        prominent, raised midrib (or 3 main veins diverging from the base)
        and sometimes with conspicuous, secondary veins; flowers small, the
        inflorescences either in the leaf axils or at the end of branches
        C   Leaves with a prominent, raised midrib and conspicuous, secondary
            veins; teeth, if present, small; flowers greenish, few to many
            in umbel-like inflorescences in the leaf axils; fruit fleshy, 5-
            15 mm long, red or black when ripe   *Rhamnus* (*see* Rhamnaceae)
        CC Leaves (rarely more than 4 cm long) with either a midrib or with
            3 main veins diverging from the base, but without conspicuous
            secondary veins; teeth, if present, either small or prominent,
            sometimes confined to the tip of the leaves; flowers usually
            blue or lavender (sometimes white), many, in panicles or corymbs
            at the ends of branches; fruit dry when mature, up to 8 mm long
                            Ceanothus (*see* Rhamnaceae)

BB Plants not conforming to all characters described in choice B
  C Leaves not toothed (rarely toothed in *Arctostaphylos*, choice D)
      D Flowers (these often appearing in winter) in short
         inflorescences at the tip of branches; corolla usually less
         than 1 cm long, urn-shaped, with small lobes, usually white
         or pink; fruit fleshy, but the pulp more commonly firm than
         soft; stems usually with a smooth, reddish brown bark;
         evergreen; leaves often stiff, sometimes grayish or bluish
         green, sometimes notched at the base
                                        Arctostaphylos (*see* Ericaceae)
      DD Plants not conforming to all characters described in choice D
         E Leaves bluish green, the blades often more than 10 cm
            long; corolla greenish yellow, mostly tubular, about 4 cm
            long, with 5 short lobes (evergreen; well established
            exotic in exposed habitats)   *Nicotiana glauca* (*Plate* 54)
                                    Tree-tobacco; sa; Solanaceae
        EE Leaves not obviously bluish green and not more than about
           6 cm long; corolla not yellow or tubular, and not so much
           as 4 cm long (in *Dirca occidentalis*, which does not have
           a corolla, the calyx is yellow and mostly tubular, but it
           has only 4 lobes and is not more than 1 cm long)
           F Leaves evergreen, dark green on the upper surface,
              white-woolly on the underside (flowers in clusters,
              usually at the end of branches; petals [5] white or
              pink; fruit red; occasional escape from cultivation)
                                  Cotoneaster (*see* Rosaceae)
         FF Leaves deciduous, not dark green on the upper surface
           and not white-woolly on the underside
           G Leaves up to 2.5 cm long; new branches not
              extremely flexible; most branches angled
              lengthwise and green; corolla urn-shaped, greenish
              white, with small lobes (fruit 1 cm long, bright
              red)   *Vaccinium parvifolium* (*Plate* 23)
                              Red Huckleberry; SCl-n; Ericaceae
          GG Leaves up to about 6 cm long; new branches
               extremely flexible; older branches not angled
              lengthwise and not green; flowers without a
              corolla, but with a yellow calyx (calyx about 1 cm
              long, with 4 short lobes; stamens [8] and style
              protruding beyond the calyx; flowers barely
              persisting until leaves appear in spring)
                                    Dirca occidentalis (*Plate* 55)
                Western Leatherwood; SFBR; 4; Thymelaeaceae
CC Leaves toothed
    D Leaf blades (these deciduous, up to about 10 cm long) with a
       prominent notch at the base, coarsely and unevenly toothed,
       and hairy; staminate flowers in catkins, pistillate flowers
       solitary or in small clusters; fruit, resembling an acorn,
       enclosed by leafy bracts
                  Corylus cornuta var. *californica* (*Plate* 74)
                              Hazelnut; SCr-n; Betulaceae
    DD Leaf blades, if with a slight indentation at the base, not
       coarsely and unevenly toothed; none of the flowers in
       catkins; fruit not resembling an acorn
       E Leaf blades (these up to 10 cm long) about twice as long
          as wide, distinctly widest below the middle, evenly and
          finely toothed; flowers, with 5 white petals about 5 mm
          long, extremely numerous in each elongated raceme; fruit
          about 1 cm long, fleshy, red or yellow (deciduous)
                  Prunus virginiana var. *demissa* (*Plate* 93)
                          Western Choke Cherry; Rosaceae
       EE Leaf blades not distinctly widest below the middle (except
          in *Holodiscus discolor*, in which they are coarsely
          toothed and also usually pinnately lobed); flowers,
          inflorescence, and fruit not as described in choice E
          F Most leaf blades (at least in their lower halves or
            two-thirds) with especially coarse teeth or toothed
            lobes (flowers with cream-white petals, less than 5 mm
            wide, in dense panicles up to more than 15 cm long;
            fruit dry, up to 1.5 mm long; widespread)
                                  Holodiscus discolor [includes
                    H. discolor var. *franciscanus*] (*Plate* 48)
                                Creambush; Rosaceae

FF Leaf blades without especially coarse teeth or toothed
   lobes
   G   Teeth usually restricted to the upper half or two-
       thirds of the leaf blades
       H   Leaf blades rounded at the base, and only
           slightly, if at all, hairy on the underside;
           flowers (these 2 cm wide, with 5 white petals
           that are slightly twisted) usually 4-6 in each
           raceme; fruit fleshy, purplish black, about 1
           cm long   *Amelanchier alnifolia* (*Plate* 48)
                                      Serviceberry; Rosaceae
       HH Leaf blades wedge-shaped at the base, usually
           distinctly hairy on the underside; flowers
           (these less than 1 cm wide, without petals) in
           clusters of 2-3; fruit an achene, the plumose
           style (usually at least 5 cm long) persisting
           (widespread) *Cercocarpus betuloides* (*Plate* 48)
                                 Mountain-mahogany; Rosaceae
   GG Teeth usually present almost all around the margin
       of the blade
       H   Leaves rarely more than 4 cm long (evergreen;
           corolla about 1 cm long, urn-shaped, with 5
           small lobes, white; flowers few in clusters
           borne in the leaf axils; fruit blue-black,
           shiny and juicy when ripe)
                             Vaccinium ovatum (*Plate* 23)
                           Evergreen Huckleberry; Ericaceae
       HH Leaf blades commonly more than 4 cm long
           I   Most leaf blades 6-8 cm wide, nearly or
               fully two-thirds as wide as long, and often
               with a slight indentation at the base;
               corolla urn-shaped, with small lobes, white
               or pinkish; fruit with several seeds
               (evergreen; leaves tough, sometimes almost
               leathery, with closely spaced small teeth;
               fruit, when ripe, purplish black, dull
               unless rubbed)
                             Gaultheria shallon (*Plate* 23)
                                           Salal; Ericaceae
           II Leaf blades rarely so much as 5 cm wide,
               usually less than two-thirds as wide as
               long, and without an indentation at the
               base; flowers with 5 white petals; fruit
               with a single, proportionately large seed
               J   Evergreen; teeth of leaf blades spine-
                   tipped; flowers numerous in each raceme
                   (fruit 1.5-2 cm long, red or blue-
                   black, resembling that of a cultivated
                   cherry)   *Prunus ilicifolia* (*Plate* 49)
                             Hollyleaf Cherry, Na-s; Rosaceae
               JJ Deciduous; teeth of leaf blades not
                   spine-tipped; flowers not often more
                   than 12 in each raceme
                   K   Racemes with up to about 12 flowers;
                       fruit rarely more than 1 cm long
                                 Prunus emarginata (*Plate* 49)
                                        Bitter Cherry; Rosaceae
                   KK Racemes with not more than 7 flowers
                       (usually only 2-4, and flowers
                       sometimes solitary); fruit 1.5-2.5
                       cm long *Prunus subcordata* (*Plate* 93)
                               Sierra Plum; Mo-n; Rosaceae

FAMILIES OF HERBACEOUS FLOWERING PLANTS

For any region that has a rich and diversified flora, it is almost impossible to construct a family key that is simple, yet infallible. The main reason for this is that easily observed features, such as leaf arrangements, leaf shape, and the number of sepals, petals, stamens, and pistils, are neither sufficiently exclusive nor necessarily adhered to by all representatives of a particular family. In the Buttercup Family (Ranunculaceae), for instance, there are plants whose flowers have petals, others whose flowers do not. Sepals, furthermore, may look like petals. When true petals are present, they may all be alike, as they are in a buttercup, or they may be of different sizes, shapes, and colors, as in a larkspur. There is sometimes a single pistil that develops into a fleshy fruit, but more commonly there are a few to many pistils that become dry fruits.

The preliminary key leads directly to certain families and genera that are so distinctive that they are easily disposed of, and also to three groups of families. For most persons who use this book to identify terrestrial plants that constitute the broad category called "wildflowers," the only key that will need to be consulted regularly is the one to Group C. The Grass Family and several families of grasslike plants, belonging to Group B, will probably be second in order of importance, except to those who have an interest in aquatic species, which are in Group A.

After you have used the preliminary key a few times, you will probably be able to skip over it and start with the relevant group key. You will be even more pleased with yourself when you are able to place many of the plants you encounter in the right family without looking at any of these keys.

An (M) after a family name indicates that it belongs to the monocotyledons. The monocotyledonous families are arranged in alphabetical order after the dicotyledonous families.

A   Plants parasitic, with slender, orange or yellow stems that twine over other plants; plants without roots and with almost no chlorophyll (therefore not green), the leaves reduced to scales; flowers small, with a 5-lobed white corolla   Cuscutaceae, Dodder Family
AA Plants, if parasitic, not as described in choice A
    B   Plants parasitic, attached to branches of trees and shrubs; leaves either broad, thick, and pale green, or reduced to scales; fruit fleshy, white or purple
                Viscaceae (formerly in Loranthaceae), Mistletoe Family
    BB Plants, if parasitic, growing out of soil (they are attached to the roots of other plants), and not as described in choice B
        C   Plants apparently lacking chlorophyll and therefore white, yellowish, pinkish, brownish, purplish, or reddish, but not green; leaves reduced to scales (either saprophytes in forest soils or parasites on roots of other species)
           D   Corolla regular, sometimes consisting of 5 free petals, sometimes of a cup-shaped or urn-shaped lower portion and several (usually 5, rarely 4 or 6) lobes; stamens usually 10 (saprophytes in forest soils)
                Ericaceae (includes Pyrolaceae), Heath Family
              Allotropa, *Hemitomes*, *Pleuricospora*, *Pityopus*, *Pyrola*
           DD Corolla irregular, either consisting of 3 free petals (in which case the lower petal is decidedly different from the upper ones, and the 3 sepals are usually colored like the petals) or with a long tube and 5 lobes, 3 forming a lower lip, 2 forming an upper lip; stamens 4 or fewer
              E   Corolla consisting of 3 separate petals, the lower one decidedly different from the upper ones; sepals 3; sepals and petals attached to the top of the fruit-forming portion of the pistil; with only 1 functional stamen; saprophytes in forest soils
                Orchidaceae, Orchid Family (M)
                Cephalanthera, *Corallorhiza*

EE Corolla consisting of a long tube with 5 lobes, 3 forming
a lower lip, 2 forming an upper lip; calyx 5-lobed;
fruit-forming portion of the pistil free within the calyx
cup and corolla tube; with 4 functional stamens;
parasites on roots of other plants
Orobanchaceae, Broom-rape Family
CC Plants with chlorophyll, therefore mostly green (even in the case
of species that are partial parasites attached to the roots of
other plants); leaves usually not scalelike (except in
Salicornia [Chenopodiaceae])
D Flowers crowded on a solitary conical inflorescence 2-4 cm
long, the base encircled by several white bracts, each more
than 1 cm long (these bracts may become reddish on the
underside); a smaller white bract (about 4-5 mm long) below
all flowers except the lowermost ones; blades of basal leaves
up to more than 15 cm long; flowering stems with a sessile
clasping leaf, from the axil of which 1-3 petioled leaves
originate; usually in low areas where the soil is alkaline or
slightly saline   Saururaceae, Lizard's-tail Family
Anemopsis californica, Yerba-mansa (*Plate* 51)
DD Plants not as described in choice D
E   Plants with flowers densely packed on a fleshy stalk
(spadix), this originating at the base of a large yellow
or white bract (spathe) that may be more than 20 cm long
Araceae, Arum Family (M)
EE Plants not as described in choice E
F   Succulent plants growing in salt marshes and some
inland alkaline habitats; branches opposite; stems
jointed, with the flowers (these very small and
without petals) crowded into a short inflorescence;
leaves reduced to scarcely noticeable scales
Chenopodiaceae, Goosefoot Family
Salicornia
FF Plants, if succulent and growing in salt marshes, not
as described in choice F
G   All leaves in a basal cluster and with conspicuous
gland-tipped hairs (these usually red) that trap
insects (flowers small, produced at the top of a
single stalk that is generally less than 20 cm
tall; corolla white; restricted to sphagnum bogs)
Droseraceae, Sundew Family
Drosera rotundifolia, Roundleaf Sundew
GG Leaves, if in a basal cluster, without gland-tipped
hairs that trap insects
H   Terrestrial plants with nearly circular leaf
blades to which the petioles attach near the
center (one of the sepals with a long spur;
petals [2 of these different from the other
3], showy, usually orange; an escape from
gardens)   Tropaeolaceae, Nasturtium Family
Tropaeolum majus, Garden Nasturtium
HH Plants, if terrestrial, without nearly circular
leaf blades to which the petioles attach close
to the center
I   Strictly aquatic plants, either floating or
partly or wholly submerged, but not
grasslike. (For grasses and grasslike
plants, go to Group B; for aquatic plants
that do not fit any choices under Group A,
go to Group C.)
Herbaceous Flowering Plants, Group A (p. 55)
II Plants primarily terrestrial, although they
may grow in wet places and may occasionally
be inundated by flooding (or by a high
tide, in the case of species that grow in
salt marshes)
J   Grasses and grasslike plants, often
tufted, with proportionately slender
leaves that may be entirely basal or
scattered along the stems;
inflorescences often crowded, the
flowers rarely with an obvious perianth
Herbaceous Flowering Plants, Group B
(p. 58)

JJ Plants not grasslike (if the leaves are
   slender but the flowers have an obvious
   corolla [or perianth, when the petals
   and sepals have the same form and
   color], go directly to Group C)
   K   Flowers in composite heads, like
       those of a sunflower, daisy,
       thistle, or dandelion. (The several
       to many small flowers are attached
       to a flat or conical receptacle, and
       partly enclosed by somewhat leaflike
       structures called phyllaries. In
       certain genera, the corolla of some
       or all flowers is drawn out into a
       petal-like structure called a ray.)
                        Asteraceae (Compositae)
                               Sunflower Family
   KK Flowers, if in dense heads, not as
       described in choice K
       Herbaceous Flowering Plants, Group C
                                        (p. 59)

Herbaceous Flowering Plants, Group A: Strictly aquatic plants, either
floating or partly or wholly submerged, but not grasslike. (Grasslike
plants, even if aquatic, are in Group B; for aquatic plants that are not
grasslike, but that do not fit any of the choices under Group A, go to
Group C.)
A   Plants strictly marine, found in bays and on rocky shores, and therefore
    exposed only at low tide; flowers in a 1-sided inflorescence located
    within a boat-shaped bract; leaves narrow, with almost perfectly
    parallel margins for nearly their entire length
                             Zosteraceae, Eel-grass Family (M)
AA Plants of freshwater or brackish-water habitats, not as described in
    choice A
    B   Plants without leaves; stems floating, rarely more than 5 mm long or
        wide, flattened or nearly spherical; roots, if present, unbranched,
        and sometimes only 1; flowers microscopic, restricted to the edges
        of stems; plants reproducing vegetatively, thus often forming small
        aggregations; sometimes found on wet soil after a drop in the water
        level   Lemnaceae, Duckweed Family (M)
    BB Plants with leaves, and not as described in choice B
        C   Leaves mostly in whorls of at least 3, or appearing to be whorled
            (In Hydrocharitaceae there are two exceptions: *Najas* has
            opposite leaves, but there is usually an additional leaf in the
            axil of each one, so there are 4 at each node; in *Elodea*, some
            of the lower leaves may be opposite)
            D   Leaves slender, not divided into lobes
                E   Leaves usually 8-10 in each whorl, extending stiffly
                    outward almost at a right angle to the stem, and with
                    smooth margins; plants upright, usually only partly
                    submerged, growing at the margins of lakes and pond
                                Hippuridaceae (formerly in Haloragaceae)
                                               Mare's-tail Family
                            Hippuris vulgaris, Mare's-tail (*Plate* 83)
                EE Leaves rarely more than 8 at each node (usually 3-6), not
                    obviously extending stiffly outward at a right angle to
                    the stem, and usually with toothed margins; plants
                    submerged   Hydrocharitaceae (includes Najadaceae)
                                               Waterweed Family (M)
            DD Leaves (at least the submerged ones) divided into slender,
                sometimes nearly hairlike lobes (plants with long, weak
                stems, mostly or wholly submerged)
                E   Leaves firm, the lobes with small teeth, thus somewhat
                    rough to the touch   Ceratophyllaceae, Hornwort Family
                                Ceratophyllum demersum, Hornwort (*Plate* 77)
                EE Submerged leaves delicate, the lobes nearly hairlike and
                    not toothed (some species with terminal portions of the
                    stems emerging from the water and with leaves different
                    from those of the submerged portions; corolla white or
                    purplish)   Haloragaceae, Water-milfoil Family
        CC Plants with opposite, alternate, or mostly basal leaves
            D   Leaves opposite (but in Najadaceae, an additional leaf
                usually originates in the axil of each leaf, so there are 4
                at each node)
                E   All leaves very slender, not more than 1 mm wide

      F Leaves, at least near their bases, with finely toothed
        margins; in fresh water   Hydrocharitaceae
           (includes Najadaceae), Waterweed Family (M)
                                         Najas
     FF Leaves with smooth margins; in brackish water
           Zannichelliaceae, Horned-pondweed Family (M)
           Zannichellia palustris, Horned-pondweed
   EE Most leaves (or at least the floating leaves, in the case
   of Callitrichaceae) more than 2 mm wide
      F Terminal portions of some stems (these bearing
        flowers) upright and extending above the surface;
        leaves with obvious petioles, the blades up to 2.5 cm
        long; flowers without petals, but with 4 sepals and 4
        stamens (these at the top of the fruit-forming portion
        of the pistil); fruit plump, about 4 mm long, slightly
        4-angled   Onagraceae, Evening-primrose Family
        Ludwigia palustris, Common Water-primrose
     FF Terminal portions of stems submerged or floating;
        leaves sessile, rarely so much as 1.5 cm long; flowers
        with neither petals nor sepals, and with only 1
        stamen; fruit flattened, breaking apart into 4 1-
        seeded units (in some species, staminate and
        pistillate flowers are separate)
             Callitrichaceae, Water-starwort Family
DD Leaves (or structures appearing to be leaves) either
alternate, basal, or arising from stems buried in mud (or, in
the case of *Lilaeopsis* [Apiaceae], arising in clusters from a
creeping stem, and consisting entirely of hollow, cross-
barred petioles [there are no blades])
   E Leaves without blades, and consisting entirely of hollow,
     cross-barred petioles (leaves arising in clusters from a
     creeping stem; flowers in small umbels)
            Apiaceae (Umbelliferae), Carrot Family
        Lilaeopsis occidentalis, Western Lilaeopsis
   EE Leaves (or structures appearing to be leaves) with blades
   (but these may be divided into slender lobes)
      F Submerged leaves (or structures appearing to be
        leaves) divided into slender, nearly hairlike lobes
        (there may also be floating leaves with broad blades)
        G Some "leaves" (these are in fact modified stems)
          with small bladders in which microscopic aquatic
          organisms are trapped; plants completely submerged
          except for the inflorescences; corolla yellow
             Lentibulariaceae, Bladderwort Family
                             Utricularia
        GG Leaves without bladders; plants sometimes with
          floating leaves (these with 3-lobed blades) as
          well as submerged leaves; corolla white
               Ranunculaceae, Buttercup Family
                    Ranunculus (in part)
     FF None of the leaves divided into slender, nearly
        hairlike lobes
        G Leaves compound, with 3 leaflets, these sometimes
          more than 7 cm long (leaves with stout petioles
          arising from prostrate stems; leaflets normally
          held well above the water level; flowers with 5
          calyx lobes and 5 corolla lobes, these white to
          pale purple, with scalelike hairs on the upper
          surface)   Menyanthaceae
          (formerly in Gentianaceae), Buckbean Family
          Menyanthes trifoliata, Buckbean (*Plate* 85)
        GG Leaves not compound
          H Leaf blades arrowhead-shaped to broadly heart-
            shaped
            I Leaf blades broadly heart-shaped, generally
              floating; flowers (with many petals or
              sepals) at least 5 cm wide, solitary and
              usually floating (the color of the flowers,
              which may be yellow, white, or pink, may
              reside in either the petals or sepals,
              depending on the genus)
                Nymphaeaceae, Waterlily Family

II Leaf blades arrowhead-shaped, raised well
   above the water level; flowers (with 3
   white petals and 3 sepals) rarely more than
   2.5 cm wide, arranged in whorls
               Alismataceae, Water-plantain Family (M)
                                          Sagittaria
HH Leaf blades not arrowhead-shaped or heart-
   shaped
   I   Leaf blades (up to about 4 cm wide) either
       almost circular (in which case the petiole
       is attached near the center, as in a
       nasturtium) or approximately kidney-shaped
       and lobed
                 Apiaceae (Umbelliferae), Carrot Family
                                          Hydrocotyle
   II Leaf blades neither almost circular nor
      kidney-shaped and lobed
      J   Leaves narrow, usually at least 7 cm
          long but not more than 0.5 cm wide;
          flowers borne singly within a rolled-up
          bract, the perianth pale yellow, with a
          tubular portion at least 1.5 cm long
          and 6 lobes about 5 mm long (stems
          slender, weak, trailing or floating;
          flowers opening at the surface)
          Pontederiaceae, Pickerel-weed Family (M)
              Heteranthera dubia, Water Stargrass
      JJ Plants, if with narrow leaves, without
         flowers as described in choice J
         K   Petals (5) yellow (leaves oval or
             elongated, scattered rather evenly
             along the stems, whether these are
             submerged, upright and only partly
             submerged, or creeping over mud;
             sepals and petals originating above
             the fruit-forming part of the
             pistil; stamens 10)
               Onagraceae, Evening-primrose Family
                                          Ludwigia
         KK Petals, if present, not 5 and not
            yellow
            L   All leaves floating (these oval,
                with many lengthwise veins and
                cross-veins) (flowers, in
                dichotomously branched
                inflorescences, with 2 white
                perianth segments and 6 stamens,
                these with purple anthers;
                occasional escape from water
                gardens)   Aponogetonaceae
                     Cape-pondweed Family (M)
                        Aponogeton distachyon
                               Cape-pondweed
            LL Only some of the leaves, or none
               of them, floating
               M   Leaves basal, usually raised
                   well above the water level;
                   petioles usually at least as
                   long as the blades; flowers
                   with 3 sepals, 3 white
                   petals (these often falling
                   early), 6 to many stamens,
                   and several to many pistils
                   that develop into 1-seeded
                   fruit (leaf blades
                   arrowhead-shaped or narrowly
                   oval)   Alismataceae
                       Water-plantain Family (M)
               MM Leaves scattered along the
                  stems, sometimes floating;
                  petioles, when present, not
                  half as long as the blades;
                  flowers not as described in
                  choice M

N   Flowering stems upright, raised well above the water level; all leaves with similar broad blades, even if some float or are submerged (flowers small, in crowded inflorescences; calyx, with 5 lobes, pink to rose; corolla absent)
Polygonaceae
Buckwheat Family
Polygonum amphibium

NN All stems trailing through the water (but the inflorescences, if present, generally raised slightly above the water level); floating leaves, if present, with broad blades, and decidedly different from the submerged leaves, which may be nearly threadlike or have substantial blades with wavy margins (flowers, when produced, with 4 lobes that are perianthlike)
Potamogetonaceae
(includes Ruppiaceae)
Pondweed Family (M)

Herbaceous Flowering Plants, Group B: Grasses and grasslike plants, often tufted, with proportionately slender leaves that may be strictly basal or scattered along the stems; inflorescences often crowded, the flowers rarely with an obvious perianth

A   Flowers in a dense, almost smooth cylindrical inflorescence in which all staminate flowers are above the pistillate flowers; plants rooted in mud or muck, but only the lower portion under water
Typhaceae (includes Sparganiaceae), Cattail Family (M)
Typha

AA Flowers not in a dense cylindrical inflorescence in which all staminate flowers are above the pistillate flowers; aquatic, terrestrial, or growing in salt marshes

B   Salt marsh plants with rather succulent basal leaves; flowers (in terminal racemes) with either 3 or 6 markedly concave perianth segments (when there are 6, they are in 2 series); stamens (3 or 6) with very short filaments; fruit 2 or 3 times as long as wide, the 3-6 divisions splitting apart at maturity
Juncaginaceae (includes Lilaeaceae), Arrow-grass Family (M)
Triglochin

BB Plants, if growing in salt marshes, without succulent leaves; flowers and fruit not as described in choice B

C   Each inflorescence consisting of several globular heads of either all staminate or all pistillate flowers, the heads of pistillate flowers soon resembling burs; plants sometimes mostly submerged, but the inflorescences held above the water surface
Typhaceae (includes Sparganiaceae), Cattail Family (M)
Sparganium

CC Inflorescences not as described in choice C

D   Leaves basal, mostly cylindrical, but the lower portion broadened into an open sheath; some pistillate flowers (these with styles at least 5 cm long) borne singly in the leaf axils, others (with short styles) mixed with staminate flowers (these with a single stamen); some flowers may have a pistil and a stamen (flowers without a perianth, but each one with a bract below it)
Juncaginaceae (includes Lilaeaceae), Arrow-grass Family (M)
Lilaea scilloides, Flowering Quillwort

DD Plants not as described in choice D

E   Each leaf with a collarlike projection (ligule) originating at the place where the blade meets the sheath; stems hollow, cylindrical (flowers partly enclosed by bracts) Poaceae (Gramineae), Grass Family (M)

    EE Leaves without a collarlike projection; stems usually
        solid, cylindrical or 3-angled
        F Stems cylindrical; leaf sheaths open or closed (if the
           sheaths are closed, then the margin of the blade has
           long hairs); successive leaves on opposite sides of
           the stem; flowers not partly enclosed by bracts
           (although there may be bracts below the inflorescence)
                            Juncaceae, Rush Family (M)
        FF Stems 3-angled or cylindrical; leaf sheaths closed
           (sometimes opening with age); successive leaves not
           quite on opposite sides of the stem (flowers partly
           enclosed by bracts)   Cyperaceae, Sedge Family (M)
Herbaceous Flowering Plants, Group C: Plants primarily terrestrial, or at
least mostly developing above the water level, unless temporarily inundated
by flooding (or, in the case of salt marsh species, inundated by a high
tide), not grasslike; flowers not in composite heads like those of
sunflowers, daisies, thistles, or dandelions (for these go directly to
Asteraceae)
A Flower parts usually in cycles of 3 or 6; main veins of leaves usually
   nearly parallel to one another (there are some exceptions to both of the
   above criteria; in some Orchidaceae, moreover, the leaves are reduced to
   scales)
   B Flowers conspicuously irregular, the lower petal (lip petal)
     distinctly different from the 2 upper ones (fruit-forming portion of
     the pistil located below the bases of the sepals and petals)
                        Orchidaceae, Orchid Family (M)
   BB Flowers regular, none of the sepals and petals (or perianth segments,
     when the 3 petals and 3 sepals have the same form and color)
     distinctly different from the others
     C Fruit-forming portion of the pistil located below the bases of
       the sepals and petals, or below the perianth tube formed by
       union of the petals and sepals   Iridaceae, Iris Family (M)
     CC Fruit-forming portion of the pistil located above the bases of
       the sepals and petals
       D Plants creeping, rooting at the nodes; leaves alternate,
         their bases forming sheaths around the stems; flowers in
         umbels, with 3 white petals and 3 green sepals (occasional
         escape from gardens)   Commelinaceae, Spiderwort Family (M)
                          Tradescantia fluminensis, Spiderwort
       DD Plants not creeping or rooting at the nodes; leaves basal, in
         whorls, or alternate; flowers, if in umbels, with 6 perianth
         segments (that is, sepals and petals similar)
         Liliaceae (including part of Amaryllidaceae), Lily Family (M)
AA Flower parts not often in cycles of 3 or 6; leaves usually with a single
   main vein that originates at the base of the blade, then branches, or
   with 3 or more main veins that diverge from the base of the blade
   B Leaves alternate or entirely basal
     C Leaves generally more than 75 cm long, with 3 primary divisions,
       each with 3-5 leaflets, these usually at least 8 cm long
       (sometimes more than 20 cm long); flowers in umbel-like clusters
       on the branches of the inflorescence, with 5 whitish petals and
       minute sepals; fruit blackish; often more than 2 m tall; growing
       along streams in shaded canyons   Araliaceae, Ginseng Family
                     Aralia californica, Elk-clover (*Plate* 6)
     CC Plants not as described in choice C
       D Leaf blades usually at least 75 cm wide, more or less
         palmately lobed, on thick, fleshy petioles arising from
         creeping underground stems (inflorescences up to more than 1
         m tall; petals absent; calyx lobes 2 or 3; fruit red)
         Gunneraceae (formerly in Haloragaceae), Gunnera Family
                        Gunnera tinctoria, Gunnera
       DD Leaf blades not so much as 75 cm wide, but even if close to
         this size, not on petioles arising from creeping stems
        E Leaves basal, with petioles up to about 40 cm long and
          with 3 diverging, fan-shaped, pale green leaflets up to
          about 8 cm long (flowers, with neither petals nor sepals,
          in a short, dense inflorescence; foliage with an aroma
          slightly reminiscent of vanilla; growing in moist
          woodland habitats)   Berberidaceae, Barberry Family
                    Achlys triphylla, Vanilla-leaf (*Plate* 15)
       EE Leaves, if basal, with a petiole much shorter than 40 cm,
         and if with 3 diverging leaflets, these not fan-shaped
         and not so much as 5 cm long

      F Plants robust, up to about 3 m tall; usually with reddish or purplish stems; leaves alternate, the blades palmately lobed, up to about 40 cm wide; fruit 2 cm wide, covered with soft spines; escaping from gardens   Euphorbiaceae, Spurge Family
             Ricinus communis, Castor-bean (*Plate* 24)
     FF Plants not as described in choice F
             Herbaceous Flowering Plants, Group C, Subkey 1
BB Leaves, at least in the upper part of the plant, distinctly opposite or whorled (Note: a pair of opposite leaves may be united in such a way that they completely surround the stem.)
  C Leaves opposite (in certain species, at least a few may be alternate)
    D Plants succulent, with leaves more than 1 cm long (these 3-angled, cylindrical, or somewhat flattened) (petals numerous, slender; stamens numerous; fruit-forming portion of the pistil united with the calyx cup, the fruit therefore developing below the 5 sepals; plants usually growing on backshores of sandy beaches or in other maritime situations, but sometimes cultivated inland)   Aizoaceae, Seafig Family
    DD Plants, if succulent, with leaves much less than 1 cm long
      E Foliage with stinging hairs (flowers small, greenish; inflorescences in the upper leaf axils, the pistillate and staminate flowers either in the same inflorescence or in separate inflorescences)   Urticaceae, Nettle Family
      EE Foliage without stinging hairs
          Herbaceous Flowering Plants, Group C, Subkey 2 (p. 67)
  CC Leaves arranged in whorls
    D Plants with a single whorl of leaves just below 1 or several flowers (there may be scalelike leaves on the lower part of the stem, or a single basal leaf that usually disappears early)
      E Whorl consisting of 3 leaves, each with 3 leaflets or lobes, the margins of these toothed; flower single, with 5 white or pale blue sepals, these not pointed at the tip (petals absent)   Ranunculaceae, Buttercup Family
           Anemone oregana, Wood Anemone (*Plate* 91)
      EE Whorl consisting of more than 3 leaves, these neither lobed nor toothed; flowers (on extremely slender peduncles) several, with 5-7 pink petals, these pointed at the tip   Primulaceae, Primrose Family
           Trientalis latifolia, Starflower (*Plate* 45)
    DD Plants usually with at least 2 whorls of leaves (many whorls in most Rubiaceae)
      E Stems cylindrical, smooth; leaf blades tough, usually more than 1 cm wide and with rather evenly toothed margins; flowers at least 1 cm wide, the 5 petals separate, white to deep pink; stamens 10; perennial (evergreen, somewhat woody; in coniferous forests)
          Ericaceae (includes Pyrolaceae), Heath Family
                    Chimaphila
      EE Stems 4-angled, often with bristles that engage the skin or stick to clothing; leaf blades not tough, rarely so much as 1 cm wide, and not toothed along the margins (sometimes they are very short, slender, and bristle-tipped); flowers much less than 1 cm wide, the corolla consisting of a short tube and 3 or 4 lobes, these white, yellowish, greenish, pinkish, or bluish; stamens 3 or 4; annual or perennial   Rubiaceae, Madder Family

Herbaceous Flowering Plants, Group C, Subkey 1: Other terrestrial plants with alternate or entirely basal leaves
A Flowers without a corolla, but usually with a calyx. (Note: in some species that have separate staminate and pistillate flowers, the latter do not have a calyx; in other species, the calyx may be colored in such a way that it resembles a corolla; an involucre, located just below the flowers of some species, may resemble a calyx.)
  B Leaves compound, with at least 3 leaflets
    C Pistillate flowers (these separate from staminate flowers) with several to many pistils (in *Thalictrum*, staminate and pistillate flowers are on separate plants; in *Isopyrum*, the sepals are commonly white, pinkish, or purplish, and thus resemble petals)
          Ranunculaceae, Buttercup Family
               Isopyrum, *Thalictrum*

CC None of the flowers with more than 1 pistil (there may be separate staminate and pistillate flowers, or some or all flowers may have stamens and a pistil)

    D Leaves mainly basal, with 11-17 pinnately arranged leaflets, each with 3-7 lobes; stamens 3-5 (dark purple); in open, sandy habitats   Rosaceae, Rose Famliy
        Acaena pinnatifida var. *californica*, California Acaena

    DD Leaves not basal, the larger ones usually with 3 or 5 pinnately arranged divisions, these with deeply lobed and coarsely toothed leaflets; stamens numerous; in moist woods (sepal tips sometimes pink)   Ranunculaceae, Buttercup Family
            Actaea rubra, Baneberry (*Plate* 91)

BB Leaves not compound, but they may be deeply lobed

  C Flowers usually with 4 sepals, 6 stamens (4 long, 2 short), and fruit usually with a partition that divides it into 2 halves that crack apart at maturity
          Brassicaceae (Cruciferae), Mustard Family

  CC Flowers not with a combination of 4 sepals, 6 stamens, and a fruit with a partition that divides it into 2 halves that crack apart at maturity

    D Leaves palmately or pinnately lobed

      E Leaves up to 20 cm long, not fan-shaped, pinnately lobed, the lobes with coarse teeth; often more than 100 cm tall (usually growing along dry stream beds; staminate and pistillate flowers mostly separate; calyx lobes green)
            Datiscaceae, Datisca Family
          Datisca glomerata, Durango-root (*Plate* 79)

      EE Leaves less than 1.5 cm long, fan-shaped, usually with 3 lobes, these not toothed; rarely more than 10 cm tall (in open or lightly shaded areas) Rosaceae, Rose Family
          Aphanes occidentalis, Western Dewcup (*Plate* 48)

    DD Leaves not lobed, but they may be toothed

      E Leaf blades heart-shaped, up to 10 cm wide, with the aroma of ginger when bruised; fruit-forming part of the pistil below the 3 sepals, which are brownish purple and drawn out into slender "tails" more than 2 cm long; low ground cover in moist woods Aristolochiaceae, Pipevine Family
          Asarum caudatum, Wild-ginger (*Plate* 6)

      EE Plants not as described in choice E

        F Plants up to more than 2 m tall; leaves up to 30 cm long (flowers, in racemes up to 20 cm long, with 10 stamens as well as a pistil; sepals white or pink; fruit about 1 cm wide, fleshy, purplish black; occasional weed)   Phytolaccaceae, Pokeweed Family
          Phytolacca americana, Pokeweed

        FF Plants rarely more than 1 m tall (mostly not more than 50 cm); leaves not more than 15 cm long

          G Fruit-forming part of the pistil fused to the calyx cup (thus the fruit develops below the sepals) (leaves succulent; stamens 7-16; pistil becoming a dry, 4-horned fruit; calyx lobes greenish yellow; branches falling down, but the tips upright)
            Aizoaceae, Seafig Family
           Tetragonia tetragonioides
          New Zealand Spinach (*Plate* 4)

          GG Fruit-forming portion of the pistil free of the calyx (calyx may be lacking in pistillate flowers)

            H Leaves and stems usually either scaly or powdery, often at least slightly succulent, or many of the leaves slender, rigid, and prickle-tipped (the plant as a whole therefore prickly) (when there are separate staminate and pistillate flowers, the latter lack a calyx, but are enclosed within a pair of bracts, these sometimes becoming fleshy and colored; some species with a strong odor)
           Chenopodiaceae, Goosefoot Family

            HH Leaves and stems generally not scaly or powdery (but they may be whitish or grayish and almost completely covered with branched hairs), and if there are slender, rigid leaves, these not prickle-tipped (in some Polygonaceae, however, the bracts that form involucres below the flower clusters may be prickly)

I Calyx of pistillate flowers a deep cup,
with 4 lobes; stigma of pistil consisting
of a tuft of slender threads;
inflorescences in the leaf axils, the
flowers few; leaves up to about 2.5 cm
long, the blade narrowly oval, pointed at
the tip; seed black, shiny

Urticaceae, Nettle Family
Parietaria judaica, Pellitory

II Plants not as described in choice I

J Fruit markedly 3-lobed, each lobe with a
single seed (staminate flowers without
a calyx and with 1 stamen)

Euphorbiaceae, Spurge Family
Euphorbia, *Croton*

JJ Fruit not 3-lobed, and producing only 1
seed

K Stems branching dichotomously (and
forming grayish mats or mounds);
plants with a strong odor;
pistillate flowers (in the axils of
lower leaves) without a calyx
(staminate flowers in corymbs at the
tip of stems; some leaves may be
opposite)

Euphorbiaceae, Spurge Family
Eremocarpus setigerus
Turkey-mullein (*Plate* 24)

KK Stems not branching dichotomously;
plants without a strong odor; all
flowers (including pistillate
flowers, if these are separate from
staminate flowers) with a calyx

L Leaves often with membranous
stipules that form sheaths
around the stems; calyx with 5
or 6 lobes or separate sepals
(these white, green, or brightly
colored); flowers generally with
a pistil and stamens; wall of
the fruit usually tightly bound
to the seed

Polygonaceae, Buckwheat Family

LL Leaves without stipules; calyx
with 2-5 separate sepals (these
thin and dry); pistillate and
staminate flowers separate, or
some flowers with both a pistil
and stamens; wall of the fruit
loosely enveloping the seed

Amaranthaceae, Amaranth Family

AA Flowers with a corolla as well as a calyx (in Papaveraceae, however, the
sepals fall when the flowers open, and they may also be united, during
the bud stage, to form a cap over the petals and other flower parts)

B Petals united to the extent that there is an obvious corolla tube

C Corolla irregular, the lobes not all of the same size or shape,
and usually distinctly 2-lipped (the corolla is not noticeably
2-lipped, however, in *Synthyris*, *Verbascum*, and *Veronica*
[Scrophulariaceae]; furthermore, it may appear, at first glance,
to be regular)

D Fruit-forming portion of the pistil free of the calyx;
flowers usually with 4 functional stamens, plus 1 sterile
stamen (5 functional stamens in *Verbascum* and only 2 in
Synthyris and *Veronica*) (a large, diversified family)

Scrophulariaceae, Snapdragon Family

DD Fruit-forming portion of the pistil united with much of the
calyx (thus the fruit develops below the sepals); flowers
with 5 functional stamens (2 shorter than the other 3 in
Downingia) (leaves either basal or the upper ones sessile;
plants low, found either on dry slopes [*Nemacladus*] or in
drying vernal pools [*Downingia*])

Campanulaceae, Bluebell Family
Downingia, *Nemacladus*

CC Corolla regular, all lobes almost exactly the same size and shape
   (if 2 upper lobes of a yellow corolla are slightly smaller than
   the 3 lower lobes, take choice C)
   D   Herbaceous vines with separate staminate and pistillate
       flowers, these with 5-7 corolla lobes (the pistillate flowers
       are solitary in the same leaf axils as those that give rise
       to racemes of staminate flowers; stamens 3; in pistillate
       flowers, the fruit-forming portion is united to the calyx and
       develops into a large fruit that often has a few to many soft
       or fairly rigid spines)   Cucurbitaceae, Gourd Family
   DD Either not vinelike or not otherwise as described in choice D
      E   Stamens in line with the corolla lobes
          F   Pistil with 5 styles; fruit a small, 1-seeded nut
              (plants mostly growing in salt marshes or on bluffs
              near the shore)   Plumbaginaceae, Leadwort Family
          FF Pistil with a single style; fruit a capsule containing
              numerous seeds   Primulaceae, Primrose Family
      EE Stamens alternating with the corolla lobes (but sometimes
         absent between certain corolla lobes)
         F   Corolla dry and membranous in texture, with 4 lobes
             (these either spreading outward, or upright and
             covering the fruit) (stamens 4 or 2; sepals 4, usually
             with membranous margins; leaves basal)
                                    Plantaginaceae, Plantain Family
         FF Corolla not dry and membranous in texture, and with 5
             lobes
             G   Fruit-forming portion of the pistil united with
                 much of the calyx, the fruit thus developing below
                 the sepals (stigma divided into 2 or more lobes)
                                    Campanulaceae, Bluebell Family
             GG Fruit-forming portion of the pistil free of the
                 calyx
                 H   Stigma 3-lobed (the separation sometimes begins
                     lower, on the style)
                                    Polemoniaceae, Phlox Family
                 HH Stigma either not divided, or divided into only
                     2 lobes
                     I   Sepals separate or nearly separate
                         J   Plants usually twining or trailing (but
                             sometimes rather compact); flowers
                             solitary; corolla generally more than 2
                             cm long, the lobes twisted together in
                             the bud stage (and usually with
                             distinct pleats after opening)
                                    Convolvulaceae, Morning-glory Family
                         JJ Plants usually upright; flowers usually
                             in groups, but sometimes solitary (when
                             grouped, often in somewhat coiled
                             inflorescences); corolla usually less
                             than 2 cm long, the lobes not twisted
                             together in the bud stage (but they may
                             overlap)
                                    Hydrophyllaceae, Waterleaf Family
                     II Sepals united for much of their length, a
                         substantial portion of the calyx therefore
                         cup-shaped or tubular
                         J   Fruit-forming portion of the pistil
                             deeply 4-lobed, and typically
                             separating into 4 1-seeded nutlets
                             (these usually roughened or bristly; in
                             some species, not all of the lobes
                             mature); flowers often in slightly
                             coiled, 1-sided inflorescences
                                    Boraginaceae, Borage Family
                         JJ Fruit-forming portion of the pistil not
                             deeply 4-lobed, and not separating into
                             4 1-seeded nutlets; flowers solitary,
                             or if in clusters, these not coiled, 1-
                             sided inflorescences (corolla lobes
                             often twisted together in the bud
                             stage, each one sometimes with a pleat
                             along the midline; fruit either a
                             fleshy berry or a dry capsule)
                                    Solanaceae, Nightshade Family

BB Petals separate or nearly so, not united to the extent that there is
   an obvious corolla tube
   C Leaves compound, with at least 3 leaflets
      D Corolla irregular (flowers with 5 petals, the 2 lower ones
         united by their edges to form a keel, and the uppermost one
         enlarged to form a banner; with 10 stamens)
                                    Fabaceae (Leguminosae), Pea Family
      DD Corolla regular, unless some of the petals do not develop
         normally (which is typical of some Ranunculaceae)
         E Fruit separating early into 2 1-seeded halves (these often
            bristly or ribbed lengthwise); inflorescence usually an
            umbel; flowers with 5 small petals; sepals, if obvious,
            reduced to small toothlike structures at the top of the
            fruit-forming part of the pistil; leaves, if bruised,
            usually aromatic   Apiaceae (Umbelliferae), Carrot Family
         EE Fruit not separating into 2 1-seeded halves, and plants
            not as described in choice E
            F Flowers with a single pistil (there may be more than 1
               style, however)
               G Stamens many more than 10; leaflets coarsely
                  toothed (flowers small, in terminal racemes; tip
                  of sepals sometimes pink; with 1-several small
                  white petals; fruit nearly spherical, about 1 cm
                  long, fleshy, usually red, sometimes white)
                                    Ranunculaceae, Buttercup Family
                                    Actaea rubra, Baneberry (Plate 91)
               GG Stamens not more than 10; leaflets not toothed
                  H Leaflets 3; stamens 10; styles 5 (leaflets
                     bilobed, diverging at the end of the petiole,
                     as in a clover; leaves with a very sour
                     taste)   Oxalidaceae, Oxalis Family
                  HH Leaflets more than 3; stamens 6; style 1
                     (flowers in a loose panicle, the buds nodding
                     before opening; sepals and petals 6, the
                     petals turned back, and each one ending in a
                     hoodlike structure; in coniferous forests)
                                    Berberidaceae, Barberry Family
                                    Vancouveria planipetala
                                    Inside-out-flower (Plate 15)
            FF Flowers with more than 1 pistil
               G Flowers with numerous pistils, these developing
                  into small, 1-seeded dry fruits (the pistils may,
                  however, be attached to a fleshy receptacle, as
                  they are in a strawberry) (stamens usually
                  numerous)
                  H Leaves with stipules at the base of the
                     petiole; sepals (5) usually alternating with
                     bracts   Rosaceae, Rose Family
                                    Fragaria, Potentilla
                  HH Leaves without stipules; sepals not alternating
                     with bracts (sometimes falling early)
                                    Ranunculaceae, Buttercup Family
                                    Ranunculus (in part)
               GG Flowers with only a few pistils, each pistil
                  producing more than 1 seed
                  H Petals red and yellow, not especially thick,
                     with long spurs; sepals somewhat pointed,
                     flat, red   Ranunculaceae, Buttercup Family
                                    Aquilegia
                  HH Petals mostly brownish red, thick, and without
                     spurs; sepals bluntly rounded, markedly
                     concave, green   Paeoniaceae, Peony Family
                                    Paeonia brownii, Western Peony (Plate 87)
   CC Leaves not compound, but they may be deeply lobed
      D Corolla irregular, the lobes of different shapes and/or sizes
         E Flowers with at least 12 stamens
            F Flowers white, not more than 1 cm long; stamens 12-15,
               these all on one side of the flower; petals 6, usually
               deeply lobed, without spurs; calyx with 4-7 lobes,
               these without spurs (escaping from gardens)
                                    Resedaceae, Mignonette Family
                                    Reseda alba, White Mignonette

FF Flowers usually not white, more than 1 cm long;
stamens usually many more than 15, not all on one side
of the flower; petals 4, in unequal pairs, those of
the upper pair with spurs; calyx composed of 5
separate sepals, the uppermost one with a spur
Ranunculaceae, Buttercup Family
Delphinium

EE Flowers with not more than 10 stamens, these usually
rather evenly (or at least symmetrically) arranged
  F Flowers with 2 sepals and 4 petals, 1 or 2 of which
have a saclike spur at the base
Papaveraceae (includes Fumariaceae), Poppy Family
Dicentra, Fumaria

  FF Flowers with 5 sepals and 3 or 5 petals, none of which
have spurs (in Polygalaceae, 2 of the 5 sepals are
colored like the petals)
    G Flowers with 3 petals, these united at the base,
and the lowermost one boat-shaped; the 2 sepals at
the sides of the flower colored pink like the
petals; stamens 8   Polygalaceae, Milkwort Family
Polygala californica, Milkwort (Plate 43)

    GG Flowers with 5 petals, these not united at the
base, the lowermost one with a pouchlike spur;
none of the sepals resembling the petals; stamens
5   Violaceae, Violet Family

DD Corolla regular, the petals more or less equal in size and
shape
  E Flowers with several to many pistils
    F Plants succulent, with thick, fleshy leaves; petals
usually yellow, at least 5 mm long, usually united at
the base; calyx lobes separate nearly to the base,
without spurs; stamens 10; pistils 5, separate or
united at the base   Crassulaceae, Stonecrop Family
    FF Plants not succulent; petals greenish yellow or white,
2-3 mm long, not united at the base; calyx consisting
of completely separate sepals, each one with a spur
near the base; stamens 5-25; pistils numerous, on a
long, slender receptacle
Ranunculaceae, Buttercup Family
Myosurus minimus, Mousetail (Plate 91)

  EE Flowers with one pistil
    F Calyx fused, for at least half the length of its
cuplike or tubular portion, to the fruit-forming part
of the pistil (thus the fruit develops below the
sepals or calyx lobes)
      G Plants succulent, with thick, fleshy leaves
(flowers with numerous slender, bright yellow
petals, these usually in more than 1 series;
stamens numerous)   Aizoaceae, Seafig Family
Conicosia pugioniformis, Conicosia

      GG Plants, if succulent, not as described in choice G
        H Flowers with 5 petals and 5 calyx lobes
          I Flowers usually with 5 or 10 stamens,
sometimes fewer than 5; petals white,
pinkish, or greenish, sometimes reddening
with age, and often divided into slender
lobes; about half to three-fourths of the
calyx cup fused to the fruit-forming part
of the pistil
Saxifragaceae, Saxifrage Family
Boykinia, Heuchera, Tellima

          II Flowers with many stamens; petals yellow,
never divided into lobes (usually with a
silky appearance); all of the calyx cup
fused to the fruit-forming part of the
pistil   Loasaceae, Blazing-star Family
        HH Flowers, if with 5 petals, not also with 5
sepals
          I Flowers with 5 petals, 2 sepals, and 7-20
stamens (petals greenish yellow; plants
succulent, prostrate; garden weed)
Portulacaceae, Purslane Family
Portulaca oleracea
Common Purslane (<i>Plate</i> 90)</div>

II Flowers with 4 petals, 4 sepals, and 4 or 8
   stamens (sepals often turned to one side
   and adhering to one another after a flower
   opens) Onagraceae, Evening-primrose Family
FF Calyx not fused to the fruit-forming portion of the
   pistil (but in Lythraceae, much of the calyx, which is
   tubular, closely envelops the pistil)
   G Calyx tubular, closely enveloping the pistil, and
      usually ribbed lengthwise (stems angular; flowers
      with 4-6 petals, 4-10 stamens, and small teeth
      alternating with the 4-6 calyx lobes)
                     Lythraceae, Loosestrife Family
   GG Calyx not tubular, not closely enveloping the
      pistil, and usually not ribbed lengthwise
      H Flowers with 2 sepals and 5 petals
                     Portulacaceae, Purslane Family
      HH Flowers, if with only 2 sepals, not with 5
         petals
         I Stamens more than 10
            J Stamens united to form a cylinder around
               the pistil (calyx 5-lobed; petals 5)
                           Malvaceae, Mallow Family
            JJ Stamens not united to form a cylinder
               around the pistil
               K Sepals 2 or 3, falling as the
                  flowers open (in *Eschscholzia*, the 2
                  sepals are united to form a cap that
                  covers the other flower parts until
                  the flower opens); petals 4 or 6
                  Papaveraceae (includes Fumariaceae)
                                      Poppy Family
               KK Sepals 5, but 2 are smaller and
                  slightly below the others, so they
                  could be mistaken for bracts; petals
                  5 (somewhat shrubby, slightly woody
                  at the base)
                        Cistaceae, Rockrose Family
         II Stamens 10 or fewer
            J Flowers with 4 petals, 4 sepals, and
               usually 6 stamens, 2 of which are
               shorter than the other 4
                           Brassicaceae (Cruciferae)
                                      Mustard Family
            JJ Flowers with more than 4 petals and 4
               sepals, and not with the arrangement of
               stamens described in choice J
               K Pistil not deeply 5-lobed and not
                  separating into 1-seeded nutlets
                  L Flowers with 5 petals (these
                     usually falling early), 5
                     sepals, 5 stamens
                                 Linaceae, Flax Family
                  LL Flowers with 13-15 petals (these
                     white to pink), 6-8 sepals, at
                     least 40 stamens (on rocky
                     slopes at higher elevations)
                        Portulacaceae, Purslane Family
                                   Lewisia rediviva
                           Bitterroot (*Plate* 45)
               KK Pistil deeply 5-lobed, the lobes
                  eventually separating as 1-seeded
                  divisions (flowers with 5 sepals, 5
                  petals)
                  L Petioles with thin stipules at
                     the base; petals usually pink or
                     purplish (sometimes red or white
                     in escaped species); each
                     division of the pistil retaining
                     a style that becomes much longer
                     than the portion containing the
                     seed and that coils and uncoils
                     in response to changes in
                     humidity
                        Geraniaceae, Geranium Family

           LL Petioles without stipules; petals
              yellow or yellow and white,
              sometimes completely white;
              pistil with a single style, this
              not persisting as the divisions
              of the pistil ripen (in moist
              places)   Limnanthaceae
                       Meadowfoam Family

Herbaceous Flowering Plants, Group C, Subkey 2: Other terrestrial plants
with opposite leaves
A Flowers without a corolla, and sometimes without a calyx (when a calyx
   is present, it may be brightly colored and thus resemble a corolla; in
   Mirabilis [Nyctaginaceae, choice C, below], united bracts forming an
   involucre below each flower could be confused with a calyx, leading the
   observer to mistake the colored calyx for a corolla)
   B Pistillate and staminate flowers separate
     C Plants with milky sap, usually not hairy; staminate and
       pistillate flowers together in the same clusters; staminate
       flowers (each with 1 stamen) around each pistillate flower, this
       usually soon raised up above the staminate flowers and often
       nodding; none of the flowers with a calyx; fruit 3-lobed, the
       lobes eventually separating (Caution: the involucre below the
       cluster of flowers could be mistaken for a calyx.)
                     Euphorbiaceae, Spurge Family
                               Euphorbia
     CC Plants without a milky sap, the leaves and stems hairy (the hairs
       on the stems sometimes 3 mm long); staminate and pistillate
       flowers in separate clusters; staminate flowers (each with
       several stamens) in terminal corymbs, pistillate flowers 1-3 in
       the axils of the lower leaves; only the staminate flowers with a
       calyx; fruit not lobed (plants grayish, strongly aromatic,
       forming mats on open ground; stems branching dichotomously;
       sometimes a few leaves are alternate)
                     Euphorbiaceae, Spurge Family
            Eremocarpus setigerus, Turkey-mullein (*Plate* 24)
   BB All flowers with a pistil and stamens (calyx sometimes brightly
     colored)
     C Calyx with a long tube, usually yellow, pink, rose, or red, the
       lower portion of the tube persisting and adhering tightly to the
       fruit as this matures (nodes usually obviously swollen; our
       native species, belonging to *Abronia*, have flowers in umbels,
       and grow on backshores of sandy beaches; a cultivated species of
       Mirabilis, with a cup of united bracts forming an involucre
       below each flower, sometimes escapes from gardens)
                     Nyctaginaceae, Four-o'clock Family
     CC Calyx either a shallow, lobed cup or consisting of separate
       sepals, usually green (but sometimes white on the upper surface,
       or tinged with rose or a related color), none of it adhering to
       the fruit
       D Stamens not necessarily the same number as the sepals; fruit
         partitioned lengthwise into 3 divisions; nodes not obviously
         swollen
            Molluginaceae (formerly in Aizoaceae), Carpetweed Family
       DD Stamens 5, the same number as the sepals; fruit not
         partitioned into divisions; nodes often obviously swollen
                    Caryophyllaceae, Pink Family
AA Flowers with a corolla as well as a calyx (but the petals or corolla
  lobes may be inconspicuous)
   B Petals separate, or nearly so, not united to the extent that there is
     an obvious corolla tube
     C Flowers in small clusters, each cluster with numerous stiff,
       sharp-tipped bracts below it; petals 5, bluish or greenish
       white; sepals 5, persisting on the dry fruit, which is covered
       with scales; leaves irregularly toothed or lobed, sometimes
       prickly; usually in places that are wet, at least during part of
       the year   Apiaceae (Umbelliferae), Carrot Family
                                  Eryngium
     CC Plants not as described in choice C
       D Leaves pinnately compound (flowers with 5 yellow petals, 10
         stamens; fruit spiny; prostrate)
                   Zygophyllaceae, Caltrop Family
             Tribulus terrestris, Puncture Vine (*Plate* 56)
       DD Leaves not compound

E   Plants somewhat shrubby, growing at the edge of salt
    marshes (flowers with 4-6 pink petals, each with a small,
    tonguelike outgrowth near the middle)
                        Frankeniaceae, Frankenia Family
                    Frankenia salina, Alkali-heath (*Plate* 28)
EE Plants, if somewhat shrubby, not growing at the edge of
    salt marshes
    F   Plants slightly woody at the base, forming a low
        ground cover in woods; leaves with 3 main veins;
        petals 5, white, soon becoming green or falling; calyx
        fused to the lower half of the fruit-forming portion
        of the pistil   Philadelphaceae
                (formerly in Saxifragaceae), Mock-orange Family
                    Whipplea modesta, Yerba-de-selva (*Plate* 87)
    FF Plants either not woody at the base or not fitting
        other criteria in choice F
        G   Flowers with 2 petals and 2 stamens (petals and
            sepals 2, on a distinct perianth tube that
            originates at the top of the fruit-forming part of
            the pistil; petals white, notched at the tip;
            fruit covered with hooked hairs; in moist, shady
            woods   Onagraceae, Evening-primrose Family
            Circaea alpina ssp. *pacifica*, Enchanter's-nightshade
        GG Flowers with more than 2 petals and 2 stamens
            H   Flowers with 4 petals, 4 sepals, and 8 stamens
                (petals and sepals sometimes on a distinct
                perianth tube that originates at the top of
                the fruit-forming part of the pistil; petals
                notched at the tip)
                            Onagraceae, Evening-primrose Family
                                        Epilobium (in part)
            HH Flowers, if with 4 petals and 4 sepals (or 4
                calyx lobes), not with 8 stamens (except
                perhaps in Lythraceae; see choice I, below)
                I   Calyx tubular, usually ribbed lengthwise,
                    and with small teeth alternating with the
                    lobes; stems angular (petals and calyx
                    lobes 4-6, stamens 4-10, the number of
                    these parts often varying on the same
                    plant)   Lythraceae, Loosestrife Family
                II Calyx not tubular, not obviously ribbed
                    lengthwise, and without small teeth
                    alternating with the lobes or separate
                    sepals; stems not angular
                    J   Stamens and petals 4 (annual, rarely so
                        much as 7 cm tall; leaves small, but
                        succulent; flowers [each one with 4
                        pistils] about 2 mm wide, clustered at
                        the end of the stems or single in the
                        leaf axils)
                                    Crassulaceae, Stonecrop Family
                                                    Crassula
                    JJ Stamens more than 4, and not necessarily
                        the same number as the petals
                        K   Sepals falling away as the flowers
                            open (flowers with 3 sepals and 6
                            petals, or 2 sepals and 4 petals;
                            stamens numerous)
                            Papaveraceae (includes Fumariaceae)
                                                Poppy Family
                        KK Sepals not falling away as the
                            flowers open (petals 5)
                            L   Sepals 2
                                    Portulacaceae, Purslane Family
                            LL Sepals 5
                                M   Stamens numerous, these
                                    usually in a few clusters;
                                    nodes not swollen
                                                    Hypericaceae
                                        St. John's-wort Family
                                MM Stamens not more than twice
                                    as many as the petals; nodes
                                    often obviously swollen
                                    Caryophyllaceae, Pink Family

BB Petals united to the extent that there is an obvious corolla tube
   (this is sometimes as long as, or longer than, the corolla lobes)
   C   Corolla regular, the lobes equal; corolla tube without spurs or
       sacs and not at all 2-lipped
       D   Pistils 2 (but the stigmas or styles may be fused for a
           time); sap sometimes milky; seeds sometimes with tufts of
           hair (in Asclepiadaceae, one of the pistils may not develop
           into a mature fruit)
           E   Stamens united to form a tube around the pistils (flowers
               complicated; corolla lobes turned back, and between them
               and the stamens is a crown with 5 concave, hoodlike
               lobes)   Asclepiadaceae, Milkweed Family
           EE Stamens not united to form a tube around the pistils
                                          Apocynaceae, Dogbane Family
       DD Pistil 1; sap not milky; seeds without tufts of hair
           E   Stamens 4, in 2 pairs   Verbenaceae, Verbena Family
           EE Stamens, if 4, not in 2 pairs
               F   Stamens in line with the corolla lobes
                                          Primulaceae, Primrose Family
               FF Stamens alternating with the corolla lobes
                   G   Style with 3 lobes   Polemoniaceae, Phlox Family
                                                   Phlox, *Linanthus*
                   GG Style either not lobed or with only 2 lobes
                       H   Calyx consisting of 5 separate sepals; corolla
                           5-lobed; style usually 2-lobed (but sometimes
                           not lobed); corolla not twisted in the bud
                           stage; leaves often deeply lobed
                                          Hydrophyllaceae, Waterleaf Family
                                                   Nemophila, *Pholistoma*
                       HH Calyx 4- or 5-lobed; corolla 4- or 5-lobed;
                           style not lobed; corolla usually twisted in
                           the bud stage; leaves not lobed
                                          Gentianaceae, Gentian Family
   CC Corolla at least slightly irregular, the lobes sometimes of
       different shapes and sizes; corolla tube sometimes with a spur
       or sac arising from its lower side, sometimes 2-lipped
       D   Corolla tube with a spur or sac on its lower side
           E   Corolla mostly yellow or blue, markedly 2-lipped; calyx
               lobes prominent, longer than the rest of the calyx;
               fruit-forming part of the pistil free of the calyx; stems
               not 4-angled (upper leaves usually alternate)
                                          Scrophulariaceae, Snapdragon Family
                                                   Linaria
           EE Corolla white to pink or rose, sometimes 2-lipped; calyx
               lobes much shorter than the rest of the calyx; fruit-
               forming portion of the pistil fused to the calyx; stems
               sometimes 4-angled   Valerianaceae, Valerian Family
       DD Corolla tube without a spur or sac on its lower side
           E   Fruit-forming part of the pistil fused to the calyx (thus
               the fruit develops below the sepals) (stems sometimes 4-
               angled; flowers in dense heads, each of which is partly
               enclosed by an involucre which is sometimes spiny)
                                          Dipsacaceae, Teasel Family
           EE Fruit-forming part of the pistil free of the calyx
               F   Stems not 4-angled Scrophulariaceae, Snapdragon Family
               FF Stems 4-angled
                   G   Flowers markedly 2-lipped; style divided near the
                       tip into 2 unequal lobes; plants usually with a
                       strong aroma, this often "minty"
                                          Lamiaceae (Labiatae), Mint Family
                   GG Flowers only slightly 2-lipped; style not divided;
                       plants without a strong aroma
                                          Verbenaceae, Verbena Family

DICOTYLEDONOUS FAMILIES

Aceraceae--Maple Family

The Maple Family consists of trees and shrubs with deciduous, opposite leaves and opposite branches. The flowers are generally in drooping racemes or dense clusters, and in our species the calyx has 5 lobes and the corolla either 5 petals or none. The flowers may be strictly staminate, strictly pistillate, or have stamens as well as a pistil. The pistil, with 2 styles, develops into a dry fruit that consists of 2 broad-winged structures called samaras. Each of these usually contains a single seed. The samaras, while still together or after separating, are scattered with the help of wind.

Besides *Acer macrophyllum* and *Acer negundo* var. *californicum*, California has *A. glabrum* and *A. circinatum*. All are excellent subjects for gardens in places that are not especially dry. *Acer circinatum*, the Vine Maple, is particularly valuable because its height rarely exceeds 6 m, and because its leaves, before falling in the autumn, often turn orange or red.

A Leaves palmately lobed (the blades up to about 25 cm wide); flowers with petals; seed-containing portion of each samara about one-fifth the length of the samara; often more than 20 m tall (widespread)
Acer macrophyllum (*Plate* 69)
Bigleaf Maple
AA Leaves pinnately compound, usually with 3 (sometimes 5) leaflets (each about 3-10 cm long); flowers without petals; seed-containing portion of each samara about one-third the length of the samara; rarely more than 15 m tall
Acer negundo var. *californicum* (*Plate* 69)
Boxelder

Aizoaceae--Seafig Family

The Seafig Family has many species, mostly natives of southern Africa, but these are apportioned to relatively few genera. Nevertheless, the group is difficult to define, partly because certain of the genera show affinities with some in the family Molluginaceae. The following summary of characters applies to our representatives.

In most species, the leaves are succulent, and the calyx is fused to the pistil. Thus the fruit, which may be fleshy or nutlike when mature, develops below the level where the sepals, petals (if present), and stamens originate. The genus *Carpobrotus* has many narrow petals, but *Tetragonia* has none, although its sepals are petal-like. Stamens tend to be numerous, although in *Tetragonia* there are not often more than 15. The pistil has several styles.

Tetragonia tetragonioides, called New Zealand Spinach and cultivated to some extent as a vegetable, has become established in many places, especially near the shores of San Francisco Bay. Two species of seafigs, with distinctively 3-angled, very succulent leaves, should be known to just about everyone who lives in the region, for they are extensively planted, especially for control of erosion on banks of highways. One of them, *Carpobrotus chilensis*, grows on coastal bluffs and the backshores of sandy beaches; it may have been on the Pacific coast before European explorers arrived, though the rest of the species of *Carpobrotus* and those of closely related genera are native to southern Africa.

A Flowers without petals (but the 3-5 sepals may be yellowish or greenish)
B Leaves alternate, up to about 5 cm long, with nearly arrowhead-shaped blades; leaves and other parts of the plant with glassy bumps (leaves succulent; sepals yellowish; usually sprawling to some extent; near the shore)
Tetragonia tetragonioides (*Plate* 4)
New Zealand Spinach; au

71

BB Leaves opposite, usually less than 3 cm long, with oval or spatula-
shaped blades; leaves and other parts of the plant without glassy
bumps
  C   Leaves succulent, spatula-shaped, up to about 2.5 cm long; sepals
rose; in saline habitats, inland as well as at the coast; up to
about 50 cm tall   *Sesuvium verrucosum*
                                                Western Sea-purslane
  CC Leaves narrowly oval, not obviously succulent, about 1 cm long;
sepals greenish; on mud banks, inland; compact, prostrate,
forming small mats   *Cypselea humifusa*
                                         Cypselea; sa; Ma, SCr
AA Flowers with numerous narrow petals
  B   Leaves differentiated into blade and petiole, the blade oval; flowers
about 1.5 cm wide (petals purple; found along the coast)
                        Aptenia cordifolia [*Mesembryanthemum cordifolium*]
                                         Iceplant; af
  BB Leaves not obviously differentiated into blade and petiole, nearly
cylindrical or 3-angled; flowers generally at least 2 cm wide (on
sand dunes and coastal bluffs)
    C   Leaves 3-angled
      D   Leaves up to 10 cm long, slightly curved, the lower edge with
fine teeth; flowers 7-10 cm wide; petals yellow when fresh,
but pink or purple after wilting (introduced for binding
sand, and now well established)
                Carpobrotus edulis [*Mesembryanthemum edule*] (*Plate* 4)
                                      Hottentot-fig; af
      DD Leaves up to 5 cm long, straight, the lower edge smooth-
margined; flowers 3-5 cm wide; petals magenta (may be native)
        Carpobrotus chilensis [*Mesembryanthemum chilense*] (*Plate* 4)
                                        Seafig; af?
    CC Leaves nearly cylindrical
      D   Leaves up to more than 10 cm long; petals yellow
                              Conicosia pugioniformis
                                  Conicosia; af; Mo, SF
      DD Leaves not often more than 1.5 cm long; petals bright pink or
rose *Drosanthemum floribundum* [*Mesembryanthemum floribundum*]
                                    Drosanthemum; af

Amaranthaceae--Amaranth Family

All of our amaranths belong to a single genus, and all of them are
annual. One or two species may have been here before the first explorers
arrived, but in general these plants are weedy immigrants. The flowers,
concentrated in dense inflorescences, lack petals, but they have 2-5
sepals, and there are also some bracts just external to the sepals. The
distinctions between the species are based to a considerable extent on the
form and proportion of the sepals and bracts of the staminate and
pistillate flowers. The fruit is dry when mature, and it contains a single
dark seed, which is shiny and disk-shaped. The seed, in fact, is perhaps
the most attractive feature of all our weedy amaranths!

Celosia argentea, called Cockscomb, is grown for its lush, usually pink
or red inflorescences, and species of *Amaranthus* have also been cultivated
to some extent. But this family is otherwise not popular with gardeners.

A   Plants low, the stems mostly lying on the ground
  B   Inflorescences up to 10 cm long, at the end of stems; blades of
larger leaves generally widest in the lower half (widespread,
sometimes growing in cracks in pavement)
                              Amaranthus deflexus (*Plate* 4)
                                    Low Amaranth; eu
  BB Inflorescences not more than 1.5 cm long, in the leaf axils; blades
of larger leaves generally widest near or above the middle
    C   Staminate flowers usually with 3 sepals, pistillate flowers with
only 1 well developed sepal; staminate flowers with 1-3 stamens
                                Amaranthus californicus
                                  California Amaranth
    CC Staminate and pistillate flowers usually with 4-5 sepals;
staminate flowers with 3-4 stamens
                              Amaranthus blitoides (*Plate* 69)
                                  Prostrate Pigweed

AA Plants upright, even if sometimes branching mostly from the base
  B   Blades of larger leaves not often so much as 3 cm long, and usually
      widest above the middle; bushy, with many branches arising at the
      base; inflorescences not more than 1.5 cm long, in the leaf axils;
      flowers with 3 sepals (stems pale)   *Amaranthus albus*
                                    Tumbleweed; sa
  BB Blades of larger leaves commonly more than 4 cm long (sometimes 10 cm
      long), and usually widest near or below the middle; not bushy,
      usually branching above, rather than at the base; inflorescences
      usually much more than 1.5 cm long, sometimes at the tip of the
      stems as well as in the leaf axils; flowers with 5 sepals (these of
      unequal length)
      C   Sepals blunt or notched at the tip, and often with a sharp
          extension
          D   Styles upright or curving inward; inflorescence green, not
              drooping (widespread)   *Amaranthus retroflexus* (*Plate* 69)
                                        Rough Pigweed; sa
          DD Styles curving downward; inflorescence usually red or
              reddish, drooping (garden flower, sometimes escaping)
                                    Amaranthus caudatus
                                          Quilete; sa
    CC Sepals narrowing gradually to a point at the tip
      D   Styles usually curving downward   *Amaranthus powellii*
                                        Powell Amaranth
      DD Styles upright
          E   Bracts of pistillate flowers about 1-1.5 times as long as
              the sepals; sepals 1.5 mm long   *Amaranthus cruentus*
                              South American Amaranth; sa
          EE Bracts of pistillate flowers nearly or fully twice as long
              as the sepals; sepals up to 2 mm long *Amaranthus hybridus*
                                    Green Amaranth; eua

Anacardiaceae--Sumac Family

Shrubs and trees of the Sumac Family have a milky or resinous, often pungent sap. The leaves, simple or compound, are alternate, and the flowers are small. In some species, they have stamens and a pistil; in others, staminate and pistillate flowers are on separate plants. The calyx has 5 lobes, and there are usually 5 petals and 5 or 10 stamens. This family includes the Mango, *Mangifera indica* (southeast Asia), *Pistacia vera* (probably Iran), which produces pistachio nuts, and the Cashew, *Anacardium occidentale* (Central and South America).

Our only native representatives are members of the genera *Toxicodendron* and *Rhus*, both with deciduous leaves that usually have 3 leaflets. The first is Poison-oak, a plant that should be avoided because contact with it brings on mild to severe dermatitis in most people. The other, called Skunkbush, is not likely to cause any problems, and is an attractive, easily grown shrub that deserves to be included in gardens that emphasize native species.

A   Trees up to 25 m tall, evergreen; leaflets numerous, peppery-smelling
    (extensively cultivated and sometimes escaping)   *Schinus molle*
                                        Peruvian Peppertree; sa
AA Shrubs not more than 3 m tall, deciduous; leaflets usually 3
    (occasionally 5), not peppery-smelling (but may be aromatic)
  B   Leaflets rounded, more or less equal; flowers and fruits whitish
      (sometimes vinelike) (Caution: contact with any part of this plant,
      even its bare stems, may result in a severe rash.) (widespread)
          Toxicodendron diversilobum [*Rhus diversiloba*] (*Plate* 4)
                                    Western Poison-oak
  BB Leaflets triangular, the terminal one larger than the 2 lateral ones;
      flowers greenish, fruits red   *Rhus trilobata*
      [includes *R. trilobata* vars. *malacophylla* and *quinata*] (*Plate* 4)
                                          Skunkbush

Apiaceae (Umbelliferae)--Carrot Family

Members of the Carrot Family have alternate leaves, and the petioles of these are usually dilated, near their bases, into sheaths that clasp the stems. The stems, morever, are often furrowed. The plants, or at least parts of them, usually have an aroma of some sort. The aromas--and the flavor of carrot, parsley, coriander, parsnip, celery, fennel, dill, and anise--are due primarily to various oils.

The flowers are nearly always concentrated in flat-topped inflorescences of the type called umbels. These are usually compound, the rays of the primary umbels giving rise to secondary umbels that consist of pedicels bearing the flowers. There are 5 petals (which may be unequal, especially in flowers that are located along the outer edge of the umbel) and 5 stamens. The fruit-forming portion of the pistil is below the level where the petals and stamens originate. If there are recognizable sepals, these will be just below the petals. The fruit consists of a pair of 1-seeded divisions that are for a time joined tightly to a central partition. They are often conspicuously ribbed, and in some species the ribs, or certain of them, are drawn out into membranous "wings."

Some members of this family are poisonous and others may irritate the skin. It is therefore prudent to handle unfamiliar plants carefully.

A   Plants somewhat resembling thistles, the bracts beneath the heads stiff, prickly (the leaves also often prickly); inflorescence a dense head, rather than an umbel (usually in wet places, such as vernal pools)
  B   Bracts with smooth margins, prickly only at the tip; coastal
                                        Eryngium armatum (*Plate* 70)
                                                     Coyote-thistle

  BB Bracts with spiny margins; Coast Ranges
    C   Styles on fruit much longer than the sepals
                                Eryngium aristulatum var. *aristulatum*
                                            Button Celery

    CC Styles on fruit about as long as the sepals
                                Eryngium aristulatum var. *hooveri*
                                Hoover Button Celery; SFBR; 4

AA Plants not resembling thistles, neither the leaves nor the bracts (if present beneath the umbels) stiff and prickly; inflorescence usually an umbel (but in *Bowlesia incana* there are just a few flowers [sometimes only 1] in each axillary inflorescence, and in *Hydrocotyle verticillata* the inflorescence consists of several whorls)
  B   Leaves reduced to hollow petioles, which are cross-barred and thus have a jointed appearance (small, creeping aquatic perennials, most commonly found in saline habitats)
    C   Leaves cylindrical or flattened, usually more than 1 mm thick, the cross-walls (4-13) well defined
                                Lilaeopsis occidentalis (*Plate* 71)
                                      Western Lilaeopsis; Sl, Ma-n
    CC Leaves cylindrical, usually less than 1 mm thick, the cross-walls
    (3-8) poorly defined   *Lilaeopsis masonii*
                                  Mason Lilaeopsis; SFBR; 1b

  BB Leaves with well developed blades
    C   All leaves clearly opposite, the blades mostly 5-lobed; fruit (and the portion of the pistil that will become the fruit) covered with star-shaped hairs; umbels, in the leaf axils, with few flowers, sometimes only 1; annuals, usually growing in shaded situations   *Bowlesia incana*
                                            Bowlesia; Sn
    CC Plants not conforming to the description in choice C
      D   Small, creeping aquatic plants with nearly circular leaf blades, these either lobed or with shallow, rounded teeth
        E   Leaf blades lobed, the petiole attached at the point of separation of two of the lobes
                                  Hydrocotyle ranunculoides (*Plate* 71)
                                          Water Pennywort

        EE Leaf blades with shallow rounded teeth, the petiole attached near the center
        F   Inflorescence a simple umbel   *Hydrocotyle umbellata*
                                        Marsh Pennywort

Adiantum aleuticum
Fivefinger Fern

Adiantum jordanii
California Maidenhair

Athyrium filix-femina var. *cyclosorum*
Lady Fern

Dryopteris arguta
Coastal Wood Fern

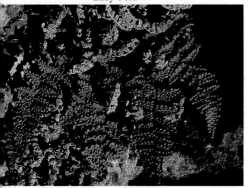

Cheilanthes covillei
Coville Lace Fern

Polystichum californicum
California Shield Fern

Polypodium californicum
California Polypody

FERNS **PLATE 1**

Pentagramma triangularis
Goldback Fern

Pteridium aquilinum var. *pubescens*
Bracken

Pellaea andromedifolia
Coffee Fern

Woodwardia fimbriata
Giant Chain Fern

Pellaea mucronata
Bird's-foot Fern

Equisetum telmateia ssp. *braunii*
Giant Horsetail

Cupressus macrocarpa
Monterey Cypress

Juniperus californica
California Juniper

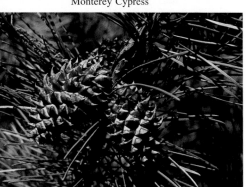

Pinus muricata
Bishop Pine

Pinus ponderosa
Pacific Ponderosa Pine

Sequoia sempervirens
Coast Redwood

Pseudotsuga menziesii
Douglas-fir

Tsuga heterophylla
Western Hemlock

GYMNOSPERMS   **PLATE 3**

Taxus brevifolia
Pacific Yew (Gymnosperm)

Torreya californica
California Nutmeg (Gymnosperm)

Carpobrotus chilensis
Seafig (Aizoaceae)

Carpobrotus edulis
Hottentot-fig (Aizoaceae)

Tetragonia tetragonioides
New Zealand Spinach (Aizoaceae)

Amaranthus deflexus
Low Amaranth (Amaranthaceae)

Rhus trilobata
Skunkbush (Anacardiaceae)

Toxicodendron diversilobum
Western Poison-oak (Anacardiaceae)

Gymnosperms; Dicotyledons: AIZOACEAE-ANACARDIACEAE **PLATE**

Heracleum lanatum
Cow Parsnip

Lomatium dasycarpum
Hog Fennel

Lomatium utriculatum
Spring-gold

Osmorhiza chilensis
Sweet-cicely

Sanicula arctopoides
Footsteps-of-spring

Sanicula bipinnatifida
Purple Sanicle

Sanicula crassicaulis
Snakeroot

Scandix pecten-veneris
Shepherd's-needle

Dicotyledons: APIACEAE **PLATE 5**

Aralia californica
Elk-clover (Araliaceae)

Aristolochia californica
Dutchman's-pipe (Aristolochiaceae)

Asarum caudatum
Wild-ginger (Aristolochiaceae)

Asclepias californica
California Milkweed (Asclepiadaceae)

Asclepias fascicularis
Narrowleaf Milkweed (Asclepiadaceae)

Achillea millefolium
Yarrow (Asteraceae)

Achyrachaena mollis
Blow-wives (Asteraceae)

Agoseris apargioides var. *eastwoodiae*
Coast Dandelion (Asteraceae)

Dicotyledons: ARALIACEAE-ASTERACEAE **PLATE**

Agoseris grandiflora
California Dandelion

Anaphalis margaritacea
Pearly Everlasting

Ambrosia chamissonis
Silvery Beachweed

Arctium minus
Common Burdock

Artemisia douglasiana
Mugwort

Arnica discoidea
Coast Arnica

Dicotyledons: ASTERACEAE   **PLATE 7**

Artemisia californica
California Sagebrush

Artemisia pycnocephala
Coastal Sagewort

Aster chilensis
Common California Aster

Baccharis pilularis
Coyotebrush

Baccharis salicifolia
Mulefat

Bidens frondosa
Sticktight

Dicotyledons: ASTERACEAE   **PLATE**

Blennosperma nanum var. *nanum*
Common Blennosperma

Calycadenia multiglandulosa
Sticky Rosinweed

Carduus pycnocephalus
Italian Thistle

Centaurea solstitialis
Yellow Star Thistle

Cichorium intybus
Chicory

Cirsium occidentale var. *venustum*
Venus Thistle

Cirsium quercetorum
Brownie Thistle

Coreopsis calliopsidea
Leafystem Coreopsis

Dicotyledons: ASTERACEAE　**PLATE 9**

Cotula coronopifolia
Brassbuttons

Crepis capillaris
Smooth Hawk's-beard

Cynara cardunculus
Cardoon

Ericameria linearifolia
Interior Goldenbush

Erigeron petrophilus var. *petrophilus*
Rock Daisy

Erigeron glaucus
Seaside Daisy

Eriophyllum confertiflorum
Golden Yarrow

Dicotyledons: ASTERACEAE   **PLATE 1**

Eriophyllum lanatum var. *achillaeoides*
Common Woolly Sunflower

Gnaphalium californicum
California Everlasting

Grindelia camporum
Great Valley Gumplant

Helenium puberulum
Rosilla

Filago californica
California Fluffweed

Gnaphalium purpureum
Purple Cudweed

Helianthus annuus
Common Sunflower

Dicotyledons: ASTERACEAE   **PLATE 11**

Helianthus gracilentus
Slender Sunflower

Hemizonia congesta ssp. *congesta*
Yellow Hayfield Tarweed

Hemizonia congesta ssp. *luzulifolia*
Hayfield Tarweed

Heterotheca grandiflora
Telegraphweed

Heterotheca sessiliflora ssp. *echioides*
Bristly Golden Aster

Holocarpha heermannii
Heermann Tarplant

Hypochaeris radicata
Rough Cat's-ear

Jaumea carnosa
Fleshy Jaumea

Dicotyledons: ASTERACEAE   **PLATE 1**

Lactuca serriola
Prickly Lettuce

Lasthenia californica
California Goldfields

Layia platyglossa
Tidytips

Madia elegans ssp. *elegans*
Elegant Madia

Picris echioides
Bristly Ox-tongue

Madia gracilis
Slender Madia

Senecio elegans
Purple Ragwort

Dicotyledons: ASTERACEAE   **PLATE 13**

Silybum marianum
Milk Thistle

Solidago californica
California Goldenrod

Solidago canadensis ssp. *elongata*
Canada Goldenrod

Solidago spathulata
Coast Goldenrod

Stephanomeria virgata
Tall Stephanomeria

Tanacetum camphoratum
Dune Tansy

Tragopogon porrifolius
Salsify (Asteraceae)

Uropappus lindleyi
Silverpuffs (Asteraceae)

Wyethia angustifolia
Narrowleaf Mule-ears (Asteraceae)

Wyethia glabra
Smooth Mule-ears (Asteraceae)

Achlys triphylla
Vanilla-leaf (Berberidaceae)

Berberis pinnata
Shinyleaf Oregon-grape (Berberidaceae)

Vancouveria planipetala
Inside-out-flower (Berberidaceae)

Dicotyledons: ASTERACEAE-BERBERIDACEAE   **PLATE 15**

Amsinckia menziesii var. *intermedia*
Intermediate Fiddleneck

Amsinckia tessellata var. *tessellata*
Devil's-lettuce

Cryptantha muricata
Prickly Cryptantha

Cryptantha micromeres
Minuteflower Cryptantha

Heliotropium curassavicum
Seaside Heliotrope

Cynoglossum grande
Hound's-tongue

Plagiobothrys nothofulvus
Common Popcornflower

Dicotyledons: BORAGINACEAE   **PLATE 1**

Arabis blepharophylla
Coast Rock Cress (Brassicaceae)

Cakile maritima
Horned Searocket (Brassicaceae)

Cardamine californica var. *californica*
Common Milkmaids (Brassicaceae)

Erysimum capitatum ssp. *capitatum*
Western Wallflower (Brassicaceae)

Hirschfeldia incana
Hirschfeldia (Brassicaceae)

Thysanocarpus curvipes
Hairy Fringepod (Brassicaceae)

Calycanthus occidentalis
Spicebush (Calycanthaceae)

Dicotyledons: BRASSICACEAE-CALYCANTHACEAE   **PLATE 17**

Lonicera hispidula var. *vacillans*
Hairy Honeysuckle

Lonicera involucrata var. *ledebourii*
Twinberry

Sambucus mexicana
Blue Elderberry

Sambucus racemosa
Red Elderberry

Symphoricarpos albus var. *laevigatus*
Common Snowberry

Symphoricarpos mollis
Creeping Snowberry

Cerastium arvense
Field Chickweed (Caryophyllaceae)

Silene californica
Indian Pink (Caryophyllaceae)

Silene gallica
Windmill Pink (Caryophyllaceae)

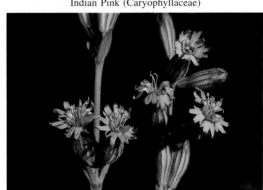
Silene scouleri ssp. *grandis*
Scouler Catchfly (Caryophyllaceae)

Spergularia rubra
Ruby Sand-spurrey (Caryophyllaceae)

Spergularia macrotheca var. *macrotheca*
Perennial Sand-spurrey (Caryophyllaceae)

Euonymus occidentalis
Western Burningbush (Celastraceae)

Paxistima myrsinites
Oregon Boxwood (Celastraceae)

Dicotyledons: CARYOPHYLLACEAE-CELASTRACEAE   **PLATE 19**

Atriplex leucophylla
Beach Saltbush (Chenopodiaceae)

Atriplex semibaccata
Australian Saltbush (Chenopodiaceae)

Salsola tragus
Russian-thistle (Chenopodiaceae)

Salicornia virginica
Virginia Pickleweed (Chenopodiaceae)

Calystegia malacophylla ssp. *pedicellata*
Hairy Morning-glory (Convolvulaceae)

Calystegia purpurata ssp. *purpurata*
Climbing Morning-glory (Convolvulaceae)

Calystegia soldanella
Beach Morning-glory (Convolvulaceae)

Cornus nuttallii
Mountain Dogwood (Cornaceae)

Cornus sericea ssp. *occidentalis*
Western Creek Dogwood (Cornaceae)

Crassula connata
Sand Pygmyweed (Crassulaceae)

Dudleya cymosa
Common Dudleya (Crassulaceae)

Sedum spathulifolium
Broadleaf Stonecrop (Crassulaceae)

Cuscuta salina var. *major*
Saltmarsh Dodder (Cuscutaceae)

Dipsacus fullonum
Wild Teasel (Dipsacaceae)

Scabiosa atropurpurea
Pincushion (Dipsacaceae)

Arbutus menziesii
Pacific Madrone (Ericaceae)

Arctostaphylos auriculata
Mount Diablo Manzanita (Ericaceae)

Arctostaphylos glandulosa
Eastwood Manzanita (Ericaceae)

Arctostaphylos pallida
Pallid Manzanita (Ericaceae)

Arctostaphylos tomentosa ssp. *crustacea*
Brittleleaf Manzanita

Arctostaphylos stanfordiana ssp. *stanfordiana*
Stanford Manzanita

Arctostaphylos uva-ursi
Bearberry

Gaultheria shallon
Salal

Rhododendron macrophyllum
Rosebay

Rhododendron occidentale
Western Azalea

Vaccinium ovatum
Evergreen Huckleberry

Vaccinium parvifolium
Red Huckleberry

Dicotyledons: ERICACEAE   **PLATE 23**

Chamaesyce maculata
Spotted Spurge (Euphorbiaceae)

Croton californicus
Croton (Euphorbiaceae)

Eremocarpus setigerus
Turkey-mullein (Euphorbiaceae)

Euphorbia peplus
Petty Spurge (Euphorbiaceae)

Ricinus communis
Castor-bean (Euphorbiaceae)

Cercis occidentalis
Western Redbud (Fabaceae)

Cytisus scoparius
Scotch Broom (Fabaceae)

Genista monspessulana
French Broom (Fabaceae)

Dicotyledons: EUPHORBIACEAE-FABACEAE   **PLATE 2**

Lathyrus littoralis
Beach Pea

Lathyrus vestitus
Woodland Pea

Lotus crassifolius
Broadleaf Lotus

Lotus humistratus
Colchita

Lotus purshianus
Spanish Lotus

Lotus scoparius
Deerweed

Lotus wrangelianus
California Lotus

Dicotyledons: FABACEAE   **PLATE 25**

Lupinus albifrons var. *albifrons*
Silver Lupine

Lupinus arboreus
Yellow Bush Lupine

Lupinus microcarpus var. *densiflorus*
Gully Lupine

Lupinus succulentus
Arroyo Lupine

Pickeringia montana
Chaparral Pea

Melilotus indica
Sour Clover

Pediomelum californicum
Indian Breadroot

Rupertia physodes
California-tea

Spartium junceum
Spanish Broom

Thermopsis macrophylla
False Lupine

Trifolium fucatum
Bull Clover

Trifolium microcephalum
Smallhead Clover

Trifolium pratense
Red Clover

Dicotyledons: FABACEAE   **PLATE 27**

Trifolium oliganthum
Fewflower Clover (Fabaceae)

Trifolium willdenovii
Tomcat Clover (Fabaceae)

Vicia gigantea
Giant Vetch (Fabaceae)

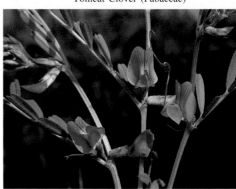

Vicia sativa ssp. *sativa*
Spring Vetch (Fabaceae)

Quercus agrifolia
Coast Live Oak (Fagaceae)

Chrysolepis chrysophylla var. *minor*
Golden Chinquapin (Fagaceae)

Frankenia salina
Alkali-heath (Frankeniaceae)

Dicotyledons: FABACEAE-FRANKENIACEAE   **PLATE**

Centaurium muehlenbergii
Monterey Centaury (Gentianaceae)

Gentiana sceptrum
King's Gentian (Gentianaceae)

Erodium botrys
Broadleaf Filaree (Geraniaceae)

Erodium cicutarium
Redstem Filaree (Geraniaceae)

Geranium bicknellii
Bicknell Geranium (Geraniaceae)

Geranium dissectum
Cutleaf Geranium (Geraniaceae)

Geranium robertianum
Herb-Robert (Geraniaceae)

Ribes aureum var. *gracillimum*
Golden Currant (Grossulariaceae)

Ribes californicum
Hillside Gooseberry (Grossulariaceae)

Ribes menziesii
Canyon Gooseberry (Grossulariaceae)

Ribes sanguineum var. *glutinosum*
Pinkflower Currant (Grossulariaceae)

Aesculus californica
California Buckeye (Hippocastanaceae)

Ribes speciosum
Fuchsia-flower Gooseberry (Grossulariaceae)

Emmenanthe penduliflora var. *penduliflora*
Whispering-bells

Hydrophyllum occidentale
Heliotrope

Phacelia californica
California Phacelia

Phacelia ramosissima var. *ramosissima*
Branched Phacelia

Eriodictyon californicum
Yerba-santa

Nemophila menziesii var. *menziesii*
Baby-blue-eyes

Phacelia ciliata
Field Phacelia

Hypericum anagalloides
Tinker's-penny (Hypericaceae)

Hypericum concinnum
Goldwire (Hypericaceae)

Lamium amplexicaule
Clasping Henbit (Lamiaceae)

Lamium purpureum
Red Henbit (Lamiaceae)

Lepechinia calycina
Pitcher Sage (Lamiaceae)

Mentha pulegium
Pennyroyal (Lamiaceae)

Monardella villosa ssp. *villosa*
Common Coyotemint (Lamiaceae)

Pogogyne serpylloides
Thymelike Pogogyne

Prunella vulgaris var. *vulgaris*
European Selfheal

Salvia columbariae
Chia

Salvia mellifera
Black Sage

Scutellaria tuberosa
Blue Skullcap

Salvia spathacea
Hummingbird Sage

Stachys bullata
California Hedgenettle (Lamiaceae)

Umbellularia californica
California Bay (Lauraceae)

Linum bienne
Narrowleaf Flax (Linaceae)

Limnanthes douglasii ssp. *douglasii*
Douglas Meadowfoam (Limnanthaceae)

Mentzelia lindleyi
Lindley Blazing-star (Loasaceae)

Linum lewisii
Western Blue Flax (Linaceae)

Mentzelia micrantha
Golden Blazing-star (Loasaceae)

Dicotyledons: LAMIACEAE-LOASACEAE   **PLATE**

Lythrum californicum
California Loosestrife (Lythraceae)

Lavatera arborea
Tree Mallow (Malvaceae)

Malva neglecta
Common Mallow (Malvaceae)

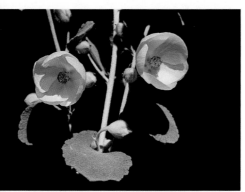

Malacothamnus fremontii
Fremont Mallow (Malvaceae)

Malva nicaeensis
Bull Mallow (Malvaceae)

Sidalcea malvaeflora ssp. *malvaeflora*
Common Checkerbloom (Malvaceae)
(A basal leaf is shown.)

Sidalcea calycosa ssp. *calycosa*
Annual Checkerbloom (Malvaceae)

Dicotyledons: LYTHRACEAE-MALVACEAE   **PLATE 35**

Abronia latifolia
Yellow Sand-verbena (Nyctaginaceae)

Abronia umbellata ssp. *umbellata*
Coast Sand-verbena (Nyctaginaceae)

Camissonia cheiranthifolia
Beach Primrose (Onagraceae)

Camissonia ovata
Suncup (Onagraceae)

Clarkia amoena ssp. *huntiana*
Hunt Clarkia (Onagraceae)

Clarkia biloba
Bilobe Clarkia (Onagraceae)

Clarkia concinna
Redribbons

Clarkia davyi
Davy Clarkia

Clarkia epilobioides
Willowherb Clarkia

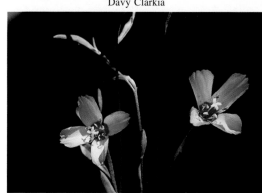

Clarkia franciscana
Presidio Clarkia

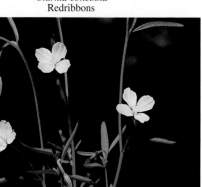

Clarkia gracilis ssp. *sonomensis*
Slender Clarkia

Clarkia purpurea ssp. *quadrivulnera*
Winecup Clarkia

Clarkia purpurea ssp. *viminea*
Large Clarkia

Clarkia rhomboidea
Tongue Clarkia

Dicotyledons: ONAGRACEAE   **PLATE 37**

Clarkia rubicunda
Godetia

Clarkia unguiculata
Elegant Clarkia

Epilobium angustifolium ssp. *circumvagu*
Fireweed

Epilobium brachycarpum
Panicled Willowherb

Epilobium canum
California Fuchsia

Oenothera deltoides ssp. *howellii*
Antioch Dunes Evening-primrose

Oenothera glazioviana
Biennial Evening-primrose

Boschniakia strobilacea
California Groundcone (Orobanchaceae)

Orobanche fasciculata
Clustered Broom-rape (Orobanchaceae)

Orobanche uniflora
Naked Broom-rape (Orobanchaceae)

Oxalis corniculata
Creeping Oxalis (Oxalidaceae)

Oxalis oregana
Redwood Sorrel (Oxalidaceae)

Oxalis rubra
Windowbox Oxalis (Oxalidaceae)

Dendromecon rigida
Bush Poppy (Papaveraceae)

Dicentra chrysantha
Golden Eardrops (Papaveraceae)

Dicotyledons: OROBANCHACEAE-PAPAVERACEAE **PLATE 39**

Dicentra formosa
Bleeding-heart

Eschscholzia caespitosa
Tufted Poppy

Eschscholzia californica
California Poppy

Fumaria officinalis
Fumitory

Stylomecon heterophylla
Wind Poppy

Platystemon californicus
Creamcups

Plantago erecta
California Plantain (Plantaginaceae)

Plantago maritima
Sea Plantain (Plantaginaceae)

Armeria maritima ssp. *californica*
Thrift (Plumbaginaceae)

Limonium californicum
California Sea-lavender (Plumbaginaceae)

Collomia grandiflora
Largeflower Collomia (Polemoniaceae)

Collomia heterophylla
Variedleaf Collomia (Polemoniaceae)

Dicotyledons: PLANTAGINACEAE-POLEMONIACEAE **PLATE 41**

Gilia achilleifolia ssp. *achilleifolia*
California Gilia

Gilia capitata ssp. *capitata*
Globe Gilia

Gilia tricolor
Bird's-eyes

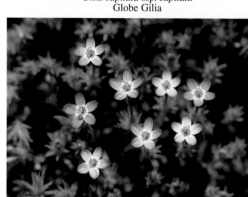

Linanthus bicolor
Bicolor Linanthus

Linanthus grandiflorus
Largeflower Linanthus

Linanthus parviflorus
Common Linanthus

Navarretia pubescens
Downy Navarretia

Phlox gracilis
Slender Phlox

Dicotyledons: POLEMONIACEAE   **PLATE 4**

Polygala californica
Milkwort (Polygalaceae)

Chorizanthe membranacea
Pink Spineflower (Polygonaceae)

Eriogonum fasciculatum
California Buckwheat (Polygonaceae)

Eriogonum latifolium
Coast Buckwheat (Polygonaceae)

Eriogonum luteolum var. *caninum*
Tiburon Buckwheat (Polygonaceae)

Eriogonum nudum var. *nudum*
Nakedstem Buckwheat (Polygonaceae)

Eriogonum umbellatum var. *bahiiforme*
Sulfurflower Buckwheat (Polygonaceae)

Eriogonum vimineum
Wicker Buckwheat (Polygonaceae)

Dicotyledons: POLYGALACEAE-POLYGONACEAE   **PLATE 43**

Polygonum amphibium var. *emersum*
Swamp Knotweed

Polygonum paronychia
Beach Knotweed

Polygonum persicaria
Lady's-thumb

Rumex acetosella
Sheep-sorrel

Rumex crispus
Curly Dock

Rumex salicifolius
Willow Dock

Calandrinia ciliata
Redmaids (Portulacaceae)

Claytonia gypsophiloides
Santa Lucia Claytonia (Portulacaceae)

Claytonia sibirica
Candyflower (Portulacaceae)

Lewisia rediviva
Bitterroot (Portulacaceae)

Anagallis arvensis
Scarlet Pimpernel (Primulaceae)

Dodecatheon hendersonii
Mosquito-bills (Primulaceae)

Trientalis latifolia
Starflower (Primulaceae)

Dicotyledons: PORTULACACEAE-PRIMULACEAE   **PLATE 45**

Aquilegia formosa
Columbine

Clematis lasiantha
Pipestems

Delphinium nudicaule
Red Larkspur

Delphinium variegatum
Royal Larkspur

Ranunculus californicus
California Buttercup

Ranunculus muricatus
Prickleseed Buttercup

Dicotyledons: RANUNCULACEAE   **PLATE 4**

Ranunculus repens
Creeping Buttercup (Ranunculaceae)

Ceanothus cuneatus
Buckbrush (Rhamnaceae)

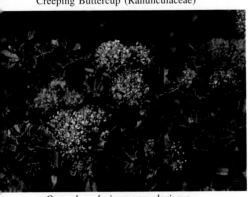

Ceanothus gloriosus var. *gloriosus*
Point Reyes Ceanothus (Rhamnaceae)

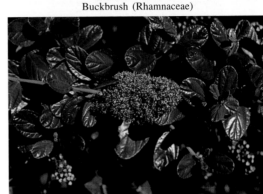

Ceanothus griseus
Carmel Ceanothus (Rhamnaceae)

Ceanothus sonomensis
Sonoma Ceanothus (Rhamnaceae)

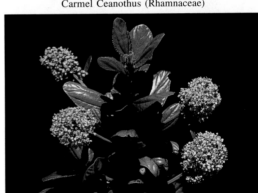

Ceanothus thyrsiflorus
Blue-blossom (Rhamnaceae)

Rhamnus californica
California Coffeeberry (Rhamnaceae)

Rhamnus crocea
Spiny Redberry (Rhamnaceae)

Dicotyledons: RANUNCULACEAE-RHAMNACEAE **PLATE 47**

Adenostoma fasciculatum
Chamise

Amelanchier alnifolia
Serviceberry

Aphanes occidentalis
Western Dewcup

Cercocarpus betuloides
Mountain-mahogany

Fragaria chiloensis
Beach Strawberry

Heteromeles arbutifolia
Toyon

Holodiscus discolor
Creambush

Dicotyledons: ROSACEAE   **PLATE 4**

Horkelia californica ssp. *californica*
California Horkelia

Physocarpus capitatus
Ninebark

Potentilla anserina ssp. *pacifica*
Pacific Cinquefoil

Potentilla glandulosa
Sticky Cinquefoil

Potentilla recta
Pale Cinquefoil

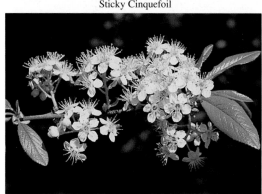
Prunus emarginata
Bitter Cherry

Prunus ilicifolia
Hollyleaf Cherry

Rubus spectabilis
Salmonberry

Dicotyledons: ROSACEAE   **PLATE 49**

Rosa californica
California Rose (Rosaceae)

Galium andrewsii
Phloxleaf Bedstraw (Rubiaceae)

Rosa gymnocarpa
Wood Rose (Rosaceae)

Galium aparine
Goosegrass (Rubiaceae)

Galium californicum
California Bedstraw (Rubiaceae)

Galium porrigens var. *porrigens*
Climbing Bedstraw (Rubiaceae)

Sherardia arvensis
Field Madder (Rubiaceae)

Ptelea crenulata
Hoptree (Rutaceae)

Anemopsis californica
Yerba-mansa (Saururaceae)

Lithophragma parviflorum
Prairie Starflower (Saxifragaceae)

Lithophragma affine
Woodland-star (Saxifragaceae)

Tellima grandiflora
Fringecups (Saxifragaceae)

Tolmiea menziesii
Piggy-back-plant (Saxifragaceae)

Antirrhinum kelloggii
Lax Snapdragon

Antirrhinum vexillo-calyculatum ssp. *breweri*
Brewer Snapdragon

Bellardia trixago
Bellardia

Castilleja attenuata
Valley-tassels

Castilleja affinis ssp. *affinis*
Common Indian-paintbrush

Castilleja exserta ssp. *exserta*
Purple Owl's-clover

Castilleja foliolosa
Woolly Paintbrush

Dicotyledons: SCROPHULARIACEAE   **PLATE 5**

Collinsia heterophylla
Chinesehouses

Castilleja wightii
Seaside Paintbrush

Mimulus aurantiacus
Bush Monkeyflower

Collinsia sparsiflora var. *sparsiflora*
Blue-eyed-Mary

Mimulus cardinalis
Scarlet Monkeyflower

Mimulus guttatus
Common Monkeyflower

Dicotyledons: SCROPHULARIACEAE   **PLATE 53**

Penstemon heterophyllus var. *heterophyllus*
Foothill Penstemon (Scrophulariaceae)

Pedicularis densiflora
Indian-warrior (Scrophulariaceae)

Scrophularia californica
Beeplant (Scrophulariaceae)

Triphysaria eriantha ssp. *eriantha*
Yellow Johnnytuck (Scrophulariaceae)

Nicotiana glauca
Tree-tobacco (Solanaceae)

Solanum sarrachoides
Hairy Nightshade (Solanaceae)

Solanum umbelliferum
Blue Nightshade (Solanaceae)

Dicotyledons: SCROPHULARIACEAE-SOLANACEAE **PLATE 5**

Fremontodendron californicum
Flannelbush (Sterculiaceae)

Dirca occidentalis
Western Leatherwood (Thymelaeaceae)

Urtica dioica ssp. *holosericea*
Hoary Nettle (Urticaceae)

Plectritis congesta
Seablush (Valerianaceae)

Plectritis macrocera
Longhorn Plectritis (Valerianaceae)

Verbena lasiostachys var. *scabrida*
Robust Verbena (Verbenaceae)

Viola adunca
Western Dog Violet (Violaceae)

Viola glabella
Stream Violet (Violaceae)

Viola ocellata
Western Heart's-ease (Violaceae)

Viola pedunculata
Johnny-jump-up (Violaceae)

Arceuthobium campylopodum
Western Dwarf-mistletoe (Viscaceae)

Phoradendron villosum
Oak Mistletoe (Viscaceae)

Vitis californica
California Wild Grape (Vitaceae)

Tribulus terrestris
Puncture Vine (Zygophyllaceae)

Dicotyledons: VIOLACEAE-ZYGOPHYLLACEAE   **PLATE 5**

Lysichiton americanum
Yellow Skunk-cabbage (Araceae)

Cyperus eragrostis
Tall Cyperus (Cyperaceae)

Iris fernaldii
Fernald Iris (Iridaceae)

Iris douglasiana
Douglas Iris (Iridaceae)

Sisyrinchium californicum
Golden-eyed-grass (Iridaceae)

Sisyrinchium bellum
Blue-eyed-grass (Iridaceae)

Luzula comosa
Wood Rush (Juncaceae)

Monocotyledons: ARACEAE-JUNCACEAE   **PLATE 57**

Allium acuminatum
Hooker Onion

Allium falcifolium
Sickleleaf Onion

Allium lacunosum
Wild Onion

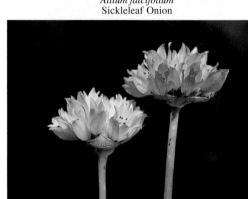

Allium serra
Serrated Onion

Allium unifolium
Clay Onion

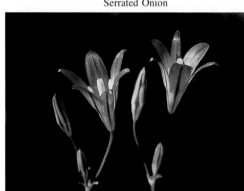

Brodiaea elegans
Elegant Brodiaea

Brodiaea terrestris
Dwarf Brodiaea

Calochortus luteus
Yellow Mariposa Lily

Monocotyledons: LILIACEAE   **PLATE**

Calochortus pulchellus
Mount Diablo Fairy-lantern

Calochortus tiburonensis
Tiburon Mariposa Lily

Calochortus umbellatus
Oakland Startulip

Calochortus venustus
Butterfly Mariposa Lily

Camassia quamash
Common Camas

Chlorogalum pomeridianum var. *divaricatum*
Soap-plant

Clintonia andrewsiana
Red Bead Lily

Monocotyledons: LILIACEAE   **PLATE 59**

Dichelostemma congestum
Ookow

Erythronium californicum
California Fawn Lily

Fritillaria affinis
Checker Lily

Fritillaria recurva
Scarlet Fritillary

Muilla maritima
Common Muilla

Lilium pardalinum ssp. *pardalinum*
Leopard Lily

Smilacina racemosa
Fat Solomon

Monocotyledons: LILIACEAE   **PLATE**

Trillium albidum
Sweet Trillium (Liliaceae)

Triteleia laxa
Ithuriel's-spear (Liliaceae)

Triteleia lugens
Uncommon Triteleia (Liliaceae)

Zigadenus fremontii
Common Star Lily (Liliaceae)

Zigadenus venenosus
Death Camas (Liliaceae)

Calypso bulbosa
Fairy-slipper

(Orchidaceae)

Cephalanthera austiniae
Phantom Orchid
(Orchidaceae)

Corallorhiza maculata
Spotted Coralroot

Corallorhiza striata
Striped Coralroot

Cypripedium californicum
California Lady's-slipper

(Orchidaceae)

(Orchidaceae)

Epipactis gigantea
Stream Orchid

Epipactis helleborine
Helleborine

Piperia unalascensis
Alaska Rein Orchid

(Orchidaceae)

(Orchidaceae)

Ammophila arenaria
European Beach Grass (Poaceae)

Arundo donax
Giantreed (Poaceae)

Monocotyledons: ORCHIDACEAE-POACEAE   **PLATE 6**

Anthoxanthum odoratum
Sweet Vernal Grass

Briza maxima
Big Quaking Grass

Bromus madritensis ssp. *rubens*
Foxtail Chess

Cynosurus cristatus
Crested Dogtail

Cynosurus echinatus
Hedgehog Dogtail

Dactylis glomerata
Orchard Grass

Leymus mollis
Dune Grass

Poa bulbosa
Bulbous Blue Grass

Monocotyledons: POACEAE   **PLATE 63**

Echinochloa crus-galli
Barnyard Grass
(Poaceae)

Elymus multisetus
Big Squirreltail
(Poaceae)

Hordeum murinum ssp. *leporinum*
Hare Barley

Lolium multiflorum
Italian Rye Grass
(Poaceae)

Lolium perenne
Perennial Rye Grass
(Poaceae)

Polypogon monspeliensis
Annual Beard Grass

Potamogeton natans
Floating Pondweed (Potamogetonaceae)

Typha latifolia
Broadleaf Cattail (Typhaceae)

Monocotyledons: POACEAE-TYPHACEAE **PLATE 6**

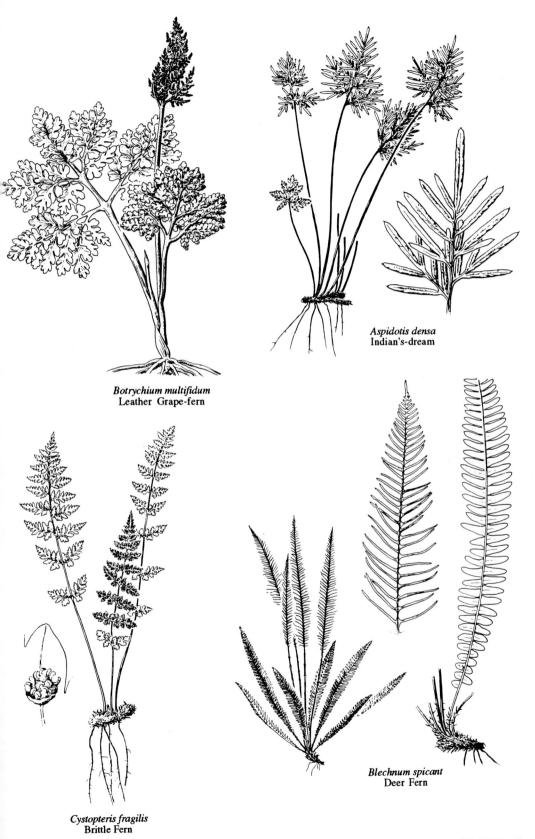

Botrychium multifidum
Leather Grape-fern

Aspidotis densa
Indian's-dream

Cystopteris fragilis
Brittle Fern

Blechnum spicant
Deer Fern

FERNS **PLATE 65**

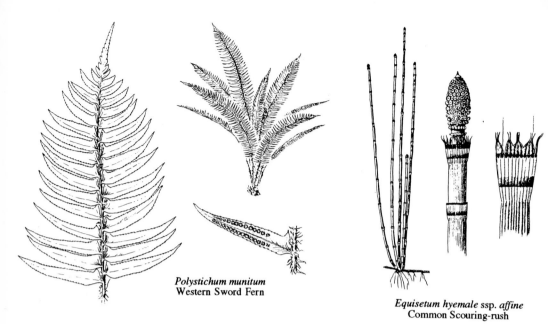

Polystichum munitum
Western Sword Fern

Equisetum hyemale ssp. *affine*
Common Scouring-rush

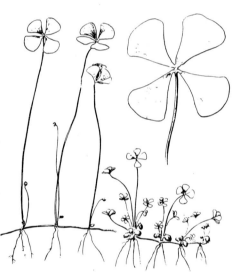

Marsilea vestita
Hairy Pepperwort

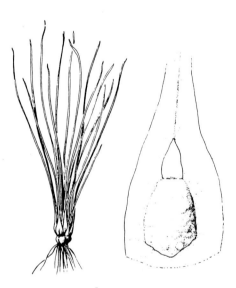

Isoetes nuttallii
Nuttall Quillwort

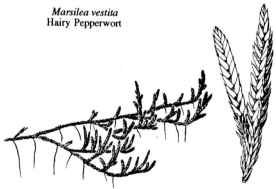

Selaginella wallacei
Little Clubmoss

Azolla filiculoides
Waterfern

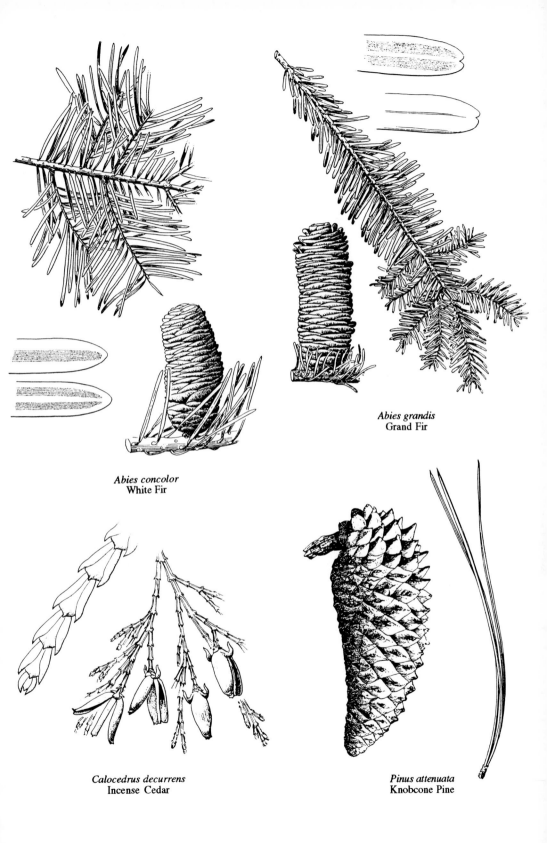

Abies grandis
Grand Fir

Abies concolor
White Fir

Calocedrus decurrens
Incense Cedar

Pinus attenuata
Knobcone Pine

GYMNOSPERMS   **PLATE 67**

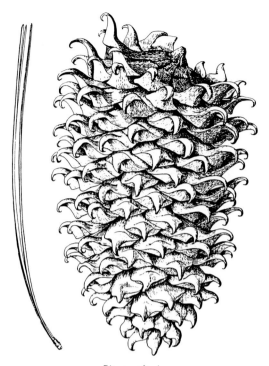

Pinus coulteri
Coulter Pine

Pinus sabiniana
Foothill Pine

Pinus ponderosa
Pacific Ponderosa Pine

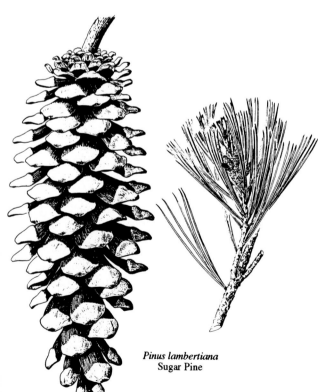

Pinus lambertiana
Sugar Pine

Pinus radiata
Monterey Pine

GYMNOSPERMS   **PLATE 68**

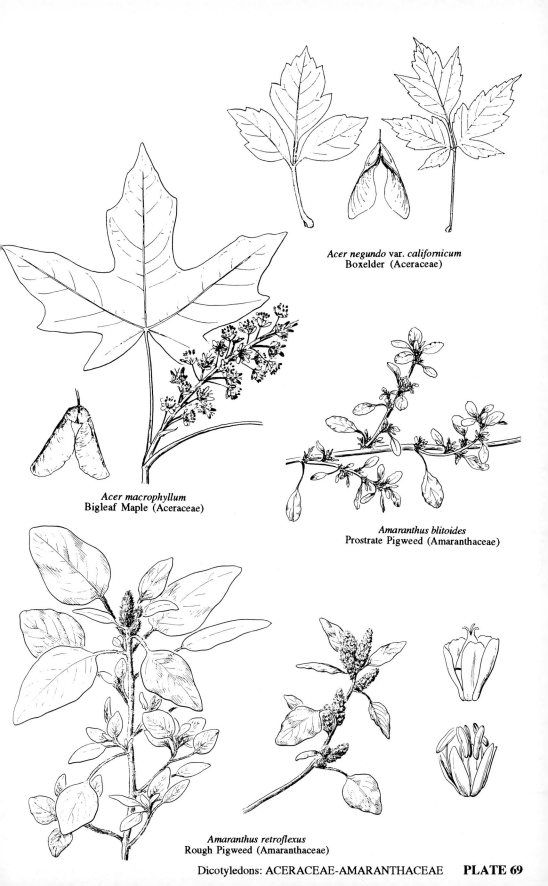

Acer negundo var. *californicum*
Boxelder (Aceraceae)

Acer macrophyllum
Bigleaf Maple (Aceraceae)

Amaranthus blitoides
Prostrate Pigweed (Amaranthaceae)

Amaranthus retroflexus
Rough Pigweed (Amaranthaceae)

Dicotyledons: ACERACEAE-AMARANTHACEAE   **PLATE 69**

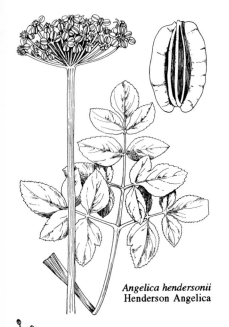

Angelica hendersonii
Henderson Angelica

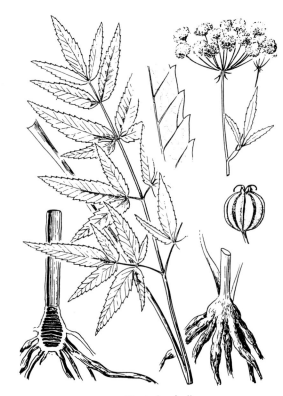

Cicuta douglasii
Douglas Water-hemlock

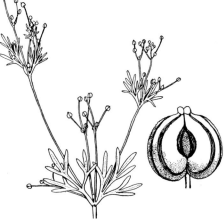

Apiastrum angustifolium
Wild Celery

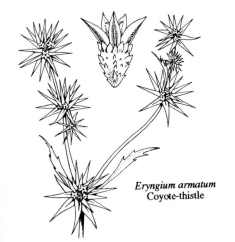

Eryngium armatum
Coyote-thistle

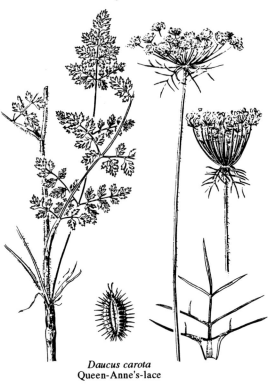

Daucus carota
Queen-Anne's-lace

Dicotyledons: APIACEAE   **PLATE 70**

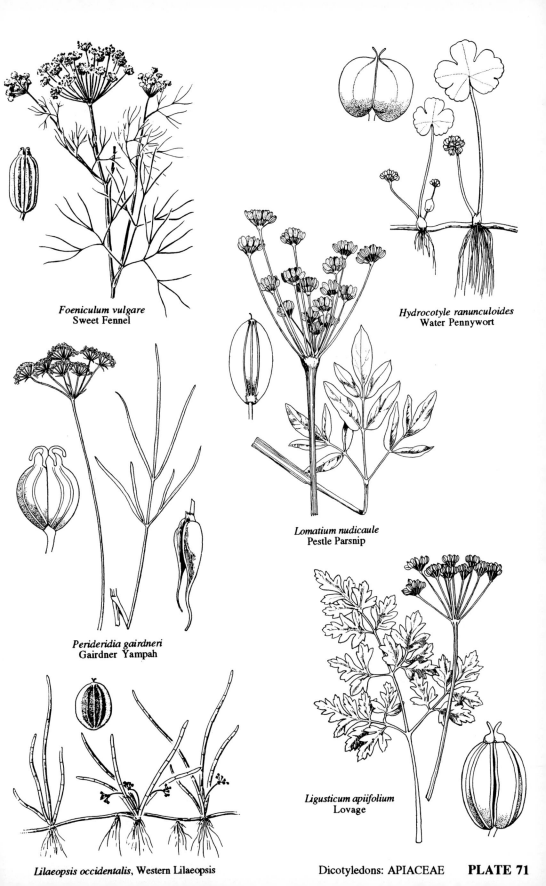

Foeniculum vulgare
Sweet Fennel

Hydrocotyle ranunculoides
Water Pennywort

Lomatium nudicaule
Pestle Parsnip

Perideridia gairdneri
Gairdner Yampah

Ligusticum apiifolium
Lovage

Lilaeopsis occidentalis, Western Lilaeopsis

Dicotyledons: APIACEAE   **PLATE 71**

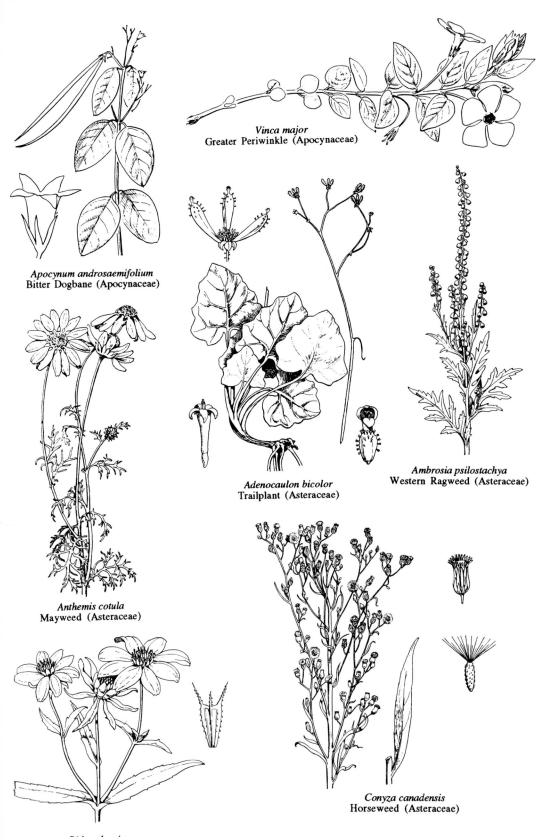

Vinca major
Greater Periwinkle (Apocynaceae)

Apocynum androsaemifolium
Bitter Dogbane (Apocynaceae)

Adenocaulon bicolor
Trailplant (Asteraceae)

Ambrosia psilostachya
Western Ragweed (Asteraceae)

Anthemis cotula
Mayweed (Asteraceae)

Conyza canadensis
Horseweed (Asteraceae)

Bidens laevis
Bur Marigold (Asteraceae)

Dicotyledons: APOCYNACEAE-ASTERACEAE **PLATE 72**

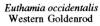

Ericameria ericoides
Mock-heather

Euthamia occidentalis
Western Goldenrod

Chamomilla suaveolens
Pineappleweed

Iva axillaris ssp. *robustior*
Povertyweed

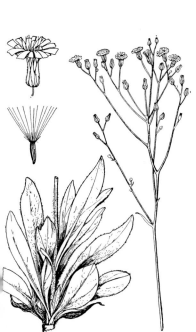

Hieracium albiflorum
Hawkweed

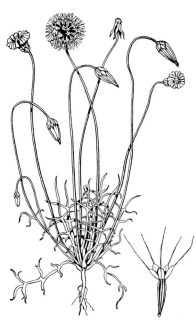

Microseris douglasii ssp. *douglasii*
Douglas Microseris

Dicotyledons: ASTERACEAE   **PLATE 73**

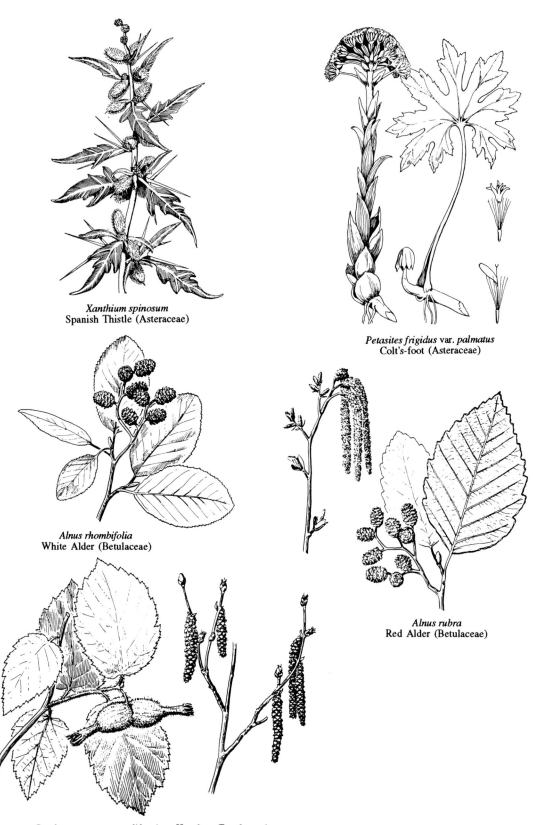

Xanthium spinosum
Spanish Thistle (Asteraceae)

Petasites frigidus var. *palmatus*
Colt's-foot (Asteraceae)

Alnus rhombifolia
White Alder (Betulaceae)

Alnus rubra
Red Alder (Betulaceae)

Corylus cornuta var. *californica*, Hazelnut (Betulaceae)

Dicotyledons: ASTERACEAE-BETULACEAE   **PLATE 74**

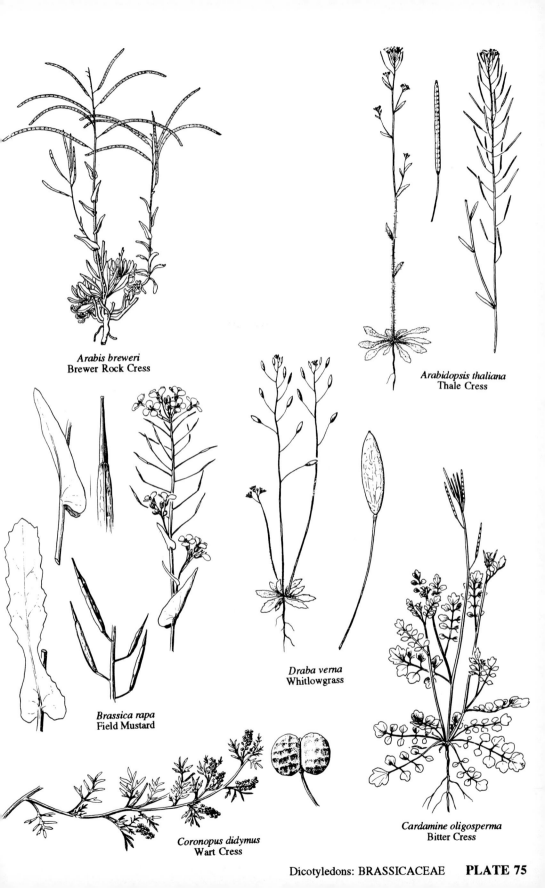

Arabis breweri
Brewer Rock Cress

Arabidopsis thaliana
Thale Cress

Brassica rapa
Field Mustard

Draba verna
Whitlowgrass

Coronopus didymus
Wart Cress

Cardamine oligosperma
Bitter Cress

Dicotyledons: BRASSICACEAE   **PLATE 75**

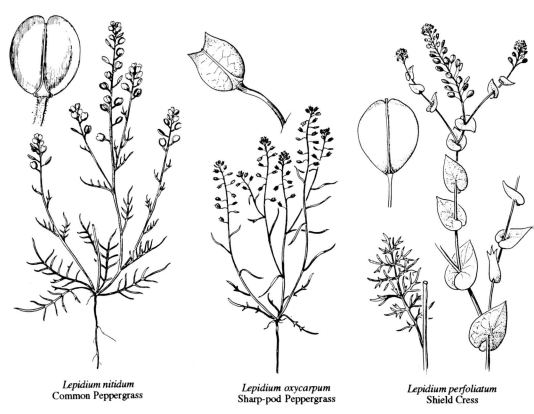

Lepidium nitidum
Common Peppergrass

Lepidium oxycarpum
Sharp-pod Peppergrass

Lepidium perfoliatum
Shield Cress

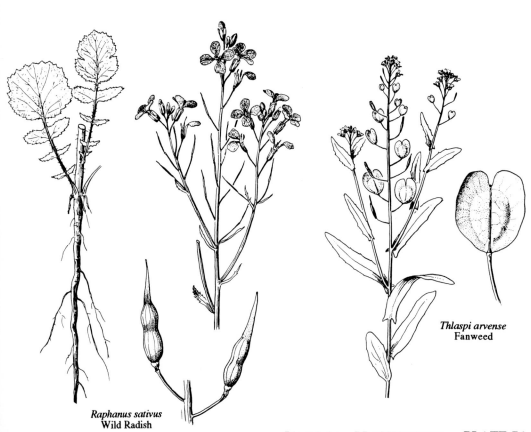

Raphanus sativus
Wild Radish

Thlaspi arvense
Fanweed

Dicotyledons: BRASSICACEAE   **PLATE 76**

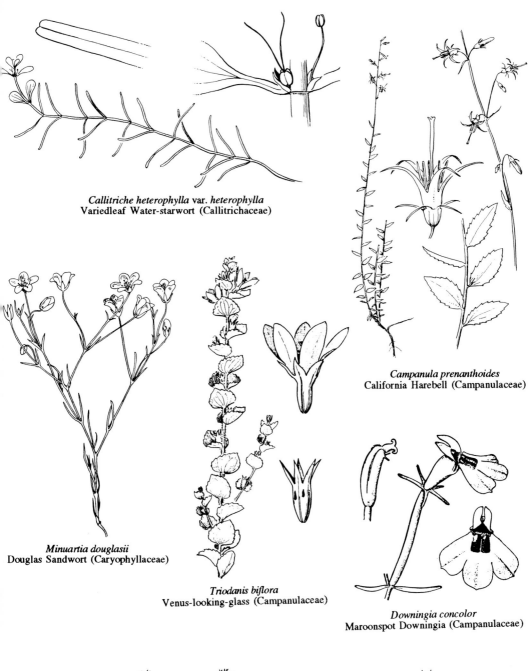

Callitriche heterophylla var. *heterophylla*
Variedleaf Water-starwort (Callitrichaceae)

Campanula prenanthoides
California Harebell (Campanulaceae)

Minuartia douglasii
Douglas Sandwort (Caryophyllaceae)

Triodanis biflora
Venus-looking-glass (Campanulaceae)

Downingia concolor
Maroonspot Downingia (Campanulaceae)

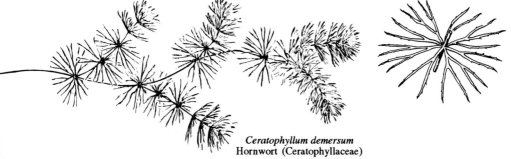

Ceratophyllum demersum
Hornwort (Ceratophyllaceae)

Dicotyledons: CALLITRICHACEAE-CERATOPHYLLACEAE   **PLATE 77**

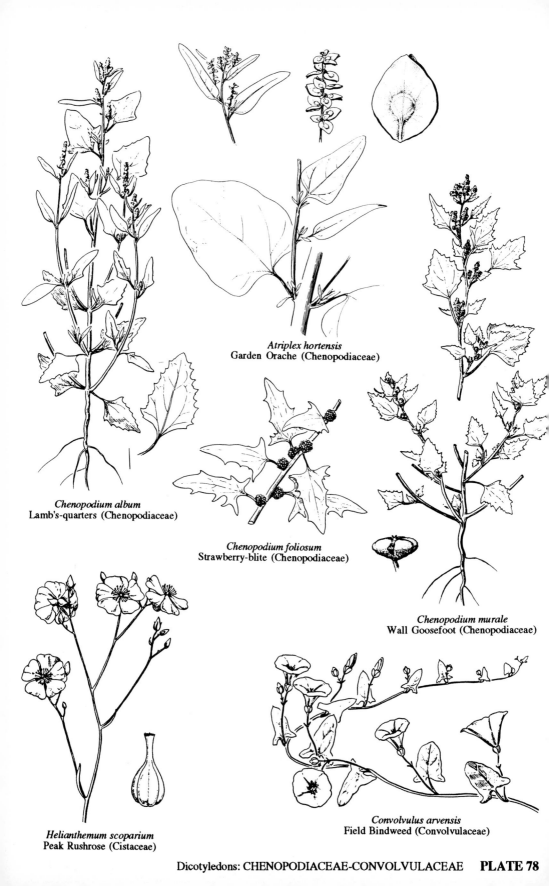

Atriplex hortensis
Garden Orache (Chenopodiaceae)

Chenopodium album
Lamb's-quarters (Chenopodiaceae)

Chenopodium foliosum
Strawberry-blite (Chenopodiaceae)

Chenopodium murale
Wall Goosefoot (Chenopodiaceae)

Helianthemum scoparium
Peak Rushrose (Cistaceae)

Convolvulus arvensis
Field Bindweed (Convolvulaceae)

Dicotyledons: CHENOPODIACEAE-CONVOLVULACEAE **PLATE 78**

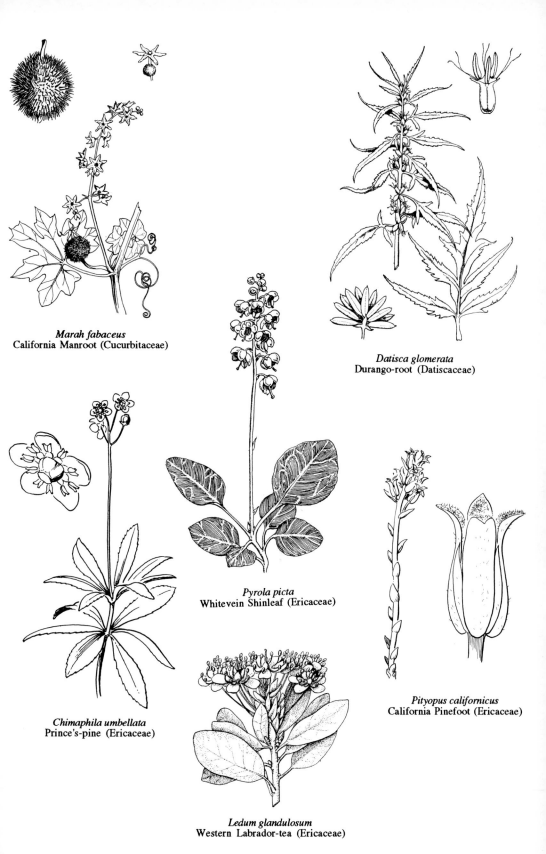

Marah fabaceus
California Manroot (Cucurbitaceae)

Datisca glomerata
Durango-root (Datiscaceae)

Pyrola picta
Whitevein Shinleaf (Ericaceae)

Chimaphila umbellata
Prince's-pine (Ericaceae)

Pityopus californicus
California Pinefoot (Ericaceae)

Ledum glandulosum
Western Labrador-tea (Ericaceae)

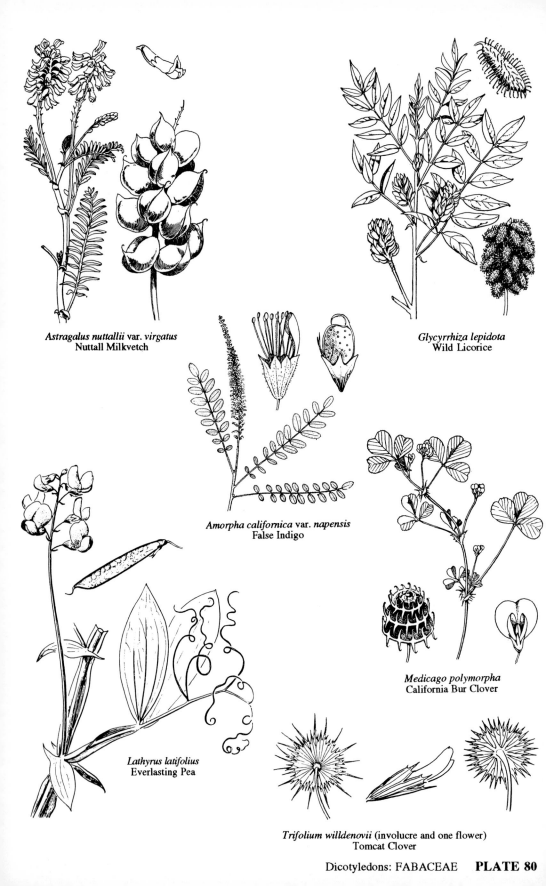

Astragalus nuttallii var. *virgatus*
Nuttall Milkvetch

Glycyrrhiza lepidota
Wild Licorice

Amorpha californica var. *napensis*
False Indigo

Medicago polymorpha
California Bur Clover

Lathyrus latifolius
Everlasting Pea

Trifolium willdenovii (involucre and one flower)
Tomcat Clover

Dicotyledons: FABACEAE   **PLATE 80**

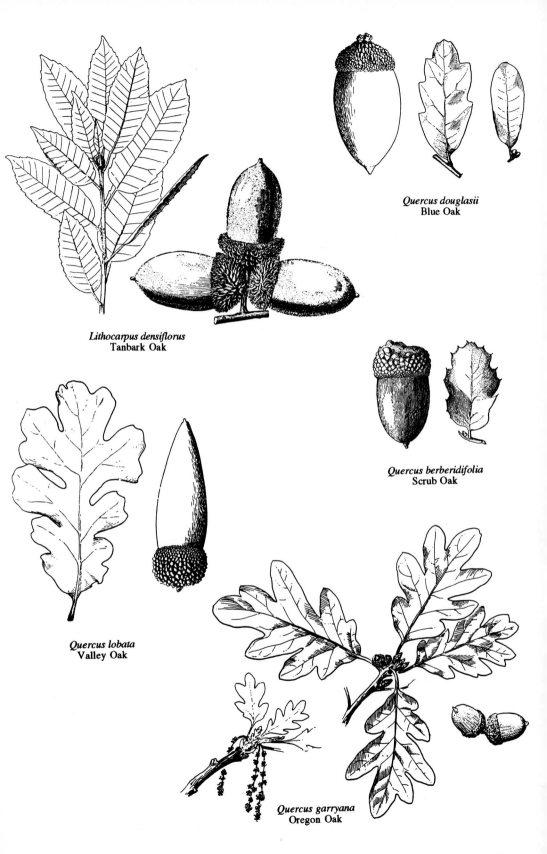

Quercus douglasii
Blue Oak

Lithocarpus densiflorus
Tanbark Oak

Quercus berberidifolia
Scrub Oak

Quercus lobata
Valley Oak

Quercus garryana
Oregon Oak

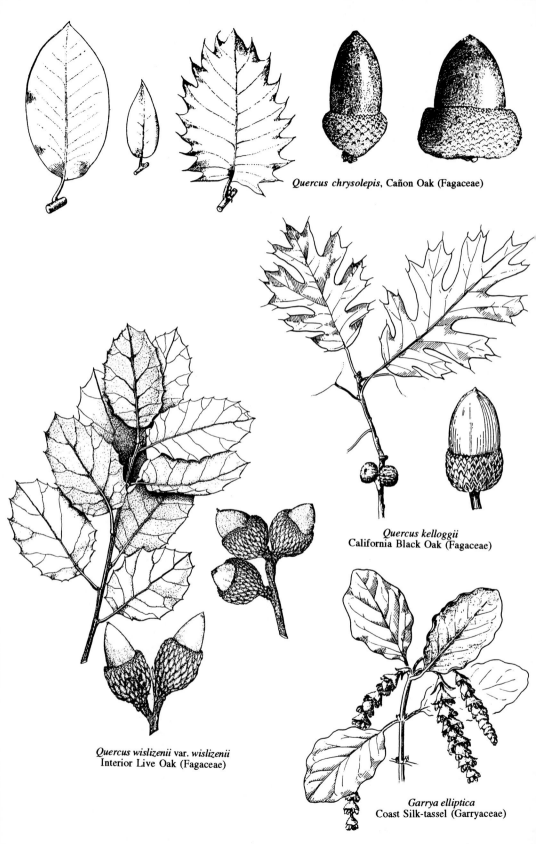

Quercus chrysolepis, Cañon Oak (Fagaceae)

Quercus kelloggii
California Black Oak (Fagaceae)

Quercus wislizenii var. *wislizenii*
Interior Live Oak (Fagaceae)

Garrya elliptica
Coast Silk-tassel (Garryaceae)

Dicotyledons: FAGACEAE-GARRYACEAE **PLATE 82**

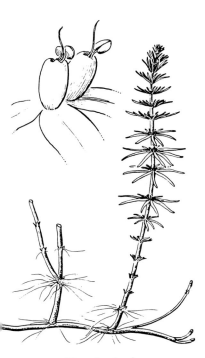

Hippuris vulgaris
Mare's-tail (Hippuridaceae)

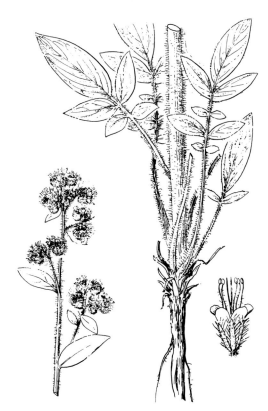

Phacelia nemoralis
Bristly Phacelia (Hydrophyllaceae)

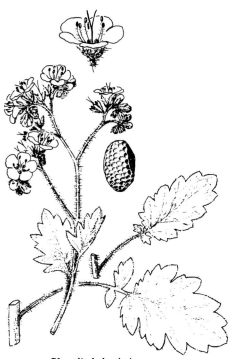

Phacelia bolanderi
Bolander Phacelia (Hydrophyllaceae)

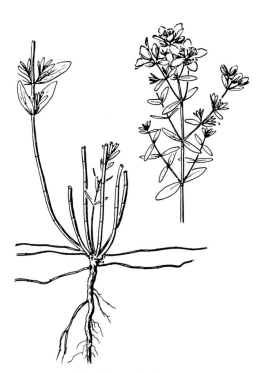

Hypericum perforatum
Klamathweed (Hypericaceae)

Dicotyledons: HIPPURIDACEAE-HYPERICACEAE   **PLATE 83**

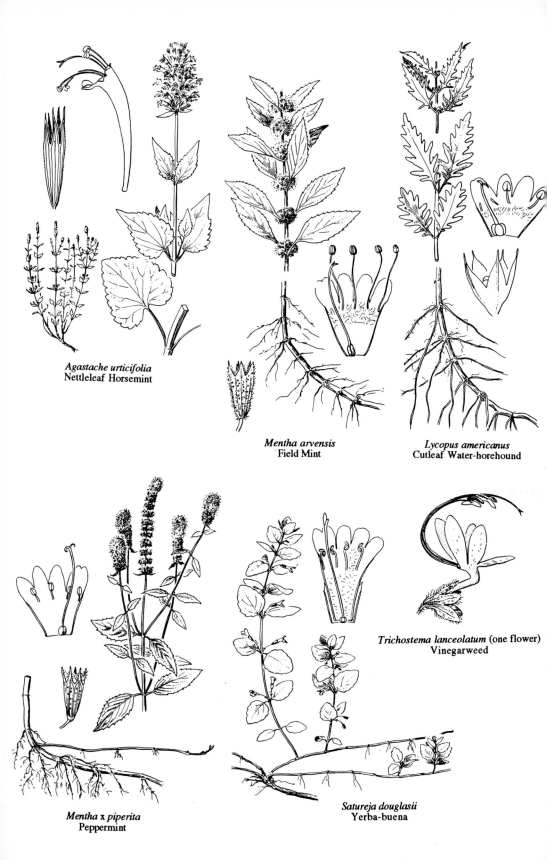

Agastache urticifolia
Nettleleaf Horsemint

Mentha arvensis
Field Mint

Lycopus americanus
Cutleaf Water-horehound

Trichostema lanceolatum (one flower)
Vinegarweed

Mentha x piperita
Peppermint

Satureja douglasii
Yerba-buena

Dicotyledons: LAMIACEAE   **PLATE 84**

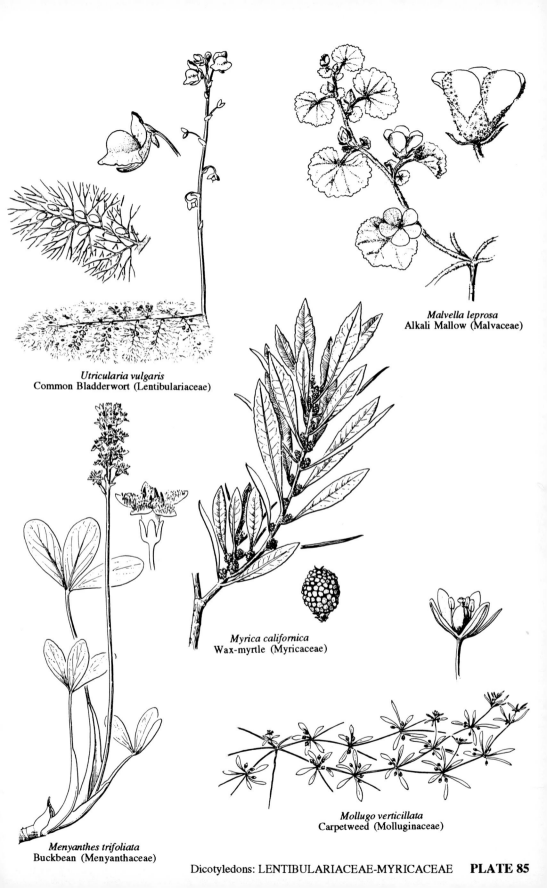

Malvella leprosa
Alkali Mallow (Malvaceae)

Utricularia vulgaris
Common Bladderwort (Lentibulariaceae)

Myrica californica
Wax-myrtle (Myricaceae)

Menyanthes trifoliata
Buckbean (Menyanthaceae)

Mollugo verticillata
Carpetweed (Molluginaceae)

Dicotyledons: LENTIBULARIACEAE-MYRICACEAE   **PLATE 85**

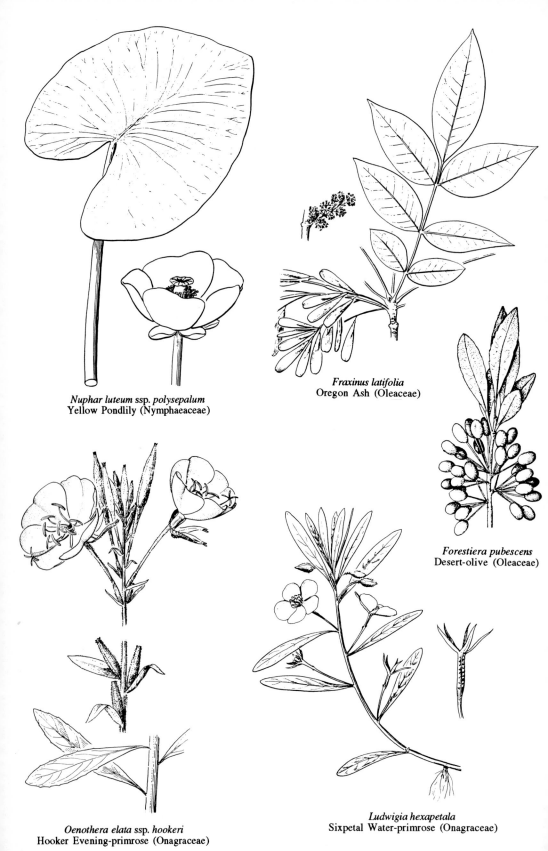

Nuphar luteum ssp. *polysepalum*
Yellow Pondlily (Nymphaeaceae)

Fraxinus latifolia
Oregon Ash (Oleaceae)

Forestiera pubescens
Desert-olive (Oleaceae)

Ludwigia hexapetala
Sixpetal Water-primrose (Onagraceae)

Oenothera elata ssp. *hookeri*
Hooker Evening-primrose (Onagraceae)

Dicotyledons: NYMPHAEACEAE-ONAGRACEAE   **PLATE 86**

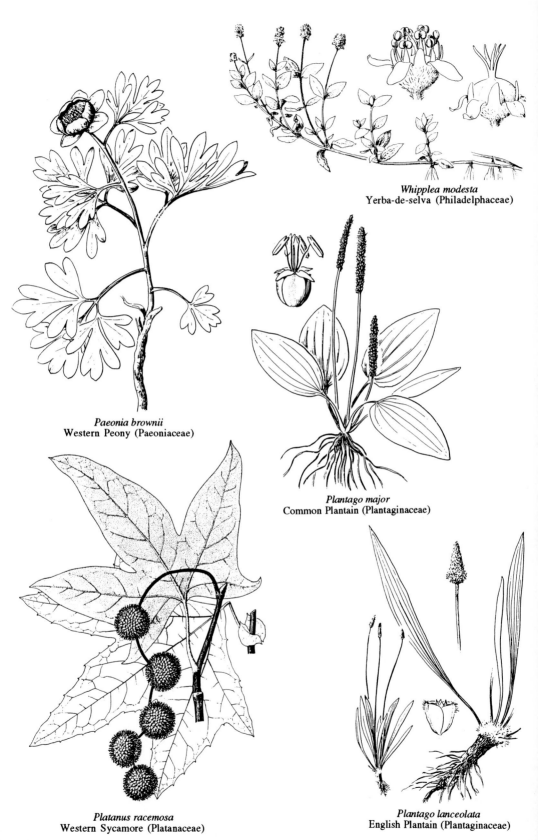

Whipplea modesta
Yerba-de-selva (Philadelphaceae)

Paeonia brownii
Western Peony (Paeoniaceae)

Plantago major
Common Plantain (Plantaginaceae)

Platanus racemosa
Western Sycamore (Platanaceae)

Plantago lanceolata
English Plantain (Plantaginaceae)

Dicotyledons: PAEONIACEAE-PLATANACEAE   **PLATE 87**

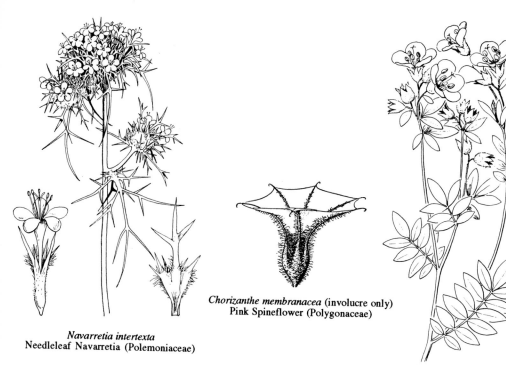

Chorizanthe membranacea (involucre only)
Pink Spineflower (Polygonaceae)

Navarretia intertexta
Needleleaf Navarretia (Polemoniaceae)

Polemonium carneum
Jacob's-ladder (Polemoniaceae)

Eriogonum nudum var. *nudum*
Nakedstem Buckwheat (Polygonaceae)

Polygonum hydropiper
Marshpepper (Polygonaceae)

Dicotyledons: POLEMONIACEAE-POLYGONACEAE   **PLATE 88**

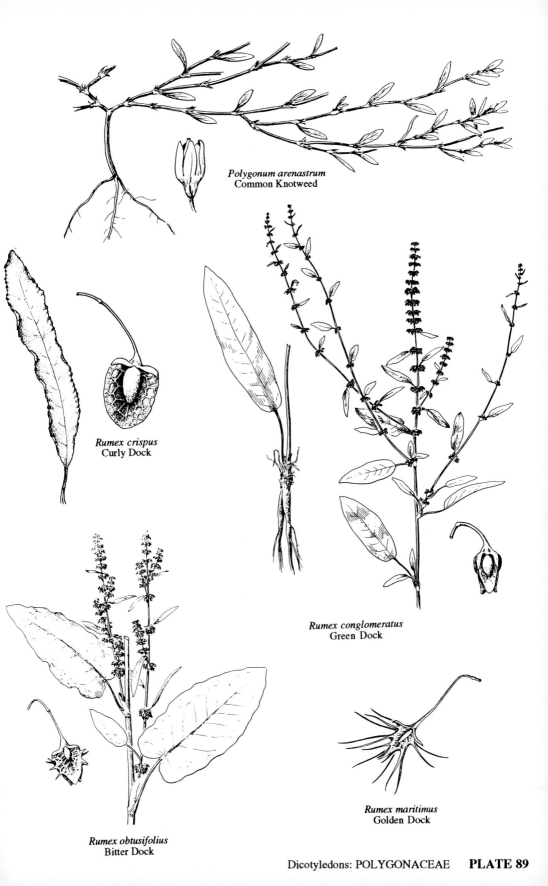

Polygonum arenastrum
Common Knotweed

Rumex crispus
Curly Dock

Rumex conglomeratus
Green Dock

Rumex maritimus
Golden Dock

Rumex obtusifolius
Bitter Dock

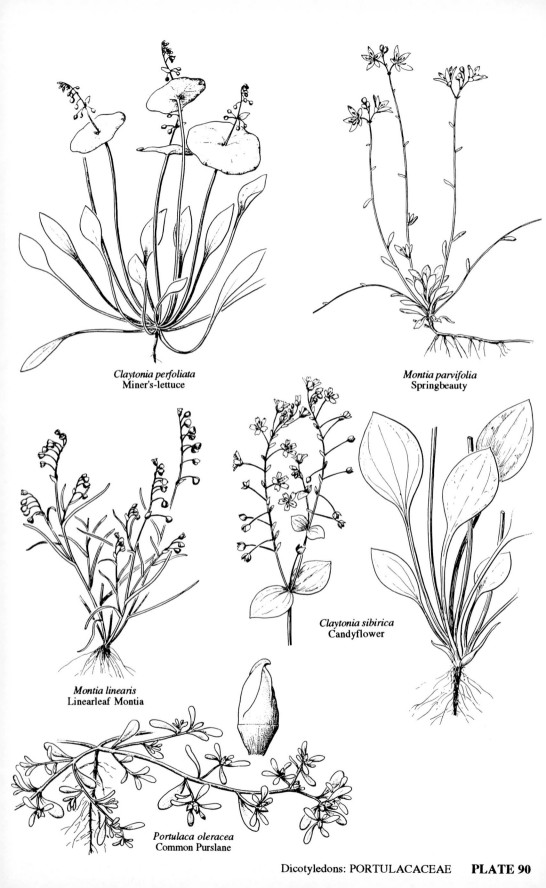

Claytonia perfoliata
Miner's-lettuce

Montia parvifolia
Springbeauty

Montia linearis
Linearleaf Montia

Claytonia sibirica
Candyflower

Portulaca oleracea
Common Purslane

Dicotyledons: PORTULACACEAE   **PLATE 90**

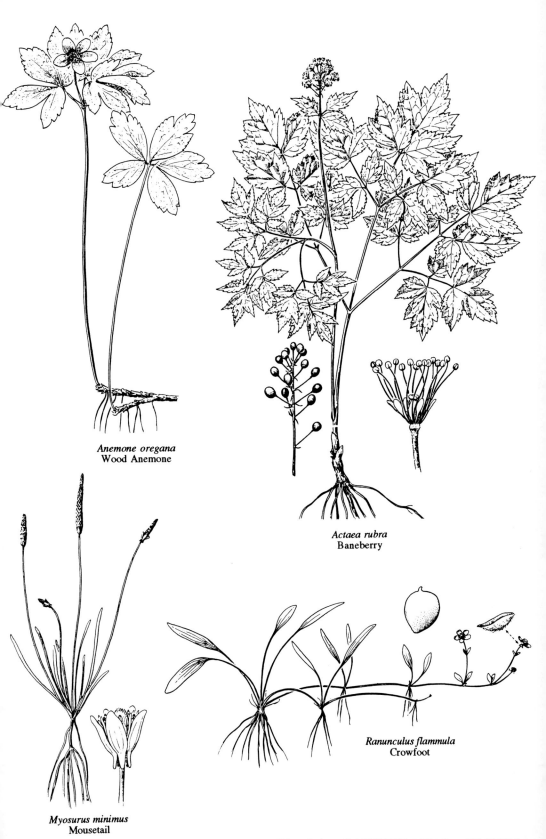

Anemone oregana
Wood Anemone

Actaea rubra
Baneberry

Myosurus minimus
Mousetail

Ranunculus flammula
Crowfoot

Dicotyledons: RANUNCULACEAE   **PLATE 91**

Ranunculus aquatilis var. *hispidulus*
Water Buttercup (Ranunculaceae)

Ranunculus hebecarpus
Downy Buttercup (Ranunculaceae)

Thalictrum fendleri var. *polycarpum*
Meadowrue (Ranunculaceae)

Malus fusca
Oregon Crabapple (Rosaceae)

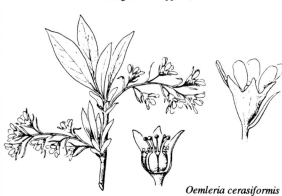

Oemleria cerasiformis
Osoberry (Rosaceae)

Dicotyledons: RANUNCULACEAE-ROSACEAE   **PLATE 92**

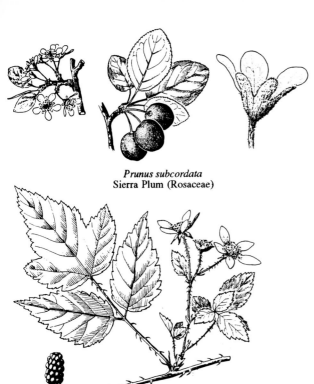

Prunus subcordata
Sierra Plum (Rosaceae)

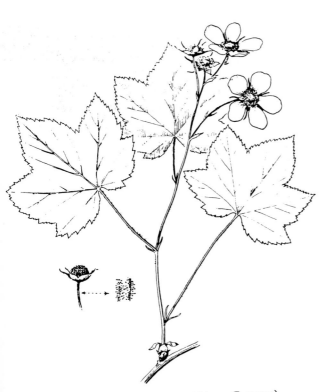

Rubus ursinus
California Blackberry (Rosaceae)

Prunus virginiana var. demissa
Western Choke Cherry (Rosaceae)

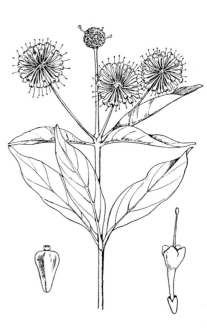

Rubus parviflorus, Thimbleberry (Rosaceae)

Cephalanthus occidentalis var. californicus
California Button-willow (Rubiaceae)

Dicotyledons: ROSACEAE-RUBIACEAE   **PLATE 93**

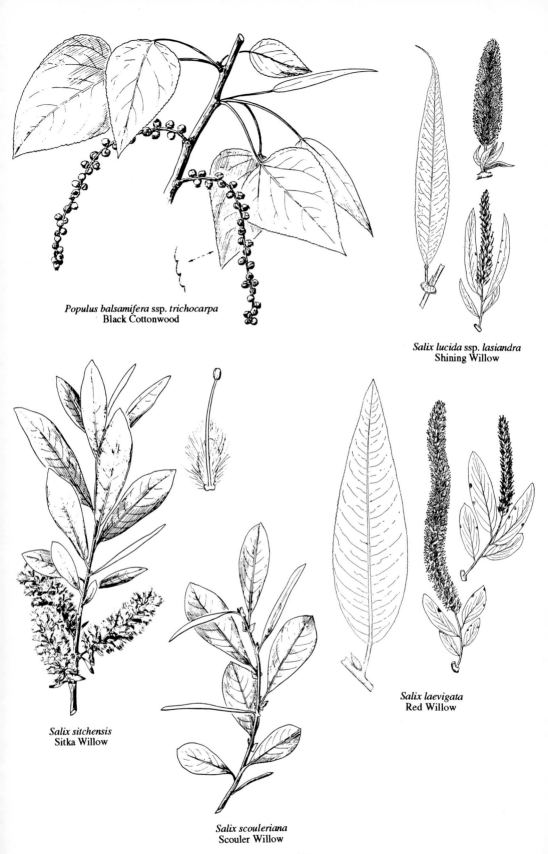

Populus balsamifera ssp. *trichocarpa*
Black Cottonwood

Salix lucida ssp. *lasiandra*
Shining Willow

Salix sitchensis
Sitka Willow

Salix laevigata
Red Willow

Salix scouleriana
Scouler Willow

Dicotyledons: SALICACEAE   **PLATE 94**

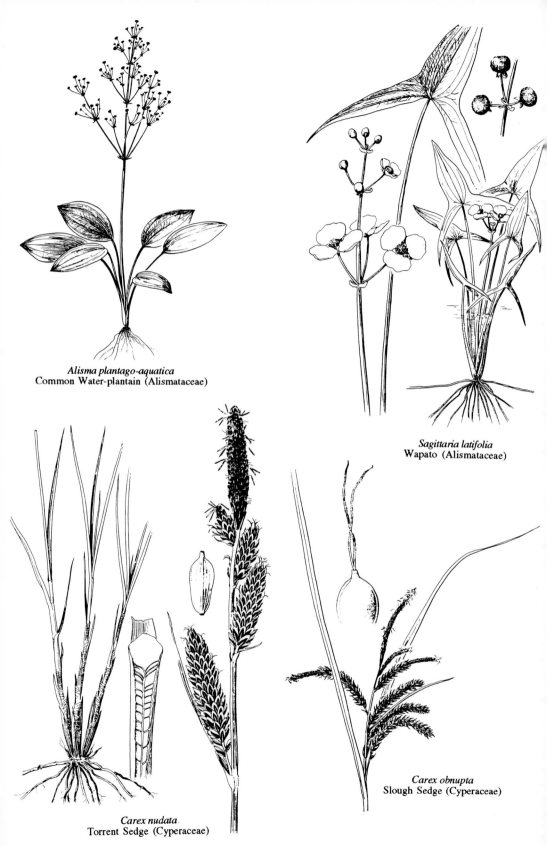

Alisma plantago-aquatica
Common Water-plantain (Alismataceae)

Sagittaria latifolia
Wapato (Alismataceae)

Carex nudata
Torrent Sedge (Cyperaceae)

Carex obnupta
Slough Sedge (Cyperaceae)

Monocotyledons: ALISMATACEAE-CYPERACEAE   **PLATE 98**

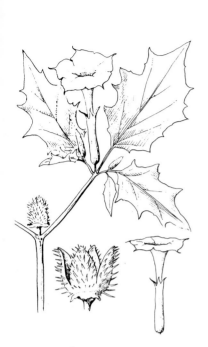

Solanum dulcamara
Bittersweet (Solanaceae)

Datura stramonium
Jimsonweed (Solanaceae)

Viola sempervirens
Evergreen Violet (Violaceae)

Solanum nigrum
Black Nightshade (Solanaceae)

Verbena bracteata
Bract Verbena (Verbenaceae)

Dicotyledons: SOLANACEAE-VERBENACEAE **PLATE 97**

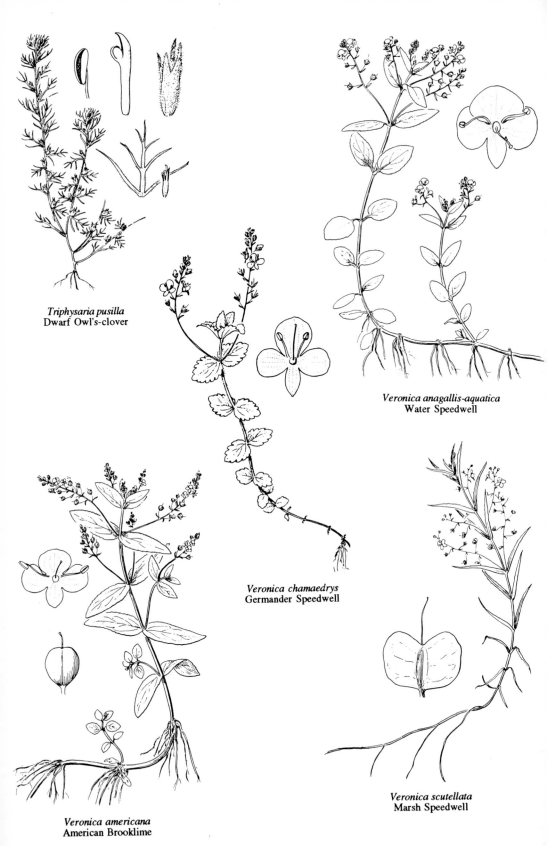

Triphysaria pusilla
Dwarf Owl's-clover

Veronica anagallis-aquatica
Water Speedwell

Veronica chamaedrys
Germander Speedwell

Veronica americana
American Brooklime

Veronica scutellata
Marsh Speedwell

Dicotyledons: SCROPHULARIACEAE   **PLATE 96**

Heuchera micrantha
Smallflower Alumroot (Saxifragaceae)

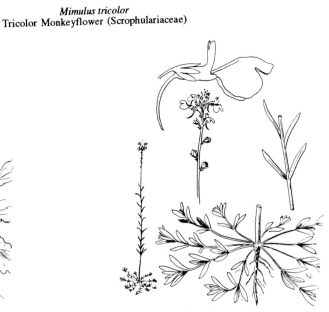

Digitalis purpurea
Foxglove (Scrophulariaceae)

Mimulus tricolor
Tricolor Monkeyflower (Scrophulariaceae)

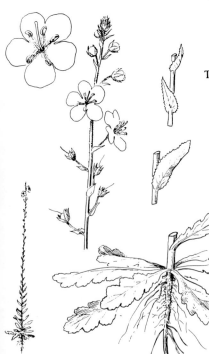

Verbascum blattaria
Moth Mullein (Scrophulariaceae)

Linaria canadensis
Blue Toadflax (Scrophulariaceae)

Dicotyledons: SAXIFRAGACEAE-SCROPHULARIACEAE **PLATE 95**

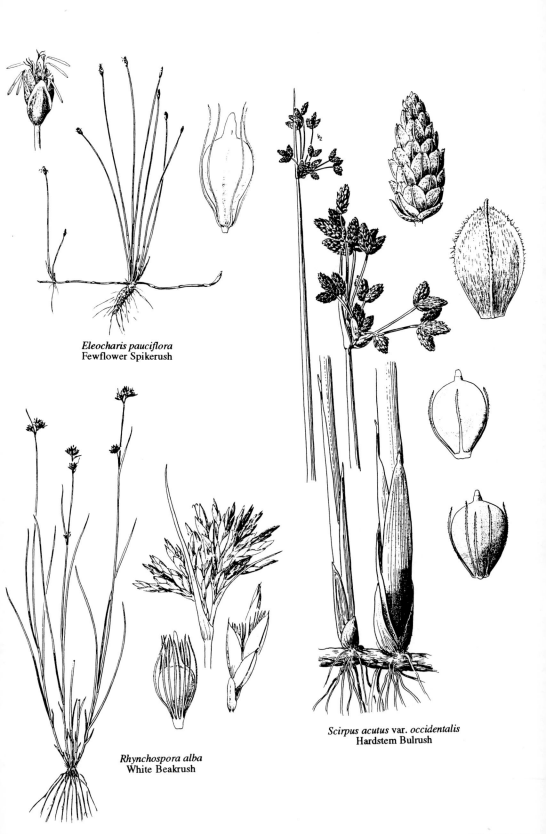

Eleocharis pauciflora
Fewflower Spikerush

Rhynchospora alba
White Beakrush

Scirpus acutus var. *occidentalis*
Hardstem Bulrush

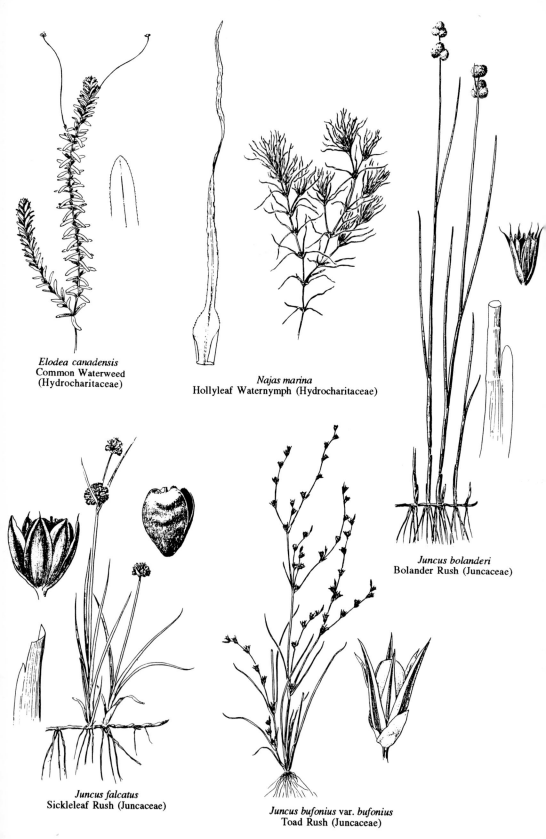

Elodea canadensis
Common Waterweed
(Hydrocharitaceae)

Najas marina
Hollyleaf Waternymph (Hydrocharitaceae)

Juncus bolanderi
Bolander Rush (Juncaceae)

Juncus falcatus
Sickleleaf Rush (Juncaceae)

Juncus bufonius var. *bufonius*
Toad Rush (Juncaceae)

Monocotyledons: HYDROCHARITACEAE-JUNCACEAE   **PLATE 100**

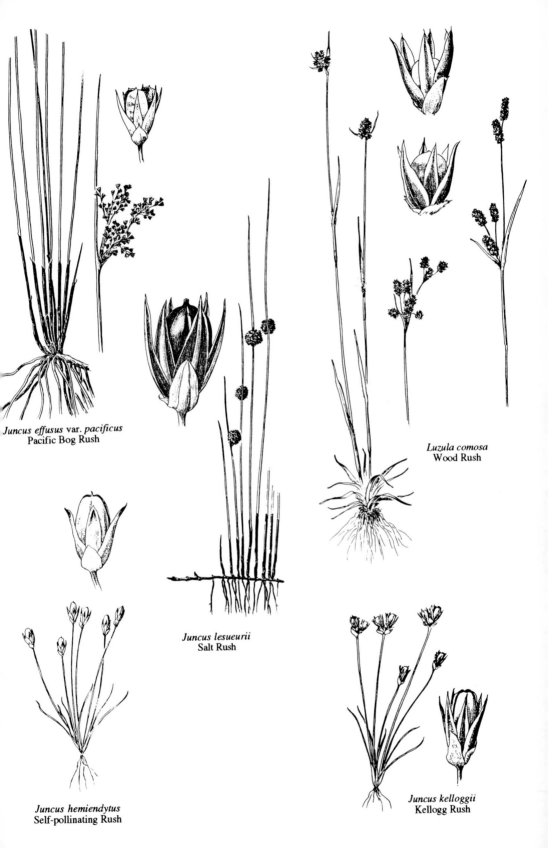

Juncus effusus var. *pacificus*
Pacific Bog Rush

Luzula comosa
Wood Rush

Juncus lesueurii
Salt Rush

Juncus hemiendytus
Self-pollinating Rush

Juncus kelloggii
Kellogg Rush

Monocotyledons: JUNCACEAE   **PLATE 101**

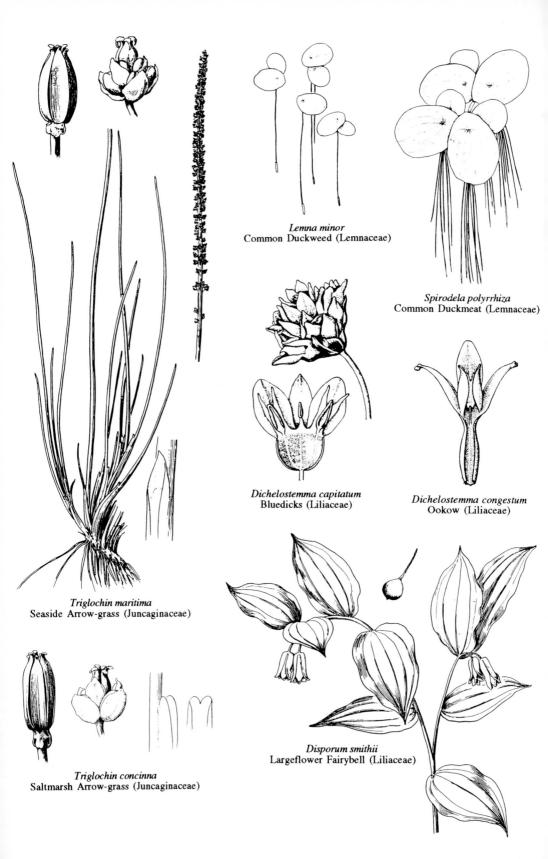

Lemna minor
Common Duckweed (Lemnaceae)

Spirodela polyrrhiza
Common Duckmeat (Lemnaceae)

Dichelostemma capitatum
Bluedicks (Liliaceae)

Dichelostemma congestum
Ookow (Liliaceae)

Triglochin maritima
Seaside Arrow-grass (Juncaginaceae)

Triglochin concinna
Saltmarsh Arrow-grass (Juncaginaceae)

Disporum smithii
Largeflower Fairybell (Liliaceae)

Monocotyledons: JUNCAGINACEAE-LILIACEAE   **PLATE 102**

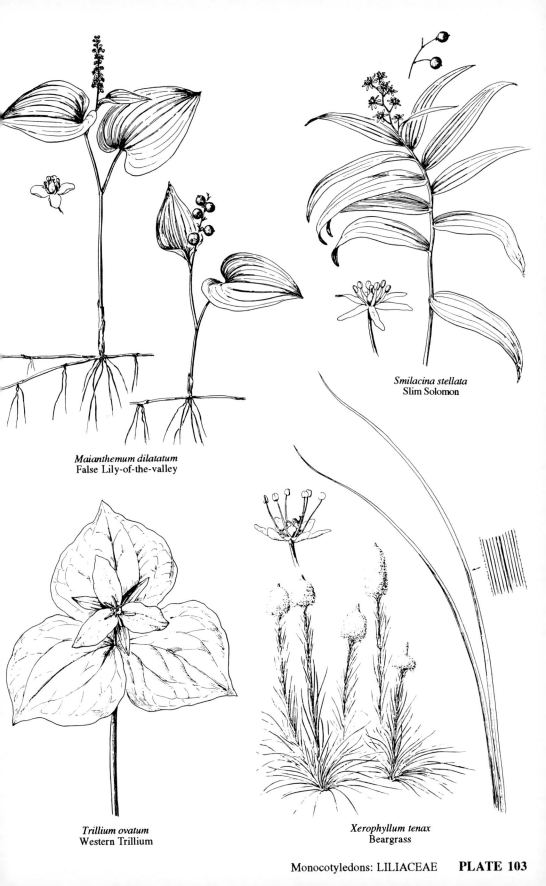

Smilacina stellata
Slim Solomon

Maianthemum dilatatum
False Lily-of-the-valley

Trillium ovatum
Western Trillium

Xerophyllum tenax
Beargrass

Monocotyledons: LILIACEAE **PLATE 103**

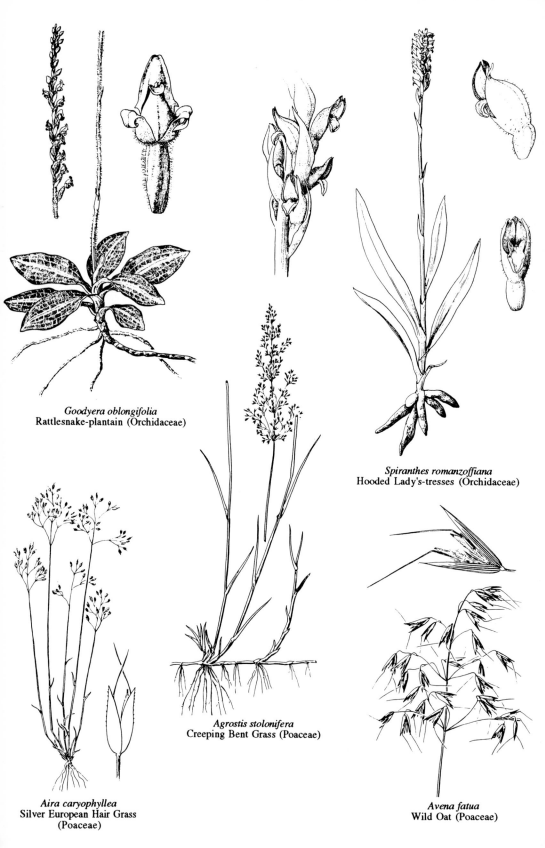

Goodyera oblongifolia
Rattlesnake-plantain (Orchidaceae)

Spiranthes romanzoffiana
Hooded Lady's-tresses (Orchidaceae)

Agrostis stolonifera
Creeping Bent Grass (Poaceae)

Aira caryophyllea
Silver European Hair Grass
(Poaceae)

Avena fatua
Wild Oat (Poaceae)

Monocotyledons: ORCHIDACEAE-POACEAE   **PLATE 104**

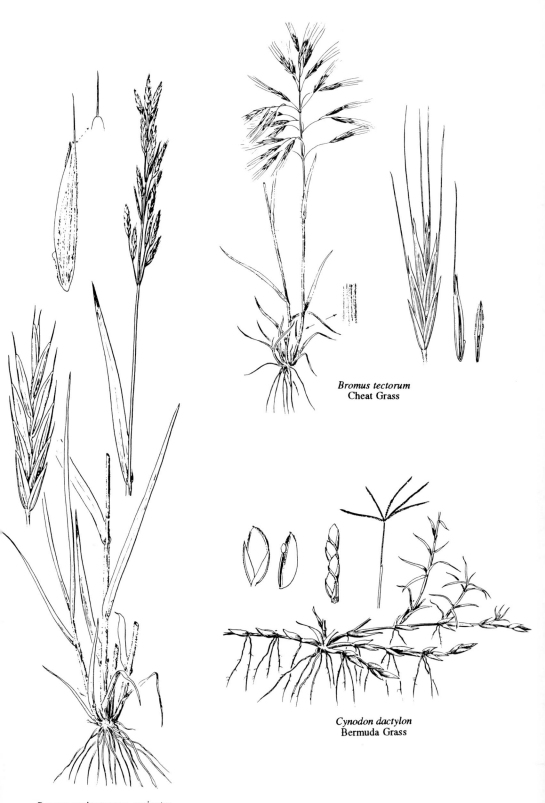

Bromus tectorum
Cheat Grass

Cynodon dactylon
Bermuda Grass

Bromus carinatus var. *carinatus*
California Brome

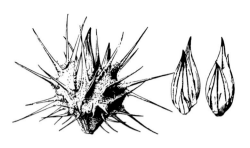

Cenchrus longispinus
Mat Sandbur

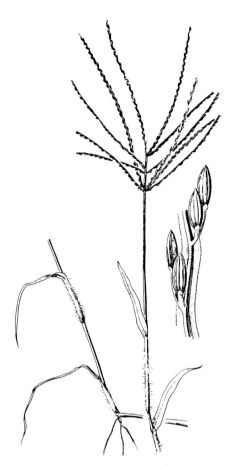

Digitaria sanguinalis
Crab Grass

Deschampsia danthonioides
Annual Hair Grass

Distichlis spicata
Salt Grass

Monocotyledons: POACEAE   **PLATE 106**

Elymus glaucus
Blue Wild Rye

Elytrigia repens
Quack Grass

Festuca rubra
Red Fescue

Monocotyledons: POACEAE   **PLATE 107**

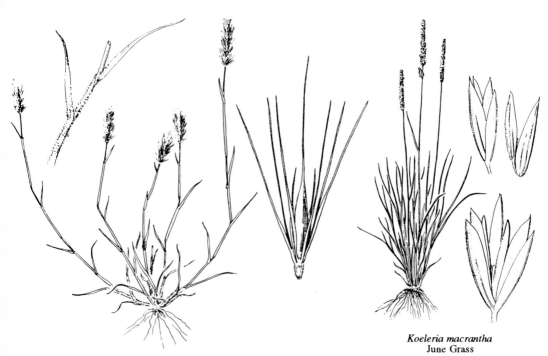

Koeleria macrantha
June Grass

Hordeum marinum ssp. *gussoneanum*
Mediterranean Barley

Leymus mollis
Dune Grass

Lolium perenne
Perennial Rye Grass

Monocotyledons: POACEAE   **PLATE 108**

Melica subulata
Alaska Onion Grass

Poa annua
Annual Blue Grass

Vulpia octoflora var. *octoflora*
Sixweeks Fescue

Vulpia bromoides
Brome Fescue

Poa pratensis
Kentucky Blue Grass

Monocotyledons: POACEAE   **PLATE 109**

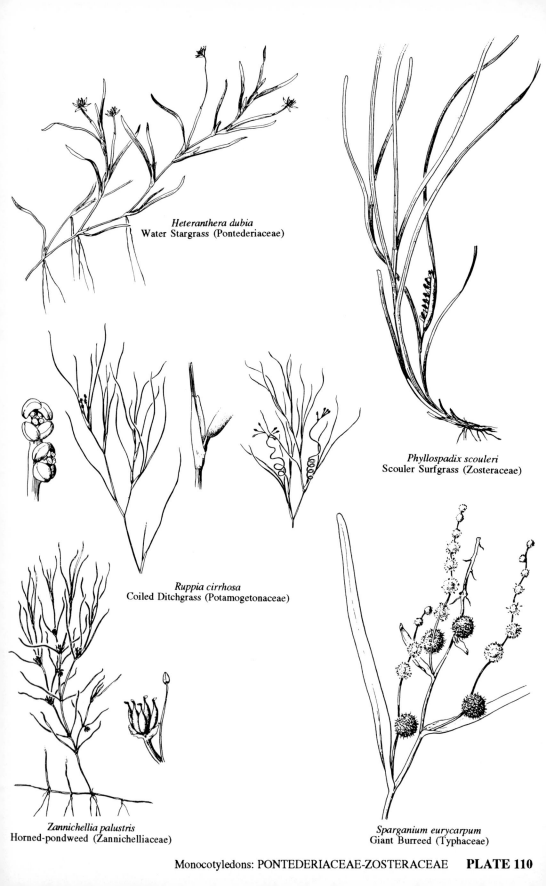

Heteranthera dubia
Water Stargrass (Pontederiaceae)

Phyllospadix scouleri
Scouler Surfgrass (Zosteraceae)

Ruppia cirrhosa
Coiled Ditchgrass (Potamogetonaceae)

Zannichellia palustris
Horned-pondweed (Zannichelliaceae)

Sparganium eurycarpum
Giant Burreed (Typhaceae)

Monocotyledons: PONTEDERIACEAE-ZOSTERACEAE   **PLATE 110**

```
            FF Inflorescence consisting of several whorls of flowers
                                        Hydrocotyle verticillata
                         [includes H. verticillata var. triradiata]
                                      Whorled Marsh Pennywort; SFBR
    DD Plants, if aquatic, not especially small, and not with nearly
       circular leaf blades
       E   Ultimate divisions of leaves cylindrical, sometimes nearly
           hairlike
           F  Leaves up to 10 cm long; corolla white; up to 0.5 m
              tall    Ciclospermum leptophyllum [Apium leptophyllum]
                                          Marsh Parsley; sa; Al
           FF Leaves up to about 30 cm long; corolla yellow; up to 2
              m tall (widespread)    Foeniculum vulgare (Plate 71)
                                            Sweet Fennel; me
       EE Ultimate divisions of leaves not cylindrical, definitely
          not hairlike (they may, however, be extremely narrow)
          F   Leaves similar to those of parsley in having small
              leaflets or deeply separated lobes, these usually much
              less than 0.5 cm wide and 1.5 cm long. (If the
              ultimate divisions, even if narrow, are several to
              many times as long as wide, go to Subkey 2.)
                                      Apiaceae, Subkey 1
          FF Leaves not like those of parsley, with relatively few
             leaflets or deeply separated lobes, these either at
             least 1 cm wide and 1.5 cm long or, if narrower,
             several to many times as long as wide. (In Perideridia
             californica, however, only the terminal leaflet and
             terminal lobe of the lateral leaflets are several
             times as long as wide.)      Apiaceae, Subkey 2 (p. 77)
```

Apiaceae, Subkey 1: Leaves similar to those of parsley in having small
ultimate leaflets or deeply separated lobes, these usually much less than
0.5 cm wide and 1.5 cm long. (If the ultimate divisions, even if narrow,
are several to many times as long as wide, go to Subkey 2.)

```
A   Plants often more than 2 m tall, the stems sometimes more than 2 cm
    thick; stems hollow, purple-spotted; leaves often 30 cm long (the deadly
    hemlock of ancient literature; may also cause a skin rash if handled;
    widespread)                                      Conium maculatum
                                               Poison-hemlock; eu
AA Plants rarely so much as 1 m tall, and not as described in choice A
   B   Fruit with a beak about 4 cm long (at least some long-beaked fruits
       will usually be found on plants that are still flowering) (main body
       of fruit about 1 cm long; corolla white; weedy annual up to 30 cm
       tall)                              Scandix pecten-veneris (Plate 5)
                                              Shepherd's-needle; me
   BB Fruit either without a beak, or the beak less than 1 cm long
      C   Fruit with stiff bristles (these sometimes hooked) or
          conspicuous, tall bumps
          D   Bracts either absent below the primary umbels or small,
              slender, and not divided into lobes (flowers of secondary
              umbels with such short pedicels that they may seem to
              originate directly from the rays of the primary umbels)
              E   Peduncle of each umbel rarely more than 2 cm long; bracts
                  usually absent below primary umbels; half of each fruit
                  with hooked bristles, the other half with rounded bumps;
                  often somewhat sprawling                Torilis nodosa
                                          Knotted Hedge Parsley; eua
              EE Peduncle of each umbel usually at least 5 cm long; bracts
                 (up to 2) usually present below primary umbels; both
                 halves of fruit with hooked bristles; usually upright
                                                         Torilis arvensis
                                          Hedge Parsley; eu; Sn
          DD Bracts below primary umbels divided into lobes or leaflets
             (in Yabea microcarpa, they resemble leaves on lower portions
             of the plant)
             E   Bracts below primary umbels so much divided that they
                 resemble leaves on lower portions of the plant (secondary
                 umbels with fewer than 10 flowers; corolla white; fruit
                 with lengthwise ribs bearing distinctly hooked bristles;
                 plants often small and inconspicuous)
                                  Yabea microcarpa [Caucalis microcarpa]
                                          California Hedge Parsley
             EE Bracts below primary umbels not so much divided that they
                resemble leaves on lower portions of the plant
                F   Spines on fruit arranged in distinct lengthwise rows
```

G   Bracts beneath the primary umbel shorter than the
     rays of the umbel, and with slender lobes that are
     not divided again; each secondary umbel usually
     with more than 15 flowers; one flower near the
     center of the umbel with a purple corolla; plants
     biennial, up to about 100 cm tall (widespread)
                              Daucus carota (*Plate* 70)
                              Queen-Anne's-lace; eu
GG   Bracts beneath the primary umbel as long as the
     rays of the umbel, and with lobes that are usually
     divided again into 2-3 lobes; each secondary umbel
     with not more than 12 flowers; none of the flowers
     distinguished by a purple corolla; annual, mostly
     less than 60 cm tall (plants growing near the
     coast are more compact and fleshier than those of
     inland situations)   *Daucus pusillus*
                              Rattlesnake-weed
FF Spines or bumps on fruit scattered, not arranged in
    distinct lengthwise rows
    G   Both halves of fruit prolonged into a beak that is
         about one-third as long as the rest of the fruit;
         fruit with slender, hooked bristles; corolla white
                              Anthriscus caucalis [*A. scandicina*]
                              Bur Chervil; eua
    GG   Neither half of the fruit prolonged into a beak
         (but the styles may persist); fruit with tall
         bumps, some of which may end in bristles or hooks;
         corolla yellow, straw-, or salmon-colored
         H   Corolla straw- or salmon-colored; bumps on
              upper part of fruit ending in bristles or
              hooks; limited to elevations above 3000'
                              Sanicula saxatilis
                              Rock Sanicle; CC(MD), SCl(MH); 1b
         HH Corolla yellow; bumps on fruit not ending in
              bristles or hooks; not limited to higher
              elevations   *Sanicula tuberosa*
                              Tuberous Sanicle
CC Fruit without stiff bristles or conspicuous, tall bumps (but it
    is ribbed lengthwise, except sometimes in *Apiastrum
    angustifolium*, and it may be hairy)
    D   Fruit about 1.5 mm long and wide, heart-shaped, owing to a
         cleft at the base, the ribs poorly developed, often
         indistinct; annual, rather delicate, freely branched, with
         leaves (some of these opposite) scattered along the stems;
         primary umbels usually sessile in the axils of upper leaves,
         the secondary umbels with only a few flowers (corolla white)
                              Apiastrum angustifolium (*Plate* 70)
                              Wild Celery; La-s
    DD Fruit more than 5 mm long, longer than wide, not at all
         heart-shaped, the ribs prominent; plants perennial, stout,
         the leaves mostly basal; primary umbels on distinct
         peduncles, the secondary umbels sometimes with many flowers
         E   All ribs of the fruit at least slightly extended as wings;
              plants not hairy; bracts of secondary umbels not more
              than 1 mm wide (corolla yellow; leaves gray-green, the
              ultimate divisions about 1 mm wide)
                              Cymopterus terebinthinus var. *californicus*
                              [*Pteryxia terebinthina* var. *californica*]
                              Cymopterus; Sn-n
         EE Only the 2 ribs at the opposite edges of the flattened
              fruit extended as wings; plants usually at least somewhat
              hairy; bracts of secondary umbels usually at least 2-3 mm
              wide, sometimes partly joined
              F   Corolla generally white or greenish, but sometimes
                   pale yellow or purple (ultimate divisions of leaves
                   not more than 7 mm long)
                   G   Corolla and body of fruit not hairy; mature fruit
                        sometimes more than 3 times as long as wide
                        (pedicels up to 14 mm long)   *Lomatium macrocarpum*
                              Sheep Parsnip; SLO-n
                   GG   Corolla and body of fruit hairy; mature fruit
                        rarely more than twice as long as wide (pedicels
                        up to 20 mm long)   *Lomatium dasycarpum* (*Plate* 5)
                              Hog Fennel

```
FF Corolla bright yellow or purple
   G   Plants usually with 2-3 leaves along the stem;
       wings of the mature fruit about as wide as the
       body of the fruit (corolla bright yellow;
       widespread)                Lomatium utriculatum (Plate 5)
                                                   Spring-gold
   GG Plants with not more than one leaf along the stem;
       wings of the mature fruit not quite so wide as the
       body of the fruit
       H   Plants either not hairy or only slightly hairy;
           corolla yellow; mature fruit less than twice
           as long as wide; ultimate divisions of leaves
           up to 15 mm long          Lomatium caruifolium
                                Carawayleaf Lomatium; Me-SLO
       HH Plants grayish, decidedly hairy; corolla
           purple; mature fruit fully twice as long as
           wide; ultimate divisions of leaves not more
           than 7 mm long (bracts fan-shaped, sometimes
           toothed; rays of primary umbels 7-12; pedicels
           2-8 mm long; on serpentine)
                           Lomatium ciliolatum var. hooveri
                                  Hoover Lomatium; Na; 4
```

Apiaceae, Subkey 2: Leaves not like those of parsley, with relatively few ultimate leaflets or deeply separated lobes, these divisions usually either at least 1 cm wide and 1.5 cm long or, if narrower, several to many times as long as wide. (In *Perideridia californica*, only the terminal leaflet, and sometimes the terminal lobes of the lateral leaflets, are several times as long as wide.)

```
A   Leaves deeply lobed in a palmate, pinnate, or bipinnate pattern, but not
    truly compound (bracts beneath both primary and secondary umbels)
    B   Lower leaves lobed in a pinnate or bipinnate pattern; corolla usually
        purple, sometimes yellow (fruit covered with hooked prickles; plant
        sometimes almost stemless; widespread)
                                       Sanicula bipinnatifida (Plate 5)
                                                   Purple Sanicle
    BB Lower leaves lobed in a palmate pattern; corolla yellow
        C   Margins of leaf lobes smooth; maturing fruit prickly, but only in
            the lower third                    Sanicula maritima
                                          Adobe Sanicle; SF-SLO; 1b
        CC Margins of leaf lobes noticeably toothed; maturing fruit prickly
            to the base
            D   Leaves yellow throughout the normal growing season
                (prostrate; leaf lobes with large, sharp marginal teeth;
                umbels compact, 1-1.5 cm wide; coastal)
                                       Sanicula arctopoides (Plate 5)
                                         Footsteps-of-Spring; Mo-n
            DD Leaves green during the normal growing season
                E   Plants typically branching well above the base; primary
                    lobes of leaves shallowly divided, the margins evenly
                    toothed (widespread)   Sanicula crassicaulis (Plate 5)
                                                       Snakeroot
                EE Plants typically branching only at the base; primary lobes
                    of leaves often deeply divided, the margins with
                    irregular, bristle-tipped teeth (coastal)
                                                Sanicula laciniata
                                              Coast Sanicle; SLO-n
AA Leaves truly compound, the primary divisions, at least, clearly separate
   to the rachis and thus considered to be leaflets
   B   Leaves with only 3 leaflets, these usually more than 10 cm long and
       deeply lobed (blades sometimes more than 40 cm wide; umbels up to
       about 25 cm wide; corolla white; robust plants of moist habitats,
       especially near the coast; contact with the leaves or stems,
       followed by exposure to sunlight, may lead to a skin irritation;
       widespread)                       Heracleum lanatum (Plate 5)
                                               Cow Parsnip; Mo-n
   BB Leaves with more than 3 primary leaflets, these arranged in a pinnate
       pattern (the leaves may also be partly or wholly bipinnate)
       C   Leaves not more than once pinnate, although some of the leaflets
           may be deeply divided
           D   Corolla yellow or red; basal leaves often more than 50 cm
               long, the leaflets sometimes more than 10 cm long; lower
               leaflets often deeply divided into 3 or more secondary lobes;
               not aquatic                      Pastinaca sativa
                                                   Parsnip; eu
```

DD Corolla white; basal leaves not often more than 30 cm long,
the leaflets generally less than 5 cm long; leaflets toothed,
but not deeply divided; in shallow water or very wet places
Berula erecta
Water Parsnip
CC Leaves at least partly bipinnate, some of the primary leaflets
being completely divided into secondary leaflets (these in turn
usually deeply divided)
  D Bracts present beneath both the primary and secondary umbels
    E Fruit with hooked bristles (corolla yellow; widespread)
Sanicula bipinnata
Poison Sanicle; Me-s

    EE Fruit without hooked bristles
      F Nearly all leaflets at least 10 times as long as wide,
         and smooth-margined
        G Styles of pistil diverging widely, but not curving
           downward along the sides of the conical elevations
           from which they originate   *Perideridia kelloggii*
Kellogg Yampah; Mo-n
        GG Styles of pistil curving downward along the sides
           of the conical elevations from which they
           originate   *Perideridia gairdneri* (*Plate* 71)
Gairdner Yampah; 4
      FF Most leaflets only a few times as long as wide, and
         with lobed or toothed margins. (In *Perideridia
         californica*, however, the terminal leaflet, and
         sometimes the terminal lobes of the lateral leaflets,
         are long and narrow, thus different from the other
         leaflets and lobes.)
        G Rays of primary umbels 5-10 (these 3-8 cm long);
           terminal leaflet, and sometimes the terminal lobes
           of the lateral leaflets, long, narrow, and smooth-
           margined, the other leaflets broader and shallowly
           lobed   *Perideridia californica*
California Yampah; CC-SLO
        GG Rays of primary umbels more than 10; all leaflets
           essentially alike, with toothed margins (some of
           them may also be lobed, however)
          H Rays of primary umbels up to 3 cm long; primary
             lateral veins of leaflets mostly directed into
             the marginal teeth; mature fruit more than 1.5
             times as long as wide, almost cylindrical;
             styles more than half as long as the fruit
             (often fully as long); stems reclining, but
             the flowering shoots upright
Oenanthe sarmentosa
Pacific Oenanthe
          HH Rays of primary umbels up to 6 cm long; primary
             lateral veins of leaflets mostly directed
             toward the clefts between the marginal teeth;
             mature fruit not much longer than wide; styles
             much less than half as long as the fruit;
             stems upright, not reclining (both species are
             poisonous; handle with care)
            I Plants of coastal salt marshes; ribs of
               fruit narrower than the darker intervals
               between them   *Cicuta maculata*
var. *bolanderi* [*C. bolanderi*]
Bolander Water-hemlock; Ma, Sl, CC
            II Plants of freshwater habitats; ribs of
               fruit broader than the darker (usually
               reddish brown) intervals between them
Cicuta douglasii (*Plate* 70)
Douglas Water-hemlock
  DD Bracts absent beneath the primary umbels, and sometimes
     absent beneath the secondary umbels
    E Ultimate divisions of leaves (that is, the leaflets or
       well separated lobes of the leaflets) slender, at least
       several times as long as wide, and often with nearly
       parallel margins for much of their length

F   Bracts, when present beneath the secondary umbels, not
    especially slender (usually not more than 4 times as
    long as wide, and sometimes widest above the middle,
    and often touching or even overlapping one another),
    often toothed; mostly around drying pools and in
    grassland (flowers usually yellow; sometimes
    purplish)   *Lomatium caruifolium*
                            Carawayleaf Lomatium; Me-SLO
FF Bracts, when present beneath the secondary umbels,
    slender, sometimes almost hairlike, not toothed;
    mostly on dry slopes
    G   Corolla yellow   *Lomatium marginatum* var. *marginatum*
                            Yellow Hartweg Lomatium; Na-n
    GG Corolla purple   *Lomatium marginatum* var. *purpureum*
                            Purple Hartweg Lomatium; Na-n
EE Ultimate divisions of leaves not especially slender, and
    not often more than 4 times as long as wide
    F   Corolla yellow or greenish yellow
        G   Some leaves present on the stem (although they may
            be reduced compared to the basal leaves)
            (secondary umbels with bracts)
            H   Rays of primary umbels 2-5; fruit elongated,
                more than 4 times as long as wide (mostly in
                woods)   *Osmorhiza brachypoda*
                            California Sweet-cicely; CC-s
            HH Rays of primary umbels more than 10; fruit not
                more than 3 times as long as wide
                                          Lomatium californicum
                                          Chu-chu-pate
        GG Leaves entirely basal
            H   Bracts absent beneath both the primary and
                secondary umbels; ultimate leaflets neither
                lobed nor obviously toothed (peduncles
                noticeably swollen just below the point where
                the rays of the primary umbels diverge)
                               Lomatium nudicaule (*Plate* 71)
                                          Pestle Parsnip
            HH Bracts present beneath the secondary umbels;
                ultimate leaflets toothed and often lobed
                I   Rays of primary umbels 10-20 (ultimate
                    leaflets often as wide as, or wider than,
                    long)   *Lomatium repostum*
                                          Napa Lomatium; Na; 4
                II Rays of primary umbels 5-10
                    J   Bracts beneath secondary umbels usually
                        longer than the flower pedicels; mature
                        fruit 4-7 mm long   *Tauschia hartwegii*
                                          Hartweg Tauschia; CC-s
                    JJ Bracts beneath secondary umbels about
                        the same length as, or shorter than,
                        the flower pedicels; mature fruit 3-5
                        mm long   *Tauschia kelloggii*
                                          Kellogg Tauschia; SCr-n
    FF Corolla white or greenish white
        G   Flowers or fruits in each secondary umbel not more
            than 10; fruit elongated, more than 4 times as
            long as wide (bracts absent below the secondary
            umbels; mostly in woods; widespread)
                               Osmorhiza chilensis (*Plate* 5)
                                          Sweet-cicely
        GG Flowers or fruits in each secondary umbel usually
            more than 20; fruit not more than 3 times as long
            as wide
            H   Rays of primary umbels 7-20, not more than 5 cm
                long
                I   Bracts absent beneath the secondary umbels;
                    mature fruit about 1.5 mm long; pedicels 1-
                    6 mm long   *Apium graveolens*
                                          Celery; eua
                II Bracts present beneath the secondary
                    umbels; mature fruit 3.5-5 mm long;
                    pedicels 5-10 mm long
                               Ligusticum apiifolium (*Plate* 71)
                                          Lovage; SM-n

HH Rays of primary umbels 25-45, up to 12 cm long
(bracts present below secondary umbels)
    I   Underside of leaves usually woolly; coastal
                        Angelica hendersonii (*Plate* 70)
                            Henderson Angelica; Mo-n
    II Underside of leaves sometimes hairy, but
        not woolly; inland   *Angelica tomentosa*
                            California Angelica; SCr-n

Apocynaceae--Dogbane Family

Commonly cultivated plants that belong to the Dogbane Family are the
Oleander, *Nerium oleander*, a shrub or small tree of Mediterranean Europe,
and 2 species of *Vinca*, called Periwinkle, which are ground covers that
sometimes become so well established along roadsides that they look wild.
The stems of *Apocynum cannabinum* have long fibers that native Americans
have used for making ropes and cords.

The family as a whole consists of perennials with opposite, simple
leaves. Both the calyx and corolla have 5 lobes, but the lobes of the calyx
are sometimes separated nearly to the base, whereas the corolla has a
substantial tube. The 5 stamens, alternating with the corolla lobes, are
attached to the tube. The pistil, with a single style, is usually free or
almost free of the calyx. As the fruit ripens, the 2 halves of it separate,
each half forming a pod that splits open after it has dried. In some
genera, including *Apocynum*, the sap is milky.

A   Flowers solitary; corolla about 4 cm wide, blue or violet, the 5 lobes
    all skewed slightly in the same direction; rooting at the nodes or at
    the tips of the stems, thus spreading and forming large masses,
    especially on somewhat shaded banks (widely cultivated as a ground
    cover, and often escaping)   *Vinca major* (*Plate* 72)
                            Greater Periwinkle; eu
AA Flowers in panicles; corolla less than 1 cm wide, greenish white, white,
    or pink, the lobes not skewed; bushy, the stems not spreading and taking
    root (pods of fruit up to 10 cm long, releasing seeds with tufts of
    white hair; the sap is somewhat poisonous)
    B   Corolla white or greenish white, less than 5 mm long; leaves often
        more than 3 times as long as wide, and not drooping
                        Apocynum cannabinum [includes
        A. cannabinum var. *glaberrimum* and *A. sibiricum* var. *salignum*]
                            Indian-hemp
    BB Corolla pink (sometimes very pale), mostly 5-10 mm long; leaves
        rarely more than twice as long as wide, usually drooping
            Apocynum androsaemifolium [includes *A. medium* var. *floribundum*,
                and *A. pumilum* vars. *pumilum* and *rhomboideum*] (*Plate* 72)
                            Bitter Dogbane; Na, Sn-n

Araliaceae--Ginseng Family

The Araliaceae, which has about 50 genera, provides many species that
are cultivated primarily for their leaves. English Ivy (*Hedera helix*), a
native of Eurasia, is known to almost everyone, but the family also
includes the numerous house plants called aralias, as well as *Fatsia
japonica*, commonly grown outdoors in milder parts of California. The
several species of ginseng, dug for their roots which have medicinal
properties, belong to the genus *Panax*. They are natives of Asia and eastern
North America.

The flowers are usually small and borne in umbels or other types of
clusters. There are generally 5 petals and 5 stamens. The sepals are small
and may be absent altogether. The fruit, which is at least slightly fleshy,
develops below the level where the petals and sepals (if present) are
attached.

The only member of the family growing wild in our area is *Aralia
californica* (*Plate* 6), called Elk-clover. It is found in moist, tree-shaded
canyons, where it reaches a height of almost 3 m. In spite of its large

size, however, it qualifies as a herb, rather than a shrub, for its stems do not become woody. The large leaves are divided ternately, and each division is pinnately compound, with 3-5 leaflets, these often more than 10 cm long and slightly notched at the base. The inflorescences, with hundreds of small flowers, are usually 30-40 cm long. The fruit is about 4 mm wide and dark when ripe.

Aristolochiaceae--Pipevine Family

Two very different-looking plants, *Aristolochia californica* and *Asarum caudatum*, represent the Pipevine Family in our region. Nevertheless, they share several features. The leaves are alternate and have heart-shaped blades. The flowers lack a corolla, but the calyx, part of which is fused to the fruiting portion of the pistil, resembles a corolla because of its coloration. In *Aristolochia californica*, furthermore, the calyx is a curved, 2-lipped tube. The fruit has 6 seed-producing chambers that correspond to 6 styles on the pistil, although this is obscure in *Aristolochia* because the styles are fused into a single column. The stamens--6 in *Aristolochia*, 12 (of 2 different lengths) in *Asarum*--are joined to the style complex.

Both of our species are interesting and useful subjects for native gardens and are generally available at plant sales. *Asarum caudatum*, called Wild-ginger, makes an excellent evergreen ground cover for a shady woodland. It cannot, however, withstand an extended drought. *Aristolochia californica*, Dutchman's-pipe, a deciduous woody vine, requires the support of shrubs over which it is allowed to climb.

A Small herbs that usually form a ground cover in moist woods; leaf blades wider than long, with a deep cleft at the base; calyx tube cup-shaped, the lobes equal and several times as long as wide, purplish brown (plants with aroma of ginger)   *Asarum caudatum* (*Plate* 6)
                                          Wild-ginger; SCl (SCM)-n
AA Woody vines that climb over shrubs; leaf blades mostly longer than wide, with only a shallow indentation at the base; calyx tube deep, curved, thus somewhat resembling the bowl of a smoker's pipe, the lobes not quite equal and not much longer than wide, greenish with purple veins
                                Aristolochia californica (*Plate* 6)
                                          Dutchman's-pipe; Mo-n

Asclepiadaceae--Milkweed Family

Milkweeds, so called because of their white sap, are perennial plants with opposite or whorled leaves. The flowers, in which both the calyx and corolla are 5-lobed, are rather intricate. The 5 stamens, attached near the base of the corolla, are united to form a tube around the pistil. Just outside this tube is a crownlike structure whose 5 lobes are called hoods. In most of the species of our only genus, *Asclepias*, there is a further complication: the inner face of each hood has an outgrowth called a horn. The fruit dries as it matures, and on cracking open it releases seeds that have tufts of long, silky hairs.

A species from the Near East, *A. syriaca*, has been used medicinally; its milky sap is thought to be helpful in cases of asthma. *Asclepias incarnata*, the Swamp Milkweed, and *A. tuberosa*, the Butterfly Weed, both from eastern North America, have been widely cultivated for their flowers, which are, respectively, reddish purple and orange. Some wildflower enthusiasts grow milkweeds, including our native *A. speciosa*, because they are the food of the caterpillars of the Monarch Butterfly (*Danaus plexippus*).

A Stems reclining, and usually undulating or zigzagging, distinctly flattened; corolla lobes not more than 2.5 mm long (flowers reddish purple; on serpentine outcrops)   *Asclepias solanoana*
                                Serpentine Milkweed; Na, Sn-n; 4

AA Stems upright, not flattened; corolla lobes usually more than 5 mm long
   (slightly shorter in *Asclepias fascicularis*) (Note: the corolla lobes
   are narrower than the calyx lobes, and, like the latter, they are turned
   back when the flowers open.)
   B   Flower color light, generally greenish, whitish, or cream, but often
       tinged with pink or purple; each hood (the hoods are saclike
       outgrowths attached to the top of the tube formed by the united
       filaments of the stamens) with a hornlike structure within it
       C   Hoods slender, extending far above the stamens and stigma; tube
           formed by the united filaments of the stamens scarcely evident
                                                      Asclepias speciosa
                                                      Showy Milkweed; Sl-n
       CC Hoods not extending appreciably above the stamens and stigma;
           tube formed by the filaments of the stamens well developed
           D   Leaves and stems dark green, at most only slightly hairy;
               leaves mostly at least 6 times as long as wide, and rarely
               more than 1 cm wide   *Asclepias fascicularis* (*Plate* 6)
                                                      Narrowleaf Milkweed
           DD Leaves and stems whitish, very hairy; leaves less than 3
               times as long as wide, and generally at least 3 cm wide
                                                      Asclepias eriocarpa
                                                      Kotolo; Me-s
   BB Flowers mostly richly colored (reddish purple or maroon); hoods
       without hornlike structures
       C   Leaves and stems whitish, owing to abundant woolly hairs; leaves
           with short petioles, not clasping the stem; hoods with a
           slitlike opening extending from the top down to about the middle
           of the outer side (widespread)   *Asclepias californica*
                       [includes *A. californica* ssp. *greenei*] (*Plate* 6)
                                                      California Milkweed
       CC Leaves and stems green, with only very short hairs; leaves
           without petioles, clasping the stem; hoods open at the top and
           on the inner side   *Asclepias cordifolia*
                                                      Purple Milkweed; Sl-n

Asteraceae (Compositae)--Sunflower Family

In California, as well as in many other parts of the world, the
Asteraceae has more species than any other family of flowering plants.
Keying an unfamiliar member of this group to genus and species may
therefore be a longer process and require more attention to details than
would be necessary to identify a mint or a saxifrage, for example. But
recognition of the family itself is easy because of the distinctive
structure of the flower heads. Each head consists of several to many small
flowers attached to a disk-shaped, conical, or concave receptacle. Below
and around this receptacle are leaflike bracts called phyllaries. When you
are served a cooked artichoke, you nibble off the more nutritious portions
of the phyllaries, then scrape away the flowers and eat the receptacle.

There are 2 main types of flowers: those in which the corolla is tubular
and usually 5-lobed, and those in which the corolla is drawn out into a
petal-like structure. In a daisy, flowers with the tubular type of corolla
occupy the central part of the head and are called disk flowers; those with
the petal-like type of corolla are distributed around the margins, and are
called ray flowers. In many members of the family there are only disk
flowers or only flowers of the ray-flower type. Thus it is convenient to
break up the large key for identifying plants of the family Asteraceae into
three main groups: those that have only ray flowers, those that have only
disk flowers, and those that have both.

Sepals as we know them in other families are absent or are replaced by
scales or hairs that form what is called a pappus. If present, the pappus
usually facilitates dispersal of the small, dry fruit by wind, animals, or
other means. The fruit, called an achene, is below the pappus and corolla.
It is fused tightly to the single seed inside it. When stamens are present,
they are attached to the corolla, and their anthers are united into a tube
around the style. Certain flowers of the head may lack stamens; others may
lack a pistil, and some may lack both.

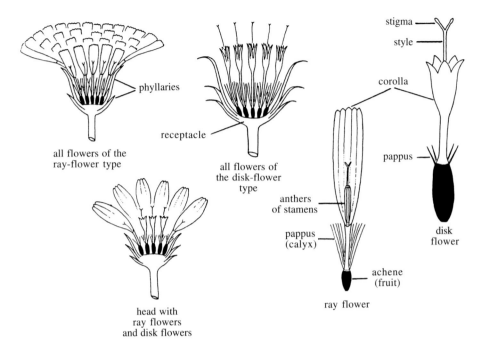

all flowers of the
ray-flower type

all flowers of
the disk-flower
type

head with
ray flowers
and disk flowers

ray flower

Of the numerous species of the family growing in our area, many are
introduced. Some of the introductions are among the most aggressive of
weeds. They fill up the hard-packed dirt of vacant lots and roadsides,
establish themselves in lawns and gardens, and create problems for farmers.
Yet the family has given us hundreds of kinds of garden flowers, food
plants, and sources of valuable oils, dyes, and medicines.

Perennial species that are good subjects for gardens devoted to
indigenous plants are sold by the California Native Plant Society, and also
by some commercial nurseries. The following are frequently available.

Baccharis pilularis,
  Coyotebrush (especially
  valuable for erosion control)
Erigeron glaucus, Seaside Daisy
Eriophyllum confertiflorum,
  Golden Yarrow

Heterotheca sessiliflora ssp.
  bolanderi, Golden Aster
Tanacetum camphoratum, Dune
  Tansy

Among the many annuals that will effectively cover bare ground and make
a colorful showing, *Coreopsis calliopsidea* (Leafystem Coreopsis), *Lasthenia
californica* (California Goldfields), and *Layia platyglossa* (Tidytips), are
of particular interest. Seeds are available at plant sales, in the stores
of botanical gardens, and from commercial sources.

The first step in identifying a member of the Asteraceae is to use the
preliminary key, which will lead you to one of the three main divisions of
the family.

A   Flowers in each head either all of the ray-flower type or all of the
    disk-flower type
    B   Flowers all of the ray-flower type; usually with milky sap
                                  Asteraceae, Division I (p. 84)
    BB Flowers all of the disk-flower type (or, if rays are present, these
        are either less than 2 mm long or do not extend more than 2 mm
        beyond the phyllaries); without milky sap
                                  Asteraceae, Division II (p. 87)
AA At least some of the flower heads with both ray flowers and disk flowers
                                  Asteraceae, Division III (p. 98)

Asteraceae, Division I: Flowers all of the ray-flower type; usually with milky sap
A Mature plants with only 1-2 flower heads at the end of unbranched stems
   B Corollas purple (widespread)   *Tragopogon porrifolius* (*Plate* 15)
                                                                Salsify; eu
  BB Corollas orange, yellow, or white
    C Pappus consisting of straight bristles at the end of a slender
       beak on the achene
        D Leaves not toothed or lobed, scattered along the flowering
           stems (annual or biennial)   *Tragopogon dubius*
                                           Yellow Salsify; eu
      DD Leaves usually toothed or lobed, and mostly basal
         E Achenes with short spines at the upper end, just below the
           slender beak (beaks 5-10 mm long; widespread)
                      Taraxacum officinale [includes *T. laevigatum*]
                                    Common Dandelion; eu
        EE Achenes without spines
          F Leaf blades with narrow lobes or teeth that are bent
            slightly toward the base of the leaves (perennial; at
            elevations above 2500')   *Agoseris retrorsa*
                                  Spearleaf Dandelion
          FF Lobes or teeth, if present on the leaf blades, not
            bent toward the base of the leaves
             G Leaf blades 5-10 cm long; peduncles of flower heads
               often more than 2-3 times as long as the leaves;
               annual   *Agoseris heterophylla*
                                     Annual Dandelion
            GG Leaf blades 10-25 cm long; peduncles of flower
               heads usually less than twice as long as the
               leaves; perennial
               H Phyllaries in 3 series, those of the uppermost
                  series taller and narrower than the others;
                  flower heads 2.5-4 cm high
                      Agoseris grandiflora (*Plate* 7)
                            California Dandelion
              HH Phyllaries in 2 series, mostly equal in size
                and shape; flower heads 1.5-2 cm high
                (coastal)
                 I Beak of achene slightly longer than the
                   rest of the achene
                    Agoseris apargioides var. *apargioides*
                            Seaside Dandelion
                 II Beak of achene half as long as the rest of
                  the achene
                    Agoseris apargioides var. *eastwoodiae*
                [*A. apargioides* var. *eastwoodae*] (*Plate* 6)
                               Coast Dandelion
  CC Pappus consisting of flattened scales that end in bristles
    (scales sometimes absent in *Microseris douglasii* ssp. *tenella*)
    D Some phyllaries 2-3 cm long; flower heads upright; achene and
       pappus together often more than 2.5 cm long (up to 4 cm long)
       (annual, up to 60 cm tall; widespread)
           Uropappus lindleyi [*Microseris lindleyi*] (*Plate* 15)
                                      Silverpuffs
    DD Phyllaries not more than 1.5 cm long; flower heads nodding in
      the bud stage, upright when fully open; achene and pappus
      together usually less than 2.5 cm long
      E Flower heads often more than 2 cm wide; perennials up to
        80 cm tall
        F Pappus white, 12-20 mm long; in the Coast Ranges
                            Microseris laciniata
                    Cutleaf Microseris; Na, Sn-n
        FF Pappus brown or yellowish, 8-13 mm long; coastal
                           Microseris paludosa
                     Coast Microseris; Sn-Mo
      EE Flower heads usually less than 1.5 cm wide (except
        sometimes in *Microseris douglasii*); annuals up to 60 cm
        tall, but mostly less
        F Leaves mostly basal, but usually some on the lower
          part of the stem, these not pinnately lobed (pappus
          white to brownish, 14-19 mm long; achenes 6-8 mm long;
          up to 40 cm tall; coastal)
             Stebbinsoseris decipiens [*Microseris decipiens*]
                    Santa Cruz Stebbinsoseris; Ma; 1b

FF Leaves entirely basal, often pinnately lobed
    G   Pappus 8-18 mm long, white to brownish, the scales
        of its lower portion 4-11 mm long; flower heads
        with up to 50 flowers; up to 30 cm tall (inland)
                                    Microseris acuminata
                            Sierra Foothills Microseris; Al-n
    GG Pappus 4-14 mm long, silvery to blackish, the
        scales of its lower portion 1-6 mm long; flower
        heads with more than 100 flowers; up to 60 cm tall
        H   Mainly coastal, in sandy soil; corollas yellow
            or orange (pappus scales 1-4 mm long; achene
            widest near the middle)   *Microseris bigelovii*
                                        Bigelow Microseris
        HH Mainly inland, in serpentine and clay soils;
            corollas yellow or white
            I   Achene widest near the tip; pappus scales
                1-6 mm long   *Microseris douglasii*
                                ssp. *douglasii* (*Plate* 73)
                                        Douglas Microseris
            II Achene widest near the middle; pappus
                scales, if present, less than 1 mm long
                            Microseris douglasii ssp. *tenella*
                            Delicate Douglas Microseris; SFBR-s
AA Mature plants with a few to many flower heads on or near the end of
branched or unbranched stems
    B   Corollas purple, blue, pink, or white
        C   Corollas bright blue (occasionally pink or white); flower heads
            at least 3 cm wide when open (they close in the afternoon)
            (flower heads scattered along the stems; up to 1 m tall;
            widespread)   *Cichorium intybus* (*Plate* 9)
                                                    Chicory; eu
        CC Corollas purple, pale blue, pink, or white; heads less than 3 cm
            wide
            D   Flower heads 2-3 cm wide, solitary at the end of branches
                (corollas white; upper leaves reduced, sessile; up to more
                than 1 m tall)   *Rafinesquia californica*
                                            California Chicory
            DD Flower heads 1-1.5 cm wide, not solitary at the end of
                branches
                E   Flower heads almost sessile along the leafless branches
                    (corollas pink or white; usually up to 1 m tall, but
                    sometimes much taller; widespread)
                                        Stephanomeria virgata (*Plate* 14)
                                                Tall Stephanomeria
                EE Flower heads on peduncles at the end of branches
                    F   Lower leaves not distinctly toothed or lobed (corollas
                        white; leaves mostly basal, the petioles winged; less
                        than 1 m tall; widespread)
                                        Hieracium albiflorum (*Plate* 73)
                                                    Hawkweed
                    FF Lower leaves toothed or lobed
                        G   Leaves not mainly basal, not hairy, the upper ones
                            sessile; corollas pale blue or cream; often more
                            than 1.5 m tall   *Lactuca biennis*
                                            Tall Blue Lettuce; na; Ma-n
                        GG Leaves mainly basal with tufts of woolly hairs on
                            the underside near the margin; corollas white or
                            pinkish; not more than 0.5 m tall
                                                Malacothrix floccifera
                                                    Woolly Malacothrix
    BB Corollas yellow or greenish yellow
        C   Leaves mostly basal (when leaves are present along the stem, only
            a few, if any, are as substantial as those at the base)
            D   Leaves bristly, rough to the touch
                E   Flower heads 2-3 cm high; leaves up to 20 cm long;
                    flowering stems sometimes branched at least once; up to
                    80 cm tall (widespread) *Hypochaeris radicata* (*Plate* 12)
                                                Rough Cat's-ear; eu
                EE Flower heads up to 1.5 cm high; leaves up to 12 cm long;
                    flowering stems not branched; up to 30 cm tall
                                    Leontodon taraxacoides [*L. leysseri*]
                                                Hawkbit; eu; SCl-n
            DD Leaves not bristly, although they may be hairy

E   Pappus either lacking or consisting of. 1-8 bristles (less
    than 0.5 m tall)
    F   Stems branching from the base; leaves with tufts of
        woolly hairs on the underside near the margin
                                          Malacothrix floccifera
                                             Woolly Malacothrix
    FF Stems not branching from the base (but there may be
        branches above); leaves not hairy
                                          Malacothrix clevelandii
                                          Cleveland Malacothrix
EE Pappus consisting of more than 25 bristles
    F   Involucre 10-16 mm high
        G   Largest leaves 15-40 cm long, pinnately lobed; up
            to 70 cm tall (at elevations above 1500')
                                          Crepis intermedia
                                    Intermediate Hawk's-beard
        GG Largest leaves 2-6 cm long, usually not lobed; up
            to 40 cm tall (opening of flower heads limited to
            periods of bright sunshine)   *Hypochaeris glabra*
                                          Smooth Cat's-ear; eu
    FF Involucre usually 5-9 mm high (sometimes 11 mm)
        G   Leaves toothed or lobed, the lobe at the end of the
            blade usually at least 2 or 3 times as long as
            wide; up to 90 cm tall (widespread)
                                          Crepis capillaris (*Plate* 10)
                                          Smooth Hawk's-beard; eu
        GG Leaves with definite lobes, the one at the end of
            the blade not much longer than wide; up to 35 cm
            tall   *Crepis bursifolia*
                                    Italian Hawk's-beard; me; SFBR
CC With at least some substantial leaves above the basal portion of
    the plant
    D   Margin of the leaves, and sometimes other parts of the plant,
        prickly or bristly (upper leaves clasping the stem)
        E   Phyllaries widest at the middle; leaves mostly toothed,
            with whitish spots (up to 80 cm tall; widespread)
                                          Picris echioides (*Plate* 13)
                                          Bristly Ox-tongue; eu
        EE Phyllaries widest at the base; leaves toothed and often
            lobed, but without whitish spots
            F   Flower heads 20-25 mm high; stems and phyllaries
                prickly; less than 0.5 m tall   *Urospermum picroides*
                                          Urospermum; eu; Al
            FF Flower heads 10-15 mm high; stems and phyllaries not
                prickly (although they may have glandular hairs); up
                to more than 1 m tall
                G   Flower heads about as wide as high; veins of leaves
                    not prickly (but the margin of the leaves are
                    prickly) (widespread)   *Sonchus asper*
                                          Prickly Sow Thistle; eu
                GG Flower heads half as wide as high; veins of leaves
                    prickly   *Lactuca virosa*
                                          Wild Lettuce; eu; SFBR
    DD Plants not prickly
        E   Upper leaves clasping the stem; often more than 1 m tall
            F   Stems (one to several originating at the base) not
                branched (except for the short peduncles of the flower
                heads, which are in narrow racemes up to 30 cm long)
                                          Lactuca saligna
                                          Willow Lettuce; eu; SFBR
            FF Upper portion of flowering stems branched
                G   Flower heads 4 times as high as wide before opening
                    (upper leaves with fine teeth, lower leaves
                    sometimes lobed; widespread)
                                          Lactuca serriola (*Plate* 13)
                                          Prickly Lettuce; eu
                GG Flower heads only 1-3 times as high as wide before
                    opening
                    H   Plants with a well developed basal cluster of
                        leaves; most stem leaves, especially in the
                        upper part of the plant, much reduced and
                        narrow (widespread)
                                          Crepis capillaris (*Plate* 10)
                                          Smooth Hawk's-beard; eu

HH Plants without a basal cluster of leaves; stem
   leaves well developed and often lobed (and
   with fine teeth; widespread) *Sonchus oleraceus*
                                 Common Sow Thistle; eu
EE Upper leaves not clasping the stem; less than 1 m tall
  F Leaves not more than 1 cm wide (pappus consisting of
    1-6 bristles; leaves sometimes toothed or lobed, not
    hairy)   *Malacothrix clevelandii*
                               Cleveland Malacothrix
  FF Some leaves over 2 cm wide
    G Pappus consisting of many more than 10 bristles
      (leaves lobed or toothed; plants with long,
      glandular hairs)   *Crepis monticola*
                     Mountain Hawk's-beard; SCl-n
    GG Pappus absent
      H Stems with glandular hairs at the base; flower
        heads 5-10 mm high; upper leaves with obvious
        teeth; lower leaves lobed as well as toothed
                         Lapsana communis
                                     Nipplewort; eu
      HH Stems without glandular hairs at the base
        (there may be hairs, however); flower heads
        10-13 mm high; upper leaves scarcely toothed;
        lower leaves lobed but not toothed
                         Rhagadiolus stellatus
                            Rhagadiolus; eu; Na

Asteraceae, Division II: Flowers all of the disk-flower type (or, if rays
are present, these are either less than 2 mm long or do not extend more
than 2 mm beyond the phyllaries); without milky sap
A Phyllaries (or the involucres, when individual phyllaries are not
  distinct) with sharp spines (see note in choice AA regarding *Ambrosia*,
  Ancistrocarphus, and *Soliva*, which are keyed there)
  B Phyllaries (or an involucre that has no distinct phyllaries) with
    hooked spines, but plants as a whole not spiny (except *Xanthium*
    spinosum, which is generally spiny)
    C Plants with trios of spines on the stems, the spines straw-
      colored and 1.5-2 cm long; leaves 3-8 cm long, dark green on the
      upper surface, grayish on the underside (rather distinctly 3-5
      lobed; widespread)   *Xanthium spinosum (Plate 74)*
                                  Spanish Thistle; eu
    CC Plants without spines on the stems; larger leaves much more than
      5 cm long, not especially darker on the upper surface than on
      the underside (usually toothed or lobed)
      D Flower heads decidedly higher than wide (sometimes nearly
        twice as high as wide); phyllaries fused, thus not distinct;
        leaves up to 20 cm long   *Xanthium strumarium*
          [includes *X. strumarium* vars. *canadense* and *glabratum*]
                                      Cocklebur
      DD Flower heads about as high as wide; phyllaries distinct;
        leaves up to 30 cm long
        E Blade of larger leaves usually somewhat pointed at the
          tip; petiole of larger leaves hollow, and without angular
          ridges   *Arctium minus (Plate 7)*
                                Common Burdock; eu
        EE Blade of larger leaves usually rounded at the tip; petiole
          of larger leaves solid, and usually with angular ridges
                             Arctium lappa
                             Great Burdock; eu
BB Phyllaries with straight spines, and plants as a whole spiny (except
  in *Cynara scolymus* and species of *Centaurea*, which are not spiny
  except for the phyllaries)
  C Leaves with white mottling (flower heads up to 5 cm wide, with
    long spines; corollas purple; up to 2 m tall; widespread)
                        Silybum marianum (Plate 14)
                                Milk Thistle; me
  CC Leaves without white mottling (although the leaf veins may be
    whitish)
    D Pappus bristles, if present, not feathery
      E Leaves without spines; basal leaves (sometimes absent at
        flowering time) lobed and sometimes toothed, the upper
        ones usually smooth-margined
        F Corollas usually purple or pink, sometimes white
          (spines on phyllaries 10-25 mm long, stout)

G   Basal leaves 15-25 cm long, with a few wide lobes,
these usually not toothed; corollas usually pink,
sometimes white; achenes with a distinct pappus
Centaurea iberica
Iberian Knapweed; me; Al
GG Basal leaves less than 10 cm long, with many narrow
lobes, these usually toothed; corollas purple;
achenes without a distinct pappus
Centaurea calcitrapa
Purple Star Thistle; eu; Sl
FF Corollas yellow (leaves less than 15 cm long, often
with a few lobes, these usually not toothed; common)
G   Spines of phyllaries purple, 5-10 mm long, slender
Centaurea melitensis
Tocalote; me
GG Spines of phyllaries yellow, 10-25 mm long, stout
Centaurea solstitialis (*Plate* 9)
Yellow Star Thistle; me
EE Leaves with spines; all leaves lobed (may also be toothed)
F   Involucre 2-4 cm high; stems without spiny wings, or
with only short ones; corollas yellow
G   Involucre 3-4 cm high, almost hidden by the leaves
below it; phyllaries not resembling the stem
leaves   *Cnicus benedictus*
Blessed Thistle; eu
GG Involucre 2 cm high, not hidden by the leaves below
it; phyllaries resembling the stem leaves
Carthamus lanatus
Woolly Distaff Thistle; me; SFBR
FF Involucre 1.5-2 cm high; stems with prominent, spiny
"wings" that extend from one leaf base nearly to the
next; corollas pink or purplish
G   Branches usually ending in only 1-5 flower heads;
outer surface of phyllaries roughened (widespread)
Carduus pycnocephalus (*Plate* 9)
Italian Thistle; me; Sn-s
GG Branches usually ending in more than 5 flower heads
(sometimes at least 20); outer surface of
phyllaries smooth   *Carduus tenuiflorus*
Slenderflower Thistle; me
DD Pappus bristles feathery
E   Stems with spiny "wings" that extend from one leaf base
nearly to the next; upper surface of leaves with hairs
that are rough to the touch (widespread) *Cirsium vulgare*
Bull Thistle; eu
EE Stems without spiny wings, or with only short ones; upper
surface of leaves without rough hairs
F   Upper half of most phyllaries with a lengthwise
resinous ridge (and with margins slightly toothed or
fringed with hairs); in wet areas
G   Leaves white-woolly (the upper surface much less so
than the underside); corollas white to purplish
red (sometimes on serpentine)   *Cirsium douglasii*
Swamp Thistle; Me-Mo
GG Upper surface of leaves scarcely hairy, the
underside woolly; corollas pale pink or rose
H   Upper leaves with many narrow lobes, those just
below upper flower heads 1-2 cm long; in
brackish marshes
Cirsium hydrophilum var. *hydrophilum*
Suisun Thistle; Sl; 1b
HH Upper leaves either smooth-margined or with a
few wide lobes, those just below upper flower
heads 2-4 cm long; on serpentine
Cirsium hydrophilum var. *vaseyi*
Mount Tamalpais Thistle; Ma(MT); 1b
FF Phyllaries without a lengthwise resinous ridge; mostly
in dry areas
G   Margin of phyllaries with fine or jagged teeth
H   Leaves mostly basal; not more than 30 cm tall
(phyllaries with lighter margins; corollas
purple or white)
Cirsium quercetorum [*C. walkerianum*] (*Plate* 9)
Brownie Thistle; SFBR-n

HH Stem leaves well developed; usually much more
than 30 cm tall
  I   Margin of phyllaries with jagged teeth;
upper portion of phyllaries sometimes
expanded (leaf spines 1-4 mm long; corollas
cream to pink)   *Cirsium remotifolium*
[includes *C. callilepis* vars.
callilepis and *pseudocarlinoides*]
Fringebract Thistle; SFBR-n
  II Margin of phyllaries with fine teeth; upper
portion of phyllaries not expanded
Cirsium occidentale var. *californicum*
[*C. californicum*]
California Thistle; CC-s
GG Margin of phyllaries not toothed
  H   Phyllaries not hairy
    I   Leaves mostly basal; not more than 30 cm
tall (phyllaries with lighter margins, not
bent back, or bent back only at the tip;
corollas purple or white)
Cirsium quercetorum
[*C. walkerianum*] (*Plate* 9)
Brownie Thistle; SF-n
    II Stem leaves well developed; up to more than
60 cm tall
      J   Phyllaries not thickened at the base,
bent back for more than half their
length; corollas white
Cirsium fontinale var. *campylon*
[*C. campylon*]
Mount Hamilton Thistle; SCl(MH); 1b
      JJ Phyllaries thickened at the base, not
bent back; corollas bluish or violet
        K   Leaves with stout spines up to 20 mm
long; corollas violet (widespread)
Cynara cardunculus (*Plate* 10)
Cardoon; me; SFBR
        KK Leaves with weak spines only 1-2 mm
long; corollas bluish (not common)
Cynara scolymus
Artichoke; me
HH Phyllaries hairy, but sometimes only on the
margin
  I   Spines on phyllaries at least 10 mm long
(corollas purple or red to white)
    J   Leaf below each flower head at least as
long as the head   *Cirsium brevistylum*
Indian Thistle
    JJ Leaf below each flower head shorter than
the head   *Cirsium andrewsii*
Franciscan Thistle; Sn-SM; 4
  II Spines on phyllaries less than 10 mm long
(usually not more than 5 mm long)
    J   Corollas white
      K   Phyllaries reddish, usually bent
back, with short hairs mainly on the
inner surface
Cirsium fontinale var. *fontinale*
Fountain Thistle; SM; 1b
      KK Phyllaries greenish, not regularly
bent back, the outer surface hairy
Cirsium cymosum
Peregrine Thistle; Mo-n
    JJ Corollas crimson, pink, pinkish purple,
or lavender (sometimes white in *Cirsium
arvense*)
      K   Flower heads 1-1.5 cm wide and high;
phyllaries somewhat hairy, but never
woolly (corollas usually pinkish
purple, sometimes white; at low
elevations; widespread)
Cirsium arvense
Canada Thistle; eu

                              KK Flower heads 2-5 cm wide and high
                                 (sometimes more); phyllaries
                                 sometimes woolly
                                 L Corollas crimson (on dry slopes
                                    at elevations of up to 8500';
                                    widespread) *Cirsium occidentale*
                                            var. *venustum* (*Plate* 9)
                                                       Venus Thistle
                                 LL Corollas usually purple (in sandy
                                    soil along the coast and on
                                    adjacent hills)
                                               Cirsium occidentale
                                       var. *occidentale* [*C. coulteri*]
                                                     Cobwebby Thistle
AA Phyllaries (or the involucres, when individual phyllaries are not
   distinct) without spines (Exceptions: in species of *Ambrosia*, there are
   spines on the involucre of pistillate flower heads, which are located in
   the leaf axils below the racemes of staminate flower heads; in
   Ancistrocarphus filagineus, each cluster of 3-4 staminate flowers,
   without an involucre, has 5 hook-tipped bracts just below it; in *Soliva*,
   the flattened achenes, which could be confused with phyllaries, end in
   spines.)
   B All leaves alternate
      C Upper half of each phyllary with 12-15 pointed marginal teeth;
         peripheral flowers much larger than those near the center of the
         head and with decidedly unequal lobes; corollas usually deep
         blue (but sometimes white, pink, or purple)   *Centaurea cyanus*
                                               Bachelor's-button; me
      CC Flower heads and corollas not as described in choice C
         D Flower heads nearly smooth, resembling buttons, about 1 cm
            wide; corollas yellow, more than 100 in each head; leaves
            irregularly lobed; stems rooting at the nodes; in wet places,
            including salt marshes   *Cotula coronopifolia* (*Plate* 10)
                                                   Brassbuttons; af
         DD Plants not in every respect as described in choice D
                              Asteraceae, Division II, Subkey (p. 91)
   BB Some or all leaves opposite
      C Leaves opposite below, alternate above
         D Leaves mostly smooth-margined, sometimes finely toothed;
            flower heads 15 mm high, solitary at the end of branches
            (corollas usually orange; ray flowers present, but the rays
            of these not extending more than 2 mm beyond the phyllaries;
            widespread)   *Achyrachaena mollis* (*Plate* 6)
                                                       Blow-wives
         DD Leaves with teeth, shallow lobes, or pinnately, bipinnately,
            or tripinnately lobed; flower heads less than 8 mm high, the
            staminate flowers in terminal racemes, the pistillate flowers
            in the leaf axils below the racemes
            E Plants forming mats; stems not erect (leaves with teeth,
               shallow lobes, or 1-3 times pinnately lobed; involucre of
               pistillate flower heads spiny; sandy seacoast habitats)
                                               Ambrosia chamissonis
                      [includes *A. chamissonis* ssp. *bipinnatisecta*] (*Plate* 7)
                                          Silvery Beachweed; Mo-n
            EE Plants not forming mats; stems erect
               F Leaves pinnately lobed, the lobes not separated nearly
                  to the midrib; involucre of pistillate flowers
                  sometimes with pronounced bumps, but without distinct
                  spines   *Ambrosia psilostachya* (*Plate* 72)
                                                    Western Ragweed
               FF Leaves bipinnately lobed, the primary lobes separated
                  nearly to the midrib; involucre of pistillate flowers
                  with hooked spines   *Ambrosia confertiflora*
                                                     Ragweed; SF
      CC All leaves opposite
         D Lower leaves with petioles
            E Leaves compound, with 3-5 leaflets; plants not glandular,
               annual (up to 120 cm tall, branched; in damp places)
                                          Bidens frondosa (*Plate* 8)
                                                    Sticktight; Sn
            EE Leaves not compound; plants glandular, perennial

       F   Corollas yellow; involucre 10-17 mm high; up to 60 cm
           tall, the stems not purplish
                          Arnica discoidea [includes *A. discoidea* vars.
                                   alata and *eradiata*] (*Plate* 7)
                                                Coast Arnica; Mo-n
       FF Corollas whitish; involucre not more than 5 mm high;
           up to 150 cm tall, the stems purplish
                          Ageratina adenophora [*Eupatorium adenophorum*]
                                   Sticky Eupatorium; mx; SFBR-s
    DD Leaves sessile
       E   Flower heads not woolly; leaves 3-10 cm long; corollas
           (yellow) with rays 1 mm long; stems weak, but usually
           upright, up to 35 cm long; in wet clay soils
                                             Lasthenia glaberrima
                                   Smooth Goldfields; Al-n
       EE Flower heads woolly; leaves not more than 2.5 cm long;
           corollas without rays; stems prostrate, up to 15 cm long;
           mostly in hard-packed soil, sometimes dry mud of vernal
           pools
           F   Leaves (12-20 mm long) often extending 5-10 mm beyond
               the flower heads; leaves 6-10 times as long as wide
                                        Psilocarphus oregonus
                                         Oregon Woolly-heads
           FF Leaves not extending more than 2-3 mm beyond the
               flower heads; leaves 1-6 times as long as wide
               G   Flower heads 6-10 mm wide; leaves up to 25 mm long
                          Psilocarphus brevissimus var. *multiflorus*
                                   Delta Woolly-marbles; SFBR; 4
               GG Flower heads 2-5 mm wide; leaves up to 15 mm long
                                        Psilocarphus tenellus
                                         Slender Woolly-heads

Asteraceae, Division II, Subkey: Flowers all of the disk-flower type;
phyllaries without spines; all leaves alternate
A   Corollas not yellow, yellowish, red, or reddish (but they may be tipped
    with red) (if purplish, try also choice AA)
    B   Shrubs, woody at least at the base (corollas whitish)
        C   Flower heads solitary in the leaf axils along the terminal
            portion of the branches (in salt marshes or alkaline places)
                               Iva axillaris ssp. *robustior* (*Plate* 73)
                                         Povertyweed; Sl, Al, CC
    CC Flower heads in clusters of several to many, at the end of short
        side branches
        D   Flower heads 12-14 mm high   *Brickellia californica*
                                         California Brickellia
        DD Flower heads about 5 mm high
            E   Leaves not more than 4 cm long, usually toothed on the
                margin and with a single vein; in dry habitats
                (widespread)   *Baccharis pilularis*
                      [includes *B. pilularis* var. *consanguinea*] (*Plate* 8)
                                                Coyotebrush
            EE Larger leaves up to 15 cm long, not distinctly toothed,
                with 3 veins (the middle vein more prominent than the
                others); in moist habitats
                F   Flower heads limited to a few branches on the
                    uppermost portion of the main stems; woody only near
                    the base; stems greenish   *Baccharis douglasii*
                                                Marsh Baccharis
                FF Flower heads on numerous branches, not necessarily on
                    the uppermost portion of the main stems; woody nearly
                    throughout; stems brownish
                                   Baccharis salicifolia [*B. viminea*] (*Plate* 8)
                                                Mulefat
BB Herbs, not woody, even at the base
    C   Leaves usually lobed, sometimes bipinnately, or with a few broad
        teeth
        D   Leaf blades nearly or fully as wide as long, either palmately
            lobed or with a few prominent teeth, decidedly whitish on the
            underside
            E   Leaves palmately lobed; pappus consisting of bristles
                    Petasites frigidus var. *palmatus* [*P. palmatus*] (*Plate* 74)
                                                Colt's-foot; SM-n
            EE Leaves with broad teeth, but not truly lobed; pappus
                absent   *Adenocaulon bicolor* (*Plate* 72)
                                                Trailplant; SCr-n

DD Leaf blades, if nearly as wide as long, pinnately or
   bipinnately divided into narrow lobes, not especially whitish
   on the underside
   E   Larger leaves pinnately lobed, the primary lobes toothed
       or again slightly lobed; upper leaves much smaller than
       the lower leaves; flower heads 15-20 mm high, at the tip
       of branches; corollas pink or lavender-blue
                                  Acroptilon repens [*Centaurea repens*]
                                             Russian Knapweed; eua
   EE Leaves bipinnately divided into narrow lobes; upper leaves
       about the same size as the lower leaves; flower heads
       less than 5 mm high, borne in the leaf axils; corollas
       greenish (achenes flattened, with broad, membranous
       wings; style of the pistil persisting as a rigid spine)
                                  Soliva sessilis [includes *S. daucifolia*]
                                             Common Soliva; sa
CC Leaves not lobed, and either without teeth, or only with small
   teeth
   D   Stems and leaves mostly green, without long, woolly hairs
       (but there may be short hairs or glandular hairs, or both)
       E   Flower heads nodding, solitary in the leaf axils (leaves
           mostly sessile; found in salt marshes or alkaline places)
                                  Iva axillaris ssp. *robustior* (*Plate* 73)
                                             Povertyweed; SL, Al, CC
       EE Flower heads not nodding, either in dense clusters or
           solitary, but if solitary not confined to the leaf axils
           F   Flower heads in dense clusters of more than 10;
               peduncles less than 1 cm long (plants with a strong
               odor)
               G   Flower clusters in the leaf axils; corollas purple;
                   lower leaves with petioles; in moist areas,
                   including saline habitats   *Pluchea odorata*
                                             Saltmarsh Fleabane; SFBR
               GG Flower clusters at the end of main stems or
                   branches; corollas white, green, or pink; all
                   leaves sessile; in dry habitats
                   H   Phyllaries white; largest leaves 10-15 mm wide;
                       involucre about as long as wide (widespread)
                                  Gnaphalium californicum (*Plate* 11)
                                             California Everlasting
                   HH Phyllaries white, green, or pink; largest
                       leaves 5-7 mm wide (most leaves less than 3 mm
                       wide); involucre longer than wide
                       (widespread)   *Gnaphalium ramosissimum*
                                             Pink Everlasting; Ma-s
           FF Flower heads solitary; peduncles 1-5 cm long
               G   Flower heads both in the leaf axils and at the end
                   of main stems or branches; leaves not in a basal
                   cluster; upper leaves not glandular or hairy,
                   reduced upward but not scalelike (in wet habitats,
                   including salt marshes)
                                  Aster subulatus var. *ligulatus* [*A. exilis*]
                                             Slim Aster; SFBR-s
               GG Flower heads at the end of main stems or branches;
                   larger leaves in a basal cluster (but these may
                   wither early); upper leaves glandular or hairy,
                   sometimes scalelike
                   H   Phyllaries glandular, but not especially hairy
                       or woolly; upper leaves glandular
                                             Lessingia ramulosa
                                             Sonoma Lessingia; SN, Sl
                   HH Phyllaries (at least the lower ones) hairy or
                       woolly, but not glandular; upper leaves hairy
                       or woolly (at least when young)
                       I   Flower heads 9-12 mm high; flowers at the
                           edge of the head considerably larger than
                           those in the center; basal leaves usually
                           persisting through the period of flowering
                                  Lessingia hololeuca var. *hololeuca*
                                             Woollyhead Lessingia; Na, Ma-SCl

```
II Flower heads 6-9 mm high; flowers at the
      edge of the head not obviously larger than
      those in the center; basal leaves usually
      withering and falling away before the
      period of flowering    Lessingia arachnoidea
                          [L. hololeuca var. arachnoidea]
                    Crystal Springs Lessingia; SM-SCr; 1b
DD Stems and one or both surfaces of leaves whitish due to long
   woolly hairs (the hairs may not be very dense, however)
   E  One leaf surface much more hairy than the other
      F  Upper surface of leaves densely woolly compared to the
         underside (flower heads solitary)
         G  Phyllaries and upper leaves glandular; basal leaves
            present, larger than the stem leaves
                          Lessingia micradenia var. micradenia
                                Tamalpais Lessingia; Ma(MT)-n; 1b
         GG Phyllaries and upper leaves not glandular; larger
            basal leaves absent after flowering
                          Lessingia micradenia var. glabrata
                                Smooth Lessingia; SCl(SCM); 1b
      FF Upper surface of leaves greenish and scarcely hairy,
         but the underside whitish and densely woolly (leaves
         2-8 cm long)
         G  Inflorescence dense, up to 2 cm wide (and 2-7 cm
            long); flower heads sometimes densely woolly;
            phyllaries not pearly white    Gnaphalium purpureum
                          [includes G. peregrinum] (Plate 11)
                                          Purple Cudweed
         GG Inflorescence spreading, up to 10 cm wide, the
            panicles slightly rounded or nearly flat-topped;
            flower heads not woolly; phyllaries pearly white
            (widespread)       Anaphalis margaritacea (Plate 7)
                                          Pearly Everlasting
   EE Leaves uniformly hairy on both surfaces, but sometimes not
      densely so (flower heads solitary or in clusters)
      F  Leafy bracts just below the clusters of flower heads
         either absent or much shorter than the clusters
         G  Flower heads solitary at the end of branches; stem
            leaves markedly reduced compared with basal leaves
            (corollas pink to lavender)
            H  Phyllaries and upper leaves glandular
                                       Lessingia ramulosa
                                Sonoma Lessingia; SN, Sl
            HH Phyllaries and upper leaves not glandular
                          Lessingia hololeuca var. hololeuca
                                Woollyhead Lessingia; Na, Ma-SCl
         GG Flower heads grouped in slightly rounded or nearly
            flat-topped clusters; stem leaves not markedly
            reduced
            H  Stems and leaves greenish yellow; phyllaries
               greenish yellow; in moist areas (leaves 2-6 mm
               wide, 20-50 mm long; corollas yellow or tan)
                          Gnaphalium stramineum [G. chilense]
                                          Cotton-batting-plant
            HH Stems and leaves whitish because of woolliness;
               phyllaries white or brownish; mostly in dry
               areas
               I  Leaves up to 4 mm wide and 40 mm long;
                  phyllaries brownish (corollas tipped with
                  red)             Gnaphalium luteo-album
                                Weedy Cudweed; eua; Sn-Mo
               II Leaves 4-10 mm wide and up to 80 mm long;
                  phyllaries white, sometimes tan
                  J  Stems not branched, except for the short
                     branches of the inflorescence (common)
                          Anaphalis margaritacea (Plate 7)
                                          Pearly Everlasting
                  JJ Stems usually extensively branched in
                     the middle part of the plant
                          Gnaphalium canescens ssp. microcephalum
                                       [G. microcephalum]
                                White Everlasting; Ma-s
      FF Tip of leafy bracts below the clusters of flower heads
         as long as, or longer than, the clusters
```

94   (D) Asteraceae (Division II, Subkey)--Sunflower Family

      G Flower heads solitary in the leaf axils; leaf
         petioles 1-1.5 cm long
                   Hesperevax sparsiflora [*Evax sparsiflora*]
                                          Erect Hesperevax
     GG Flower heads in clusters at the end of branches;
         leaves without petioles
      H Leaves 10-50 mm long and 3-5 mm wide (sometimes
         larger) (in moist habitats)
          I Leaves 20-50 mm long; involucre 5-6 mm
             high, greenish-yellow
                   Gnaphalium stramineum [*G. chilense*]
                                     Cotton-batting-plant
         II Leaves 10-30 mm long; involucre 3 mm high,
             brownish   *Gnaphalium palustre*
                                          Lowland Cudweed
     HH Leaves usually not more than 15 mm long or 3 mm
         wide (clusters of flower heads usually less
         than 1 cm wide)
          I Tip of leafy bracts below the clusters of
             flower heads extending more than 3 mm
             beyond the clusters
             J Staminate flowers in clusters of 3-4,
                each cluster without phyllaries but
                with 5 hook-tipped bracts just beneath
                it (each flower in the pistillate heads
                enclosed by a woolly, boat-shaped
                phyllary) *Ancistrocarphus filagineus*
                              [*Stylocline filaginea*]
                       Woolly Fish-hooks; Me-s
            JJ Individual flower heads, regardless of
                their composition, without specialized
                bracts that end in hooked spines
                K Clusters of flower heads rather
                   widely spaced, not hiding the stems;
                   up to 30 cm tall (widespread)
                                       Filago gallica
                                        Fluffweed; me
               KK Clusters of flower heads very
                   crowded, hiding the stems; prostrate
                   (often on dried mud of vernal pools,
                   but also in other dry habitats)
                                   Psilocarphus oregonus
                                    Oregon Woolly-heads
         II Tip of leafy bracts below the clusters of
             flower heads not extending beyond the
             clusters
             J Phyllaries, and also the bracts below
                the clusters of flower heads, scarcely
                visible because of woolliness
                K Flower heads 3.5-5 mm long
                                  Micropus amphibolus
                              [*Stylocline amphibola*]
                   Mount Diablo Cottonweed; Ma-Al; 4
               KK Flower heads 2-4 mm long
                                  Micropus californicus
                                   Slender Cottonweed
            JJ Phyllaries, and also the bracts below
                the clusters of flower heads, visible
                K Tips of corollas purplish, visible
                   above the phyllaries (widespread)
                              Filago californica (*Plate* 11)
                                   California Fluffweed
               KK Corollas hidden by the phyllaries
                                  Stylocline gnaphaloides
                                    [*S. gnaphalioides*]
                       Everlasting Neststraw; Ma
AA Corollas yellow, yellowish, red, or reddish, sometimes purplish
    B Largest leaves at or near the base of the plant (stem leaves few and
       reduced or numerous and scalelike) (phyllaries often dark-tipped)
       C Leaves sessile (lower leaves elliptic, densely hairy; upper
          leaves scalelike, scarcely hairy)   *Lepidospartum squamatum*
                                          Scalebroom; SCl-s
      CC Some leaves with petioles

D   Petioles of basal leaves winged; plants whitish (in wet
    places)   *Senecio hydrophilus*
                              Alkali-marsh Butterweed; SFBR-n
DD Petioles of basal leaves not winged; plants not whitish
                                                     Senecio aronicoides
                                   California Butterweed; SM-n
BB Stems with well developed leaves (some of them sessile, however)
  C   Involucre very glandular
    D   Phyllaries widest at the middle, tapering abruptly at the
        tip; pappus either absent or not evident
      E   Involucre 4-6 mm high; glands on phyllaries obvious (at
          elevations above 3500')   *Madia glomerata*
                                             Mountain Madia
      EE Involucre 2-3 mm high; glands on phyllaries not obvious
          without a hand lens (they are among long hairs)
          (widespread)   *Madia exigua*
                                          Threadstem Madia
    DD Phyllaries widest at the base and tapering gradually to the
        tip; pappus obvious, consisting of many bristles
      E   Leaves and stems hairy
              Erigeron petrophilus var. *petrophilus* (*Plate* 10)
                                            Rock Daisy; Mo-n
      EE Upper leaves and stems not hairy
        F   Leaves 2-6 mm wide; lower stems and leaves hairy or
            glandular   *Erigeron petrophilus* var. *viscidulus*
                                [*E. inornatus* var. *viscidulus*]
                                 Klamath Rock Daisy; Ma(PR); 4
        FF Leaves less than 2 mm wide; lower stems and leaves not
            hairy or glandular *Erigeron reductus* var. *angustatus*
                                 [*E. inornatus* var. *angustatus*]
                                  Northern Rayless Daisy; SM-n
  CC Involucre not glandular
    D   Involucre less than 5 mm high
      E   Flower heads solitary at the end of branches that are
          usually at least 3 cm long (leaves sessile, clasping)
        F   Leaves bipinnately lobed
          G   Flower heads 2-5 mm wide   *Cotula australis*
                              Southern Brassbuttons; au
          GG Flower heads 6-10 mm wide
            H   Flower heads 6-9 mm high (widespread)
                                          Chamomilla suaveolens
                              [*Matricaria matricarioides*] (*Plate* 73)
                                       Pineappleweed; na
            HH Flower heads 10-11 mm high
                                         Chamomilla occidentalis
                                 [*Matricaria occidentalis*]
                                      Valley Pineappleweed
        FF Leaves either not lobed or only irregularly lobed,
            definitely not bipinnately lobed (flower heads 8-10 mm
            wide)
          G   Pappus absent or consisting of only 3 bristles
              (leaves not hairy)
            H   Involucre 3-5 mm high; stems branched at the
                base if at all
                     Pentachaeta exilis [*Chaetopappa exilis*]
                                       Meager Pentachaeta
            HH Involucre up to 3 mm high; stems branched above
                the base
                Pentachaeta alsinoides [*Chaetopappa alsinoides*]
                                          Tiny Pentachaeta
          GG Pappus consisting of many bristles
            H   Stems and leaves hairy   *Erigeron petrophilus*
                              var. *petrophilus* (*Plate* 10)
                                            Rock Daisy; Mo-n
            HH Upper stems and leaves not hairy (but lower
                stems and leaves may be hairy)
              I   Leaves 2-6 mm wide; lower stems and leaves
                  hairy or glandular
                     Erigeron petrophilus var. *viscidulus*
                              [*E. inornatus* var. *viscidulus*]
                              Klamath Rock Daisy; Ma(PR); 4

II Leaves less than 2 mm wide; lower stems and
leaves not hairy or glandular
Erigeron reductus var. *angustatus*
[*E. inornatus* var. *angustatus*]
Northern Rayless Daisy; SM-n
EE Flower heads many, in elongated racemes or extensive
panicles whose branches are mostly less than 2 cm long
F Flower heads in elongated racemes, these about 2 cm
wide; flower heads without any rays
G Larger leaves either not lobed or with a few lobes
or coarse teeth, but most of these divisions not
separated nearly to the midrib (leaves densely
hairy beneath)
H Involucre hairy (widespread)
Artemisia douglasiana (*Plate* 7)
Mugwort
HH Involucre not hairy (coastal)
Artemisia suksdorfii
Coastal Mugwort; Sn-n
GG Larger leaves lobed, most of the primary lobes
separated nearly to the midrib
H Leaves not hairy; lobes of the larger leaves
regularly toothed; annual or biennial, up to
30 cm tall   *Artemisia biennis*
Biennial Sagewort; eu
HH Leaves hairy; lobes of the larger leaves not
toothed; perennial, up to more than 70 cm tall
I Stems and leaves usually whitish; hairs
silky, not matted; involucre 4-5 mm high,
hairy; woody only at the base; up to 70 cm
tall (in coastal habitats)
Artemisia pycnocephala (*Plate* 8)
Coastal Sagewort; Mo-n
II Stems and leaves grayish; hairs matted;
involucre 2-3 mm high, not hairy; woody
almost throughout; up to 100 cm tall
(widespread)
Artemisia californica (*Plate* 8)
California Sagebrush; Ma, Na-s
FF Flower heads in spreading panicles, these often 10-15
cm wide; flower heads sometimes with very small rays
G Lower leaves up to 30 mm long; upper leaves mostly
not more than 5 mm long; pappus bristles reddish
(outer disk flowers sometimes much larger than
those near the center)
H Outer margin of phyllaries densely hairy, but
not glandular
Lessingia tenuis [*L. germanorum* var. *parvula*]
Spring Lessingia; SCl-SLO; 4
HH Outer margin of phyllaries not densely hairy,
but glandular   *Lessingia glandulifera*
[*L. germanorum* var. *glandulifera*]
Valley Lessingia; SFBR-s
GG Lower leaves more than 60 mm long; upper leaves at
least 10 mm long; pappus bristles whitish or tan
H Vines (but not woody), the stems up to 6 m
long; leaf blades about as long as wide
(involucre 3-4 mm high, not hairy; in moist
coastal habitats)   *Senecio mikanioides*
German-ivy; af
HH Plants upright, the stems up to 2 m tall; leaf
blades at least twice as long as wide
I Leaves with lobes or teeth of rather
uniform size; lower leaves up to 6 cm long;
up to 1 m tall (involucre hairy, 2-3 mm
high)   *Conyza coulteri*
Coulter Horseweed; SCl-s
II Leaves either with smooth margins or with a
few irregular lobes or teeth; lower leaves
up to 10 cm long; up to 2 m tall
J Phyllaries without a dark midvein;
involucre hairy, 4-5 mm high (extremely
small rays present) *Conyza bonariensis*
South American Horseweed; sa; SFBR-s

JJ Phyllaries with a dark midvein;
   involucre not hairy or only scarcely
   hairy, 3-4 mm high
   K Hairs usually present on both leaf
      surfaces, and also on the margin;
      very small rays present (widespread)
         Conyza canadensis (*Plate* 72)
                              Horseweed
   KK Hairs present only on the leaf
      margin; rays absent *Conyza bilboana*
                     Rayless Horseweed
DD Involucre at least 5 mm high
   E Some leaves (often only the lower ones, however) toothed
      or pinnately lobed
      F Pappus lacking or consisting of scales
         G Leaves pinnately lobed; stems not woody, sprawling
            with the tips rising (pappus consisting of scales;
            on coastal dunes) *Tanacetum camphoratum* (*Plate* 14)
                                       Dune Tansy; SF
         GG Leaves toothed or irregularly lobed but not
            pinnately lobed; stems woody, at least at the
            base, upright
            H Leaves evenly toothed, up to 3.5 cm long;
               pappus lacking; established in coastal areas
               Santolina chamaecyparisus [*S. chamaecyparissus*]
                              Lavender-cotton; me; Sn
            HH Leaves irregularly lobed, 2-7 cm long; pappus
               consisting of very small scales; on coastal
               bluffs   *Eriophyllum staechadifolium* [includes
                  E. staechadifolium var. *artemisiaefolium*]
                        Seaside Woolly Sunflower; SCr-Me
      FF Pappus consisting of bristles
         G Shrubs, woody at the base; leaves often in bunches
                           Isocoma menziesii var. *vernonioides*
                        [*Haplopappus venetus* ssp. *vernonioides*]
                              Coast Goldenbush; SF-s
         GG Herbs, not woody; leaves not in bunches
            H All leaves sessile
               I Pappus bristles white; plants with 5 or
                  fewer branches, sometimes not branched
                  J Leaves not hairy; up to 25 cm tall
                                    Senecio aphanactis
                           Rayless Ragwort; Sl-s; 2
                  JJ Leaves sparsely to densely hairy on the
                     underside; up to 80 cm tall
                                    Senecio sylvaticus
                           Wood Groundsel; eu; Ma
               II Pappus bristles reddish; plants with
                  multiple branches that are again branched
                  (outer disk flowers sometimes raylike)
                  J Stems and leaves not glandular; on sand
                     dunes and in other coastal habitats
                                    Lessingia germanorum
                        San Francisco Lessingia; SFBR-s; 1b
                  JJ Stems and leaves glandular; inland
                     K Outer margin of phyllaries densely
                        hairy, but not glandular
                                       Lessingia tenuis
                           [*L. germanorum* var. *parvula*]
                           Spring Lessingia; SCl-SLO; 4
                     KK Outer margin of phyllaries not
                        densely hairy, but glandular
                                    Lessingia glandulifera
                           [*L. germanorum* var. *glandulifera*]
                              Valley Lessingia; SFBR-s
            HH Lower leaves with petioles
               I Leaves finely toothed (1-1.5 cm wide)
                           Erechtites minima [*E. prenanthoides*]
                        Australian Burnweed; au; Mo, Ma-n
               II Leaves with regularly spaced lobes
                  J Phyllaries black-tipped; involucre with
                     small, black-tipped bracts at the base
                     (widespread)   *Senecio vulgaris*
                           Common Groundsel; eua

```
                      JJ Phyllaries not black-tipped; involucre
                         without small bracts at the base
                         K  Involucre and stem only slightly if
                            at all hairy; small rays present
                                        Senecio sylvaticus
                                   Wood Groundsel; eu; Ma
                         KK Lower part of involucre and upper
                            stem woolly; rays absent (coastal)
                                   Erechtites glomerata [E. arguta]
                                   Bushman's Burnweed; au; SLO-n
             EE Leaves with smooth margins
                F  Phyllaries keeled (woody shrubs with a strong odor)
                            Chrysothamnus nauseosus ssp. mohavensis
                                   Rubber Rabbitbrush; SCl(MH)-s
                FF Phyllaries not keeled
                   G  Pappus consisting of scales
                      H  Plants not woody; flower heads globose (in
                         wetlands)        Helenium puberulum (Plate 11)
                                                             Rosilla
                      HH Plants woody at the base; flower heads conical
                         I  Bark not whitish; on coastal bluffs
                                   Eriophyllum staechadifolium [includes
                            E. staechadifolium var. artemisiaefolium]
                                   Seaside Woolly Sunflower; SCr-Me
                         II Bark whitish; on inland hillsides
                                            Eastwoodia elegans
                                   Yellow Mock Aster; Al-s, e
                   GG Pappus consisting of bristles
                      H  Flower heads 4-8 mm high; up to 300 cm tall
                         I  Largest leaves up to 10 cm long; not woody,
                            annual             Conyza bonariensis
                                   South American Horseweed; sa; SFBR-s
                         II Largest leaves up to 6 cm long; woody, at
                            least at the base, perennial
                            J  Leaves 1-4 cm long, often in clusters;
                               involucre 5-7 mm high; woody only at
                               the base; coastal
                                   Isocoma menziesii var. vernonioides
                               [Haplopappus venetus ssp. vernonioides]
                                            Coast Goldenbush; SF-s
                            JJ Leaves 3-6 cm long, rarely in clusters;
                               involucre 4-5 mm high; stems woody;
                               inland (common)  Ericameria arborescens
                                        [Haplopappus arborescens]
                                                        Goldenfleece
                      HH Flower heads 10-15 mm high; up to 60 cm tall
                         I  Upper leaves sparse, some of them reduced
                            to a length of 1 cm or less
                                   Heterotheca oregona var. scaberrima
                                   [Chrysopsis oregona var. scaberrima]
                                   Sticky Oregon Golden Aster; Al, SCl-Mo
                         II Upper leaves crowded, 2-5 cm long
                            J  Stems and leaves with stout, long hairs;
                               not confined to stream beds; inland
                                        Heterotheca oregona var. rudis
                                        [Chrysopsis oregona var. rudis]
                                   Inland Oregon Golden Aster; SF-n
                            JJ Stems and leaves without hairs; in dry
                               stream beds; coastal
                                        Heterotheca oregona var. oregona
                                        [Chrysopsis oregona var. oregona]
                                        Oregon Golden Aster; Ma-n
```

Asteraceae, Division III: At least some of the flower heads with both ray
flowers and disk flowers
```
A  Ray flowers white, pink, blue, violet, purple, or lavender
   B  Ray flowers usually pink, blue, violet, purple, or lavender, but
      sometimes white (perennial, or sometimes biennial in Erigeron
      philadelphicus)
      C  Tips of phyllaries all at the same level
         D  Flower heads with more than 100 rays
            E  Ray flowers pinkish or white, the rays less than 1 mm
               wide; flower heads 6-15 mm wide; not confined to coastal
               locations              Erigeron philadelphicus
                                               Philadelphia Daisy
```

         EE Ray flowers purple, blue, or white, the rays 2-3 mm wide;
            flower heads 15-35 mm wide; on coastal bluffs
                                  Erigeron glaucus (*Plate* 10)
                                        Seaside Daisy; SLO-n
      DD Flower heads with not more than 60 rays
         E   Involucre more than 8 mm high; ray flowers purple or
             reddish purple
             F   Leaves 10-20 cm long, toothed; involucre 12-15 mm
                 high; rays 30-60; on open areas including burns
                                     Hulsea heterochroma
                                     Redray Hulsea; SCl-s
             FF Leaves 3-8 cm long, the lower ones deeply lobed;
                 involucre about 10 mm high; rays about 13; established
                 along the coast   *Senecio elegans* (*Plate* 13)
                                           Purple Ragwort; af
         EE Involucre 4-7 mm high; ray flowers usually blue
             F   Leaves (and sometimes also the stems) hairy
                                  Erigeron foliosus var. *foliosus*
                                         Leafy Daisy; SFBR
             FF Leaves and stems scarcely hairy (when hairs are
                 present, they are rather widely spaced)
                                  Erigeron foliosus var. *hartwegii*
                                         Hartweg Daisy; Mo-n
   CC Tips of phyllaries at 2 or more different levels
      D   Leaves mostly grayish green, due to a fine, white woolliness
          (this is sometimes shed); ray flowers not producing achenes
          (ray flowers pink, purple, or white; mainly coastal)
          E   Flower heads 2-3.5 cm wide
                       Lessingia filaginifolia var. *californica* [includes
                  Corethrogyne californica vars. *californica* and *obovata*]
                                         California Aster; Mo-n
          EE Flower heads rarely more than 2.5 cm wide
                       Lessingia filaginifolia var. *filaginifolia* [includes
                  Corethrogyne filaginifolia vars. *hamiltonensis* and
                                  virgata, and *C. leucophylla*]
                                         Common Lessingia; Al-s
      DD Leaves not grayish green and not woolly (there may be sparse
          fine hairs, however); ray flowers producing achenes
          E   Lower leaves 2-3 times as long as wide, usually distinctly
              toothed (4-10 cm long); involucre 6-9 mm high; rays 10-
              15, white to pale violet; in dry forests *Aster radulinus*
                                         Broadleaf Aster; SLO-n
          EE Most leaves 6-8 times as long as wide, smooth-margined or
              toothed; involucre 5-7 mm high; rays 20-35, violet or
              purple; usually in open areas
              F   Plants not hairy; in salt marshes around Suisun Bay or
                  saline habitats around San Francisco Bay *Aster lentus*
                  [includes *A. chilensis* vars. *lentus* and *sonomensis*]
                                  Suisun Marsh Aster; SF, Sn, Na, SCl
              FF Tip of peduncles hairy (leaves and stems sometimes
                  also hairy); mostly outer Coast Ranges, but sometimes
                  in moist areas of the inner Coast Ranges (widespread)
                                         Aster chilensis
                  [includes *A. chilensis* var. *invenustus*] (*Plate* 8)
                                  Common California Aster
BB Ray flowers usually white, but sometimes pink, reddish, or purple (if
   your specimen does not agree well with any of the following choices,
   try choice B, above)
   C   Plants with glandular hairs, or the phyllaries sticky because of
       glandular secretions
       D   Glands on phyllaries saucer-shaped; flower heads at the end
           of branches or scattered along the stems (annual)
           E   Glands yellow; rays 5-11; up to 180 cm tall
                                  Blepharizonia plumosa
                                  Big Tarweed; Al-e; 1b
           EE Glands black; rays 1-5; not more than 70 cm tall (ray
               flowers white to rose or reddish)
               F   Phyllaries with only a few hairs; rays 1 or 2, divided
                   into 3 lobes that are separate nearly to the base of
                   each ray; flower heads less than 0.5 cm wide; leaves
                   up to 5 cm long   *Calycadenia pauciflora*
                                  Smallflower Rosinweed; Sn-n

FF Phyllaries densely hairy on the margin; rays 2-5,
   divided into 3 lobes that are separate for about half
   the length of each ray flower heads up to 2 cm wide;
   leaves up to 8 cm long   *Calycadenia multiglandulosa*
                    [includes *C. hispida* ssp. *reducta* and
                            C. multiglandulosa sspp.
                       cephalotes and *robusta*] (*Plate* 9)
                              Sticky Rosinweed; Me-SCl
DD Glands on phyllaries globular; flower heads mostly at the end
   of branches
   E   Achene of each ray flower completely enclosed by the
       adjacent phyllary
       F   Leaves opposite below, alternate above; perennial
           (involucre up to 5 mm high; rays up to 4.5 mm long,
           each divided into 3 lobes)   *Holozonia filipes*
                                   Hareleaf; Na, Ma-SCl
       FF All leaves alternate; annual
           G   Rays 6-15 mm long; involucre up to 10 mm high; on
               sandy soil, inland   *Layia glandulosa*
                                     White Layia; CC-s
           GG Rays up to 2.5 mm long; involucre up to 7.5 mm
               high; only on coastal sand dunes   *Layia carnosa*
                                   Beach Layia; SF; 1b
   EE Achene of each ray flower only partly enclosed by the
       adjacent phyllary (lower leaves about as wide as the
       upper leaves; annual)
       F   Flower heads in both the leaf axils and at the end of
           branches   *Hemizonia congesta* ssp. *clevelandii*
                                Cleveland Spikeweed; Sn, Na-n
       FF All flower heads at the end of branches
                            Hemizonia congesta ssp. *luzulifolia* [includes
           H. luzulaefolia sspp. *luzulaefolia* and *rudis*] (*Plate* 12)
                                   Hayfield Tarweed; La-s
CC Plants without glandular hairs or sticky secretions
   D   Flower heads more than 1.5 cm wide
       E   Leaves all basal (ray flowers white to purple; widespread)
                                        Bellis perennis
                                   English Daisy; eu
       EE Stem leaves well developed
           F   Leaves pinnately divided into lobes less than 1 mm
               wide
               G   Flower heads 2.5-3.5 cm wide; stems reddish
                            Chamaemelum fuscatum [*Anthemis fuscata*]
                                        Dogfennel; me; Sn
               GG Flower heads 1.5-2 cm wide; stems green
                   (widespread)   *Anthemis cotula* (*Plate* 72)
                                        Mayweed; eu
           FF Leaves either toothed or pinnately divided, but the
               lobes at least 3 mm wide
               G   Flower heads 2.5-5 cm wide, solitary
                   Leucanthemum vulgare [*Chrysanthemum leucanthemum*]
                                        Ox-eye Daisy; eu
               GG Flower heads usually less than 2 cm wide, clustered
                   Tanacetum parthenium [*Chrysanthemum parthenium*]
                                        Feverfew; eu
   DD Flower heads less than 1.5 cm wide
       E   Leaf blades about 10 cm in diameter, palmately lobed (some
           flower heads without ray flowers)
           Petasites frigidus var. *palmatus* [*P. palmatus*] (*Plate* 74)
                                        Colt's-foot; SLM-n
       EE Leaf blades not more than 6 cm long and 1 cm wide, either
           not divided or divided pinnately
           F   Leaves divided pinnately into lobes less than 1 mm
               wide (flower heads 4-6 mm high; rays 2-5 mm long,
               white to pink; widespread)
                       Achillea millefolium [includes *A. borealis* sspp.
                            arenicola and *californica*] (*Plate* 6)
                                        Yarrow
           FF Leaves not divided into lobes (but they may be
               toothed)
               G   Leaves alternate (lower leaves 1-3.5 cm long; up to
                   20 cm tall)   *Pentachaeta bellidiflora*
                                      [*Chaetopappa bellidiflora*]
                          Whiteray Pentachaeta; Ma, SM, SCr; 1b

```
    GG Leaves opposite
        H  Flower heads 5-10 mm wide; leaves and stems
           covered with short hairs (in freshwater
           marshes)            Eclipta prostrata [E. alba]
                                      False Daisy; SFBR
        HH Flower heads 3-4 mm wide; leaves and stems
           either not hairy, or the leaves with only a
           few hairs on the veins   Galinsoga parviflora
                               Littleflower Quickweed; sa; SF
AA Ray flowers completely or partly yellow, greenish yellow, orange, red,
   or a closely related color
   B  Leaves mostly basal, with stem leaves absent or reduced (in species
      of Lagophylla, keyed in choice BB, the leaves are sometimes lacking)
      C  Flower heads at least 4 on each main stem (rays fewer than 10)
         D  Phyllaries apparently in 1 series, the tip of all of them at
            the same level
            E  Petiole of lower leaves winged; plants whitish; in damp
               places (leaf blades up to 20 cm long, smooth-margined or
               toothed; phyllaries 4-9 mm long, the tips black)
                                            Senecio hydrophilus
                                  Alkali-marsh Butterweed; SFBR-n
            EE Petiole not winged; plants not whitish; in dry areas
               F  Lower leaves up to 10 cm long, usually deeply lobed,
                  and the lobes toothed; phyllaries 8-12 mm long, the
                  tips green (sometimes on serpentine)
                                            Senecio eurycephalus
                                       Cutleaf Butterweed; Sn-n
               FF Lower leaves up to 20 cm long, smooth-margined or
                  toothed; phyllaries 4-8 mm long, the tips black
                                            Senecio aronicoides
                                    California Butterweed; SM-n
         DD Phyllaries in 2 or more series, the tips at different levels
            E  Flower heads 1.5-3 cm wide; rays 5-12 mm long
                        Pyrrocoma racemosa [Haplopappus racemosus]
                                      Racemose Pyrrocoma; Na
            EE Flower heads less than 1 cm wide; rays 2-3 mm long
               F  Leaves toothed; phyllaries oblong, blunt (mostly
                  coastal)            Solidago spathulata (Plate 14)
                                         Coast Goldenrod; Mo-n
               FF Leaves not toothed; phyllaries slender, pointed
                                            Solidago guiradonis
                                       Guirado Goldenrod; 4
      CC Flower heads usually solitary at the end of main stems
         D  Leaves much more than 2 cm wide
            E  Leaves pinnately to bipinnately lobed; pappus absent
                                        Balsamorhiza macrolepis
                                   Bigscale Balsamroot; SF-e; 1b
            EE Leaves smooth-margined or toothed, but not lobed; pappus
               absent or consisting of a crown of scales
               F  Outer phyllaries enlarged and extending beyond the
                  disk and the rays
                  G  Leaves (at least most of them) with a woolly or
                     cottony coating; on sunny slopes
                                            Wyethia helenioides
                                            Gray Mule-ears
                  GG Leaves without a woolly or cottony coating,
                     although they may be somewhat hairy and rough to
                     the touch; on shady slopes (widespread)
                                            Wyethia glabra (Plate 15)
                                            Smooth Mule-ears
               FF Outer phyllaries not enlarged and not extending beyond
                  the disk and rays
                  G  Phyllaries with hairs about 2 mm long, these mainly
                     concentrated at the margin (the phyllaries are
                     otherwise scarcely hairy except at the base);
                     pappus consisting of a crown of scales
                     (widespread)        Wyethia angustifolia (Plate 15)
                                            Narrowleaf Mule-ears
                  GG Phyllaries with hairs less than 1 mm long, these
                     not confined to the margin (they are especially
                     abundant at the tip); pappus absent
                                            Helianthella californica
                                   California Helianthella; SCl-n
         DD Leaves not more than 1.5 cm wide (sometimes much less)
```

E   Phyllaries without membranous margins, those of the upper
    and lower series similar
    F   Length of peduncle above last stem leaves at least
        twice as long as these leaves
                                Chaenactis glabriuscula var. *lanosa*
                                    Yellow Pincushion; SF-s
    FF Length of peduncle above last stem leaves less than
        twice as long as these leaves (sometimes on
        serpentine) *Chaenactis glabriuscula* var. *heterocarpha*
        [*C. glabriuscula* var. *gracilenta* and *C. tanacetifolia*]
                                    Inland Yellow Pincushion; Na-s
EE Phyllaries of the upper series with membranous margins
    (and often yellowish and/or marked with darker lines),
    those of the lower series green and leaflike
    F   Leaves rarely with more than 2 lateral lobe
                                    Coreopsis douglasii
                                    Douglas Coreopsis; SCl-s
    FF Leaves usually with at least 4 lateral lobes, these
        sometimes divided again
        G   Stems leafy for about one-third or one-half their
            length; pappus scales 4 mm long
                                Coreopsis calliopsidea (*Plate* 9)
                                    Leafystem Coreopsis; Al-s
        GG Leaves all basal; pappus scales lacking or just 1
            mm long
            H   Leaves up to 10 cm long, with lobes 4-8 mm
                long, these not crowded; pappus not apparent;
                found at elevations of 100-3000'
                                    Coreopsis stillmanii
                                    Stillman Coreopsis; CC-s
            HH Leaves up to 5 cm long, with lobes 2-3 mm long,
                these crowded; pappus scales 1 mm long; found
                at elevations of 1800-4500'
                                    Coreopsis hamiltonii
                        Mount Hamilton Coreopsis; CC(MD), SCl(MH); 1b
BB Leaves present along the stem, not insignificant compared to the
    basal leaves (*Coreopsis calliopsidea*, which may have well developed
    leaves in the lower half of the stem, is keyed in choice B)
    C   All leaves alternate (*Madia*, keyed under choice CC, sometimes has
        alternate leaves)   Asteraceae, Division III, Subkey (p. 105)
    CC Some or all leaves opposite (*Eriophyllum*, keyed under choice C,
        sometimes has opposite leaves)
        D   Plants with glandular hairs (often only in the upper
            portion), or the phyllaries sticky because of glandular
            secretions (leaves opposite below, alternate above, with long
            soft hairs, and usually with a strong odor; achene of each
            ray flower completely enclosed within a phyllary)
            E   Rays less than 5 mm long
                F   Disk flowers not more than 10; ray flowers 1-8
                    G   Flower heads 2-4 mm high; leaves 1-2 cm long; up to
                        15 cm tall   *Madia minima*
                                    Small Madia; SCl(MH)
                    GG Flower heads 5-9 mm high; leaves up to 9 cm long;
                        up to 70 cm tall
                        H   Ray flowers 1-3; leaves up to 9 cm long; at
                            elevations of 3500-8800'   *Madia glomerata*
                                    Mountain Madia
                        HH Ray flowers 3-7; leaves up to 7 cm long; at
                            elevations up to 500'   *Madia anomala*
                                    Plumpseed Madia; CC-n
                FF Disk flowers 15 to many; ray flowers 5-15
                    G   Leaves lemon-scented, the largest up to 9 cm long
                        (plants glandular only in the upper half, but
                        hairy throughout; involucre 6-8 mm high)
                                    Madia citriodora
                                    Lemonscent Madia; Na-n
                    GG Leaves not lemon-scented (but with a resinous
                        odor), the largest more than 9 cm long
                        H   Plants glandular and hairy throughout;
                            involucre up to 15 mm high; leaves more than
                            10 mm wide at the base (widespread)
                                    Madia sativa [includes *M. capitata*]
                                    Coast Madia

        HH Plants glandular only in the upper half, but
           hairy throughout; involucre up to 9 mm high;
           leaves 5 mm wide at the base (widespread)
                      Madia gracilis (*Plate* 13)
                           Slender Madia
EE Rays 6-15 mm long
  F Pappus consisting of scales
    G Leaves opposite well up the stem; up to 65 cm tall;
       perennial   *Madia madioides*
                          Woodland Madia
    GG Leaves alternate except for the lowest pair; up to
       25 cm tall; annual
       H Leaves evenly spaced on stems; plants generally
          branched; flower heads nodding in bud stage;
          ray flowers 5-8; on volcanic ash *Madia nutans*
                    Nodding Madia; Na, Sn; 4
       HH Leaves mainly in crowded clusters; plants
          mostly not branched; flower heads not nodding
          in bud stage; ray flowers 3-6; on serpentine
                          Madia hallii
                  Hall Madia; Na; 1b
  FF Pappus absent (or at least not obvious)
    G Rays 3-8 mm long, barely extending beyond the
       phyllaries (black-tipped glandular hairs usually
       present on the phyllaries and stem; widespread)
                      Madia gracilis (*Plate* 13)
                           Slender Madia
    GG Rays 6-16 mm long, extending beyond the phyllaries
       H Anthers yellow (spring-flowering, March-June;
          found at elevations below 3000'; rays usually
          yellow, without a maroon blotch)
          I Lower leaves upright along the stem; flower
             heads many; annual up to 90 cm tall; on
             grassy slopes   *Madia radiata*
                  Showy Madia; CC-s; 1b
          II Lower leaves spreading horizontally; flower
             heads few; perennial up to 65 cm tall; in
             woods   *Madia madioides*
                      Woodland Madia
       HH Anthers black or purple-black (stems often
          reddish or dark purple; on dry slopes)
          I Basal cluster of leaves well developed;
             glands on upper stems and leaves usually
             obvious without a hand lens; up to more
             than 1 m tall, fall-flowering, August-
             November (at elevations up to 3000';
             widespread) *Madia elegans* ssp. *densifolia*
                        Common Madia
          II Basal cluster of leaves either lacking or
             not well developed; glands on upper stems
             and leaves not obvious without a lens; less
             than 1 m tall, spring- or summer-flowering,
             March-August
             J Rays usually yellow, without a maroon
               blotch at the base; spring-flowering,
               March-June; at elevations below 3000'
                    Madia elegans ssp. *vernalis*
                        Springtime Madia
            JJ Rays usually yellow, often with a maroon
              blotch at the base; summer-flowering,
               June-August; at elevations of 3000-
              8000'
                Madia elegans ssp. *elegans* (*Plate* 13)
                        Elegant Madia
DD Plants without glandular hairs (but they may have other
  hairs)
  E Leaves opposite on the lower portion of the stems, but
    alternate above
    F Leaves not rough to the touch, but covered with soft
       hairs
       G Phyllaries united, forming a cup; in grassland
                         Monolopia major
                 Cupped Monolopia; SF-Mo

GG Phyllaries not united; in wooded or shrubby areas
Monolopia gracilens
Woodland Monolopia; CC(MD)-s
FF Leaves rough to the touch because of bumps or short
hairs
G   Leaves broad, rarely so much as twice as long as
wide; disk flowers reddish (flower heads 2.5-3.5
cm wide; widespread)   *Helianthus annuus*
[includes *H. annuus* ssp. *lenticularis*] (*Plate* 11)
Common Sunflower
GG Most leaves 4 times as long as wide; disk flowers
yellowish or brownish
H   Phyllaries densely hairy
I   Rays 8-13 mm long; bracts below the flower
heads extending beyond the phyllaries
Lagophylla minor
Lesser Hareleaf; Na-n
II Rays 3-5 mm long; bracts below the flower
heads usually not extending beyond the
phyllaries
J   Flower heads in dense clusters, these
1.5-6 cm wide   *Lagophylla ramosissima*
ssp. *congesta* [*L. congesta*]
Rabbitfoot; SCr-n
JJ Flower heads single or in clusters not
more than 1.5 cm wide
Lagophylla ramosissima ssp. *ramosissima*
Common Hareleaf
HH Phyllaries not densely hairy (but they may have
some hairs)
I   Lower phyllaries longer than the combined
height of the involucre and disk; mostly in
wet places
J   Flower heads 2.5-3.5 cm wide, solitary
Helianthella castanea
Diablo Helianthella
JJ Flower heads 1.8-2.5 cm wide, in
panicles   *Helianthus californicus*
California Sunflower; Na-s
II Lower phyllaries shorter than the combined
height of the involucre and disk; mostly in
dry places
J   Phyllaries with long hairs (visible
without a lens) on the margin; leaves
narrow, with one main vein from the
base   *Helianthus ciliaris*
Blueweed; na; Al-s
JJ Phyllaries with short hairs, these not
confined to the margin; leaves wide,
with 3 main veins from the base
Helianthus gracilentus (*Plate* 12)
Slender Sunflower; CC-s
EE All leaves opposite
F   Leaves toothed
G   Leaves triangular, the blades not so much as 2
times as long as wide; petiole of lower leaves 4-7
cm long   *Arnica cordifolia*
Heartleaf Arnica
GG Leaves elliptic, the blades 2-4 times as long as
wide; petiole of lower leaves 1-2 cm long
H   Rays not more than 2 cm long, pale yellow;
flower heads nodding as they age *Bidens cernua*
Nodding Bur Marigold; SF-n
HH Rays 1.5-3 cm long, bright yellow; flower heads
not nodding as they age
Bidens laevis (*Plate* 72)
Bur Marigold; Ma-s
FF Leaves smooth-margined or lobed, but not toothed
(leaves up to 8 mm wide, usually much less)
G   Phyllaries in 2-3 series, the tips not all at the
same level (leaves succulent; in salt marshes)
Jaumea carnosa (*Plate* 12)
Fleshy Jaumea; SFBR-n

GG Phyllaries in 1 series, the tips all at the same
level
  H Phyllaries united at their bases to form a cup
(the tips, however, are free)
    I Rays shorter than the height of the
involucre, scarcely visible
                Lasthenia glaberrima
                Smooth Goldfields; Al-n
    II Rays longer than the height of the
involucre
      J Achenes not hairy, but the surface may
be roughened (mostly in vernal pools
and alkaline flats) *Lasthenia glabrata*
              Yellowray Goldfields
      JJ Achenes with short hair
              Lasthenia ferrisiae
            Ferris Goldfields; CC-Sl0
HH Phyllaries free from one another
  I Phyllaries 3-6; flower heads higher than
wide, ray flowers 1-3 *Lasthenia microglossa*
            Smallray Goldfields; CC-s
  II Phyllaries at least 8; flower heads usually
about as high as wide, ray flowers 10-11
    J Leaves 4-8 mm wide, either with several
lobes (these mainly at the tip) or with
long hairs on the margin
      K Leaves with several lobes
concentrated at the tip, without
hairs; basal leaves absent (somewhat
succulent; coastal)
              Lasthenia maritima
          [*L. minor* ssp. *maritima*]
          Maritime Goldfields; Mo-n
      KK Leaves not lobed, with long hairs on
the margin; basal leaves prominent
            Lasthenia macrantha
          Seacoast Goldfields; Ma-n
    JJ Leaves rarely more than 3 mm wide,
either not lobed or with a few narrow
lobes (these not mainly at the tip)
      K Leaves less than 1 mm wide (and
usually divided pinnately or
bipinnately into slender lobes);
rays 3-6 mm long (often near drying
pools)   *Lasthenia fremontii*
        Fremont Goldfields; Sl, CC
      KK Leaves (and the lobes, if these are
present), at least 2 mm wide; rays
4-10 mm long
        L Leaves not lobed; on grassy
slopes (widespread)
            Lasthenia californica
      [*L. chrysostoma* sspp. *chrysostoma*
and *hirsutula*] (*Plate* 13)
              Goldfields
        LL Lower leaves lobed; in moist
areas   *Lasthenia minor*
         Woolly Goldfields; Sn-s

Asteraceae, Division III, Subkey: Flower heads with both ray flowers and
disk flowers; ray flowers completely or partly yellow, greenish yellow,
orange, red, or a closely related color; leaves present along the stem, all
alternate
A Plants woody, at least at the base
  B Leaves neither toothed nor lobed
    C Underside of leaves densely hairy   *Eriophyllum staechadifolium*
        [includes *E. staechadifolium* var. *artemisiaefolium*]
          Seaside Woolly Sunflower; SCr-Me
    CC Underside of leaves not densely hairy
      D Rays 13-18, up to about 15 mm long (leaves 10-40 mm long and
up to 2.5 mm wide)   *Ericameria linearifolia*
        [*Haplopappus linearifolius*] (*Plate* 10)
          Interior Goldenbush; La-s
      DD Rays 11 or fewer, not more than 5 mm long

       E   Rays 2-6; leaves 4-12 mm long, almost cylindrical, about 1
         mm wide, and in clusters (on backshores of beaches)
            Ericameria ericoides [*Haplopappus ericoides*] (*Plate* 73)
                               Mock-heather
      EE Rays 8-11; leaves up to 50 mm long, flat, mostly wider
         than 1 mm, not in clusters   *Gutierrezia californica*
                  California Matchweed; SFBR-n + e
BB Leaves toothed or lobed
  C   Leaves toothed, rarely lobed, not hairy; phyllaries glandular
    (flower heads 3-5 cm wide; at borders of salt marshes and on
    seaside bluffs)
      D   Lower leaves with petioles; not often more than 50 cm tall,
         woody only near the base (widespread)   *Grindelia stricta*
            var. *platyphylla* [includes *G. stricta* ssp. *venulosa*]
                        Pacific Gumplant; Ma-n
      DD All leaves sessile; up to 150 cm tall, woody almost
         throughout, except for stems produced in the last year
            Grindelia hirsutula var. *hirsutula* [includes *G. humilis*]
                        Hairy Gumplant; SFBR
  CC Leaves lobed, woolly-hairy on the underside; phyllaries not
    glandular
      D   Involucre not more than 5 mm wide; flower heads in clusters
         of 5 or more (usually more than 10), each head on a peduncle
         not more than 1 cm long
         E   Rays (6-8) 6-9 mm long   *Eriophyllum latilobum*
                  San Mateo Woolly Sunflower; SM; 1b
         EE Rays 2-5 mm long
           F   Phyllaries 5-6; rays 4-6; on dry slopes in the Coast
              Ranges (widespread)   *Eriophyllum confertiflorum*
              [includes *E. confertiflorum* var. *laxiflorum*] (*Plate* 10)
                        Golden Yarrow
           FF Phyllaries 8-11; rays 6-9; on coastal bluffs
                    Eriophyllum staechadifolium
            [includes *E. staechadifolium* var. *artemisiaefolium*]
               Seaside Woolly Sunflower; SCr-Me
      DD Involucre at least 10 mm wide; flower heads mostly solitary
         on peduncles at least 2 cm long
         E   Rays 5-9, golden (involucre 5-7 mm high; sometimes on
           serpentine)   *Eriophyllum jepsonii*
                Jepson Woolly Sunflower; CC-SB; 4
         EE Rays 8-15, yellow
           F   Involucre 5-8 mm high; leaves up to 1.5 cm wide, the
              unlobed portion nearest the petiole 2-3 mm wide; in
              Coast Ranges (widespread)
              Eriophyllum lanatum var. *achillaeoides* (*Plate* 11)
                Common Woolly Sunflower; SCl, Mo-n
           FF Involucre 8-10 mm high; leaves up to 2 cm wide, the
              unlobed portion nearest the petiole about 5 mm wide;
             coastal bluffs   *Eriophyllum lanatum* var. *arachnoideum*
                   Coastal Woolly Sunflower
AA Plants not woody
  B   Plants with glandular hairs, or phyllaries sticky because of
    glandular secretions
    C   Upper leaves often clustered, ending abruptly in an open gland
      D   Anthers yellow
         E   Stems and leaves in upper part of plant densely hairy;
           each phyllary with 25-50 slender, gland-tipped structures
                  Holocarpha heermannii (*Plate* 12)
                 Heermann Tarplant; CC-SCl(MH)
         EE Stems and leaves in upper part of plant not densely hairy;
           each phyllary with 5-20 stout, gland-tipped structures
         Holocarpha obconica [includes *H. obconica* ssp. *autumnalis*]
                San Joaquin Tarplant; Al, CC-s
      DD Anthers black
         E   Ray flowers 3-7; flower heads scattered in a raceme
                         Holocarpha virgata
                 Virgate Tarplant; Na, SCl
         EE Ray flowers 8-16; flower heads densely clustered
                     Holocarpha macradenia
           Santa Cruz Tarplant; Ma, Al, SCr; 1b
  CC Upper leaves not clustered, and not ending in an open gland
    (there may be glands elsewhere on the plant, however)
    D   Leaves (not more than 5 mm wide) smooth-margined

E Achene of each ray flower partly enclosed by the adjacent
   phyllary (leaves 6-15 cm long and 0.5 cm wide; rays often
   3-lobed; plants sometimes glandular only on the lower
   portion, but usually hairy throughout)
                           Hemizonia congesta ssp. *congesta*
                   [includes *H. lutescens* and *H. multicaulis* sspp.
                           multicaulis and *vernalis*] (*Plate* 12)
                           Yellow Hayfield Tarweed; Me-Ma, Al, Sn
EE Achenes of individual ray flowers not enclosed by
   phyllaries
   F Leaves up to 5 cm long; very small, globular glands
      present on the stems and leaves; involucre 7-10 mm
      high; rays not 3-lobed; perennial (plant hairy all
      over and also camphor-scented)
          Heterotheca sessiliflora ssp. *echioides* [*Chrysopsis
          villosa* vars. *camphorata* and *echioides*] (*Plate* 12)
                           Bristly Golden Aster; Sn, Al, SCl(MH)-s
   FF Leaves up to more than 8 cm long; prominent, saucer-
      shaped glands present at the tip of the upper leaves
      and sometimes on the phyllaries; involucre 4-7 mm
      high; rays often 3-lobed (sometimes with a red dot at
      the base); annual
      G Leaves and stems usually not hairy (but the basal
         portions of the lower leaves have bristly
         margins); saucer-shaped glands usually not present
         on the phyllaries   *Calycadenia truncata*
                           Rosinweed; Me-SCl
      GG Leaves and stems usually hairy; saucer-shaped
         glands usually present on the phyllaries
              Calycadenia multiglandulosa [includes *C. hispida*
              ssp. *reducta* and *C. multiglandulosa* sspp.
                  cephalotes and *robusta*] (*Plate* 9)
                           Sticky Rosinweed; Me-SCl
DD Leaves toothed, lobed, or compound (sometimes, however, the
   leaves of species of *Grindelia* are smooth-margined)
   E Achene of each ray flower completely enclosed by the
      adjacent phyllary
      F Stems with black dots or streaks
         G Rays usually much more than 5 mm long; anthers
            black   *Layia gaillardioides*
                           Woodland Layia
         GG Rays less than 4 mm long; anthers yellow
                           Layia hieracioides
                           Tall Layia; Al, CC(MD)
      FF Stems without black dots or streaks (rays much more
         than 5 mm long)
         G Anthers black (widespread)   *Layia platyglossa*
            [includes *L. platyglossa* ssp. *campestris*] (*Plate* 13)
                           Tidytips; Me-s
         GG Anthers yellow
            H Pappus consisting of 16-21 bristles; often on
               serpentine soils   *Layia septentrionalis*
                           Colusa Layia; Sn; 1b
            HH Pappus consisting of 10-15 bristles; on sandy
               soil, inland   *Layia glandulosa*
                           White Layia; CC-s
   EE Achene of each ray flower only partly, or not at all,
      enclosed by the adjacent phyllary
      F Phyllaries in one series (and partly enclosing each of
         the ray-flower achenes); lower leaves lobed
         G Leaves with spinelike tips or lobes
                           Hemizonia fitchii
                           Fitch Spikeweed; Me-SLO
         GG Leaves without spinelike tips
            H Rays 5; anthers yellow   *Hemizonia kelloggii*
                           Kellogg Spikeweed; SFBR-s
            HH Rays at least 8; anthers black or tan
               I Rays 18-32, without purple veins; anthers
                  black   *Hemizonia corymbosa*
                           Coast Spikeweed; Me-Mo

II Rays 8-13, usually with purple veins;
anthers tan
Hemizonia congesta ssp. *congesta* [includes
H. lutescens and *H. multicaulis* sspp.
multicaulis and *vernalis*] (*Plate* 12)
Yellow Hayfield Tarweed; Me-Ma, Al, Sn
FF Phyllaries in more than 1 series; lower leaves
toothed, but not lobed
  G   Tip of outer phyllaries curled back, sometimes so
  much that they form loops
    H   Plants slightly succulent; upper stems somewhat
    hairy; in coastal areas, including borders of
    salt marshes (sometimes prostrate; widespread)
Grindelia stricta var. *platyphylla*
[includes *G. stricta* ssp. *venulosa*]
Pacific Gumplant
    HH Plants not succulent; upper stems not hairy;
    inner Coast Ranges
      I   Lower leaves up to 15 cm long, not sessile;
      rays 12-25   *Grindelia nana*
Idaho Gumplant; Na-n
      II Lower leaves up to 8 cm long, sessile; rays
      18-35
Grindelia camporum [includes *G. camporum*
var. *parviflora* and *G. procera*] (*Plate* 11)
Great Valley Gumplant
  GG Tip of outer phyllaries straight, spreading
  outward, or curved, but not curled back
    H   Upper stems mostly hairy; rays (16-35) up to 20
    mm long (in Coast Ranges and salt marshes)
Grindelia hirsutula var. *hirsutula*
[includes *G. humilis*]
Hairy Gumplant; Na-Mo
    HH Upper stems usually not hairy; rays up to 15 mm
    long
      I   Lower leaves up to 18 cm long, not sessile;
      rays 30-40 (on ocean bluffs)
Grindelia hirsutula var. *maritima*
[*G. maritima*]
San Francisco Gumplant; SF; 1b
      II Lower leaves up to 8 cm long, sessile; rays
      18-35
Grindelia camporum [includes *G. camporum*
var. *parviflora* and *G. procera*] (*Plate* 11)
Great Valley Gumplant
BB Plants not glandular (but they may be hairy)
  C   Leaves not toothed or lobed (species of *Layia*, keyed under choice
  CC, sometimes have leaves without teeth or lobes)
    D   Phyllaries obscured by dense hairs
      E   Flower heads in dense clusters, these 1.5-6 cm wide
Lagophylla ramosissima ssp. *congesta* [*L. congesta*]
Rabbitfoot; SCr-n
      EE Flower heads single or in clusters not more than 1.5 cm
      wide   *Lagophylla ramosissima* ssp. *ramosissima*
Common Hareleaf
    DD Phyllaries not obscured by dense hairs (but some hairs may be
    present)
      E   Flower heads 4-5 cm wide (rays about 20 mm long)
Helianthella castanea
Diablo Helianthella; 1b
      EE Flower heads less than 2.5 cm wide
        F   Rays 2-3 mm long
          G   Leaves not often more than 1 mm wide; flower heads
          solitary at the tip of branches (usually not more
          than 10 flower heads on each plant); pappus
          consisting of scales; annuals, not more than 30 cm
          tall   *Rigiopappus leptocladus*
Rigiopappus
          GG Leaves usually at least 3 mm wide, flower heads
          numerous, concentrated in cymes or racemes; pappus
          consisting of bristles; perennials, up to more
          than 100 cm tall

H   Leaves usually not more than 5 mm wide, with
dark, glandular pits; flowers in cymes
Euthamia occidentalis
[*Solidago occidentalis*] (Plate 73)
Western Goldenrod
HH Leaves generally at least 10 mm wide (and
usually widest above the middle), without
pits; flowers in racemes (widespread)
Solidago californica (Plate 14)
California Goldenrod
FF Rays at least 5 mm long
G   Leaves with dark glandular pits
H   Leaves about 2 mm wide, clustered; rays 7-10
(these 6-12 mm long)   *Helenium amarum*
Fineleaf Sneezeweed; na; Ma
HH Leaves 10-40 mm wide, not clustered; rays 13-20
I   Rays 4-10 mm long; inflorescence usually
with more than 10 flower heads; annual
Helenium puberulum (Plate 11)
Rosilla
II Rays 13-25 mm long; inflorescence with
fewer than 10 flower heads; perennial
Helenium bigelovii
Bigelow Sneezeweed
GG Leaves without dark glandular pits
H   Leaves up to 1.5 cm long and less than 5 mm
wide, scarcely if at all hairy; pappus absent;
in hard-packed soil of dried vernal pools
(involucre 6-8 mm high; stigmas of ray flowers
red)   *Blennosperma bakeri*
Baker Blennosperma; Sn; 1b
HH Leaves up to 5 cm long and at least 5 mm wide,
hairy; pappus of bristles; in sandy soil
I   Leaves up to 5 cm long; leafy bracts not
present below the flower heads; involucre
7-10 mm high   *Heterotheca sessiliflora*
ssp. *echioides* [*Chrysopsis villosa* vars.
camphorata and *echioides*] (Plate 12)
Bristly Golden Aster; Sn, Al, SCl(MH)-s
II Leaves not more than 4 cm long; leafy
bracts present directly below the flower
heads; involucre 10-15 mm high (usually
coastal)
J   Leaves and stems densely hairy, the
upper leaves more than 1 cm long, with
petioles; branching extensively from
the base
Heterotheca sessiliflora ssp. *bolanderi*
[*Chrysopsis villosa* var. *bolanderi*]
Bolander Golden Aster; SFBR-Me
JJ Leaves and stems not densely hairy, the
upper leaves not more than 1 cm long,
sessile; usually not branching from the
base (but they may branch above)
Heterotheca sessiliflora ssp.
sessiliflora [*Chrysopsis villosa*
var. *sessiliflora*]
Sessile Golden Aster; Me-s
CC Some leaves toothed or lobed
D   Achene of each ray flower enclosed by a phyllary
E   Leaves without spinelike tips (phyllaries with stout hairs
on the margins)   *Layia chrysanthemoides*
[includes *L. chrysanthemoides* ssp. *maritima*]
Smooth Layia; Me-Mo
EE Leaves with spinelike tips
F   Flower heads 5-10 mm high; pappus of 3-5 scales
G   Leaves and stems glandular and with long hairs;
rays 3-6 mm long (often near marshes)
Hemizonia parryi ssp. *parryi*
Pappose Spikeweed; SM-n
GG Leaves and stems not glandular and without long
hairs; rays 2.5-3 mm long
Hemizonia parryi ssp. *congdonii*
Congdon Spikeweed; Al, CC-s; 1b

```
FF Flower heads 3-6 mm high; pappus absent
   G  Leaves rough to the touch; usually much more than
      10 cm tall; inland, in dry habitats
                           Hemizonia pungens ssp. pungens
                                  Common Spikeweed; Ma
   GG Leaves not rough to the touch; less than 10 cm
      tall; around San Francisco Bay, in salt marshes as
      well as in dry habitats
                          Hemizonia pungens ssp. maritima
                                Saltmarsh Spikeweed; SFBR-s
DD Achene of each ray flower not enclosed by a phyllary
   E  Flower heads more than 1.5 cm wide (leaves toothed)
      F  Rays curled at the tip
         G  Stem and leaves hairy; pappus red (widespread)
                            Heterotheca grandiflora (Plate 12)
                                        Telegraphweed
         GG Stem and leaves not hairy; pappus tan or brown
                    Prionopsis ciliatus [Haplopappus ciliatus]
                                        Prionopsis; na; SFBR
      FF Rays not curled at the tip
         G  Rays golden yellow, 3-4 mm wide, with a deep notch
            at the tip; leaves mostly bipinnately lobed
                                   Chrysanthemum segetum
                                        Corn Daisy; me
         GG Rays reddish yellow or purple, 1 mm wide, usually
            somewhat frayed at the tip (but without a distinct
            notch); leaves coarsely toothed
                                   Hulsea heterochroma
                                    Redray Hulsea; SCl-s
   EE Flower heads less than 1.5 cm wide
      F  Leaves not lobed, but some may have small teeth
         G  Stems and leaves not hairy; involucre 5-6 mm high
            (mostly along the coast)
                           Solidago spathulata (Plate 14)
                                  Coast Goldenrod; Mo-n
         GG Stems hairy, and leaves sometimes hairy, at least
            on one surface; involucre 3-5 mm high
            H  Leaves 1-5.5 cm long and less than 0.5 cm wide,
               the upper ones not reduced    Lagophylla minor
                                  Lesser Hareleaf; Na-n
            HH Lower leaves 5-12 cm long and 1-3.5 cm wide,
               the upper ones sometimes much reduced
               I  Inflorescence usually more than 2 (often 4)
                  times as long as wide; leaves hairy on both
                  surfaces; basal and lower leaves toothed,
                  upper leaves smooth-margined, much reduced
                  (widespread)
                            Solidago californica (Plate 14)
                                    California Goldenrod
               II Inflorescence usually not more than 2 times
                  as long as wide; leaves either not hairy or
                  hairy only on the lower surface; all leaves
                  either toothed or smooth-margined, similar
                  in size (widespread)
                  Solidago canadensis ssp. elongata (Plate 14)
                                    Canada Goldenrod
      FF Leaves lobed
         G  Stems and sometimes also the phyllaries and leaves
            with woolly hairs (the hairs may be very dense)
            H  Lobes of leaves regularly toothed or divided
               again into small secondary lobes; pappus
               absent or consisting of scales
                                   Anthemis tinctoria
                                 Golden Marguerite; me; Al
            HH Lobes of leaves not toothed or divided again;
               pappus consisting of bristles
               Senecio flaccidus var. douglasii [S. douglasii]
                                  Bush Groundsel; Me-s
         GG Stems, phyllaries, and leaves without woolly hairs
            H  Lower leaves with 1-3 lobes, these 10-15 mm
               long                    Blennosperma bakeri
                                  Baker Blennosperma; Sn; 1b
            HH Lower leaves with 7-15 lobes, these not more
               than 5 mm long
```

I   Involucre 7 mm high; achenes 3-4.5 mm long;
    up to 30 cm tall; in sandy soil on Pt.
    Reyes   *Blennosperma nanum* var. *robustum*
                   Point Reyes Blennosperma; Ma(PR); 1b
II Involucre 5-6 mm high; achenes 2.5-3 mm
    long; up to 20 cm tall; on wet clay soil
    (widespread)
               Blennosperma nanum var. *nanum* (*Plate* 9)
                                   Common Blennosperma

Berberidaceae--Barberry Family

The Barberry Family, consisting of shrubs and perennial herbs, has only
a few local species. The basic flower formula is as follows: 6 sepals,
often petal-like, in 2 circles; 6 petals, also in 2 circles; 6 stamens;
fruit dry or fleshy, with only 1 compartment. But there are exceptions, one
of them being *Achlys triphylla*, Vanilla-leaf. Its flowers lack sepals and
petals, and have 6-13 stamens.

Many exotic species of *Berberis* have long been in cultivation. There are
4 western species with which gardeners interested in native plants should
be familiar. Two of them--*B. pinnata* and *B. nervosa*--grow wild in our area.
The former prefers a sunny situation, the latter requires shade. *Berberis
aquifolium*, called Oregon-grape (it is the State Flower of Oregon), is also
an excellent subject for a sunny garden. *Vancouveria planipetala*, called
Inside-out-flower, is a deciduous herb found mostly in redwood forests. It
spreads by underground stems and makes an attractive ground cover. The
species mentioned are likely to be available at sales of native plants, and
at some nurseries.

A   Shrubs; leaves pinnately compound; marginal teeth of leaflets spine-
    tipped
    B   Stems not branched; leaves 25-45 cm long; leaflets usually 11-21,
        each with 12-24 teeth; usually in shaded habitats   *Berberis nervosa*
                                        Longleaf Oregon-grape; Mo-n
    BB Stems branched; leaves 5-15 cm long; leaflets usually 5-11, each with
        16-40 teeth; usually in sunny habitats   *Berberis pinnata* (*Plate* 15)
                                        Shinyleaf Oregon-grape
AA Herbs; leaves compound, but not in a pinnate pattern; marginal teeth, if
    present on leaflets, not spine-tipped
    B   Leaves with 3 approximately fan-shaped leaflets arising at the top of
        the petiole, the broad end of the leaflets coarsely toothed;
        inflorescence dense, narrow, the flowers less than 5 mm long, with
        neither calyx nor corolla   *Achlys triphylla* (*Plate* 15)
                                        Vanilla-leaf; Sn-n
    BB Leaves with 3 primary divisions, each of these divided again into 3
        long-stalked leaflets; inflorescence a loose panicle, the flowers 6-
        8 mm long, with 6 sepals (in 2 series) and 6 petals, these turned
        back   *Vancouveria planipetala* (*Plate* 15)
                                        Inside-out-flower; Mo-n

Betulaceae--Birch Family

The Birch Family is represented in our region by a hazelnut and 2
species of alders, all of which have alternate, deciduous leaves. In
alders, both the pistillate and staminate flowers are in catkins; the
pistillate catkins, especially after becoming woody, resemble the cones of
coniferous trees. In hazelnuts, only the staminate flowers are in catkins;
the pistillate flowers are produced singly or in clusters, but as a rule
only 1 matures as an acornlike nut.

Alders, like many other plants, especially members of the Pea Family,
have nitrogen-fixing bacteria in nodules on their roots. These bacteria,
with help from their hosts, are able to utilize atmospheric nitrogen gas in
synthesis of amino acids and proteins. Much of the nitrogen "fixed" by the
bacteria becomes incorporated into the trees. Thus alder seedlings may
flourish on soils that are poor in ammonia and nitrates, which are the

usual sources of nitrogen for plants. The symbiotic association accounts, in part, for the success of some species on land that has been deforested. When the leaves decay, their organic nitrogen is converted by various soil bacteria into ammonia and nitrates. By enriching the soil, and also by making the soil more acid, alders prepare the land for colonization by other trees, such as Douglas Fir. It is a little ironic that the increased acidity makes the soil unsuitable for succeeding generations of alder seedlings.

A   Leaves slightly notched at the base, the 2 sides of the basal portion
    often not quite equal; staminate flowers in catkins, pistillate flowers
    (each with 2 red stigmas) single or in small clusters; pistillate
    flowers developing into nuts about 15 mm long, each nut enclosed by a
    leafy involucre   *Corylus cornuta* var. *californica* (*Plate* 74)
                                                                   Hazelnut
AA Leaves not notched at the base, and the 2 sides of the basal portion
    equal; both staminate and pistillate flowers in catkins; pistillate
    catkins, when mature, resembling conifer cones (they consist of numerous
    bracts, each of which has a pair of pistils above it); fruit flattened,
    about 2.5 mm long
    B   Margin of leaf blades rolled, with coarse teeth that are usually
        slightly serrated; underside of leaves slightly rusty-gray; fruit
        (attached to the axis of the catkins) with prominent membranous
        margins that are at least half as wide as the thickened portion;
        mostly near the coast   *Alnus rubra* [*A. oregona*] (*Plate* 74)
                                                            Red Alder; SCr-n
    BB Margin of leaf blades not rolled, the teeth small and more or less
        the same size; underside of leaves not at all rusty-gray; fruit with
        thin margins, but these not membranous; mostly along streams away
        from the coast (widespread)   *Alnus rhombifolia* (*Plate* 74)
                                                                White Alder

Boraginaceae--Borage Family

Some of the Old World species of the Borage Family have long been used as kitchen flavorings, herbal remedies, and sources of dyes. A few of them, such as Borage (*Borago officinalis*) and Comfrey (*Symphytum asperum*), are found growing as escapes in California, but they will be close to civilization and thus not likely to be mistaken for native species.

With few exceptions, the foliage and stems of our local representatives have bristly hairs. The inflorescences are often one-sided and at least slightly coiled, in general form resembling the scroll of a violin. The calyx and corolla are generally 5-lobed. The 5 stamens, attached to the corolla tube, alternate with the corolla lobes. The fruiting portion of the pistil is usually deeply divided into 4 lobes (sometimes only 2 lobes) that typically separate into dry, 1-seeded nutlets. In some species, however, the nutlets do not develop equally, and only one may reach maturity. There are various other deviations from the formula of flower structure, but after seeing a few members of the family, one will generally be able to recognize a borage even if it does not meet all of the qualifications set forth above. Separating species of *Plagiobothrys* and *Cryptantha*, or even telling these two genera apart, may require use of a dissecting microscope.

A   Corolla orange or yellow (in *Myosotis discolor*, under choice AA, the
    corolla changes from yellow to blue as the flowers age)
    B   Corolla yellow
        C   Corolla throat constricted and partly obstructed by somewhat
            saclike elevations; all stamens attached at the same level below
            the constriction (usually in moist areas) *Amsinckia lycopsoides*
                                                     Bugloss Fiddleneck; SLO-n
        CC Corolla throat wide open, not obstructed by any elevations;
            stamens attached somewhat irregularly, not all at the same level
            D   Corolla 4-7 mm long, the tube barely, if at all, extending
                beyond the calyx lobes
                        Amsinckia menziesii var. *menziesii* [includes *A. retrorsa*]
                                                            Rancher's-fireweed

DD Corolla 7-10 mm long, the tube extending well beyond the
   calyx lobes
   E   Corolla slightly 2-lipped, and its tube bent (Coast
       Ranges)   *Amsinckia lunaris*
                                     Bentflower Fiddleneck; 4
   EE Corolla perfectly regular, and its tube straight
                            Amsinckia menziesii var. *intermedia*
                                 [*A. intermedia*] (*Plate* 16)
                                     Intermediate Fiddleneck

BB Corolla orange
   C   Corolla 14-18 mm long (and 8-10 mm wide) (inland)
       D Nutlets with a smooth surface, shiny   *Amsinckia grandiflora*
                            Largeflower Fiddleneck; CC, Al; 1b
       DD Nutlets with a rough surface
                  Amsinckia eastwoodiae [*A. intermedia* var. *eastwoodae*]
                                     Valley Fiddleneck
   CC Corolla not more than 12 mm long (nutlets with a rough surface)
       D   Corolla tube with 20 dark veins below the level where the
           stamens are attached (the veins are visible externally);
           surface of nutlets resembling a cobblestone pavement (the
           bumps are crowded, contiguous) (plants very bristly-hairy; in
           sandy or gravelly areas, inner Coast Ranges)
           E   Corolla 8-12 mm long and 2-6 mm wide; anthers pressed
               against the stigma
                       Amsinckia tessellata var. *tessellata* (*Plate* 16)
                                     Devil's-lettuce; CC-s
           EE Corolla 12-16 mm long and 6-10 mm wide; anthers diverging
               away from the stigma
                       Amsinckia tessellata var. *gloriosa* [*A. gloriosa*]
                                 Largeflower Devil's-lettuce; CC-s
       DD Corolla tube with only 10 dark veins below the level where
           the stamens are attached; surface of nutlets not resembling a
           cobblestone pavement (the bumps are not crowded together)
           E   Corolla usually 1-1.5 cm long; 2 or 3 calyx lobes, due to
               fusion, much shorter than the others, which are separate
               nearly to the base (usually around salt marshes and in
               sandy habitats at the coast)   *Amsinckia spectabilis*
                                     Coast Fiddleneck
           EE Corolla not often more than 1 cm long; all calyx lobes
               separate nearly to the base
               F   Corolla slightly 2-lipped, and its tube bent (Coast
                   Ranges)   *Amsinckia lunaris*
                                     Bentflower Fiddleneck; 4
               FF Corolla perfectly regular, and its tube straight
                            Amsinckia menziesii var. *intermedia*
                                 [*A. intermedia*] (*Plate* 16)

AA Corolla primarily blue, cream, white, or white tinged with yellow or
   pink, but not distinctly yellow or orange (in *Myosotis discolor*, under
   choice B, the corolla changes from yellow to blue as the flowers age)
   B   Corolla primarily blue or purple
       C   Blades of lower leaves oval, up to about 15 cm long; corolla
           bright blue, with a ring of white lobes at the throat and
           usually with some purple in the throat (flowers in a loose
           raceme; nutlets covered with short prickles; generally in
           somewhat shaded habitats)   *Cynoglossum grande* (*Plate* 16)
                                     Hound's-tongue
       CC Blades of largest leaves not more than 10 cm long, and usually
           much smaller; corolla not blue with white lobes (introduced,
           sometimes escaping from gardens)
           D   Corolla 2 cm wide (blue) (stamens protruding from the
               corolla)   *Borago officinalis*
                                     Borage; me; SF
           DD Corolla not more than 1.2 cm wide
               E   Corolla 12 mm wide (purple or dark blue); inflorescence
                   tightly coiled; shrubby, 1-2 m tall   *Echium candicans*
                                     Viper's-bugloss; af; SF, Ma
               EE Corolla not more than 10 mm wide; inflorescence not
                   tightly coiled; herbaceous, less than 1 m tall (usually
                   much less) (in moist areas)
                   F   Corolla about 5-8 mm wide, light blue, with a ring of
                       white lobes at the throat; largest leaves about 2 cm
                       wide; up to 50 cm tall, perennial (established along
                       the coast)   *Myosotis latifolia*
                                     Forget-me-not; af; Mo-n

FF Corolla about 4 mm wide, at first pale yellow, later
blue; largest leaves less than 1 cm wide; up to 30 cm
tall, annual   *Myosotis discolor* [*M. versicolor*]
                Yellow-and-blue Scorpion-grass; eu; Sn-n
BB Corolla primarily white, or white tinged with yellow (sometimes
becoming yellow or pink on drying)
C   Leaves either all alternate, or mainly basal, in which case there
are a few reduced upper leaves (in *Plagiobothrys tener*, under
choice CC, the leaves are sometimes mostly basal)
D   Basal cluster of leaves well developed, upper leaves much
reduced (widespread)
E   Corolla 6-8 mm wide (sometimes only 3 mm) (calyx 2-3 mm
long; sap purple; hairs on the lobes of the calyx rust-
colored, those on the rest of the calyx mostly white)
                Plagiobothrys nothofulvus (*Plate* 16)
                    Common Popcornflower
EE Corolla not more than 4 mm wide
F   Calyx 5 mm or more long; corolla 3-4 mm wide; sap
purple; hairs on calyx mostly rust-colored
                        Plagiobothrys fulvus
            [includes *P. fulvus* var. *campestris*]
                    Hairy Popcornflower; SCl-n
FF Calyx 3-5 mm long; corolla 2-3 mm wide; sap usually
not purple; hairs on calyx white or rust-colored
                        Plagiobothrys tenellus
                    Delicate Popcornflower
DD Basal cluster of leaves absent or not well developed, upper
leaves not markedly reduced
E   Corolla with a purple ring at the throat (the purple may
be preceded by yellow); leaves slightly succulent, not
hairy; perennial (stems and leaves slightly grayish owing
to a coating of a bluish wax; usually growing in moist,
somewhat saline places at the coast, but occasionally in
similar habitats inland)   *Heliotropium curassavicum*
                [*H. curassavicum* var. *oculatum*] (*Plate* 16)
                        Seaside Heliotrope
EE Corolla without a purple ring in the throat; leaves not
succulent, usually at least slightly hairy; annual
F   Calyx lobes with hooked bristles on the margin;
corolla not extending beyond the calyx; mostly less
than 20 cm tall   *Pectocarya pusilla*
                    Little Pectocarya; Mo-n
FF Calyx lobes without hooked bristles on the margin;
corolla extending beyond the calyx; mostly more than
20 cm tall
G   Midrib of calyx lobes and leaves purple-stained;
corolla white, drying to pink
                        Plagiobothrys infectivus
                    Dye Popcornflower; SLO-n
GG Plants not purple-stained; corolla white, drying to
yellow   *Lithospermum arvense*
                    Gromwell; eua; SFBR
CC All leaves, or at least the lower ones, opposite (sometimes the
lower leaves on *Heliotropium curassavicum*, under choice C
[above], are opposite)
D   Stems and leaves with densely crowded long hairs; calyx lobes
with white or straw-colored hairs which are usually longer
than the lobes; stems usually branched near the tip into 2-5
leafless flower racemes (sometimes also branched at the
base); side of nutlets on which the attachment scar is
located without a keel, but with a distinct lengthwise
groove, this reaching the attachment scar, which is close to
the base
E   Bristles on calyx distinctly curved, sometimes hooked
(hairs on stems lying flat, pointing upwards; mostly on
dry slopes)   *Cryptantha flaccida*
                    Flaccid Cryptantha
EE Bristles on calyx not distinctly curved or hooked
F   Calyx around maturing nutlets usually less than 2 mm
long
G   Nutlets 4, one usually noticeably larger than the
other and at least one roughened (often on burned
areas)   *Cryptantha micromeres* (*Plate* 16)
                    Minuteflower Cryptantha; Ma-s

GG Nutlet solitary, smooth (on dry slopes)
$$Cryptantha\ microstachys$$
Hairy Cryptantha
FF Calyx around maturing nutlets at least 2 mm long
  G Many hairs on stems pointing outward, rather than
lying flat against the stems
    H Corolla not more than 1.5 mm in diameter;
nutlets smooth (calyx around maturing nutlets
3-6 mm long, the bristles 2-2.5 mm long;
mostly on hillsides near the coast)
Cryptantha torreyana
[includes *C. torreyana* var. *pumila*]
Torrey Cryptantha; Ma-n
    HH Corolla 2-7 mm in diameter; nutlets roughened
      I   Calyx lobes not wider at the base than near
the middle; nutlets smooth
Cryptantha clevelandii
[includes *C. clevelandii* var. *florosa*]
Cleveland Cryptantha; Ma(PR)
      II Calyx lobes widest at or near the base; at
least one nutlet roughened
        J Calyx around maturing nutlets 4-6 mm
long; nutlets elongated; style of
pistil not extending beyond the tip of
the nutlets   *Cryptantha intermedia*
Intermediate Cryptantha; La, Na, SLO
        JJ Calyx around maturing nutlets 2-4 mm
long; nutlets rounded; style of pistil
extending beyond the tip of the nutlets
(mostly in gravelly or rocky habitats;
widespread)
Cryptantha muricata (*Plate* 16)
Prickly Cryptantha; CC-s
GG Most hairs on stems pointing upward and lying flat
against the stems
  H At least one nutlet roughened
    I   Corolla less than 1 mm long; calyx around
maturing nutlet 4-5 mm long; up to 15 cm
tall, the stems sometimes falling; in open
sandy habitats at elevations below 2500'
Cryptantha hooveri
Hoover Cryptantha; CC-s; 4
    II Corolla 1-1.5 mm long; calyx around
maturing nutlets 3-7 mm long; up to 40 cm
tall, the stems not falling; in woods at
elevations above 2500' *Cryptantha simulans*
Pine Cryptantha
  HH Nutlets smooth (calyx around maturing nutlets
2-4 mm long)
    I   Corolla 2-2.5 mm wide; nutlet solitary; on
serpentine soil (inner Coast Ranges)
Cryptantha hispidula
Napa Cryptantha; La, Na
    II Corolla not more than 1.5 mm wide; nutlets
usually 4; not on serpentine soil
      J Corolla 1-1.5 mm wide; in sandy habitats
near the shore   *Cryptantha leiocarpa*
Coast Cryptantha
      JJ Corolla less than 1 mm wide; on slopes,
mostly away from the coast
Cryptantha nemaclada
Colusa Cryptantha; SLO-n
DD Stems and leaves without densely crowded long hairs, often
scarcely hairy; calyx lobes with rust-colored hairs at the
tip, the hairs very much shorter than the lobes (there may
also be short white hairs on the lower part of the lobes);
stems branched near the base, each branch with some leaves in
the lower part and a leafless raceme above; side of nutlets
on which the attachment scar is located with a prominent
lengthwise keel (this may extend from the tip to the
attachment scar)
  E Corolla 5-12 mm wide

F   Leaves all opposite, up to 5 mm wide (plants with long
    hairs, those on the calyx lobes yellowish; perennial,
    up to 30 cm tall, the stems often falling down)
                        Plagiobothrys mollis var. *vestitus*
                        Petaluma Popcornflower; Sn; 1a(1888)
FF Leaves opposite below, alternate above, generally much
    less than 5 mm wide
    G   Nutlets 2.5-3 mm long with spines about 1 mm long
        (mostly in the dried mud of vernal pools)
                                Plagiobothrys greenei
                        Green Popcornflower; Sn-n
    GG Nutlets less than 2 mm long, usually without spines
        (the nutlets of *Plagiobothrys stipitatus*, however,
        may be up to 2.5 mm long)
        H   Stems and leaves usually not hairy; known only
            from around sulphur springs near Calistoga,
            Napa County (attachment scar of nutlets deeply
            concave and about one-half as long as the
            nutlets)   *Plagiobothrys strictus*
                        Calistoga Popcornflower; Na; 1b
        HH Stems and leaves at least slightly hairy; not
            confined to sulphur springs
            I   Attachment scar of nutlets at the base (if
                the scar is almost at the base, but
                nevertheless on the side, take choice II)
                (in alkaline habitats or vernal pools)
                                Plagiobothrys stipitatus
                        Valley Popcornflower; Na, Sn-n
            II Attachment scar of nutlets not at the base
                J   Keel on nutlets not in a groove (in wet
                    habitats; corolla 3-7 mm wide)
                                Plagiobothrys tener
                        Inland Popcornflower; Na-n
                JJ Keel on nutlets within a groove (mostly
                    coastal and in moist areas; stems
                    upright or falling)
                    K   Corolla 6-10 mm wide; pedicels
                        usually longer than the calyx
                                Plagiobothrys chorisianus
                                        var. *chorisianus*
                        Choris Popcornflower; SF-SLO
                    KK Corolla 5-6 mm wide; pedicels
                        usually shorter than the calyx
                                Plagiobothrys chorisianus
                                        var. *hickmanii*
                        Hickman Popcornflower
EE Corolla less than 5 mm wide
    F   Nutlets with prominent spines, these covered with
        small barbs
        G   Corolla about 1 mm wide; nutlets 1.5-2 mm long;
            spines short, the bases of these connected by
            transverse ridges   *Plagiobothrys hystriculus*
                        Bearded Popcornflower; Sl; 1a(1892)
        GG Corolla 2.5-4 mm wide; nutlets 2.5-3 mm long;
            spines long, the bases of these not connected by
            transverse ridges (mostly on clay or dried mud)
                                Plagiobothrys greenei
                        Green Popcornflower; Sn-n
    FF Nutlets without spines
        G   Stems succulent, somewhat inflated, and hollow
            (corolla about 3 mm wide; around salt marshes or
            saline habitats inland)   *Plagiobothrys glaber*
                        Hairless Popcornflower; SF; 1a(1954)
        GG Stems not succulent, and not inflated or hollow
            H   Attachment scar of nutlets deeply concave and
                one-third to one-half as long as the nutlets
                (corolla up to 5 mm wide; plants usually not
                hairy; known only from around sulphur springs
                near Calistoga, Napa County)
                                Plagiobothrys strictus
                        Calistoga Popcornflower; Na; 1b
            HH Attachment scar of nutlets not deeply concave
                and not more than one-fifth as long as the
                nutlets

I   Keel on nutlets within a groove
  J   Attachment scar of nutlets about as wide
     as long; corolla up to 3.5 mm wide; in
     coastal locations
              Plagiobothrys reticulatus
     var. *rossianorum* [includes *P. diffusus*]
           Northern Popcornflower; Ma, SF
  JJ Attachment scar of nutlets narrow;
     corolla up to 2 mm wide; in mud flats
     and vernal pools
             Plagiobothrys undulatus
         Southern Popcornflower; Me-s
II Keel on nutlets not in a groove (in mud
  flats or vernal pools)
  J   Racemes with few, if any, bracts (in wet
     habitats, mostly away from the coast)
              Plagiobothrys tener
         Inland Popcornflower; Na-n
  JJ Racemes with bracts either at the base
     or throughout
     K   Bracts at the base of the raceme;
        attachment scar of nutlets somewhat
        out of line with the keel
             Plagiobothrys bracteatus
          Bracted Popcornflower
     KK Bracts along the entire length of
        the raceme; attachment scar of
        nutlets clearly in line with the
        keel   *Plagiobothrys trachycarpus*
        Roughseed Popcornflower; CC-s

Brassicaceae (Cruciferae)--Mustard Family

Plants of the Mustard Family are easy to recognize when in flower, for
there are few exceptions to the following formula: 4 sepals, 4 petals, 6
stamens (4 long, 2 short), and a pistil that is partitioned lengthwise into
2 divisions, each with its own row of seeds.

This large family provides us not only with mustard, but with radishes,
cabbage, cauliflower, broccoli, bok choy, kohlrabi, watercress, and other
food plants. It also gives us numerous valuable ornamentals: *Aubrieta
deltoidea*, the Purple Rock Cress, *Erysimum capitatum*, Common Wallflower,
and *Matthiola incana*, called Stock, are just a few of them.

Not many California species are sufficiently showy to be widely
cultivated, although some kinds of *Erysimum*, *Arabis*, and *Draba* are grown in
rock gardens.

A   Fruit markedly flattened (only slightly flattened in *Cardaria*), usually
  not more than twice as long as wide (except in *Draba verna*, in which it
  is 3-4 times as long as wide)
  B   Fruit either flattened parallel to the partition that separates the 2
    halves (in which case the partition is visible externally along the
    narrow edges of the fruit) or not divided into 2 halves (in which
    case a prominent "vein" will be seen only along 1 edge, if at all)
    C   Fruit 3 or 4 times as long as wide (leaves not lobed, all in a
      basal cluster; petals white, deeply notched, about 2.5 mm long;
      one of the first herbaceous plants to bloom in late winter;
      rarely more than 10 cm tall; widespread)   *Draba verna* (*Plate* 75)
                  Whitlowgrass; SM, SCl-n
    CC Fruit less than twice as long as wide
      D   Petals 1.5-2 cm long, deep purple; mature fruit broadly oval,
        usually 3-4 cm long and nearly as wide, raised up on a stalk
        (this 1-1.5 cm long) above the level where the sepals and
        petals were attached   *Lunaria annua*
                Honesty; eu; Ma, Na
     DD Petals less than 1 cm long, not purple (but they may be
      tinged with purple); mature fruit much less than 2 cm long,
      and not raised up on a stalk
      E   Fruit without an abrupt swelling in the central portion
        (fruit with neither perforations nor marginal incisions;
        petals white)

F   Fruit densely covered with hooked hairs (petals about
    1.5 mm long)   *Athysanus pusillus*
                                           Dwarf Athysanus
FF Fruit, if hairy, without any hooked hairs
    G   Leaves entirely basal, some of them lobed; flowers
        solitary; fruit up to 8 mm long; in moist,
        gravelly habitats, mostly at elevations above
        2000'   *Idahoa scapigera*
                                          Flatpod; SCl(MH)
    GG Leaves scattered along the stem, none of them
        lobed; flowers in rather dense racemes; fruit less
        than 3 mm long; widespread in disturbed lowlands
        (flowers sweet-smelling, produced almost
        throughout the year)   *Lobularia maritima*
                                      Sweet Alyssum; me
EE Fruit with an abrupt swelling in the central portion
   (where the single seed is located)
   F   Petals pale yellow; racemes stout, with densely
       crowded flowers; fruit divided into 2 halves; fruit
       without incisions or perforations *Alyssum alyssoides*
                                           Yellow Alyssum; eua; Al
   FF Petals white or pinkish; racemes very slender, with
       rather widely spaced flowers; fruit not divided into 2
       halves; outer portion of the fruit sometimes with
       incisions or perforations
       G   Outer portion of fruit with sharply demarcated,
           slender ridges that extend outward from the
           swollen central portion; margin of fruit without
           prominent incisions (mature fruit 8-10 mm wide; in
           moist habitats)   *Thysanocarpus radians*
                                         Ribbed Fringepod; SCl-n
       GG Outer portion of fruit without distinct ridges (but
           there may be indistinct, broad ridges, and also
           perforations); margin of fruit sometimes with
           prominent incisions
           H   Most leaves (except those at the base) clasping
               the stems; basal leaves usually forming a
               cluster, and also usually hairy (mature fruit
               3-8 mm wide, the outer portion sometimes
               perforated; widespread) *Thysanocarpus curvipes*
               [includes *T. curvipes* var. *elegans*] (*Plate* 17)
                                        Hairy Fringepod
           HH Leaves not clasping the stems; basal leaves not
               often forming a distinct cluster, and usually
               not hairy (outer portion of fruit often
               perforated)   *Thysanocarpus laciniatus*
                                        [includes *T. laciniatus* var. *crenatus*]
                                        Narrowleaf Fringepod; CC-s
BB Fruit flattened at right angles to the partition that separates the 2
   halves (the location of the partition is visible externally on both
   flattened surfaces of the fruit)
   C   Fruit nearly triangular, widest at the top (where there is a
       shallow notch) and narrowing evenly to the pedicel (widespread)
                                        Capsella bursa-pastoris
                                        Shepherd's-purse; eu
   CC Fruit not nearly triangular, both margins mostly rounded
       D   Style of pistil not more than 0.5 mm long; fruit (less than 4
           mm long) divided lengthwise into 2 bulging halves (these with
           a networklike surface pattern or conspicuous pointed bumps)
           E   Fruit notched at the tip, the short style therefore
               entirely within the notch; surface of both halves of
               fruit with a networklike pattern (widespread)
                                        Coronopus didymus (*Plate* 75)
                                        Wart Cress; eu
           EE Fruit not notched at the tip, the style therefore
               projecting beyond the tip; surface of both halves with
               pointed bumps   *Coronopus squamatus*
                                           Swine Cress; eu; SFBR
       DD Style of pistil commonly more than 0.5 mm long; fruit not
           divided lengthwise into 2 bulging halves (except sometimes in
           Cardaria draba, in which the style is about 1 mm long, and in
           which the fruit has neither a networklike pattern nor pointed
           bumps)
           E   Upper leaves clasping the stem

F   Upper leaves heart-shaped, usually less than twice as
    long as wide; petals yellow; fruit not swollen, both
    surfaces flat   *Lepidium perfoliatum* (*Plate* 76)
                                        Shield Cress; eua
FF Upper leaves not heart-shaped, usually 2 or 3 times as
    long as wide; petals white or yellowish; fruit swollen
    G   Fruit slightly notched at the tip; style of pistil
        less than 1 mm long; petals white or slightly
        yellowish (mostly at elevations above 2000')
                                        Lepidium campestre
                                        Field Cress; eu
    GG Fruit not notched at the tip; style of pistil 1-1.5
        mm long; petals white
        H   Fruit obviously hairy, not constricted into 2
            halves   *Cardaria pubescens*
                                        Whitetop; as
        HH Fruit not obviously hairy, when mature often
            constricted into 2 halves   *Cardaria draba*
                                        Heartpod Hoary Cress; eua
EE Leaves not clasping the stem
    F   Fruit so deeply notched that the lobes on both sides
        of the notch are nearly as long as the rest of the
        fruit (notch narrow, the edges of the lobes bordering
        it almost parallel for most of their length) (flowers
        and fruit in congested racemes; plants low, branching
        from the base; in alkaline soils)   *Lepidium latipes*
                                        Dwarf Peppergrass
    FF Fruit not so deeply notched that the lobes on both
        sides of the notch are more than one-third as long as
        the rest of the fruit
        G   Sepals persisting until the fruit is nearly or
            fully mature   *Lepidium strictum*
                                        Prostrate Cress
        GG Sepals falling away early, or at least withering,
            about the same time as the petals and stamens
            H   Lobes of the fruit sharp-pointed (in alkaline
                soils)   *Lepidium oxycarpum* (*Plate* 76)
                                        Sharp-pod Peppergrass; Sn-SB
            HH Lobes of the fruit rounded, or at least not
                sharp-pointed
                I   Mature fruit 10-15 mm long
                                        Thlaspi arvense (*Plate* 76)
                                        Fanweed; eu
                II Mature fruit not more than 6 mm long
                    J   Pedicels distinctly flattened, at least
                        twice as wide as thick
                        K   Fruit not hairy, often slightly
                            concave on one side, convex on the
                            other (widespread)
                                        Lepidium nitidum (*Plate* 76)
                                        Common Peppergrass
                        KK Fruit usually at least slightly
                            hairy, at least along the margin,
                            not convex on one side and concave
                            on the other
                            L   Fruit with a pronounced
                                networklike pattern (in alkaline
                                soil)   *Lepidium dictyotum*
                                        Alkali Peppergrass; CC
                            LL Fruit without a pronounced
                                networklike pattern
                                        Lepidium lasiocarpum
                                        Hairypod Peppergrass; Ma
                    JJ Pedicels not flattened to the extent
                        that they are twice as wide as thick
                        K   Lower leaves more than 10 cm long,
                            the margin of the blade either
                            smooth or evenly toothed; up to more
                            than 1 m tall *Lepidium latifolium*
                                        Slender Perennial Peppergrass
                                        eua; Sn, SCl

KK Lower leaves not more than 8 cm
long, the margin either irregularly
toothed or pinnately lobed; usually
less than 0.5 m tall
L   Petals if present, shorter than
the sepals; fruit 1.5 mm long
Lepidium pinnatifidium
Eurasian Cress; eua; SF
LL Petals as long as or longer than
the sepals; fruit 2.5-4 mm long
M   Pedicels not flattened,
sparsely hairy; stems
without rigid hairs
Lepidium virginicum
var. *virginicum*
Tonguegrass; na
MM Pedicels flattened, hairy;
stems with rigid hair
Lepidium virginicum
var. *pubescens*
Hairy Tonguegrass
AA Fruit usually not flattened (except in species of *Streptanthus*), more
than 4 times as long as wide
B   Fruit divided into 2 units, the lower of these shorter than the upper
one (the upper unit may be conspicuously beaked)
C   Petals yellow; upper portion of the fruit nearly globular and
with about 5 distinct, lengthwise ridges, tapering abruptly to a
slender beak that is about as long as the upper portion
Rapistrum rugosum
Wild Turnip; eu; SF
CC Petals pale purple to nearly white; upper portion of fruit more
nearly flattened or ovoid rather than globular, somewhat
angular, tapering gradually to a short, stout beak (on
backshores of sandy beaches)
D   Leaf blades usually with a wavy margin and toothed, or
sometimes shallowly lobed; fruit widest in the upper unit;
petals about 6 mm long
Cakile edentula [includes *C. edentula* ssp. *californica*]
Searocket; na
DD Leaf blades without a wavy margin, pinnately lobed, the lobes
sometimes separated nearly to the midrib; fruit widest where
the two units are joined; petals about 10 mm long
Cakile maritima (Plate 17)
Horned Searocket; eu; Me-Mo
BB Fruit not divided into 2 slightly unequal units (there may be
constrictions, however, between the seeds)
C   Upper portion of pistil (usually just the style) developing into
a conspicuous beak that is at least one-sixth as long as the
rest of the fruit
D   Fruit not divided lengthwise by a partition
E   Fruit up to 4 cm long, with 4-8 seeds; fruit with
distinct, regularly spaced lengthwise grooves and ridges,
and conspicuous constrictions between successive seeds;
petals usually yellow, fading to white
Raphanus raphanistrum
Jointed Charlock; me
EE Fruit up to 2 cm long, with 2-3 seeds; fruit without
distinct grooves or conspicuous constrictions; petals
usually white, pink, rose, or purplish, with darker veins
(widespread)   *Raphanus sativus* (Plate 76)
Wild Radish; me
DD Fruit divided lengthwise by a partition (not conspicuously
constricted between successive seeds)
E   Petals white or pale yellow, with reddish purple veins
Eruca vesicaria ssp. *sativa* [*E. sativa*]
Garden-rocket; eu
EE Petals decidedly yellow
F   Fruit covered with conspicuous, flattened hairs
Sinapis alba [*Brassica hirta*]
White Mustard; eua
FF Fruit not covered with conspicuous, flattened hairs
G   Upper leaves sessile and clasping the stem (common)
Brassica rapa [*B. campestris*] (Plate 75)
Field Mustard; eu

GG Upper leaves, if sessile, not clasping the stem
   H   Beak 4-angled, up to about two-thirds as long
      as the rest of the fruit
      Sinapis arvensis [includes *Brassica kaber* vars.
                  kaber and *pinnatifida*]
                          Charlock; eu
   HH Beak cylindrical or slightly flattened, but not
      4-angled, and not so much as half as long as
      the rest of the fruit
      I   Maturing fruit not pressed to the stem of
         the inflorescence; fruit 30-40 mm long, the
         beak 5-8 mm long (petals 7-8 mm long;
         annual)   *Brassica juncea*
                     Indian Mustard; eua
      II Maturing fruit pressed to the stem of the
         inflorescence; fruit 10-20 mm long, the
         beak 1-4 mm long (widespread)
         J   Petals usually 7-8 mm long; fruit up to
           2 cm long, the beak almost uniformly
           slender; annual   *Brassica nigra*
                      Black Mustard; eu
         JJ Petals usually 5-6 mm long; fruit rarely
           more than 1.2 cm long, the lower two-
           thirds of the beak much thicker than
           the slender tip; biennial or perennial
                   Hirschfeldia incana
           [*Brassica geniculata*] (*Plate* 17)
                    Hirschfeldia; eu
CC Upper portion of pistil either not developing into a beak, or the
  beak less than one-sixth as long as the rest of the fruit
  D   Sepals not purplish; fruit not flattened
                  Brassicaceae, Subkey (p. 123)
  DD Sepals sometimes purplish, at least in bud or when flowers
    first open; fruit usually slightly flattened parallel to the
    partition (in *Caulanthus coulteri* var. *lemmonii*, however,
    they are not obviously flattened) (petals usually at least
    partly purple, or with one or more purple veins, the upper
    portion often ruffled and sometimes little if any wider than
    the lower portion; calyx usually "pinched in" slightly near
    the top, where the sepals diverge, so that it is vase-shaped)
    E   Stigma obviously 2-lobed, the lobes 1.5-3 mm long (leaves
      sessile, clasping the stem; sepals 7-15 mm long, unequal,
      dark purple before flowers open, then becoming paler;
      petals usually 1-1.5 cm long)
                  Caulanthus coulteri var. *lemmonii*
                  Lemmon Caulanthus; Al-s
    EE Stigma not obviously 2-lobed
      F   Calyx slightly irregular, 1 sepal extending outward
        farther than the other 3, and somewhat separate from
        them
        G   Stems and leaves (especially the margin and
          underside of the midrib) with scattered hairs
          (sometimes on serpentine)
          H   Flowers not mostly on one side of the
            inflorescence (sepals dark purple)
              Streptanthus glandulosus ssp. *glandulosus*
                  Common Jewelflower; Sn-SLO
          HH Flowers mostly on one side of the inflorescence
           I   Sepals reddish purple; petals purple
              Streptanthus glandulosus ssp. *pulchellus*
              Mount Tamalpais Jewelflower; Ma(MT); 1b
           II Sepals purple, rose, white, or greenish
             yellow; petals white to purple
              Streptanthus glandulosus ssp. *secundus*
                Oneside Jewelflower; Me-Ma
        GG Stems and leaves without obvious hairs, even when
          examined with a hand lens
          H   Sepals very dark purple (on serpentine)
                  Streptanthus niger
              Tiburon Jewelflower; Ma; 1b
          HH Sepals whitish to pale lavender
           I   Sepals greenish white; on serpentine
              Streptanthus albidus ssp. *albidus*
             Metcalf Canyon Jewelflower; SCl(MH); 1b

      II Sepals pale lavender; sometimes on
        serpentine
            Streptanthus albidus ssp. *peramoenus*
        Most Beautiful Jewelflower; Al, CC, SCl; 1b
FF Calyx regular
  G Leaves and stems with bristly hairs, especially
    near the base of the plant; not often so much as
    20 cm tall (in rocky areas)
    H Petals 6-8 mm long, purplish, with white
      margins; fruit 4-7 cm long; bristly hairs
      crowded   *Streptanthus hispidus*
        Mount Diablo Jewelflower; CC (MD); 1b
    HH Petals 10 mm long, reddish purple; fruit 1.5-2
      cm long; bristly hairs widely scattered
      (usually separated by a distance about equal
      to their length)   *Streptanthus callistus*
        Mount Hamilton Jewelflower; SCl(MH); 1b
  GG Leaves and stems without bristly hairs, even near
    the base of the plant; generally at least 30 cm
    tall
    H Lower portion of inflorescence with a few
      broad, leaflike bracts that clasp the stem
      (lower leaves usually less than 3 times as
      long as wide; petals yellowish, with purple
      veins; up to 100 cm tall)
      I Sepals 5-8 mm long; petals 6-8 mm long;
        usually biennial
          Streptanthus tortuosus var. *tortuosus*
              Mountain Jewelflower; Sn-n
      II Sepals 8-10 mm long; petals 11-13 mm long;
        perennial (sometimes on serpentine)
      Streptanthus tortuosus var. *suffrutescens*
             Shrubby Mountain Jewelflower
    HH Inflorescence without broad, leaflike bracts
      (often on serpentine)
      I Lower leaves, including basal leaves,
        usually more than 3 times as long as wide
        (petals 8-10 mm long, the upper pair white,
        the lower pair purple with white margins;
        up to 70 cm tall)   *Streptanthus barbiger*
             Bearded Streptanthus; Me-Na
      II Lower leaves not often more than 3 times as
        long as wide
        J Leaves clasping the stem (sepals mainly
          purple, not hairy; up to 60 cm tall)
          K Flowers mostly on one side of the
            inflorescence; leaves yellow-green,
            up to 5 cm long; fruit markedly
            curved, approximately C-shaped when
            mature; up to 40 cm tall
            Streptanthus breweri var. *hesperidis*
              Yellow-green Streptanthus; Na
          KK Flowers not just on one side of the
            inflorescence; leaves blue-green, up
            to 12 cm long; fruit only slightly
            curved
            Streptanthus breweri var. *breweri*
              Brewer Streptanthus; SB-n
      JJ Lower leaves not clasping the stem
        K Lower leaves up to 2.5 cm long,
          mostly 2-3 times as long as wide;
          annual, up to 18 cm tall (petals
          white, with purple veins; sepals
          green or purple with white tips, not
          hairy)   *Streptanthus batrachopus*
          Tamalpais Jewelflower; Ma(MT); 1b
        KK Lower leaves up to more than 3 cm
          long, rarely so much as twice as
          long as wide; biennial, up to more
          than 45 cm tall

In most of our genera, the corolla lobes are perfectly equal, but in *Downingia*, the corolla is strongly 2-lipped, with 3 lower lobes and 2 smaller upper lobes. The calyx lobes and stamens of *Downingia* are also not quite equal.

A Corolla regular, all lobes similar in size and shape
   B Flowers sessile (but the lower portion of the calyx, enveloping part of the pistil, may be so slender as to resemble a pedicel)
      C Calyx lobes extending far beyond the corolla (angles of stems hairy)   *Githopsis specularioides*
                                                                 Bluecup
      CC Calyx lobes not extending much beyond the corolla
         D Lobes of corolla not much longer than wide, and about half as long as the rest of the corolla; angles of stem not hairy; most leaves about as long as wide, the margin distinctly toothed (in moist shaded areas)   *Heterocodon rariflorum*
            Heterocodon
         DD Lobes of corolla more than twice as long as wide, and about as long as the rest of the corolla (note: this feature applies to flowers that open; some flowers, especially those on the lower portion of the stem, do not); angles of stem hairy; most leaves at least slightly longer than wide, the margin with bristles but usually not distinctly toothed   *Triodanis biflora* (*Plate* 77)
            Venus-looking-glass
   BB Flowers with distinct pedicels
      C Style of pistil extending far beyond the corolla; plants fairly stout, at least near the base, usually more than 50 cm tall
            Campanula prenanthoides (*Plate* 77)
            California Harebell
      CC Style of pistil not extending beyond the corolla; plants rather delicate, not often more than 30 cm tall
         D Corolla rarely more than 6 mm long, only slightly, if at all, longer than the calyx (sometimes on serpentine or burned areas)
            E Most leaves more than 3 times as long as wide; pedicels rarely more than 5 mm long; fruit distinctly longer than wide   *Campanula griffinii* [*C. angustiflora* var. *exilis*]
               Griffin Harebell; SN, Ma, SCr
            EE Most leaves only 2-3 times as long as wide; pedicels often more than 10 mm long; fruit nearly spherical
               Campanula angustiflora
               Eastwood Campanula; SN, Ma, SCr
         DD Corolla usually at least 8 mm long (sometimes more than 12 mm long), much longer than the calyx
            E Larger leaves 1.5-2 cm long (and about twice as long as wide); perennial, with weak stems (these often sprawling) up to 30 cm long; limited to swampy habitats near the coast   *Campanula californica*
               Swamp Harebell; Me-Ma; 1b
            EE Leaves rarely more than 1 cm long, mostly more than twice as long as wide; annual, up to about 15 cm tall; in dry habitats
               F Lower and upper leaves several times as long as wide; small bractlike leaves below flowers decidedly alternate   *Campanula exigua*
                  Chaparral Harebell; CC(MD), SCl(MH); 4
               FF Lower leaves not often so much as twice as long as wide, the upper leaves usually 3-4 times as long as wide; small bractlike leaves below flowers opposite or nearly so   *Campanula sharsmithiae*
                  Sharsmith Harebell; Al, SCl(MH); 1b
AA Corolla 2-lipped, the 2 upper lobes smaller and also more distinctly separate from one another than the 3 lower lobes
   B Anthers of the stamens separate from one another; corolla less than 3 mm long
      C Corolla 1.5-2.5 mm long, white or purplish; base of fruit rounded (on serpentine)   *Nemacladus montanus*
            Mountain Nemacladus; Na, La
      CC Corolla less than 1.5 mm long, white; base of fruit shaped like an inverted cone   *Nemacladus capillaris*
            Common Nemacladus; SCl-n, w

BB Anthers of the stamens (but not necessarily the filaments) joined
    together to form a tube; corolla generally more than 5 mm long (in
    drying mud of vernal pools and at the edge of ponds and other wet
    places)
    C Corolla 2.5-4 mm long, about as long as, or shorter than, the
      calyx (corolla either with white upper lobes and tricolored
      [blue, white, yellow] lower lobes, or entirely white)

Downingia pusilla
Dwarf Downingia; SN-s; 2

   CC Corolla more than 5 mm long, longer than the calyx
      D Lower lip of corolla with 3 purple spots alternating with 2
        larger yellow spots that are within a white blotch (upper
        lip, and tip of lobes of lower lip, purple)

Downingia pulchella
Flatface Downingia; Mo-n

     DD Lower lip without 3 purple spots (but there may be a single
       purple spot)
       E Margin of upper lobes of corolla usually with a fringe of
         hairs (use hand lens); lower lip of corolla with a large
         purple spot near the base (and sometimes also a white
         blotch); rest of corolla blue

Downingia concolor (*Plate* 77)
Maroonspot Downingia; La-Mo

       EE Margin of upper lobes of corolla without a fringe of
         hairs; lower lip of corolla without a large purple spot
         near the base (but there may be a little purple at the
         throat) (lower lip with a white blotch within which is a
         yellow spot, or 2 somewhat separate yellow spots); rest
         of corolla pale blue or lavender, sometimes nearly white

Downingia cuspidata
Cuspidate Downingia; SLO-n

Caprifoliaceae--Honeysuckle Family

All members of the honeysuckle family are perennial, and most are
decidedly woody. Our species include vines, shrubs, and small trees. The
leaves are opposite, and in honeysuckles (*Lonicera*) they are sometimes
joined at the base in such a way that the pairs form disks around the stem.
The calyx is usually small, with 4 or 5 lobes. The corolla, often with a
substantial tube, may have 4 or 5 essentially equal lobes, or it may be
markedly 2-lipped, in which case the upper lip has 4 lobes and the lower
lip consists of a single large lobe. The stamens are attached to the tube,
and they alternate with the lobes. The fruiting portion of the pistil is
below the level where the corolla and calyx lobes originate. The fruit,
fleshy in all our species, has 2 to 5 seed-forming divisions.

This family has contributed many garden plants, mostly European,
Asiatic, and North American species of *Lonicera* and *Viburnum*. Our native
species are suitable for a wild garden provided they are given the space
they need. Two of them, *Lonicera involucrata* var. *ledebourii* and *Sambucus
racemosa*, will do well only if they are in soil that is moist. Species of
Symphoricarpos, snowberry, especially *S. albus* var. *laevigatus*, spread
aggressively by vegetative means but can be controlled if one is watchful.

Two species whose distribution begins north of our region should be
considered for inclusion in gardens. One is *Lonicera ciliosa*, a honeysuckle
with clusters of yellow, orange, or red flowers. The other is *Linnaea
borealis*, called Twinflower, a delicate creeper than will form a ground
cover in shaded habitats with abundant humus.

A Leaves pinnately compound
  B Inflorescence not as tall as wide; fruit bluish (nearly black if the
    powdery coating is rubbed off; widespread)

Sambucus mexicana [includes *S. caerulea*] (*Plate* 18)
Blue Elderberry

  BB Inflorescence at least as tall as wide; fruit red (restricted to wet
    habitats, such as edges of marshes, streams, and ditches)

Sambucus racemosa [*S. callicarpa*] (*Plate* 18)
Red Elderberry; SM-n

AA Leaves not compound (but they may be toothed or lobed)
    B Leaves with coarse, evenly spaced teeth (corolla white; shrubs up to
       4 m tall; fruit red)   *Viburnum ellipticum*
                                Oval-leaf Viburnum; Sn, CC
    BB Leaves without regularly spaced teeth (although *Symphoricarpos* may
       have shallow, irregular lobes)
       C Flowers borne in pairs above broad bracts that become red; leaves
          usually pointed at the tip, 3-7 cm long (corolla yellow; fruit
          black when ripe; upright shrubs of moist habitats)
                     Lonicera involucrata var. *ledebourii* (*Plate* 18)
                                  Twinberry; La-SLO
      CC Flowers not in pairs above bracts; leaves rounded, not more than
        4 cm long (except in *Lonicera hispidula* var. *vacillans*)
        D Corolla cream-colored or yellowish (fruit red; leaves 1-3 cm
          long; shrubs, often twining or sprawling on other plants)
          E Leaves of one or more upper pairs often united; corolla
            lobes equal or nearly so   *Lonicera interrupta*
                            Chaparral Honeysuckle
          EE Leaves of upper pairs not united; corolla lobes unequal
                      Lonicera subspicata var. *denudata*
                      [*L. subspicata* var.*johnstonii*]
                   Southern Honeysuckle; CC(MD); SCl(MH)
        DD Corolla pink or purplish
          E Leaves 3.5-8 cm long, those of one or more upper pairs
            united; corolla lobes unequal; fruit red (sprawling or
            twining on other plants; widespread)
                      Lonicera hispidula var. *vacillans* (*Plate* 18)
                                Hairy Honeysuckle
          EE Leaves 1-4 cm long, none of the pairs united; corolla
            lobes equal; fruit white
            F Plants often more than 100 cm tall, upright; leaves
               not hairy on the underside; corolla with a nectar
               gland below only 1 lobe; fruit usually 10 mm long
               (widespread)   *Symphoricarpos albus* var. *laevigatus*
                            [*S. rivularis*] (*Plate* 18)
                             Common Snowberry
            FF Plants rarely more than 50 cm tall, usually sprawling;
               leaves usually hairy on the underside; corolla with
               nectar glands below all 5 lobes; fruit about 8 mm long
                     Symphoricarpos mollis (*Plate* 18)
                           Creeping Snowberry; Me-s

Caryophyllaceae--Pink Family

    We are indebted to the Caryophyllaceae not only for cultivated pinks,
carnations, Sweet-William, and Baby's-breath, but also for some attractive
wildflowers. The most striking species native to our region is *Silene
californica*, the Indian Pink. The family is a large one, and it has given
us many weeds, some of which have become well established in habitats where
they may seem to be part of the original flora.

    In plants of the Pink Family, the leaves are opposite, and the nodes
where they originate are usually swollen. The flowers typically have 5
sepals (or calyx lobes, when the lower part of the calyx forms a cup or
tube) and 5 petals; the petals are nearly always split into 2 or more
lobes. A few species lack petals, at least on some flowers. The number of
stamens is generally 5 or 10, and the fruit, which is dry when mature, has
from 1 to many seeds.

A Petals either absent or reduced to very small scales (leaves narrow,
   stiff, up to about 1 cm long; perennial, forming low, tufted growths)
   (Note: petals are very small or absent in certain species of *Cerastium*
   and *Stellaria*, but these plants are otherwise not as described here.)
   B Sepals all alike, each one narrowing to a stiff bristle; petals
     completely absent; on grassy hillsides   *Paronychia franciscana*
                          California Whitlow-wort; sa; SF-Sn
   BB Sepals unequal, the 3 outer ones larger than the 2 inner ones, and
     also tipped with stouter and slightly longer bristles; petals
     present, but reduced to very small scales; on dunes and other sandy
     habitats close to the shore   *Cardionema ramosissimum*
                                     Sandmat

AA Petals usually present and obvious (except, as noted in choice A, in
   certain species of *Cerastium* and *Stellaria*)
   B Lower part of the calyx a cup or tube; petals with a narrow,
      stalklike base (and usually with a pair of scalelike outgrowths on
      the inner face)
      C Petals usually bright red, sometimes deep pink
                                     Silene californica (*Plate* 19)
                                                       Indian Pink
      CC Petals white or pink
         D Calyx with at least 18 prominent ribs   *Silene multinervia*
                                           Manynerve Catchfly; Sn-s
         DD Calyx with 10 ribs, these not always prominent
            E Petals inconspicuous, the exposed portion only about twice
               as long as the calyx lobes (upper internodes usually with
               sticky bands; upper leaves often more than 10 times as
               long as wide)   *Silene antirrhina*
                                                     Sleepy Catchfly
            EE Petals conspicuous, the exposed portion at least 3 times
               as long as the calyx lobes
               F Flowers in nearly one-sided racemes; petals rounded at
                  the tip, not lobed; widely naturalized annual
                                            Silene gallica (*Plate* 19)
                                                   Windmill Pink; eu
               FF Flowers in panicles or racemes, but these not one-
                  sided; petals deeply divided into 2 primary lobes,
                  each of these sometimes with a small secondary lobe at
                  the side; native perennial
                  G Lower leaves up to 3 cm wide (petals greenish white
                     to rose; each primary petal lobe with a small
                     secondary lobe at the side; on bluffs close to the
                     coast)   *Silene scouleri* ssp. *grandis* (*Plate* 19)
                                                Scouler Catchfly; SM-n
                  GG Lower leaves rarely more than 1.2 cm wide
                     H Largest leaves 2-6 cm long; calyx usually
                        purplish; petals pink or rose; coastal (in
                        sandy areas)   *Silene verecunda* ssp. *verecunda*
                                       San Francisco Campion; SF-SCr; 1b
                     HH Largest leaves 5-9 cm long; calyx usually
                        greenish; petals white, greenish or rose;
                        inland (at elevations above 2000')
                                            Silene verecunda ssp. *platyota*
                                      Cuyamaca Catchfly; La, CC(MD)-s
   BB Sepals separate; petals without a narrow, stalklike base
      C Leaves with membranous stipules (not more than 40 cm tall)
         D Leaves not appearing to be in whorls or bundles (they are
            either in pairs, or in pairs accompanied by one to a few
            smaller leaves; corolla usually shorter than the calyx)
            E Stamens 2-5 (leaves 2-4 cm long, the stipules 2-4 mm long;
               petals white or pink; fruit up to 6 mm long; in salt
               marshes and inland saline or alkaline habitats)
                                                    Spergularia marina
                                               Saltmarsh Sand-spurrey
            EE Stamens at least 6
               F Leaves up to 2 cm long, the stipules 2-4 mm long;
                  petals white or pink; stamens 6-10; fruit 3-5 mm long,
                  the seeds slightly roughened; upright (in sand dunes
                  and other coastal habitats)   *Spergularia bocconii*
                                           Boccone Sand-spurrey; me
               FF Leaves up to 5 cm long, the stipules 3-6 mm long;
                  petals white; stamens 9-10; fruit 4.5-8 mm long, the
                  seeds smooth; upright or prostrate (around salt
                  marshes and other saline habitats) *Spergularia media*
                                          Middlesize Sand-spurrey; eu; Ma
         DD Leaves crowded in the axils, thus appearing to be in whorls
            E Pistil with 5 styles (petals white, about 5 mm long)
                                                    Spergula arvensis
                                                       Stickwort; eu
            EE Pistil with 3 styles
               F Petals 4-7 mm long; sepals 5-10 mm long; stamens 10
                  (leaves 10-35 mm long, the stipules 5-10 mm long)
                  G Petals pink; on seaside bluffs and around salt
                     marshes
                        Spergularia macrotheca var. *macrotheca* (*Plate* 19)
                                            Perennial Sand-spurrey

GG Petals white; around inland freshwater marshes and
   in alkaline habitats
                           Spergularia macrotheca var. *longistyla*
                               Whiteflower Sand-spurrey; Na-Al
FF Petals 2-5 mm long; sepals 3-5 mm long; stamens 6-10
   G   Petals pink; leaves 6-12 mm long, the stipules up
       to 5 mm long; stem below the inflorescence not
       glandular-hairy (the pedicels, however, are
       glandular-hairy)   *Spergularia rubra* (*Plate* 19)
                                     Ruby Sand-spurrey; eu
   GG Petals white; leaves 10-40 mm long, the stipules up
      to 8 mm long; stem below the inflorescence very
      glandular-hairy (in sandy soil, usually near the
      coast)   *Spergularia villosa*
                               Villous Sand-spurrey; sa
CC Leaves without stipules
   D   Petals deeply divided into 2 lobes (petals often absent in
       Stellaria crispa, and often much reduced or absent in
       Stellaria borealis ssp. *sitchana* and *Cerastium glomeratum*)
       E   Pistil with 5 styles
           F   Most leaves narrow, at least 6 times as long as wide;
               petals generally slightly more than 1 cm long; found
               in rocky or grassy habitats (perennial; widespread)
                                       Cerastium arvense (*Plate* 19)
                                         Field Chickweed; Mo-n
           FF Leaves broad, rarely more than 3 times as long as
              wide; petals less than 1 cm long; found in lawns,
              gardens, and waste places
              G   Flower pedicels usually much longer than the calyx;
                  petals up to 8 mm long, usually slightly longer
                  than the sepals; bracts usually with membranous
                  margins; short-lived perennial (often flowering
                  the first year), matted, but with flowering stems
                  rising   *Cerastium fontanum* ssp. *vulgare*
                               Large Mouse-ear Chickweed; eu
              GG Flower pedicels usually no longer than the calyx;
                 petals (when present) up to 4.5 mm long, not
                 longer than the sepals; bracts entirely green;
                 annual, upright   *Cerastium glomeratum*
                               Mouse-ear Chickweed; eu
       EE Pistil with 3 styles (rarely 4)
          F   Stems with a single lengthwise row of hairs (leaves
              oval; usually in somewhat shaded places; annual;
              widespread)   *Stellaria media*
                                 Common Chickweed; me
          FF Stems often hairy, but without a single lengthwise row
             of hairs
             G   Leaves clasping the stem, up to 4 cm long (in moist
                 habitats among sand dunes; perennial)
                                       Stellaria littoralis
                                 Beach Chickweed; SF-n
             GG Leaves not clasping the stem (but they may be
                nearly sessile), rarely more than 2 cm long
                H   Most leaves narrow, several to many times as
                    long as wide; annual (in grassy habitats)
                                       Stellaria nitens
                                     Shiny Chickweed
                HH Leaves broad, usually less than twice as long
                   as wide; perennial
                   I   Margin of leaves wavy (petals often absent;
                       in moist areas)   *Stellaria crispa*
                                   Chamisso Chickweed; Ma-n
                   II Margin of leaves not wavy
                      J   Petals if present, very small, shorter
                          than the sepals; in moist places up to
                          6000' *Stellaria borealis* ssp. *sitchana*
                                   [*S. sitchana* var. *bongardiana*]
                                   Northern Chickweed; Ma-n
                      JJ Petals up to 5 mm long, as long as, or
                         longer than the sepals; established in
                         lawns   *Stellaria graminea*
                                   Lesser Chickweed; eu; SF
DD Petals either not lobed, or with only a shallow notch

E   Styles 4-5, equal to the number of sepals, and alternating
    with them
    F   Sepals 4 (rarely 5); petals less than 2 mm long,
        shorter than the sepals (and sometimes absent);
        largest leaves (in the cluster at the base of the
        plant) not often more than 2 cm long; prostrate,
        rooting at the nodes   *Sagina procumbens*
                                        Arctic Pearlwort; Ma-n
    FF Sepals 5; petals 3 mm long, about as long as the
        sepals; largest leaves (in the basal cluster) up to 3
        cm long; up to 12 cm tall (on ocean bluffs and sand
        dunes)
                Sagina maxima ssp.*crassicaulis* [*S. crassicaulis*]
                                        Beach Pearlwort; Mo-n
EE Styles 3, fewer than the sepals, and opposite them
    F   Leaves rarely more than 5 mm long; petals 3-4 mm long
        (annual, up to about 10 cm tall; widespread)
                Minuartia californica [*Arenaria californica*]
                                        California Sandwort
    FF Most leaves more than 8 mm long; petals up to 6 mm
        long
        G   Largest leaves about 1 cm long; growing on dry,
            rocky or gravelly hillsides; upright, rarely more
            than 15 cm tall; annual (sometimes on serpentine)
                Minuartia douglasii [includes *Arenaria douglasii*
                    vars. *douglasii* and *emarginata*] (*Plate* 77)
                                        Douglas Sandwort
        GG Largest leaves about 4 cm long; sprawling, the
            stems sometimes more than 50 cm long, rooting at
            the nodes; perennial
            H   Flowers solitary, the petals 5-6 mm long; in
                swampy habitats   *Arenaria paludicola*
                                        Marsh Sandwort; SF; 1b
            HH Flowers in cymes, the petals 2-4 mm long; on
                shaded slopes at elevations above 1500'
                Moehringia macrophylla [*Arenaria macrophylla*]
                                        Largeleaf Sandwort; SCl-n

Celastraceae--Staff-tree Family

   The Celastraceae has two representatives in our region. They are
attractive and interesting shrubs that merit a place in gardens in which
native plants are emphasized.
   Both species have opposite leaves, but they are otherwise very
different. *Paxistima myrsinites* (*Plate* 19), called Oregon Boxwood, is
evergreen, with oval leaves approximately 1.5-2 cm long, and 4-petaled
flowers. The petals are only about 1 mm long, but they are of an unusual
reddish brown color. *Paxistima* grows slowly and compactly, reaching a
height of about 1 m. It is found mostly in partly shaded habitats above
2000 feet. In cultivation, it requires good drainage, and can stand
considerable drought once it is well established.
   Euonymus occidentalis (*Plate* 19), called Western Burningbush, is
deciduous, with leaves commonly at least 4-5 cm long. The branchlets are
distinctive in that they are conspicuously angled. The strange flowers have
5 rounded petals, about 3-4 mm long. They are brownish purple, dotted
delicately with white. When the fruits split open, some red tissue from the
inside adheres to the seeds, as it does in European and Asiatic relatives
that are widely cultivated in gardens. *Euonymus* grows tall, sometimes
reaching a height of more than 5 m. It is found primarily in wooded
ravines, and requires considerable moisture.

Ceratophyllaceae--Hornwort Family

   Hornworts could be confused with bladderworts (Lentibulariaceae) or
water-milfoils (Haloragaceae). The plants are of about the same size and of
similar form: stems more than 1 m long, leaves divided into slender lobes
and arranged in whorls about 2 cm in diameter. The leaves of hornworts,

however, are stiff and somewhat rough to the touch, owing to small
serrations. The only species in all of California is *Ceratophyllum demersum*
(*Plate* 77), called Hornwort.

Chenopodiaceae--Goosefoot Family

The Goosefoot Family consists largely of herbaceous plants, although it
does include shrubs and near-shrubs. The foliage is often mealy, owing to
the presence of small scales, and some species are decidedly succulent. The
flowers, usually clustered, are small and without petals, and in certain
genera there are staminate and pistillate flowers, which may be on separate
plants. Each fruit, generally dry at maturity, contains a single seed.

This family has given us the beet (whose cultivated forms include the
sugar beet and Swiss chard) and spinach. Various other members of the group
have provided food or medicines for aboriginal peoples.

Many representatives of the family live in saline or alkaline habitats,
and certain of them--notably species of *Salicornia* and *Salsola*--have been
important sources of sodium salts. Huge piles of these plants were
incinerated to produce the ash needed for making soap and glass.

A   Stems (at least those of recent growth) succulent, jointed; leaves
reduced to scales
  B   Branches alternate; somewhat woody, often more than 100 cm tall;
limited to inland alkaline soils   *Allenrolfea occidentalis*
Iodinebush; CC
  BB Branches opposite; mostly nonwoody, rarely so much as 50 cm tall; in
coastal salt marshes and saline habitats inland
    C   Inflorescences without flowers in the portion near the tip
Salicornia subterminalis
Glasswort; SF-s
    CC Inflorescences with flowers throughout their length
      D   Inflorescence usually slightly thinner than most stems;
plants branching from the base; perennial
Salicornia virginica (*Plate* 20)
Virginia Pickleweed
      DD Inflorescence usually slightly thicker than most stems;
plants branching well above the base; annual
        E   Inflorescence 4-6 mm wide; plants branching above the
middle   *Salicornia bigelovii*
Bigelow Pickleweed
        EE Inflorescence 2-4 mm wide; plants branching throughout
(usually turning bright red in autumn) *Salicornia europaea*
Slender Glasswort; Ma, Sn
AA Stems neither especially succulent nor jointed; leaves obvious, not
scalelike, although they may be slender
  B   Leaves spine-tipped (lower leaves slender, up to 5 cm long; upper
leaves much shorter, terminating in a stout prickle; plants bushy,
up to 1 m tall; widespread)   *Salsola tragus* (*Plate* 20)
Russian-thistle; eua
  BB Leaves not spine-tipped
    C   Leaves succulent, cylindrical, usually at least 5 times as long
as wide (somewhat shrubby; in alkaline and saline habitats)
      D   Flowers 2-3 mm wide, usually throughout the plant; scars of
fallen leaves generally persisting as obvious bumps; in
coastal salt marshes   *Suaeda californica*
California Seablite; SF-s
      DD Flowers 1-1.5 mm wide, usually restricted to upper branches;
scars of fallen leaves generally not persisting as obvious
bumps; in coastal salt marshes and inland alkaline habitats
Suaeda moquinii
Bush Seepweed; SCl-s
    CC Leaves not especially succulent, the blades not often so much as
4 times as long as wide
      D   Woody shrub (leaf blades oval to arrowhead-shaped, up to 5 cm
long, grayish, with fine scales on both surfaces; pistillate
flowers "sandwiched" between a pair of bracts; up to more
than 1.5 m tall; in coastal and inland saline habitats)
Atriplex lentiformis [includes *A. lentiformis* ssp. *breweri*]
Big Saltbush; SF-s, e

DD Plants either nonwoody or only slightly woody at the base
                                        Chenopodiaceae, Subkey
Chenopodiaceae, Subkey: Stems not jointed; leaves neither scalelike, spine-
tipped, nor succulent, the blades rarely so much as 4 times as long as wide
A Flowers with either stamens or a pistil, but not both (furthermore,
   staminate and pistillate flowers may be on separate plants); pistillate
   flowers "sandwiched" between a pair of bracts
   B Bracts enclosing fruits usually at least 10 mm wide (bracts often
      nearly circular; annual, up to more than 1.5 m tall)
                                        Atriplex hortensis (*Plate* 78)
                                        Garden Orache; eua; SF-SCl
   BB Bracts enclosing fruits rarely more than 8 mm wide
      C Bracts enclosing fruits fleshy, red (leaves about 2 cm long, with
         faintly toothed or wavy margins and a scaly underside; prostrate
         perennial; forming a ground cover in waste places, especially
         where saline or alkaline) *Atriplex semibaccata* (*Plate* 20)
                                        Australian Saltbush; au; CC
      CC Bracts enclosing fruits not fleshy or red
         D Blades of larger leaves usually decidedly broader near the
            base than near the middle, thus roughly triangular or
            arrowhead-shaped in general outline (annual)
            E Bracts enclosing fruits united for nearly their entire
               length (leaves and stems grayish, densely scaly; plants
               up to more than 60 cm tall)
                                        Atriplex argentea var. *mohavensis*
                                        [*A. argentea* ssp. *expansa*]
                                        Silver Saltbush; SF-s
            EE Bracts enclosing fruits united for not more than half
               their length
               F Underside of leaves densely scaly, grayish; usually in
                  moist, saline or alkaline habitats (Note: this species
                  is not clearly distinct from *A. patula* [under choice
                  DD, below].)   *Atriplex triangularis*
                                        Spearscale; Al, La
               FF Underside of leaves at most only slightly scaly, not
                  obviously grayish; mostly in dry habitats, especially
                  waste places (leaves becoming red; bracts hard when
                  fruit matures)   *Atriplex rosea*
                                        Redscale; eua
         DD Leaf blades oval or elongated, not obviously broader near the
            base than near the middle
            E Low, somewhat mat-forming perennials, rarely more than 30
               cm tall (on coastal bluffs and backshores of beaches)
               F Bracts enclosing fruits united for almost half their
                  length; pistillate and staminate flowers usually in
                  separate clusters   *Atriplex leucophylla* (*Plate* 20)
                                        Beach Saltbush
               FF Bracts enclosing fruits almost completely separate;
                  pistillate and staminate flowers usually in the same
                  cluster   *Atriplex californica*
                                        California Saltbush; Ma-s
            EE Upright annuals, usually more than 40 cm tall (mostly at
               the edge of salt marshes or on other saline soils) (Note:
               the 2 varieties below intergrade; furthermore, *A.
               triangularis* [under choice D, above] is not clearly
               distinct from them.)
               F Leaves slightly lobed or with small marginal teeth;
                  bracts enclosing fruits usually with prominent
                  projections (and often conspicuously widest near the
                  base)   *Atriplex patula* var. *obtusa*
                                        Common Orache; SFBR-n
               FF Leaves without lobes or marginal teeth; bracts
                  enclosing fruits usually without obvious projections
                                        Atriplex patula var. *patula*
                                        Spear Orache
AA All of the flowers with stamens and a pistil; none of the flowers
   "sandwiched" between a pair of bracts
   B Leaf blades usually about 10 times as long as wide, rarely so much as
      4 mm wide, sometimes with short, soft hairs; inflorescences in the
      leaf axils, with not more than 7 flowers (along roadsides and in
      waste places) *Kochia scoparia* [includes *K. scoparia* var. *subvillosa*]
                                        Summer-cypress; eua; CC, SF

BB Leaf blades not more than 4 times as long as wide, the larger leaves
   generally at least 10 mm wide, powdery, scaly, or glandular, but not
   hairy; inflorescences in leaf axils or at the tip of stems, often
   with numerous flowers
   C   Leaves and stems glandular, usually strong-scented (leaves often
       pinnately lobed, although the lobes are shallow; in waste places
       and along roadsides)
       D   Calyx with 3-5 small teeth, not lobed   *Chenopodium multifidum*
                                    Cutleaf Goosefoot; sa; SFBR-e, s
       DD Calyx deeply 5-lobed (the lobes separate nearly to the base)
           E   Largest leaf blades less than 3 cm long
                                              Chenopodium pumilio
                                              Small Goosefoot; au
           EE Largest leaf blades more than 3 cm long (sometimes more
              than 6 cm long)
              F   Flowers on short but distinct pedicels; inflorescence
                  branches noticeably curved   *Chenopodium botrys*
                                              Jerusalem-oak; eu
              FF Flowers sessile; inflorescence branches straight
                                              Chenopodium ambrosioides
                                              Mexican-tea; sa
   CC Leaves and stems not glandular (but the leaves often mealy or
      powdery, especially on the underside), sometimes somewhat
      unpleasantly scented
      D   Most flowers in globular clusters on the long, leafless
          terminal portion of main stems (below this, however, some
          flower clusters are in leaf axils); perennial (in open areas,
          especially in sand or clay)   *Chenopodium californicum*
                                              California Goosefoot
      DD Flowers not primarily in nearly globular clusters on the
         terminal portion of main stems (there are small, often
         branched inflorescences originating from many leaf axils);
         annual
         E   Most or all fruits wider than long, somewhat pumpkin-
             shaped; calyx with 5 lobes (mostly in waste places)
             F   Upper surface of leaf blades shiny, dark green, the
                 underside mealy or powdery (leaves unpleasantly
                 scented)   *Chenopodium murale* (*Plate* 78)
                                              Wall Goosefoot; eu
             FF Upper surface of leaf blades dull green, the underside
                powdery white
                G   Leaves unpleasantly scented; raised keel on each
                    calyx lobe about one-fourth as wide as the lobes
                                              Chenopodium berlandieri
                                 [includes *C. berlandieri* var. *zschackei*]
                                              Pitseed Goosefoot; Ma
                GG Leaves not unpleasantly scented; raised keel on
                   each calyx lobe much less than one-fourth as wide
                   as the lobe (common)   *Chenopodium album* (*Plate* 78)
                                              Lamb's-quarters; eu?
         EE Most or all fruits longer than wide, egg-shaped; calyx
            with 3 lobes (Note: when there are some pumpkin-shaped
            fruits, the calyx enclosing these has 4 or 5 lobes.)
            F   Calyx fleshy, often becoming reddish (mostly in saline
                habitats, including the edge of salt marshes)
                                              Chenopodium rubrum
                                              Red Goosefoot
            FF Calyx not fleshy, rarely if ever becoming reddish
               G   Ripe fruit of egg-shaped type projecting well
                   beyond the calyx (in open places, especially in
                   sand or gravel)
                            Chenopodium foliosum [*C. capitatum*] (*Plate* 78)
                                              Strawberry-blite; eu
               GG Ripe fruit of egg-shaped type completely enclosed
                  by the calyx
                  H   Young leaves densely covered with white powder
                      (they later lose much or all of the powder);
                      calyx lobes rounded at the tip (in moist
                      habitats)
                            Chenopodium macrosperum var. *halophilum*
                                 [*C. macrosperum* var. *farinosum*]
                                              Coast Goosefoot; sa?

HH Young leaves with little or no white powder;
   calyx lobes pointed at the tip (mostly on
   saline soils)   *Chenopodium chenopodioides*
            South American Goosefoot; sa; Me, Na, Sn

Cistaceae--Rock-rose Family

The flowers of rock-roses usually have 5 sepals (sometimes 3), 5 petals,
numerous stamens, and a pistil whose fruiting portion becomes a 3-lobed dry
capsule. The petals often fall away early. The leaves are simple, sometimes
alternate, sometimes opposite. Most members of the family are at least
somewhat shrubby. We have only 1 native genus, but a few of the cultivated
rock-roses from the Mediterranean region of Europe have become naturalized
in habitats that are to their liking.

A   Leaves 1-2.5 cm long, shed early; petals yellow, 4-7 mm long; up to 30
    cm tall (leaves up to 3 mm wide; in sandy or gravelly areas)
       Helianthemum scoparium [includes *H. scoparium* var. *vulgare*] (*Plate* 78)
                                            Peak Rushrose; Me-s
AA Leaves 3-7 cm long, persistent; petals not yellow, 20-30 mm long; up to
    100 cm tall (naturalized in a few areas)
    B   Flowers solitary; sepals 3; leaves with 3 main veins from the base
                                            Cistus ladanifer
                                            Gum Cistus; me
    BB Flowers in panicles; sepals 5; leaves with 5 main veins from the base
       C   Petals white   *Cistus monspeliensis*
                                            Montpelier Rock-rose; me
       CC Petals mauve, pale purple, or other nonwhite colors
                          Cistus creticus [*C. villosus* var. *corsicus*]
                                            Crete Rock-rose; me

Convolvulaceae--Morning-glory Family

Morning-glories and their relatives are mostly climbing or trailing
plants with alternate leaves. The flowers have 5 free (or nearly free)
sepals, a barely 5-lobed, usually funnel-shaped corolla, and 5 stamens that
are attached to the corolla tube. The corolla is twisted when in the bud
stage. In our species, the pistil is partitioned into 2 divisions, within
each of which 2 seeds develop.

Various morning-glories of the genera *Ipomoea* and *Convolvulus* are grown
for ornament, and *Dichondra donnelliana* is used for lawns. The family has
also given us the Sweet Potato, which is a species of *Ipomoea*, as well as
Convolvulus arvensis, the Field Bindweed, one of the most objectionable
weeds ever introduced to North America.

A   Corolla less than 1 cm long, deeply divided into lobes
    B   Leaf blades kidney-shaped, no longer than wide (1-2 cm wide); corolla
        (whitish) about 3 mm long, the lobes rounded (a small, creeping
        ground cover, native, but widely cultivated and thus escaping into
        areas where it is not indigenous)   *Dichondra donnelliana*
                                            Dichondra; Mo-n
    BB Leaf blades oval or elongated; corolla about 6 mm long, the lobes
        pointed
       C   Leaf blades oblong, up to 5 cm long; corolla pink; with long,
           trailing stems; annual   *Convolvulus simulans*
                                            Smallflower Morning-glory; CC-s; 4
       CC Leaf blades oval, about 1 cm long; corolla white; more or less
           tufted; perennial (in alkaline soil)   *Cressa truxillensis*
                          [includes *C. truxillensis* var. *vallicola*]
                                            Alkali-weed
AA Corolla more than 1 cm long, extended well beyond the calyx, not divided
    into distinct lobes
    B   Stigma of pistil nearly globular (sometimes lobed) (corolla white,
        pink, purple, or blue, 4-6 cm long; leaves and stems noticeably
        hairy; leaf blades heart-shaped, up to 12 cm long) (widely
        cultivated and frequently escaping)   *Ipomoea purpurea*
                                            Garden Morning-glory; sa
    BB Stigma of pistil elongated (or at least oblong), divided into 2 lobes

C   Without a pair of broad bracts resembling sepals located directly
    below these; bracts narrow and well separated from the sepals
    D   Corolla rarely more than 2.5 cm long; major portion of leaf
        blades (above the basal lobes) often widest near the middle;
        stems mostly prostrate (sometimes climbing), arising from a
        deep root (corolla white or pink; widespread)
                                    Convolvulus arvensis (*Plate* 78)
                                            Field Bindweed; eu
    DD Corolla usually at least 3 cm long; major portion of leaf
        blades (above the basal lobes) rarely, if ever, widest near
        the middle; generally climbing into shrubs, but sometimes
        trailing
        E   Leaves and stems (especially the underside of the leaf
            blades) somewhat hairy (use a hand lens)
                            Calystegia occidentalis [includes *C. polymorpha*]
                                    Western Morning-glory; Na-n
        EE Leaves and stems not hairy
            F   Leaves with rounded lobes
                                    Calystegia purpurata ssp. *saxicola*
                                        Marin Morning-glory; Me-Ma
            FF Leaves with pointed, angular lobes
                                    Calystegia purpurata ssp. *purpurata*
                    [includes *C. purpurata* ssp. *solanensis*] (*Plate* 20)
                                        Climbing Morning-glory
CC With a pair of broad bracts resembling sepals located directly
    below these; bracts partly covering the sepals
    D   Corolla pink, rose, or purple, even when freshly opened
        (sometimes white in *Calystegia sepium* ssp. *limnophila*);
        leaves and stems either not hairy or only slightly so;
        coastal
        E   Leaf blades kidney-shaped, about as long as wide, slightly
            succulent; growing on backshores of sandy beaches
                                    Calystegia soldanella (*Plate* 20)
                                            Beach Morning-glory
        EE Leaf blades arrowhead-shaped, much longer than wide, not
            succulent; growing in salt marshes and other moist,
            saline habitats   *Calystegia sepium* ssp. *limnophila*
                                            Hedge Bindweed; SF
    DD Corolla white to cream when freshly opened (but sometimes
        turning pink as it ages); leaves and stems decidedly hairy;
        in dry, gravelly habitats away from the coast
        E   Leaves and stems hairy, but not woolly; bracts up to 3
            times as long as wide   *Calystegia subacaulis*
                                        Hill Morning-glory; Sn, Na-SLO
        EE Leaves and stems woolly; bracts not often more than 2.5
            times as long as wide
            F   Plants compact, without any stems radiating away from
                the rootstock; bracts rounded at the tip (in rocky
                soil, including serpentine)   *Calystegia collina*
                                        Woolly Morning-glory; Me-SCl
            FF Plants with long, prostrate stems, these radiating
                away from the rootstock; bracts usually tapering to a
                point
                            Calystegia malacophylla ssp. *pedicellata* (*Plate* 20)
                                        Hairy Morning-glory; Al(MH)

Cornaceae--Dogwood Family

All members of the Dogwood Family native to North America have
deciduous, opposite leaves. In certain species, however, successive pairs
of leaves may be so close together that they appear to form a whorl. The
flowers are small; each has 4 petals and 4 stamens; the lower part of the
4-lobed calyx is joined to the pistil and thus contributes to the fleshy,
2-seeded fruit.

This family has given us some attractive garden subjects, including
Cornus florida, the so-called Flowering Dogwood of eastern North America,
Cornus kousa of Asia, and some Asiatic species of *Aucuba*. Of the 3 species
native to our region, *Cornus sericea* ssp. *occidentalis* is perhaps the most
amenable to cultivation here, though it requires considerable moisture.
Cornus nuttallii is a spectacular tree when in flower and when its foliage

turns red in autumn. Although it grows well in the mountains and in the
Pacific Northwest, it is not likely to succeed in lowland gardens.
Futhermore, it is subject to an extremely debilitating bacterial disease.
The Bunchberry, *Cornus canadensis*, a mostly herbaceous, creeping ground
cover that is widespread in North America, does not grow wild in California
south of Mendocino County. It is amenable to cultivation, however, in
woodland habitats that are kept almost constantly moist by rain or fog, or
that can be frequently watered.

A   Flowers in a dense hemispherical head, directly beneath which are 4-6
    (rarely 7) large white bracts (at first, these have a yellowish tinge,
    and they may turn pinkish as they age); fruit bright red; trees up to 25
    m tall; mostly in woods or at the edge of woods
                                    Cornus nuttallii (Plate 21)
                                    Mountain Dogwood; SC1, Na
AA Flowers in cymes, these not associated with large white bracts; fruit
    white or bluish; shrubs or small trees; mostly close to streams, ponds,
    or ditches
    B   Leaf blades dark or bright green on the upper surface, mostly 5-8 cm
        long, with 4-7 secondary veins on both sides of the midrib, the
        underside hairy
                Cornus sericea ssp. *occidentalis* [*C. occidentalis*] (Plate 21)
                                    Western Creek Dogwood
    BB Leaf blades gray-green on the upper surface, 3-5 cm long, with only
        3-4 secondary veins on both sides of the midrib, not hairy
                                    Cornus glabrata
                                    Brown Dogwood

Crassulaceae--Stonecrop Family

Most Stonecrops and their relatives are fleshy-leaved perennials whose
flowers have a calyx with 5 lobes or 5 separate sepals, 5 petals, either 5
or 10 stamens, and a pistil consisting of 5 divisions; the divisions may
separate at maturity. The genera *Crassula* and *Parvisedum*, however, are
represented in our region by annuals, and the flowers of the two species of
Crassula do not fit the typical formula, for the parts are usually in fours
instead of fives.
    The family provides us with numerous plants that are easily grown in
rockeries, dish gardens, and pots. Various stonecrops (*Sedum*) and
houseleeks (*Sempervivum*) are especially popular. The nearly ubiquitous
Sempervivum tectorum, a European species called Hen-and-chickens because of
its habit of growth, used to be planted on tile roofs because it was
thought to provide protection from lightning. While it may not help in that
regard, its juice may be useful for curing ringworm and for removing warts.

A   None of the leaves so much as 1 cm long; flowers with not more than 5
    stamens; petals not more than 3 mm long; annual, less than 7 cm tall
    (except *Parvisedum pentandrum*)
    B   Flowers in cymes at the end of branches; upper leaves alternate,
        lower leaves opposite (the lower leaves are shed early, but the
        places where they were attached will still be visible); flowers with
        5 petals, these at least 2 mm long; up to 13 cm tall (in rocky
        habitats, including serpentine)   *Parvisedum pentandrum*
                                    Parvisedum; La-SB
    BB Flowers (either single or clustered) in the leaf axils; all leaves
        opposite; flowers usually with 4 petals, these not more than 1.5 mm
        long; rarely more than 6 cm tall
        C   Flowers in clusters; upright, usually forming dense masses and
            becoming reddish as they age; in open fields and on burned areas
            (widespread)   *Crassula connata* [*C. erecta*] (Plate 21)
                                    Sand Pygmyweed
        CC Flowers single; sprawling, with stem tips rising; in muddy areas
            that dry out in late spring and summer   *Crassula aquatica*
                                    Pygmyweed
AA Largest leaves generally at least 1 cm long; flowers with 10 stamens;
    petals more than 5 mm long; perennial, the flowering stems up to 35 cm
    tall

B Plants forming mats in which there are several to many stems with
   crowded leaves, as well as the stalks of the inflorescences; leaves
   commonly not more than 2 cm long (sometimes longer in *Sedum
   spathulifolium*); petals completely separate and spreading outward,
   the corolla thus star-shaped (in rocky areas)
   C Leaves on vegetative stems nearly cylindrical, widest near the
      base, not whitish, about 1 cm long
                         Sedum radiatum [*S. stenopetalum* ssp. *radiatum*]
                                       Narrowpetal Stonecrop; Mo-n
   CC Leaves on vegetative stems flattened, widest near the apex,
      usually whitish, 1-2 cm long (widespread)
                                   Sedum spathulifolium (Plate 21)
                                     Broadleaf Stonecrop; SCr-n
BB Plants not branching extensively, all the leaves, other than those on
   the stalk of the inflorescence, concentrated in a compact basal
   cluster; largest leaves generally at least 3 cm long; corolla
   somewhat tubular, the petals joined together for part of their
   length, and spreading outward only near the tips, if at all
   C Reduced leaves on stalk of inflorescence nearly triangular,
      nearly or fully as wide as long (strictly limited to coastal
      cliffs)   *Dudleya farinosa*
                                                    Powdery Dudleya
   CC Reduced leaves on stalk of inflorescence slender, mostly 2 or 3
      times as long as wide (inland as well as near the coast)
      D Corolla generally bright yellow or reddish; basal leaves
         often more than 2 cm wide (sometimes 4 cm) (widespread)
                                         Dudleya cymosa (Plate 21)
                                          Common Dudleya; SCl-n
      DD Corolla pale yellow; basal leaves rarely so much as 2 cm wide
         (in rocky areas)
                           Dudleya setchellii [*D. cymosa* ssp. *setchellii*]
                                Santa Clara Valley Dudleya; CC-SB; 1b

Cucurbitaceae--Gourd Family

The Gourd Family is a generous one, for it has given us cucumbers,
watermelons, cantaloupes, squashes, and pumpkins. Important medicinal
substances are also derived from certain species. Plants of this group are
mostly climbing or trailing, and usually have tendrils originating at the
leaf axils.

In native representatives, which belong to the genus *Marah*, there is a
single pistillate flower in each inflorescence that otherwise consists of
staminate flowers. In flowers of both types, the tiny calyx lobes are
located between the 5 lobes of the corolla. In staminate flowers, there are
3 stamens, which are partly joined to one another; in pistillate flowers,
the greater portion of the pistil is below the combined calyx-corolla tube,
and it develops into a fruit with a few large seeds. It should be noted
that flowers of other genera deviate in one way or another from the pattern
characteristic of *Marah*. The calyx lobes, for instance, are usually
prominent, and there are generally 5 stamens, although there may appear to
be only 3, due to fusion.

A Fruit a red berry (occasionally naturalized)   *Bryonia dioica*
                                              White Bryony; eu; SF
AA Fruit a dry gourdlike capsule
   B Leaves mostly more than 10 cm wide (up to 30 cm); corolla white, the
      pistillate flower 15-17 mm wide (spines on fruit, if present, up to
      about 6 mm long, weak; in the Coast Ranges)   *Marah oreganus*
                                           Coast Manroot; SCl-n
   BB Leaves mostly less than 10 cm wide; corolla greenish or cream, the
      pistillate flower 8-15 mm wide
      C Corolla greenish, the staminate flower 5-7 mm wide; spines on
         fruit 1-2 mm long, weak (inland)   *Marah watsonii*
                                             Taw Manroot; Sn, Sl
      CC Corolla cream, the staminate flower 7-12 mm wide; spines on fruit
         5-25 mm long, rigid (widespread)
             Marah fabaceus [includes *M. fabaceus* var. *agrestis*] (Plate 79)
                                         California Manroot; Ma-Mo

Cuscutaceae--Dodder Family

Dodders, often included in the Morning-glory Family, are rootless and have almost no chlorophyll, and their leaves are reduced to tiny scales. All of them parasitize other flowering plants, whose tissues they penetrate by specialized branches of their slender, twining stems. The small flowers typically have 5 calyx and corolla lobes, and 5 stamens; the stamens are attached to the corolla tube in line with the clefts between the lobes. The pistil, with 2 styles, develops into a small dry fruit with 1-4 seeds. Most dodders are of a distinctly orange or yellow color, and are thus easily seen on the plants they parasitize.

A   Stems bright orange; restricted to salt marshes and alkaline areas
    B   Flowers 2-3 mm long; parasitic on plants of the family Chenopodiaceae in inland saline habitats   *Cuscuta salina* var. *salina*
                                                      Inland Dodder
    BB Flowers 3-4.5 mm long; parasitic on *Salicornia* (Chenopodiaceae) and some other plants in coastal salt marshes
                              Cuscuta salina var. *major* (*Plate* 21)
                                          Saltmarsh Dodder
AA Stems yellow (except in *Cuscuta subinclusa*); in many habitats, both moist and dry, but rarely in salt marshes or alkaline areas
    B   Stems orange; flowers 5-6 mm long (pedicels, if present, shorter than the flower; parasitic on various trees [*Quercus*, *Salix*, *Prunus*] and shrubs [*Ceanothus*, *Rhus*])   *Cuscuta subinclusa*
                                          Canyon Dodder
    BB Stems yellow; flowers mostly less than 5 mm long
        C   Calyx lobes pointed
            D   Flowers sessile; coastal (often parasitic on species of *Grindelia* and *Solanum*, but not confined to them)
                      Cuscuta californica var. *breviflora* [*C. occidentalis*]
                                        Western Dodder
            DD Flowers with short pedicels; not restricted to the coast (parasitic on many native shrubs and herbs)
                            Cuscuta californica var. *californica*
                                    California Dodder
        CC Calyx lobes blunt
            D   Flowers 3-4 mm long; calyx shorter than the corolla (parasitic on *Medicago sativa* and various other plants)
                                    Cuscuta indecora
                                      Common Dodder
            DD Flowers 1.5 mm long; calyx as long as the corolla (parasitic on species of *Trifolium*, *Medicago sativa*, and various members of Asteraceae)   *Cuscuta pentagona* [includes *C. campestris*]
                                          Field Dodder

Datiscaceae--Datisca Family

Datisca glomerata (*Plate* 79), called Durango-root, reaches a height of about 1 m, and is the only species of the Family Datiscaceae in California. Although it is perennial, it is not woody. The large leaf blades, with toothed margins, are irregularly and somewhat raggedly lobed in a pinnate pattern. The flowers, clustered in the leaf axils, are small and have no petals. They are, moreover, of 2 types: strictly staminate flowers, with 8-12 stamens, and flowers that have a pistil with 3 forked styles and also a few stamens. The fruit, about 8 mm long and somewhat 3-angled, develops below the level of the 3 short, toothlike sepals. *Datisca* is most commonly found in stream beds that are dry in summer.

Dipsacaceae--Teasel Family

In our region, the Teasel Family is represented by only 3 species, all introduced from Europe. These herbaceous plants have opposite leaves, and their small flowers are concentrated in dense heads whose bases are surrounded by bracts. (In *Scabiosa*, however, the bracts are inconspicuous and mostly hidden by the flowers.) A unique feature of the flowers of this family is a cuplike or funnel-like structure that encloses the calyx. The calyx, in turn, is fused to the pistil, which becomes a 1-seeded dry fruit. There are usually 4 stamens. In *Dipsacus*, the corolla has 2 larger lobes and 2 smaller lobes; in *Scabiosa*, most flowers have a regular 5-lobed corolla, but those nearest the base of the head have an upper lip of 2 lobes and a lower lip of 3 lobes.

A   Plants not especially stiff, mostly less than 50 cm tall, bushy, without prickles; flowers in heads that are usually at least as wide as high, without spinelike bracts (corolla pink, rose, purple, or white; thoroughly naturalized in some areas)   *Scabiosa atropurpurea* (*Plate* 22) Pincushion; me
AA Plants stiff, up to more than 100 cm tall, usually with a single main stem, this prickly; flowers concentrated in a thimble-shaped head, with spinelike bracts beneath it as well as beneath individual flowers
    B   Bracts beneath the inflorescence curving upwards, and some of them longer than the inflorescence; bracts beneath individual flowers ending in a straight spine   *Dipsacus fullonum* (*Plate* 22) Wild Teasel; eu; SFBR
    BB Bracts beneath the inflorescence directed outwards, generally shorter than the inflorescence; bracts beneath individual flowers ending in a downcurved spine (inflorescences have long been used in textile mills for raising the nap on woollen cloth)   *Dipsacus sativus* Fuller's Teasel; eu

Droseraceae--Sundew Family

Sundews live in moist habitats, especially sphagnum bogs. Their leaves are covered with hairs whose glandular tips trap small insects. The insects are then slowly digested, and the soluble products absorbed by the plant. Our only species is *Drosera rotundifolia*, Roundleaf Sundew, which occurs northward from Sonoma county, as well as in the Sierra Nevada. Its leaves, resembling flattened spoons, form a tight cluster. The sticky hairs are reddish. The small flowers, with white or pinkish petals, are on a tall, upright stem.

Elaeagnaceae--Oleaster Family

Elaeagnus angustifolius [*E. angustifolia*], called Oleaster, is probably native to Asia, but is common in Europe and North America. Through self-seeding, it sometimes becomes so well established that it may appear to be native. It is a shrub or tree up to about 10 m tall. The twigs and leaves are silvery, owing to the presence of scalelike hairs. The flowers, in clusters, have no petals, but they are yellowish and sweet-scented. Some of the flowers may be strictly staminate; others will have a pistil as well as stamens. The oval fruits, about 2 cm long, are yellowish or brownish and are covered with silvery scales similar to those on other parts of the plant. In Europe, they are dried and used to flavor cakes.

A species native to California (mostly east of the Sierra Nevada) is Buffalo-berry, *Shepherdia argentea*. The silvery leaves and red berries of this shrub or small tree make it an attractive subject for some gardens.

Ericaceae (includes Pyrolaceae)--Heath Family

The Heath Family consists of trees, shrubs, and perennial herbs. Familiar garden plants belonging to this large group are rhododendrons, azaleas, heaths and heathers, and blueberries. The flowers are regular, and usually have a 5-lobed calyx and a corolla that is either 5-lobed or that consists of 5 separate petals. There are 5 or 10 stamens, and these are distinctive in that their anthers release pollen through a neat pore at the tip, instead of simply splitting open. The fruit, sometimes fleshy at maturity, sometimes dry, is usually partitioned into 5 seed-producing divisions. In a few genera, such as *Vaccinium*, which includes blueberries and huckleberries, the fruit-forming portion of the pistil is below the level where the other flower parts originate.

Probably all plants of this family have fungi associated with their roots. The fungi contribute to the symbiotic relationship by absorbing nutrients and making these available to their plant hosts. Phosphorus is a particularly important element that enters the plants by way of the fungi. Several of our representatives of the Heath Family lack chlorophyll, and are saprophytic.

Many shrubs of this family are excellent subjects for gardens in which native plants are used. For our region, species of *Arctostaphylos* are especially good. Some of them, such as *A. uva-ursi*, are low ground covers, but most reach heights of more than 1 m. A wide variety of these plants are available at sales held periodically by the California Native Plant Society and botanical gardens, as well as at nurseries. *Rhododendron occidentale*, the Western Azalea, *Vaccinium ovatum*, the Evergreen Huckleberry, and *Gaultheria shallon*, called Salal, are superior shrubs for situations where the soil is reasonably moist throughout the summer, and where there is at least some protection from extremely hot sun.

Arbutus menziesii, the much-admired Madrone, is difficult to transplant successfully, even when it is only about 10 or 20 cm tall. Perhaps this is due, in part, to its root fungi not taking kindly to disturbance. After you have prepared a hole in which to plant a small Madrone, remove it from the pot without letting any soil fall away from the roots. It will need sun, and as with all newly transplanted plants, regular moisture until the young tree is well established.

You can also try growing the Madrone from seed. Collect the fleshy fruits that have fallen in the autumn, crush them to squeeze out the seeds, and press these into the soil where you want a tree to grow. A year or two later, thin the crop so that the strongest seedling will have no competition. Remember that the Madrone will grow to be large, that its fallen leaves can cause problems on a roof, and that it is susceptible to some diseases caused by fungi. Maybe you'll be better off without it.

A   Plants with only scalelike leaves, and these not green (the plants
    completely lack chlorophyll) (in shaded habitats)
    B   Stems with red and white stripes; inflorescence 5-25 cm long; up to
        50 cm tall (at elevations above 2000')   *Allotropa virgata*
                                                    Sugarstick; Na, Sn-n
    BB Stems without red and white stripes; inflorescence not more than 6 cm
       long (except in *Pyrola*); not often more than 20 cm tall
        C   Flowers in loose racemes 6-10 cm long; style of pistil projecting
            out of the flower
                    nongreen form of *Pyrola picta* [*P. picta* forma *aphylla*]
                                                            Na, Sn-n
        CC Flowers in congested heads 1.5-6 cm long; style of pistil not
           projecting out of the flower
            D   Inside of corolla not hairy; leaves often fringed with hairs
                (up to 20 cm tall)   *Pleuricospora fimbriolata*
                                                    Fringed Pinesap; SM-n
            DD Inside of corolla hairy; leaves not fringed with hairs
                E   Plants white to pink, up to 12 cm tall; stigmas obvious
                    (yellow); petals united at the base   *Hemitomes congestum*
                                                            Gnomeplant; Mo-n
                EE Plants dull white, up to 20 cm tall; stigmas not obvious;
                   petals completely separate
                                            Pityopus californicus (Plate 79)
                                            California Pinefoot; La, Ma; 4
AA Plants with broad leaves, these green (the plants have chlorophyll)
    B   Plants primarily herbaceous, either not woody or only slightly woody
        at the base, not more than 30 cm tall (in shaded habitats)
        C   Leaves basal, with white or yellowish veins; petioles 2-7 cm
            long; inflorescence 6-10 cm long (petals yellowish white,
            greenish white, or purplish)
                    Pyrola picta [includes *P. picta* forma *aphylla*] (Plate 79)
                                            Whitevein Shinleaf; Na, Sn-n

CC Leaves in whorls on the stem, the veins not obviously white or
   yellowish (except sometimes in *Chimaphila menziesii*); petioles
   absent or not more than 0.5 cm long; inflorescence not more than
   5 cm long
   D   Leaves not more than 3.5 cm long, sometimes with white veins;
       flowers 1-3 in each inflorescence; petals white, turning pink
                                               Chimaphila menziesii
                                                            Pipsissewa
   DD Leaves 3-7 cm long, the veins not white; flowers 3-7 in each
       inflorescence; petals pink   *Chimaphila umbellata*
                [includes *C. umbellata* var. *occidentalis*](*Plate* 79)
                                                        Prince's-pine
BB Trees or shrubs, woody, often more than 100 cm tall (but low and
   creeping in a few species)
   C   Petals either separate, or united for about half their length, in
       which case the lower portion of the corolla is funnel-shaped and
       the corolla as a whole is at least 2 cm long
       D   Petals (5-8 mm long) separate, white; up to 1.5 m tall
           (leaves 3-6 cm long, evergreen, leathery; found in swamps and
           bogs)   *Ledum glandulosum*
                 [includes *L. glandulosum* ssp. *columbianum*] (*Plate* 79)
                                            Western Labrador-tea; SCr-n
       DD Petals (these at least 25 mm long) united for about half
           their length (the corolla thus with a funnel-shaped lower
           portion and 5 lobes), white, cream, or rose; often more than
           2 m tall
           E   Leaves evergreen, 6-20 cm long, smooth-margined; stamens
               10; corolla mostly pink or rose (in shady coastal
               habitats)   *Rhododendron macrophyllum* (*Plate* 23)
                                                           Rosebay; Mo-n
           EE Leaves deciduous, 3-8 cm long, with finely toothed
               margins; stamens 5; corolla white or cream, often pink-
               tinged and with considerable yellow on the uppermost lobe
               (flowers fragrant; usually on stream banks and in other
               moist habitats)   *Rhododendron occidentale* (*Plate* 23)
                                                     Western Azalea; SCr-n
   CC Petals united for more than half their length, the portion of the
       corolla below the lobes shaped like a cup or urn, and the
       corolla as a whole not more than 1 cm long
       D   Flowers in panicles or racemes that are at least 5 cm long;
           leaves (these evergreen and leathery) usually more than 7 cm
           long, not hairy (in shaded habitats)
           E   Trees up to 40 m tall; leaves elliptic, usually rounded at
               the tip, and usually not toothed: flowers in panicles;
               fruit red to orange; bark reddish brown, the outer layer
               generally peeling off, exposing a smooth surface
                                               Arbutus menziesii (*Plate* 22)
                                                        Pacific Madrone
           EE Shrubs up to 2 m tall (the stem sometimes vinelike);
               leaves abruptly pointed at the tip and toothed on the
               margin; flowers in racemes; fruit dark purple; bark not
               obviously peeling   *Gaultheria shallon* (*Plate* 23)
                                                                  Salal
       DD Flowers solitary or in panicles, corymbs, or racemes that are
           rarely more than 3 cm long; leaves not more than 7 cm long,
           often hairy
           E   Flowers in panicles at the end of branches; leaves often
               more than 2 cm long (leathery, evergreen, usually smooth-
               margined); bracts present on the flower pedicels; in open
               areas, including chaparral   Ericaceae, Subkey (p. 144)
           EE Flowers borne in the leaf axils; leaves about 2 cm long;
               bracts usually absent from the pedicels; in shaded areas
               F   Leaves finely toothed, leathery, evergreen; new
                   branches without prominent ridges; flowers in corymbs
                   or racemes; fruit dark blue-black (coastal)
                                               Vaccinium ovatum (*Plate* 23)
                                                     Evergreen Huckleberry
               FF Leaves usually not toothed, thin, deciduous; new
                   branches with prominent ridges; flowers solitary;
                   fruit red   *Vaccinium parvifolium* (*Plate* 23)
                                                    Red Huckleberry; SCl(SCM)-n

Ericaceae, Subkey: Flowers solitary or in panicles that are rarely more than 3 cm long; petals united for more than half their length, the portion of the corolla below the lobes shaped like a cup or urn, and the corolla as a whole not more than 1 cm long; bracts present on pedicels; leaves not more than 7 cm long, leathery, usually smooth-margined; evergreen trees or shrubs   *Arctostaphylos*
A Bracts on pedicels considerably shorter than the pedicels
   B Stomates not visible (even with a hand lens) because of a whitish
      coating on the leaves   *Arctostaphylos glauca*
                                          Bigberry Manzanita; CC(MD)-s
   BB Stomates visible (with a hand lens) on one or both surfaces of the
      leaves (they look like white pin-pricks; view a leaf with the light
      coming from your side)
      C Leaves with stomates only on the lower surface
         D Leaves 4-6 cm long, hairy; up to more than 100 cm tall (calyx
            and corolla with 5 lobes; branchlets with short hairs)
            Arctostaphylos tomentosa ssp. *rosei* [*A. crustacea* var. *rosei*]
                              Lake Merced Manzanita; SF
         DD Leaves up to 2.5 cm long, not hairy, or only with some hairs
            on the veins and margin; prostrate or with stem tips rising
            up to about 15 cm (but sometimes *Arctostaphylos nummularia* is
            upright and considerably taller than 15 cm)
            E Calyx and corolla with 5 lobes; branchlets with short
               hairs   *Arctostaphylos uva-ursi* [includes
                A. *uva-ursi* vars. *coactilis* and *marinensis*] (*Plate* 23)
                                   Bearberry; SM-n
            EE Calyx and corolla with 4 lobes; branchlets with long hairs
               (sometimes up to 2 m tall)   *Arctostaphylos nummularia*
                     [includes A. *nummularia* var. *sensitiva*]
                    Fort Bragg Manzanita; Ma(MT)-SCl(SCM)
      CC Leaves with stomates visible on both surfaces
         D Leaves less than 2.5 cm long; petioles less than 5 mm long
            (pedicels 3-5 mm long; bracts 1-2 mm long)
            E Inflorescences, each with 10-15 flowers, not limited to
               the tip of branches (corolla 4-5 mm long; fruit 5-6 mm
               wide)   *Arctostaphylos densiflora*
                          Vine Hill Manzanita; Sn; 1b
            EE Inflorescences, each with 5-9 flowers, limited to the tip
            of branches (restricted to serpentine)
             F Corolla 4-5 mm long; fruit 4-5 mm wide
                    Arctostaphylos hookeri ssp. *ravenii*
                        Presidio Manzanita; SF; 1b
            FF Corolla 5-7 mm long; fruit 6-8 mm wide
                Arctostaphylos hookeri ssp. *franciscana*
                Franciscan Manzanita; SF; 1a(1942)
         DD Leaves usually more than 2.5 cm long; petioles 5-12 mm long
            E Petioles (and sometimes branchlets) glandular and hairy
              (glands may not be numerous, however)
              F Fruit usually 8-10 mm wide, decidedly wider than long,
                 reddish brown (on serpentine)
                Arctostaphylos bakeri [*A. stanfordiana* ssp. *bakeri*]
                           Baker Manzanita; 1b
              FF Fruit usually 5-7 mm wide, not markedly wider than
                long, whitish, tan, or brown *Arctostaphylos hispidula*
                      [*A. stanfordiana* ssp. *hispidula*]
                    Howell Manzanita; Sn; 4
           EE Petioles and branchlets not glandular (but they may be
            hairy)
            F Branchlets usually hairy
              G Corolla 7-8 mm long, white or pink; leaves up to 6
                cm long; mature fruit 8-12 mm long, red; up to 5 m
                tall (widespread)
                    Arctostaphylos manzanita ssp. *manzanita*
                       Parry Manzanita; CC-n
             GG Corolla 4-7 mm long, white; leaves about 3 cm long;
               mature fruit 5-8 mm long, dark brown; up to 1 m
               tall   *Arctostaphylos manzanita* ssp. *laevigata*
                  Contra Costa Manzanita; CC(MD); 1b
            FF Branchlets not hairy
              G Leaves rounded, whitish; corolla white or tinged
                with pink, 8-9 mm long; up to 4 m tall
                          Arctostaphylos glauca
                    Bigberry Manzanita; CC(MD)-s

GG Leaves elliptic, not whitish; corolla pink, 5-6 mm
   long; prostrate or up to 2 m tall
   H Plants prostrate *Arctostaphylos stanfordiana*
      ssp. *decumbens* [*A. stanfordiana* var. *repens*]
                Rincon Manzanita; Sn; 1b
   HH Plants up to 2 m tall
             Arctostaphylos stanfordiana
      ssp. *stanfordiana* (*Plate* 23)
           Stanford Manzanita; Na, Sn
AA Bracts on pedicels at least as long as the pedicels
 B Stomates not visible (even with a hand lens) because of dense hairs
   C Leaves sessile (flowers white or pink)
     D Leaves 3-7 cm long, dark to pale green, often with marginal
       teeth on the lower half *Arctostaphylos andersonii*
            Santa Cruz Manzanita; SF-SCr; 1b
     DD Leaves 2-5 cm long, gray-green, usually without marginal
       teeth (except perhaps at the base)
           Arctostaphylos auriculata (*Plate* 22)
           Mount Diablo Manzanita; Al, CC(MD); 1b
  CC Leaves with petioles
     D Branchlets with densely crowded cottony hairs
       E New branchlets not glandular
          Arctostaphylos canescens ssp. *canescens*
         [includes *A. canescens* var. *candidissima*]
             Hoary Manzanita; SCr-n
       EE New branchlets glandular
          Arctostaphylos canescens ssp. *sonomensis*
           Sonoma Manzanita; Sn; 1b
     DD Branchlets with long, bristly hairs
       E Leaves (4-5 cm long) sparsely hairy on both surfaces, the
         base notched (corolla white or pale pink; base of trunk
         not enlarged) *Arctostaphylos regismontana*
           King Mountain Manzanita; SM(MM); 4
       EE Leaves densely hairy on at least one surface, the base not
         notched
         F Leaves pale gray-green, up to 6 cm long, with dense
           matted hair on both surfaces; base of trunk not
           enlarged *Arctostaphylos columbiana*
            Douglas-fir Manzanita; Sn-n
         FF Leaves dark green, up to 4.5 cm long, densely hairy on
           the underside, less hairy on the upper surface; base of
           trunk enlarged *Arctostaphylos tomentosa* ssp. *crinita*
           [*A. tomentosa* var. *tomentosiformis*]
            Hairy Manzanita; SM-SLO
 BB Stomates visible on one or both surfaces of the leaves (with a hand
   lens they look like white pin-pricks; view a leaf with the light
   coming from your side)
   C Leaves with stomates only on the underside
     D Leaves 1-2.5 cm long, with petioles up to 5 mm long;
       branchlets with short hairs (coastal)
        Arctostaphylos uva-ursi [includes *A. uva-ursi* vars.
           coactilis and *marinensis*] (*Plate* 23)
             Bearberry; SM-n
     DD Leaves 2-4 cm long, with petioles 5-8 mm long; branchlets
       with bristly hairs
        Arctostaphylos tomentosa ssp. *crustacea* [*A. crustacea*
      var. *crustacea* and *A. glandulosa* var. *campbellae*] (*Plate* 23)
           Brittleleaf Manzanita; CC, SF-s
  CC Leaves with stomates on both surfaces
     D Leaves sessile
       E Leaves not hairy except at the bases of the veins on the
         underside *Arctostaphylos pallida*
         [*A. andersonii* var. *pallida*] (*Plate* 22)
            Pallid Manzanita; Al, CC; 1b
       EE Leaves with hairs usually scattered on the upper surface,
         on the margin, and on the veins on the underside
         F Corolla 3-5 mm long; pedicels 3-5 mm long; prostrate,
           or only the tips of the branches rising
            Arctostaphylos imbricata
           [*A. andersonii* var. *imbricata*]
         San Bruno Mountain Manzanita; SM(SBM); 1b
        FF Corolla 6-9 mm long; pedicels 5-6 mm long; upright
            Arctostaphylos montaraensis
           Montara Manzanita; SM(MM); 1b

```
        DD Leaves with petioles
           E  Base of trunk enlarged
              F  Leaf blades flat, the upper and lower surfaces
                 similar; petioles 5-10 mm long
                                        Arctostaphylos glandulosa
                         [includes A. glandulosa var. cushingiana] (Plate 22)
                                        Eastwood Manzanita; Ma, Na, Sn
              FF Upper surface of leaf blades convex, darker than the
                 lower surface; petioles not often more than 5 mm long
                                     Arctostaphylos tomentosa ssp. crustacea
                                     [includes A. crustacea var. crustacea and
                                      A. glandulosa var. campbellae] (Plate 23)
                                        Brittleleaf Manzanita; CC, SF-s
           EE Base of trunk not enlarged
              F  Branchlets with short hairs; leaves rarely so much as
                 3 cm long; less than 1 m tall (on serpentine)
                                     Arctostaphylos hookeri ssp. montana
                                            [A. pungens var. montana]
                                  Mount Tamalpais Manzanita; Ma(MT); 1b
              FF Branchlets usually with long hairs; most leaves at
                 least 3 cm long; usually at least 1 m tall
                 G  Hairs on the margin of the leaves (and sometimes on
                    the midrib) larger than the scattered hairs (if
                    any) on the leaf surfaces; leaves rarely so much
                    as half as wide as long    Arctostaphylos virgata
                                        Marin Manzanita; Ma; 1b
                 GG Hairs (if any) on the leaves all about the same
                    size; leaves often more than half as wide as long
                    H  Leaf blades notched at the base; petioles about
                       3 mm long          Arctostaphylos regismontana
                                        King Mountain Manzanita; SM(MM);  4
                    HH Leaf blades not notched at the base; petioles
                       3-6 mm long         Arctostaphylos columbiana
                                        Douglas-fir Manzanita; Sn-n
```

Euphorbiaceae--Spurge Family

The Spurge Family, with both shrubby and herbaceous representatives, has given us some interesting garden and greenhouse subjects, including poinsettias. Certain species from southern Africa resemble cacti and are popular with collectors of succulents. There are a few weeds in the group, as well as plants that have long been exploited as sources of dyes, oils, and other substances. One of the more valuable species is *Ricinus communis*, called Castor-bean, whose seeds yield castor oil, widely used in medicine; this oil is also used as fuel for lamps, and in the manufacture of soap, candles, varnish, and polishes. The plant as a whole is very poisonous, but when seeds are pressed, the toxic substance remains in the pulp. Species of *Euphorbia* and *Chamaesyce* have a milky sap that may irritate the skin, and the seeds of some are so toxic that extracts of them have been used for coating arrowheads and for killing fish.

Flowers of spurges are small. None of ours has petals, and in *Euphorbia* and *Chamaesyce* there is no calyx, either, although some bracts below a group of flowers may form what looks like a calyx. When a true calyx is present in an individual flower, it is inconspicuous and 5-lobed. Staminate flowers may have only 1 stamen, which is the case in *Euphorbia* and *Chamaesyce*, or up to more than a dozen; pistillate flowers, sometimes surrounded by several staminate flowers, have a single pistil. The fruit consists of 3 divisions that separate at maturity, each releasing 1 or 2 seeds.

```
A  Flowers with sepals but no petals; pistillate and staminate flowers not
   in the same cluster and sometimes on separate plants
   B  Leaves opposite (leaf blades 2-5 cm long, with fine teeth; staminate
      flowers in dense terminal inflorescences, pistillate flowers in the
      leaf axils of separate plants; up to 30 cm tall)    Mercurialis annua
                                        Mercury; eu; SM
   BB Leaves alternate
```

C Leaves 10-40 cm long, with 5-11 palmate lobes (shrub, up to 3 m
   tall)   *Ricinus communis* (*Plate* 24)
                                                    Castor-bean; eu
CC Leaves not more than 6 cm long, not lobed
   D Leaves and stems covered with long, stiff hairs; annual,
      branching dichotomously, forming mats or nearly hemispherical
      masses up to 20 cm tall (leaves with 3 prominent veins; seeds
      an important part of the diet of doves and quail;
      widespread)   *Eremocarpus setigerus* (*Plate* 24)
                                                   Turkey-mullein
   DD Leaves and stems covered with matted hairs; perennial,
      branching, but not dichotomously, in a spreading mass, or up
      to 100 cm tall (leaves grayer beneath than above; in sandy
      soil; widespread)   *Croton californicus* (*Plate* 24)
                                                      Croton; SF-s
AA Flowers without sepals or petals; pistillate and staminate flowers
   together within a toothed or lobed cup (there may be 4 glands
   alternating with the teeth or lobes of the cup, or there may be leaves
   that resemble petals below the cup)
   B Some leaves alternate (these may be confined to the lower part of the
      stem)
      C Leaves of flowering stems wedge-shaped; low, with stem tips
         rising up to 50 cm (leaves finely toothed; annual)
                                              Euphorbia helioscopia
                                                    Wartweed; eu; SF
      CC Leaves of flowering stems more or less oval or oblong, not at all
         wedge-shaped; upright
         D Leaves below the flowering stems with petioles about half as
            long as the leaves (annual garden weed, up to 35 cm tall)
                                           Euphorbia peplus (*Plate* 24)
                                                    Petty Spurge; eu
         DD Leaves below the flowering stems without distinct petioles
            (but they narrow gradually to where they join the stem)
            E Leaves without fine teeth (although the tip may be
               pointed); up to 60 cm tall (stem leaves up to 3.5 cm
               long, oblong; floral leaves rounded; usually annual)
                                              Euphorbia crenulata
                                                    Chinese-caps
            EE Leaves with fine teeth; up to 45 cm tall
               F Stems hairy; stem leaves 4-6 cm long; perennial
                                              Euphorbia oblongata
                                 European Euphorbia; eu; CC, SCl, SM
               FF Stems not hairy; stem leaves 1-3 cm long; annual
                                              Euphorbia spathulata
                                                 Spatulateleaf Spurge
   BB All leaves opposite (annual)
      C Leaves 5-14 cm long (toothed, sessile, and not hairy) (leaves
         beneath flowers 2-6 cm long; up to 1 m tall) *Euphorbia lathyris*
                                                    Caper Spurge; eu
      CC Leaves not more than 3 cm long
         D Tip of leaves with one or more teeth
            E Fruit hairy only on the angles (prostrate, the stems up to
               25 cm long)   *Chamaesyce prostrata* [*Euphorbia prostrata*]
                                                    Prostrate Spurge
            EE Fruits either not hairy or rather uniformly hairy
               F Plants not hairy; leaves 3-14 mm long; fruit not
                  hairy; prostrate or up to 35 cm tall (widespread)
                      Chamaesyce serpyllifolia [*Euphorbia serpyllifolia*]
                                                    Thymeleaf Spurge
               FF Plants hairy (but sometimes only on the tip of young
                  branches); leaves 4-17 mm long; fruit hairy;
                  prostrate   *Chamaesyce maculata*
                  [includes *Euphorbia maculata* and *E. supina*] (*Plate* 24)
                                                   Spotted Spurge; na
         DD Tip of leaves not toothed (prostrate, the stems up to 25 cm
            long)
            E Leaves up to 7 mm long, rounded
                              Chamaesyce serpens [*Euphorbia serpens*]
                                                   Serpent Spurge; sa; SF
            EE Leaves up to 11 mm long, oblong

F   Staminate flowers 4 in each very small cup
                    Chamaesyce prostrata [*Euphorbia prostrata*]
                                        Prostrate Spurge; sa
FF Staminate flowers many in each cup
                    Chamaesyce ocellata [*Euphorbia ocellata*]
                                        Valley Spurge

Fabaceae (Leguminosae)--Pea Family

The Pea Family, as represented in our area, consists mostly of
herbaceous plants. It does, however, include some shrubs and small trees.

The flowers usually have 5 sepals, 5 petals, 10 stamens, and a pistil
that is not divided into compartments, so it has a single row of seeds. For
most of our representatives, the arrangement of petals is distinctive. The
two lower ones are fused, along the edge where they touch, into a structure
called the keel; the two lateral petals, which often stand out at a 90
degree angle to the keel, are called wings, and the upper petal, usually
the largest, forms the banner. In most of our genera, the filaments of 9 of
the stamens are fused into a tube, but the uppermost one is free. The
fruit, when dry, usually splits open rather forcefully, thus effectively
scattering its seeds. In some genera, including *Acacia* and *Amorpha*, the
flowers deviate markedly from the pattern typical of the family.

Many of our species are introduced. Some were brought to California on
purpose; seeds of others were carried in accidentally. All of the medicks
(*Medicago*) and sweet clovers (*Melilotus*), and a large proportion of vetches
(*Vicia*) and clovers (*Trifolium*), are of exotic origin. So are some shrubby
species that have colonized extensive areas, especially where the native
vegetation has been disturbed. These are Scotch Broom (*Cytisus scoparius*),
French Broom (*Genista monspessulana*), Gorse (*Ulex europaea*), and Spanish
Broom (*Spartium junceum*). All of these are attractive in one way or
another, but they are so aggressive that they interfere with the recovery
of natural vegetation. Elimination of all of them is advised.

Particularly good subjects for gardens are some of the annual and
perennial lupines (*Lupinus*), which can be grown from seed. In the category
of large shrubs or small trees, the Western Redbud (*Cercis occidentalis*)
should be considered, for its flowers, produced freely in late winter or
early spring, are of a beautiful reddish purple color.

A   Trees, generally over 2 m tall (woody throughout)
    B   Leaves not compound (except sometimes the juvenile leaves of *Acacia
        longifolia* and *A. melanoxylon*)
        C   Leaf blades rounded, but with a notch at the base and sometimes
            with a slight indentation at the tip; flowers 8-12 mm long,
            reddish purple, irregular   *Cercis occidentalis* (Plate 24)
                                                Western Redbud; Sl-n
        CC "Leaves" (they are actually just flattened petioles) oblong and
            usually curved, without notches or indentations; flowers less
            than 3 mm long, cream-colored or yellow, regular
            D   Flowers cream-colored, in bunches along the raceme
                                                Acacia melanoxylon
                                                Blackwood Acacia; au
            DD Flowers yellow, solitary along the raceme   *Acacia longifolia*
                                                Golden Wattle; au
    BB Leaves compound
        C   Leaves bipinnately compound, the leaflets not more than 1.5 cm
            long
            D   Leaves with 3-6 pairs of primary divisions (each primary
                division with 12-20 pairs of leaflets, these 5-7 mm long;
                flowers less than 5 mm long)   *Acacia baileyana*
                                                Cootamundra Wattle; au; SFBR
            DD Leaves usually with more than 7 pairs of primary divisions
                E   Each primary division of the leaf with 15-35 pairs of
                    leaflets, these 5-15 mm long; flowers less than 5 mm
                    long, yellow, in several rounded clusters; fruit with
                    conspicuous constrictions   *Acacia decurrens*
                                                Green Wattle; au

EE Each primary division of the leaf with 7-10 pairs of
   leaflets, these less than 8 mm long; flowers over 35 mm
   long, yellow, orange, and red, not in clusters; fruit
   without obvious constrictions (stamens red, 8-9 cm long)
                                        Caesalpinia gilliesii
                                        Bird-of-paradise; sa; CC
CC Leaves pinnately compound, the leaflets usually at least 2 cm
   long
   D   Leaflets (usually 11-17 and of an odd number) oval, rounded
       at the tip, not hairy; base of petiole with a pair of spines;
       corolla whitish, irregular; deciduous   *Robinia pseudoacacia*
                                                 Black Locust; na
   DD Leaflets (usually 8-16 and of an even number [there is no
       terminal leaflet]) about 3 times as long as wide, pointed at
       the tip, hairy; petiole without spines; corolla yellow, only
       slightly irregular; evergreen
                            Senna multiglandulosa [*Cassia tomentosa*]
                                                 Senna; mx; Ma
AA Shrubs or herbs, usually not more than 2 m tall (flowers irregular)
   B   Shrubs with spiny branches, spinelike leaves, or nearly leafless
       branches (in which case the leaves, when present, are much-reduced)
       C   Branches not spiny, but the leaves, when present, much reduced
           (corolla yellow; branches green)   *Spartium junceum* (*Plate* 27)
                                               Spanish Broom; me
       CC Branches spiny, or the leaves stiff and spinelike
          D   Corolla yellow; leaves stiff, spinelike
                                      Ulex europaea [*U. europaeus*]
                                                 Gorse; eu
          DD Corolla not yellow; branches spiny
             E   Flowers 8-9 mm long, 4-6 in each cluster, the corolla
                 lavender-red; leaves not compound
                                 Alhagi pseudalhagi [*A. camelorum*]
                                                 Camelthorn; as; CC
             EE Flowers 16-19 mm long, solitary, the corolla rose-purple;
                leaves usually palmately compound, with 3 leaflets, but
                sometimes not compound   *Pickeringia montana* (*Plate* 26)
                                                 Chaparral Pea; Me-s
   BB Shrubs or herbs without spiny branches, spinelike leaves, or nearly
      leafless branches (leaves compound, with 3 to many leaflets)
      C   Leaves palmately compound (leaflets attached to the end of the
          petiole, and without stalks)
          D   Leaflets 4 to many (flowers in showy racemes)
              E   Calyx with prominent glands; leaflets 4-5
                  Pediomelum californicum [*Psoralea californica*] (*Plate* 27)
                                                 Indian Breadroot
              EE Calyx without glands; leaflets 4 to many
                                        Fabaceae, Subkey 1 (p. 151)
          DD Leaflets 3
             E   Flowers 17-19 mm long, in racemes 15-25 cm long (corolla
                 yellow)   *Thermopsis macrophylla* (*Plate* 27)
                                                 False Lupine
             EE Flowers not more than 12 mm long, in heads or umbels
                                        Fabaceae, Subkey 2 (p. 153)
      CC Leaves pinnately compound (leaflets attached to the side of the
         petiole as well as to the end of it, and with stalks [but these
         may be short])
         D   Leaflets either 2, or more than 3
             E   Leaves with a tendril or with a short projection about 3
                 mm long   Fabaceae, Subkey 3 (p. 155)
             EE Leaves without a tendril or other projection
                                        Fabaceae, Subkey 4 (p. 156)
         DD Leaflets 3 (there may also be a pair of stipules)
            E   Glands present on the calyx and leaves (they may not be
                visible, however, on young leaves)
                F   Calyx lobes up to 5 mm long
                    G   Calyx not hairy, the lobes more or less equal in
                        length
                           Rupertia physodes [*Psoralea physodes*] (*Plate* 27)
                                                 California-tea
                    GG Calyx hairy, with one lobe 2-3 mm longer than the
                       others (in moist areas)
                           Hoita macrostachya [*Psoralea macrostachya*]
                                                 Leather-root
                FF Calyx lobes 10-15 mm long

G   Calyx lobes equal in length, about 1 cm long (moist
    areas)   *Hoita orbicularis* [*Psoralea orbicularis*]
                                       Roundleaf Hoita
GG One calyx lobe obviously longer than the others,
    and about 1.5 cm long
                       Hoita strobilina [*Psoralea strobilina*]
                                   Loma Prieta Hoita; CC-SC1
EE Glands not present on the calyx or elsewhere
   F   Shrubs (*Lotus scoparius* is sometimes only slightly
       woody, however) (widespread)
       G   Flowers in racemes at the end of branches
                                       Genista monspessulana
                          [*Cytisus monspessulanus*] (*Plate* 24)
                                           French Broom; me
       GG Flowers single or in clusters in the leaf axils
           H   Flowers 7-10 mm long, almost sessile; stems not
               angled, usually sparingly leafy; rarely more
               than 1 m tall   *Lotus scoparius* (*Plate* 25)
                                                   Deerweed
           HH Flowers up to 20 mm long, not sessile; stems
               obviously angled, usually densely leafy; often
               more than 2 m tall (extremely invasive)
                                   Cytisus scoparius (*Plate* 24)
                                           Scotch Broom; eu
   FF Herbs
       G   Corolla white or blue
           H   Leaflets not toothed (corolla whitish with some
               pink, 5 mm long)   *Lotus purshianus* (*Plate* 25)
                                               Spanish Lotus
           HH Leaflets toothed at least at the tip
               I   Corolla white, about 3 mm long; teeth not
                   confined to the upper half of leaflets
                                       Melilotus alba [*M. albus*]
                                       White Sweet Clover; eua
               II Corolla blue, about 10 mm long; teeth
                   present only in the upper half of leaflets
                                               Medicago sativa
                                               Alfalfa; eu
       GG Corolla yellow, or yellow and red
           H   Leaflets toothed; corolla yellow
               I   Flowers 2-5 in each cluster (fruit coiled,
                   sometimes spiny)
                   Medicago polymorpha [includes *M. polymorpha*
                               var. *confina*] (*Plate* 80)
                                       California Bur Clover; me
               II Flowers numerous in each cluster
                   J   Leaflets fan-shaped, 1 cm long; fruit
                       coiled   *Medicago lupulina*
                                               Black Medick; eu
                   JJ Leaflets elongate, 2-2.5 cm long; fruit
                       not coiled
                       Melilotus indica [*M. indicus*] (*Plate* 26)
                                               Sour Clover; me
           HH Leaflets not toothed; corolla yellow or yellow
               with red (flowers in the leaf axils, solitary
               or up to several on each peduncle)
               I   Flowers solitary, almost sessile
                   (widespread)
                   J   Calyx lobes equal to the calyx tube;
                       plants scarcely hairy
                                   Lotus wrangelianus (*Plate* 25)
                                           California Lotus
                   JJ Calyx lobes twice as long as the calyx
                       tube; plants densely hairy
                                   Lotus humistratus (*Plate* 25)
                                               Colchita
               II Flowers usually more than 1 on peduncles at
                   least 5 mm long
                   J   Flowers more than 8-12 on each peduncle,
                       the corolla 12 mm long (mainly yellow,
                       but with some red); peduncle 5-15 cm
                       long (stipules leaflike)
                                       Lotus uliginosus
                                       Wetland Deerweed; eu; Ma

JJ Flowers 1-7 on each peduncle, the
   corolla usually less than 12 mm long;
   peduncle not more than 3 cm long
   K Calyx and stem uniformly hairy (use
      a hand lens)
      L Stem hairs not obvious without a
         hand lens; calyx teeth 1-2 mm
         long; corolla 7-9 mm long, red
         or yellow and red; stipules
         reduced to glands (in rocky or
         sandy areas)   *Lotus benthamii*
                        Bentham Lotus; Sn-s
      LL Stem hairs obvious; calyx teeth
         3-4 mm long; corolla 4 mm long,
         yellow; stipules leaflike
                           Lotus angustissimus
                  Slenderpod Deerweed; eu; Sn
   KK Calyx and stem not hairy, or only
      scarcely so
      L Corolla 8-12 mm long, yellow (or
         the banner sometimes red);
         stipules leaflike
                            Lotus corniculatus
                      Bird's-foot Deerweed; eu
      LL Corolla 6-7 mm long, yellow with
         red tinges; stipules reduced to
         glands (on coastal hills)
                                  Lotus junceus
                           Rush Lotus; Me-SLO

Fabaceae, Subkey 1: Shrubs or herbs; leaves palmately compound, with 4 to
many leaflets; flowers in showy racemes; calyx without glands   *Lupinus*
A Keel without hairs (use a hand lens and look at the edge of the keel
   after pushing back the wings)
   B Banner hairy on back (be sure that you are not looking at the calyx)
      (coastal)
      C Hairs not confined to the midvein of banner; woody almost
         throughout, up to 2 m tall; in sandy areas   *Lupinus chamissonis*
                                      Chamisso Bush Lupine; Ma-s
      CC Hairs confined to midvein of banner; woody only at the base, less
         than 0.5 m tall; on rocky hills   *Lupinus albifrons* var. *collinus*
                                      Bay Area Silver Lupine; SFBR
   BB Banner not hairy
      C Petioles 10-13 cm long; leaflets 5-8 cm long (robust perennial up
         to 1.5 m tall; in moist areas)   *Lupinus polyphyllus*
                                           Swamp Lupine; SCr-n
      CC Petioles not more than 6 cm long; leaflets not more than 3 cm
         long
         D Racemes up to 30 cm long, usually with many more than 25
            flowers (stems up to 80 cm tall, but sometimes prostrate,
            with ascending tips)   *Lupinus formosus*
                                              Summer Lupine
         DD Racemes not more than 12 cm long, with fewer than 25 flowers
            E Petioles up to 3 cm long; corolla (10-12 mm long)
               uniformly pale yellow to blue or lilac   *Lupinus adsurgens*
                                              Silky Lupine; SCl-n
            EE Petioles 3-6 cm long; corolla blue to lavender, the banner
               with a light spot
               F Flowers 10-14 mm long, the pedicels 4-8 mm long; woody
                  at base (on rocky, coastal hills)
                                   Lupinus albifrons var. *collinus*
                                   Bay Area Silver Lupine; SFBR
               FF Flowers less than 10 mm long, the pedicels 1-3 mm
                  long; not woody   *Lupinus pachylobus*
                                   Bigpod Lupine; SCl-n
AA Keel with hairs (but sometimes only on the lower or upper margin)
   B Banner hairy on back
      C Plants woody, at least at the base
         D Plants woody nearly throughout, 1-2 m tall, not forming mats
            (widespread)   *Lupinus albifrons* var. *albifrons* (Plate 26)
                                              Silver Lupine
         DD Plants woody only at the base, less than 0.5 m tall, forming
            mats (on rocky, coastal hills)
                                   Lupinus albifrons var. *collinus*
                                   Bay Area Silver Lupine; SFBR

```
CC Plants not woody (less than 1 m tall)
   D  Leaflets 3-5 (near the coast)
             Lupinus tidestromii [includes L. tidestromii var. layneae]
                                        Tidestrom Lupine; Ma-Mo; 1b
   DD Leaflets 6-8
      E  Keel with hairs on both upper and lower margins; at
         elevations of 2000-4000'            Lupinus sericatus
                                Cobb Mountain Lupine; Na, Sn, La; 1b
      EE Keel with hairs only on upper margin; from sea level up to
         3000'
         F  Plants rather succulent, branching only from base
                                               Lupinus affinis
                                            Sky Lupine; SCr-n
         FF Plants not succulent, branching all along the stem
                     Lupinus nanus [includes L. nanus ssp. latifolius
                                    and L. vallicola ssp. apricus]
                                            Douglas Lupine
BB Banner not hairy on back
   C  Petioles less than 4 cm long; leaflets less than 25 mm long and 6
      mm wide
      D  Keel with hairs on lower as well as upper margins (corolla
         yellow)                               Lupinus luteolus
                                            Butter Lupine
      DD Keel without hairs on lower margin
         E  Plants not woody
            F  Banner 10-15 mm long; keel almost hidden by the wings
                     Lupinus nanus [includes L. nanus ssp. latifolius and
                                    L. vallicola ssp. apricus]
                                            Douglas Lupine
            FF Banner less than 8 mm long; keel not hidden by the
               wings (widespread)            Lupinus bicolor
               [includes L. bicolor sspp. tridentatus and umbellatus,
                     L. bicolor var. trifidus, and L. polycarpus]
                                            Miniature Lupine
         EE Plants woody, at least at base
            F  Plants upright, 1-2 m tall, woody (except recent
               growth); leaflets 5-12 (banner yellow; wings usually
               yellow, but sometimes blue, or of mixed colors;
               coastal)                        Lupinus arboreus
                     [includes L. arboreus var. eximius] (Plate 26)
                                            Yellow Bush Lupine
            FF Plants low or prostrate, less than 0.5 m tall, woody
               at the base; leaflets 7-10
               G  Leaflets equally hairy on both surfaces; corolla
                  blue to lavender, banner with a light spot; up to
                  50 cm tall, forming mats (on rocky coastal hills)
                                  Lupinus albifrons var. collinus
                                    Bay Area Silver Lupine; SFBR
               GG Leaflets more hairy on the underside than on the
                  upper surface; corolla uniformly yellow, white,
                  pink, or blue; prostrate or with the stem tips
                  rising                       Lupinus variicolor
                                    Varied Lupine; SLO-n
   CC Petioles over 6 cm long; leaflets at least 30 mm long and 10 mm
      wide
      D  Keel with hairs on both lower and upper margins
         E  Corolla purple, pink, or white
                                  Lupinus succulentus (Plate 26)
                                    Arroyo Lupine; Me-s
         EE Corolla yellowish                  Lupinus luteolus
                                            Butter Lupine
      DD Keel with hairs confined to either the upper or lower margin
         E  Corolla yellow or white, sometimes pink-tinged or with
            darker veins (racemes up to 20 cm long; banner 14 mm
            long)      Lupinus microcarpus var. densiflorus [includes
               L. densiflorus vars. aureus and densiflorus] (Plate 26)
                                            Gully Lupine
         EE Corolla red, dark violet, pink, or blue
            F  Keel with hairs only on lower margin; flowers not in
               whorls (racemes up to 25 cm long; corolla bluish or
               red)                           Lupinus hirsutissimus
                                    Stinging Lupine; SM-s
            FF Keel with hairs only on upper margin; flowers usually
               in whorls
```

                    G   Banner (8-12 mm long) with a light spot (racemes up
                        to 24 cm long)   *Lupinus nanus* [includes *L. nanus*
                        ssp. *latifolius* and *L. vallicola* ssp. *apricus*]
                                                          Douglas Lupine
                    GG Banner without a light spot
                        H   Corolla dark violet or pink (banner 12-14 mm
                            long); racemes up to 15 cm long; less than 0.5
                            m tall   *Lupinus microcarpus* var. *microcarpus*
                                                            [*L. subvexus*]
                                                     Chick Lupine; La-s
                        HH Corolla blue or purple; racemes up to 45 cm
                            long; up to more than 1 m tall
                            I   Flowers 10-14 mm long; plants scarcely
                                hairy   *Lupinus latifolius* var. *latifolius*
                                                          Broadleaf Lupine
                            II Flowers 13-16 mm long; plants densely hairy
                                           Lupinus latifolius var. *dudleyi*
                                                     Dudley Lupine; SM(MM)

Fabaceae, Subkey 2: Shrubs or herbs; leaves palmately compound, with 3
leaflets; flowers not more than 12 mm long, in heads or umbels   *Trifolium*
A   Flower heads without bracts at the base (but there may be stem leaves or
    a ringlike structure)
    B   Flowers with pedicels 1-2 mm long
        C   Corolla yellow
            D   Flowers 2-4 mm long, about 10 in a head; petioles less than 1
                cm long   *Trifolium dubium*
                                                    Shamrock; eu; SFBR-n
            DD Flowers 8-12 mm long, about 4 in a head; petioles 2-10 cm
                long   *Trifolium subterraneum*
                                          Subterranean Clover; eu; Sn
        CC Corolla white to purple
            D   Peduncle and calyx hairy (leaflets notched at the tip)
                                               Trifolium bifidum var. *bifidum*
                                                      Pinole Clover; Mo-n
                                       and *Trifolium bifidum* var. *decipiens*
                                                          Deceiving Clover
            DD Peduncle and calyx not hairy
                E   Pedicels about 5 mm long; corolla 6-10 mm long (in moist
                    areas)
                    F   Stipules 10-25 mm long; corolla pink *Trifolium hybridum*
                                                          Alsike Clover; eu
                    FF Stipules 4-10 mm long; corolla white to pinkish
                                                            Trifolium repens
                                                       White Clover; eua
                EE Pedicels 1-3 mm long, corolla 5-7 mm long
                    F   Stipules less than 10 mm long; peduncle 2-6 cm long
                                                         Trifolium gracilentum
                                                           Pinpoint Clover
                    FF Stipules 15-30 mm long; peduncle 5-15 cm long
                                                          Trifolium ciliolatum
                                                              Tree Clover
    BB Flowers sessile
        C   Flower heads sessile, with some leaves and stipules directly
            below them
            D   Flower heads often in pairs; corolla and calyx about the same
                length (6 mm); prostrate, with stem tips rising
                                                          Trifolium macraei
                                                      Doublehead Clover
            DD Flower heads solitary; corolla (10-20 mm) twice as long as
                calyx; up to 60 cm tall   *Trifolium pratense* (*Plate* 27)
                                                         Red Clover; eu
        CC Flower heads (solitary) on peduncles arising from the uppermost
            group of leaves
            D   Corolla hidden by the calyx, which may be more than 1 cm long
                            Trifolium albopurpureum var. *olivaceum* [*T. olivaceum*]
                                                            Olive Clover
            DD Corolla at least as long as the calyx
                E   Corolla 12-15 mm long, calyx at least 10 mm long
                    F   Corolla purple with white tips; flower heads 2.5 cm
                        long and about as wide   *Trifolium amoenum*
                                          Showy Indian Clover; Ma-Sl; 1b
                    FF Corolla crimson; flower heads about 4 cm long and half
                        as wide   *Trifolium incarnatum*
                                                  Crimson Clover; eu; Mo-n

EE Corolla and calyx both 5-10 mm long
   F Corolla 6 mm long, almost hidden by the calyx (calyx
      teeth feathery)
                Trifolium albopurpureum var. *albopurpureum*
                                Rancheria Clover
   FF Corolla 7-10 mm long, definitely longer than the calyx
     G Flower heads narrow, with more than 10 flowers;
        calyx teeth feathery   *Trifolium albopurpureum*
                    var. *dichotomum* [*T. dichotomum*]
                    Branched Indian Clover; SCl-n
    GG Flower heads globular, with fewer than 10 flowers;
       calyx teeth not feathery (ringlike structure
       present at base of the head)
                Trifolium depauperatum var. *depauperatum*
                      Dwarfsack Clover; Al-n
AA Flower heads with bracts at the base
  B Bracts of flower heads completely separate
    C Corolla 12-25 mm long (cream or yellow, turning pink with age and
      sometimes becoming inflated; in moist areas)   *Trifolium fucatum*
      [includes *T. fucatum* vars. *gambelii* and *virescens*] (*Plate* 27)
                                  Bull Clover
   CC Corolla about 6 mm long
    D Bracts about as long as the calyx (in grassy areas)
                   Trifolium depauperatum var. *truncatum*
                      [*T. amplectens* var. *truncatum*]
                  Common Palesack Clover; La-s
    DD Bracts shorter than the calyx
     E Plants found in relatively dry grassy areas; fruit up to 4
       mm long, without an obvious stalklike base
                 Trifolium depauperatum var. *amplectens*
                    [*T. amplectens* var. *amplectens*]
                Fusedbract Palesack Clover; La-s
     EE Plants found in saline and alkaline areas; fruit up to 3
       mm long, with a distinct stalklike base (very rare)
                 Trifolium depauperatum var. *hydrophilum*
                    [*T. amplectens* var. *hydrophilum*]
                      Saline Clover; SLO-n
 BB Bracts of flower heads at least partly joined, the lobes often toothed
   C United bracts forming a cup-shaped structure
    D Lobes of bracts not toothed
                     Trifolium microcephalum (*Plate* 27)
                      Smallhead Clover
    DD Lobes of bracts toothed
     E Corolla (purple) definitely longer than the calyx; bract
       cup 1.5-2 cm wide (in moist areas)
                Trifolium barbigerum var. *andrewsii* [*T. grayi*]
                    Gray Clover; Me-Mo
     EE Corolla shorter than the calyx, bract cup about 1 cm wide
      F Corolla purple; prostrate (coastal, in moist areas)
                  Trifolium barbigerum var. *barbigerum*
                      Bearded Clover
      FF Corolla white or pink; up to more than 10 cm tall (in
        grassland)   *Trifolium microdon*
                  Valparaiso Clover; SLO-n
   CC United bracts forming a flat disk
    D Corolla 12-15 mm long
     E Plants hairy and glandular; banner (this may have a dark
       spot) and wings equally pale (in moist areas)
                        Trifolium obtusiflorum
                      Creek Clover
     EE Plants not hairy or glandular; banner (this may have a
       dark spot) lighter than the wings
      F Bracts without regularly spaced lobes, but with teeth
        1-3 mm long (in grassland)
        Trifolium willldenovii [*T. tridentatum*] (*Plates* 28, 80)
                        Tomcat Clover
      FF Bracts lobed, and with teeth on the lobes (in moist
        areas; widespread)   *Trifolium wormskioldii*
                      Cow Clover
   DD Corolla 5-10 mm long
    E Corolla definitely longer than the calyx (calyx teeth not
      hairy; in moist areas)
              Trifolium variegatum [includes *T. appendiculatum*]
                    Whitetip Clover

     EE Corolla shorter than the calyx or protruding from it for
        only 1-2 mm
        F Corolla dark purple, shorter than the calyx; calyx
           hairy (in moist areas)
                              Trifolium barbigerum var. *barbigerum*
                                                  Bearded Clover
        FF Corolla mostly lavender with white tips (but the keel
           purple), slightly longer than the calyx; calyx not
           hairy   *Trifolium oliganthum* (*Plate* 28)
                                              Fewflower Clover; SLO-n

Fabaceae, Subkey 3: Leaves pinnately compound with either 2 leaflets or
more than 3 leaflets; leaves with a tendril or with a short projection
about 3 mm long
A Flowers 1-3 cm long; wings free from keel except at the base; base of
   each wing with a crescent-shaped ridge
   B Leaflets 2
      C Flowers solitary, 10-13 mm long; corolla red   *Lathyrus cicera*
                                              Flatpod Peavine; eu; SM
      CC Flowers not solitary, 20-30 mm long; corolla purplish, pink, or
         reddish pink (sometimes white in *Lathyrus latifolius*)
         D Inflorescence with 1-3 flowers (coastal)   *Lathyrus tingitanus*
                                                     Tangier Pea; eu
         DD Inflorescence with 5-15 flowers
                                      Lathyrus latifolius (*Plate* 80)
                                               Everlasting Pea; eu
   BB Leaflets more than 8
      C Tendril reduced to a projection about 3 mm long; stems not
         winged; flowers less than 2 cm long, not more than 10 in each
         raceme
         D Flowers 8-13 mm long, 1-2 in each raceme; in open woods
                                                    Lathyrus torreyi
                                               Redwood Pea; SCr-n
         DD Flowers 12-18 mm long, usually 2-6 (sometimes 10) in each
            raceme; on backshores of sandy beaches
                                      Lathyrus littoralis (*Plate* 25)
                                                 Beach Pea; Mo-n
      CC Tendril well developed and coiling; stems winged; flowers about 2
         cm long, 15-20 in each raceme
         D Corolla white to lavender, fading to yellow (in woods or
            under shrubs)   *Lathyrus vestitus*
                      [includes *L. vestitus* ssp. *bolanderi*] (*Plate* 25)
                                               Woodland Pea; SFBR-n
         DD Corolla crimson, the banner sometimes paler (near water)
            E Plants not hairy; flowers uniformly bright pink
                                      Lathyrus jepsonii var. *jepsonii*
                                            Delta Tule Pea; SFBR; 1b
            EE Plants hairy; banner usually paler than the rest of the
               flower   *Lathyrus jepsonii* var. *californicus*
                                                Bluff Pea; SLO-n
AA Flowers often less than 1 cm long (if so, the following 2 characters
   need not be checked); wings joined at least partly to keel; wings
   without a ridge at the base
   B Flowers sessile on the stem, or on a peduncle less than 1 cm long
      C Corolla yellow or white
         D Flowers 6 mm long; corolla whitish   *Lens culinaris*
                                               Lentil; eua; SF
         DD Flowers over 15 mm long; corolla yellow and white
            E Flowers 15-18 mm long; leaflets 14-18   *Vicia pannonica*
                                            Hungarian Vetch; eu; Sn
            EE Flowers 20-25 mm long; leaflets 8-16   *Vicia lutea*
                                              Yellow Vetch; eu; SFBR
      CC Corolla violet, purple, or white with a purple spot
         D Flowers 10-18 mm long (uniformly purple)
                                      Vicia sativa ssp. *nigra* [*V. angustifolia*]
                                               Narrowleaf Vetch; eu
         DD Flowers 18-35 mm long
            E Racemes with 3-6 flowers; corolla white, with purple on
               the wings; leaflets 6-7, 15 mm wide; tendril less than
               1.5 cm long   *Vicia faba*
                                               Horsebean; eua
            EE Racemes with 1-2 flowers; corolla purple, with violet on
               the wings; leaflets 8-16, 2-3 mm wide; tendril 2-3 cm
               long   *Vicia sativa* ssp. *sativa* (*Plate* 28)
                                                 Spring Vetch; eu

BB Flowers clustered on a peduncle at least 2.5 cm long
   C   Racemes with 1-2 flowers, these 2-7 mm long
       D   Calyx lobes about as long as the calyx tube   *Vicia tetrasperma*
                                                              Tare; eu
       DD Calyx lobes much shorter than the calyx tube
           E   Leaflets 2-3 cm long
                           Vicia ludoviciana [*V. exigua* var. *exigua*]
                                                         Slender Vetch
           EE Leaflets 1-2 cm long   *Vicia hassei* [*V. exigua* var. *hassei*]
                                                         Hasse Vetch; Ma, Al-s
   CC Racemes with more than 3 flowers, these 10-18 mm long
       D   Racemes with 3-10 flowers, these not grouped on one side;
           leaflets 8-16 (widespread)   *Vicia americana* [includes
                                              V. americana ssp. *oregana*, vars.
                                         linearis and *truncata*, and *V. californica*]
                                                         American Vetch
       DD Racemes usually with more than 10 flowers (often fewer,
           however, in *Vicia benghalensis*), these grouped on one side;
           leaflets usually 16-24
           E   Upper leaflets of each leaf usually considerably smaller
               than the lower ones
               F   Stems and leaves obviously hairy; tip of petals dark
                   blue or violet, rest of corolla whitish
                                             Vicia villosa ssp. *villosa*
                                                         Hairy Vetch; eu
               FF Stems and leaves scarcely, if at all, hairy; tip of
                   petals not markedly different from rest of corolla,
                   which is usually dull red or orange-red, but sometimes
                   yellowish or purplish (in moist areas)
                                             Vicia gigantea (*Plate* 28)
                                                         Giant Vetch; SLO-n
           EE All leaflets more or less equal
               F   Leaflets up to 3.5 cm long and 10 mm wide; racemes
                   with 3-10 flowers; corolla rose-purple
                                             Vicia benghalensis
                                                 Purple Vetch; eu; SFBR
               FF Leaflets up to 1.5 cm long and 4 mm wide; racemes
                   usually with more than 10 flowers; corolla light blue
                   to purple   *Vicia villosa* ssp. *varia* [*V. dasycarpa*]
                                                 Winter Vetch; eu; Ma

Fabaceae, Subkey 4: Leaves pinnately compound with either 2 leaflets or
more than 3 leaflets, and without a tendril or short projection
A   Leaflets 4-7 (except *Lotus stipularis*); flowers solitary or in umbels
    B   Stipules obvious, similar in size to the leaflets
        C   Stipules green and leaflike; leaflets 9-19 (corolla red and
            white, about 1 cm long)   *Lotus stipularis*
                                                 Stipulate Lotus; Mo-n
        CC Stipules not green, not leaflike; leaflets 4-7
            D   Corolla white or pink, 8-9 mm long (at edge of woods)
                                             Lotus aboriginus [*L. aboriginum*]
                                                 Roseflower Lotus; Sn-n
            DD Corolla with some yellow, 9-14 mm long
                E   Stipules 3 mm long; corolla mostly greenish yellow, with
                    some red; in dry areas   *Lotus crassifolius* (*Plate* 25)
                                                 Broadleaf Lotus; SLO-n
                EE Stipules 8 mm long; banner yellow, wings and keel rose; in
                    moist coastal areas   *Lotus formosissimus*
                                                 Witch's-teeth; Mo-n
    BB Stipules not obvious (small red or black glands at the petiole base)
        C   Corolla white or reddish
            D   Corolla 7-9 mm long, reddish, with dark veins (coastal)
                                                       Lotus benthamii
                                                 Bentham Lotus; Sn-s
            DD Corolla 4-5 mm long, pinkish to red   *Lotus micranthus*
                                                              Hill Lotus
        CC Corolla yellow, sometimes becoming red as flowers age
            D   Peduncles over 10 mm long
                E   Peduncle 4-8 cm long (each with at least 2 flowers);
                    corolla more than 15 mm long   *Lotus grandiflorus*
                                                 Largeflower Lotus; Me-s
                EE Peduncle 1-4 cm long; corolla less than 9 mm long
                    F   Calyx and stem scarcely hairy (on coastal hills)
                                                       Lotus junceus
                                                 Rush Lotus; Me-SLO

```
          FF Calyx and stem hairy
              G   Wings longer than keel (leaflets about 6 mm long;
                  flowers 1-3 in each cluster)        Lotus strigosus
                          [includes L. strigosus var. hirtellus]
                                        Hairy Lotus; Ma-s
              GG Wings not longer than keel (primarily coastal)
                  H   Leaflets 5-8; flowers 1-5 in each cluster;
                      corolla yellow              Lotus salsuginosus
                                        Hookbeak Lotus; SCl-s
                  HH Leaflets 4-5; flowers 3-7 in each cluster;
                      corolla usually reddish, with darker veins,
                      but sometimes yellowish or white tinged with
                      red or purple                 Lotus benthamii
                                        Bentham Lotus; Sn-s
      DD Peduncles up to 5 mm long, sometimes absent
          E   Flowers on peduncles 2-5 mm long in the leaf axils
              F   Plants densely hairy (at least on the new stems and
                  leaves); flowers 4-10 in each raceme; corolla 3-5 mm
                  long; calyx lobes 5 mm long; mostly prostrate, forming
                  mats; in moist areas  Lotus heermanii var. orbicularis
                                  [L. heermanii var. eriophorus]
                                        Southern Lotus; SCr-s
              FF Plants scarcely hairy; flowers 1-5 in each raceme;
                  corolla 6-7 mm long; calyx lobes 1 mm long; mostly
                  upright, or at least the stem tips rising; in dry
                  areas                              Lotus junceus
                                                    Rush Lotus
      EE Flowers nearly sessile in the leaf axils (in dry areas;
          widespread)
              F   Flowers 1-5 in each raceme, the corolla 7-10 mm long;
                  somewhat woody, up to more than 1 m tall (plants
                  scarcely hairy)          Lotus scoparius (Plate 25)
                                                    Deerweed
              FF Flowers solitary, the corolla 5-7 mm long; not woody,
                  prostrate, with stem tips rising
                  G   Calyx lobes equal to the calyx tube; plants
                      scarcely hairy       Lotus wrangelianus (Plate 25)
                                        California Lotus
                  GG Calyx lobes twice as long as the calyx tube; plants
                      densely hairy        Lotus humistratus (Plate 25)
                                                    Colchita
AA Leaflets 7-40; flowers in racemes (but some racemes are so short they
  may appear to be umbels)
  B   Leaves gland-dotted; fruit not inflated
      C   Flowers with only 1 petal (the banner); fruit without hooked
          bristles; shrub, up to 2 m or more tall
                          Amorpha californica var. napensis (Plate 80)
                                        False Indigo; Ma
      CC Flowers with 5 petals; fruit with many hooked bristles; herb, up
          to 1 m tall (in moist areas)    Glycyrrhiza lepidota (Plate 80)
                                        Wild Licorice
  BB Leaves not gland-dotted; fruit of some species obviously inflated
      C   Leaflets 17-40 (flowers from 15 to more than 100 in each raceme)
          D   Flowers usually less than 1 cm long; fruit 6-8 mm long
              E   Racemes dense, 6 cm long and 2 cm wide; leaflets very
                  hairy (near salt marshes, sand dunes)
                                        Astragalus pycnostachyus
                                        Marsh Milkvetch; Me-SM
              EE Racemes open, 10-15 cm long and 1 cm wide; leaflets only
                  slightly hairy (in moist areas, often on serpentine)
                                        Astragalus clevelandii
                                        Cleveland Milkvetch; Na; 4
          DD Flowers 1-1.5 cm long; fruit 20-30 mm long
              E   Fruits on stalks 20-40 mm long; flowers 15-45 in each
                  raceme; inland              Astragalus asymmetricus
                                        San Joaquin Milkvetch; Sl, CC-s
              EE Fruits on stalks 3 mm long; flowers 40-125 in each raceme;
                  coastal               Astragalus nuttallii var. virgatus
                                  [A. nuttalli var. virgatus] (Plate 80)
                                        Nuttall Milkvetch; Ma-SM, Al
      CC Leaflets 7-17
          D   Flowers 20-95 in each raceme (in moist ares, often on
              serpentine)                   Astragalus clevelandii
                                        Cleveland Milkvetch; Na; 4
```

DD Flowers 4-15 in each raceme
 E   Flowers 3-6 mm long, in racemes not more than 1 cm long;
     corolla white, tinged with violet
  F   Racemes 8-10 mm long; fruit 3 mm long, roundish, not
      twisted at the tip; sometimes prostrate
                                  Astragalus didymocarpus
                                  Twoseed Milkvetch; CC-s
  FF Racemes 4-8 mm long; fruit 4-5 mm long, flattened,
      twisted at the tip; upright   *Astragalus gambelianus*
                                  Gambel Milkvetch
 EE Flowers 7-12 mm long, in racemes over 1 cm long; corolla
    lilac or purple, sometimes white with a purple spot at
    the tip of each petal
  F   Leaflets narrow, 10-14 mm long and about 2 mm wide
      (fruit 10-15 mm long, somewhat hairy; in moist,
      alkaline soil)   *Astragalus tener*
                                  Alkali Milkvetch; Sl, SF, SB; 1b
  FF Leaflets not especially narrow
   G   Leaflets 10 mm long; fruit 15-30 mm long (somewhat
       hairy) (often on serpentine)
                                  Astragalus rattanii var. *jepsonianus*
                                  [*A. rattani* var. *jepsonianus*]
                                  Jepson Milkvetch; Na; 1b
   GG Leaflets 6 mm long; fruit 6-15 mm long
    H   Fruit densely hairy; flowers 6-9 in each raceme
        (the racemes resemble umbels) (often on
        serpentine)   *Astragalus breweri*
                                  Brewer Milkvetch; Ma; 4
    HH Fruit somewhat hairy; flowers more than 11 in
       each raceme
                                  Astragalus lentiginosus var. *idriensis*
                                  Freckled Milkvetch; SCl(MH)

Fagaceae--Oak Family

The Oak Family consists of trees and shrubs with alternate leaves. The
staminate flowers, each with at least several stamens, are in elongated
catkins; the pistillate flowers, each of which produces a single-seeded
nut, may be solitary (in oaks), in small clusters (in beeches), or at the
base of the staminate catkins (in chestnuts and chinquapins). Both
staminate and pistillate flowers have a calyx with several lobes.

Our representatives of this family are oaks (*Lithocarpus* and *Quercus*)
and chinquapins (*Chrysolepis*). In oaks, the fruit is an acorn, consisting
partly of a 1-seeded nut derived from the pistil and partly of a scaly cup
formed by closely associated bracts that were below the flower. In
chinquapins, comparable bracts contribute to a spiny bur that encloses 1-3
nuts.

Oaks and chinquapins can be propagated from acorns. All of our native
species, some of which are evergreen shrubs, will be of interest to persons
whose gardens or natural areas provide a habitat for indigenous plants.

A   Leaves with golden scales on the underside, at least when young (older
    leaves sometimes olive-yellow on the underside); fruit consisting of 1-3
    nuts enclosed by a spiny bur; staminate catkins upright, the pistillate
    flowers at the base (leaves smooth-margined, but the margins may be
    wavy, the tips tapered)
 B   Leaves not folded, mostly 7-9 cm long; sometimes more than 30 m tall;
     at elevations up to 1500' *Chrysolepis chrysophylla* var. *chrysophylla*
                                  Giant Chinquapin; Ma-n
 BB Leaves often folded along the midrib (or wavy along the margins),
    mostly 3-4 cm long; rarely more than 8 m tall (trees, but somewhat
    shrubby); at elevations up to 6000'
                                  Chrysolepis chrysophylla var. *minor* (Plate 28)
                                  Golden Chinquapin
AA Leaves not golden on the underside (except in *Quercus chrysolepis*);
   fruit an acorn (a nut partly enclosed within a cup); staminate catkins
   hanging down (except in *Lithocarpus densiflorus*)
 B   Shrubby, up to about 3 m high, but usually less (evergreen)

C Leaves (up to 3 cm long) convex, the lower surface with star-shaped hairs (use hand lens); leaves toothed, the teeth usually spiny (petioles less than 5 mm long; often on serpentine)
Quercus durata
Leather Oak

CC Leaves flat or wavy, but not convex, the lower surface without star-shaped hairs (there may be other hairs, however); leaves toothed or nearly smooth
  D Leaves 2-6 cm long; petioles 5-10 mm long; acorns at least twice as long as wide; scales of acorn cup overlapping
Quercus wislizenii var. *frutescens*
Dwarf Interior Live Oak; La-s

  DD Leaves not more than 3 cm long; petioles 1-4 mm long; acorns less than twice as long as wide; scales of acorn cup not distinctly overlapping (they are more like bumps than scales)   *Quercus berberidifolia* (*Plate* 81)
Scrub Oak

BB Trees, up to more than 20 m tall, but often much shorter and shrublike
  C Leaves deeply lobed in a somewhat pinnate pattern; deciduous
    D Lobes of leaves with sharp tips, and sometimes with a few teeth that also have sharp tips; depth of acorn cup nearly or fully half the length of the nut *Quercus kelloggii* (*Plate* 82)
California Black Oak

    DD Lobes of leaves more or less rounded, smooth-margined; depth of acorn cup usually much less than half the length of the nut
      E Leaves bluish green above, paler beneath (widespread)
Quercus douglasii (*Plate* 81)
Blue Oak

      EE Leaves dark green above
        F Leaves 5-11 cm long, not leathery; petioles less than 1.5 cm long; nuts commonly 3-4 cm long and about 3 times as long as wide (away from the immediate coast)
Quercus lobata
Valley Oak

        FF Leaves 9-17 cm long, leathery; petioles 1.5-2.5 cm long; nuts not more than 2.5 cm long, and not more than twice as long as wide
Quercus garryana (*Plate* 81)
Oregon Oak; SCl, Ma-n

CC Leaves not obviously lobed, although they may be toothed; evergreen (except *Quercus douglasii*)
  D Largest leaves up to more than 12 cm long (with blunt tips and usually with hairs on the underside); catkins mostly upright (with an unpleasant odor); cup of acorn shallow (about one-fourth as long as the nut) and with its scales drawn out into slender projections; petioles 1-2 cm long (leaves with more than 8 prominent lateral veins on both sides, each one ending at a marginal tooth; mostly in coastal forests; extract of the bark long used for tanning leather; common)   *Lithocarpus densiflorus* [*L. densiflora*] (*Plate* 81)
Tanbark Oak

  DD Largest leaves usually not more than 9 cm long; catkins hanging down; cup of acorn not necessarily shallow, its scales not drawn out into slender projections; petioles 2-15 mm long
    E Leaves bluish green on the upper surface (paler bluish green and slightly hairy on the underside), sometimes slightly lobed; deciduous; cup of acorn about one-fourth as long as the nut, its diameter less than that of the widest part of the nut (mostly away from the coast; widespread)   *Quercus douglasii* (*Plate* 81)
Blue Oak

    EE Leaves dark green on the upper surface, the margins toothed or smooth; evergreen; cup of acorn at least one-third as long as the nut
      F Underside of young leaves with golden hairs, the underside of older leaves grayish; acorn cup 17-30 mm wide (densely hairy on the scales and inside the cup) (leaves flat or wavy, the tips tapered, usually smooth-margined)   *Quercus chrysolepis* (*Plate* 82)
Cañon Oak

FF Underside of leaves neither with golden hairs, or
   grayish; acorn cup not more than 18 mm wide
   G Leaves usually smooth-margined, the tips tapered
      (underside of leaves not hairy)
                              Quercus parvula var. *shrevei*
                                         Shreve Oak; SFBR
   GG Leaves usually with small teeth, but sometimes
      smooth-margined, the tips blunt (widespread)
      H Leaves flat, the underside not hairy; nut
         usually about 3 times as long as wide
            Quercus wislizenii var. *wislizenii* (*Plate* 82)
                                       Interior Live Oak
      HH Leaves generally convex, the underside often
         with tufts of hair on the veins; nut usually
         about twice as long as wide (the most abundant
         evergreen oak in our region; in some years,
         the foliage is severely damaged by
         caterpillars of the California Oak Moth,
         Phryganidia californica)
                              Quercus agrifolia (*Plate* 28)
                                          Coast Live Oak

Frankeniaceae--Frankenia Family

Plants belonging to the Frankeniaceae have opposite, sessile or nearly
sessile leaves. Two pairs of leaves may be so close together, however, as
to suggest a whorl of four leaves. The calyx is tubular, with short lobes,
and the petals are attached to the inside of the tube near its top. Each
petal has a small scalelike structure on its upper face. The fruit is dry
at maturity. Some species are rather inconsistent with respect to the
number of petals and stamens. This is the case in *Frankenia salina* [*F.
grandifolia*] (*Plate* 28), called Alkali-heath, a much-branched shrubby plant
that grows at the edge of salt marshes from Marin and Solano counties south
to Baja California. It usually has 5 pink petals, but there are often 6,
and the number of stamens ranges from 4 to 7. *Frankenia* reaches a height of
about 30 cm and may form rather dense thickets.

Garryaceae--Silk-tassel Family

Silk-tassel shrubs, all in the genus *Garrya*, have opposite leaves that
are evergreen and rather tough. The flowers are concentrated in
inflorescences that resemble catkins, and pistillate and staminate flowers
are on separate plants. Within an inflorescence of pistillate flowers,
there is a single flower above each little bract. The fruit, when ripe, is
black or purple and somewhat fleshy inside, but has a hard, dry covering.
Staminate flowers, with 4 stamens, are borne in threes above each pair of
bracts. The long inflorescences and evergreen leaves make these plants very
attractive garden subjects. *Garrya elliptica* is the most widely cultivated
species.

A Leaf margins wavy (underside of leaves densely hairy)
   B Individual hairs on underside of leaves scarcely distinguishable with
      a hand lens; leaf blades 6-8 cm long; petioles 6-12 mm long
                              Garrya elliptica (*Plate* 82)
                                        Coast Silk-tassel
   BB Individual hairs on underside of leaves distinguishable with a hand
      lens; leaf blades 2.5-6 cm long; petioles 4-8 mm long
                              Garrya congdonii [*G. congdoni*]
                                      Congdon Silk-tassel; SB-n
AA Leaf margins flat (blades 2-5 cm long)
   B Fruit not hairy, or only slightly so when young; underside of leaves
      scarcely hairy   *Garrya fremontii*
                                      Fremont Silk-tassel; Mo-n
   BB Fruit densely hairy; underside of young leaves densely hairy (but the
      hairs usually disappear as the leaves age)
                    Garrya flavescens [includes *G. flavescens* var. *pallida*]
                                      Ashy Silk-tassel; Al-s

Gentianaceae--Gentian Family

Gentians and their relatives have opposite, sessile leaves. Both the calyx and corolla have either 4 or 5 lobes (usually 5), and the number of stamens is the same. The stamens are attached to the corolla tube. In the bud stage, the corolla lobes overlap one another with a slight twist. The dry fruit splits lengthwise into 2 halves.

Various exotic species--mostly European and Asiatic species of the genus *Gentiana*--have long been in cultivation. Most highly prized are certain alpine species grown in rock gardens. Some of the native montane gentians from California and other western states are strikingly beautiful, but they demand very special conditions. The genus *Centaurium* is represented by several rather attractive pink-flowered species, all annuals that are easily grown from seed. One of them, *C. erythraea*, was introduced from Europe, and has become weedy in the Pacific Northwest. It should perhaps be avoided.

A Corolla primarily blue (but sometimes with greenish dots or streaks
   within)
  B Leaves less than twice as long as wide; folds within the corolla
   (these are located at the base of the clefts between lobes) fringed
   with a few slender teeth; all flowers sessile
                                 Gentiana affinis var. *ovata* [*G. oregana*]
                                       Prairie Gentian; Ma-n
  BB Most leaves about 3 times as long as wide; folds within the corolla
   not fringed with slender teeth; some flowers (especially those
   originating from lower axils) on distinct pedicels (mostly in
   sphagnum bogs; coastal)   *Gentiana sceptrum* (*Plate* 29)
                                          King's Gentian; Sn

AA Corolla pink (sometimes white) or yellow
  B Corolla deep yellow (corolla about 6 mm long, including the lobes,
   smaller than the broad, cup-shaped calyx, closing in the afternoon;
   less than 5 cm tall; in grassy places)   *Cicendia quadrangularis*
                                               Cicendia
  BB Corolla pink (sometimes white)
     C Flowers on such short pedicels (less than 1 mm long) that they
      appear to be sessile (corolla tube 6-12 mm long, the lobes
      usually 3-5 mm long) (Note: if the corolla lobes are more than 5
      mm long, go to choice CC, even if the pedicels are very short.)
              Centaurium muehlenbergii [includes *C. floribundum*] (*Plate* 29)
                                     Monterey Centaury; Mo-n
     CC Some or all flowers on distinct pedicels
        D Corolla lobes not more than twice as long as wide, and not
         more than half as long as the tube; anthers not often more
         than 2 mm long (mostly in areas close to the coast, and often
         in moist habitiats)   *Centaurium davyi*
                                           Davy Centaury; SLO-n
        DD Corolla lobes more than twice as long as wide, and more than
         half as long as the tube; anthers usually 3-6 mm long
           E Corolla lobes usually less than 10 mm long; stigma without
            distinct lobes (Note: this species is not definitely
            distinct from *C. venustum*.)   *Centaurium trichanthum*
                                        Alkali Centaury; SM-n
           EE Corolla lobes mostly more than 10 mm long; stigma with
            distinct lobes, each on a short stalk *Centaurium venustum*
                                              Canchalagua

Geraniaceae--Geranium Family

Geraniums and their relatives usually have regular flowers (cultivated pelargoniums, with slightly irregular flowers, are exceptions) with 5 separate sepals and petals, and with 10 or 15 stamens. Some of the stamens may not have anthers, however, and the filaments are more or less joined to one another at the base. The 5 styles of the pistil form a long beak above the 5-lobed portion within which seeds develop. When the fruit matures, the lobes separate, and each one, containing a single seed, gets one of the styles. In *Erodium*, the styles become coiled like corkscrews as they dry.

When wet, however, they straighten out, and the action of uncoiling literally drills the seeds into the soil.

This family has contributed numerous weeds to our flora. Some of the species of *Erodium* are so well established in otherwise nearly wild areas that they may seem to be native.

A With 10 anther-bearing stamens
  B Leaves pinnately compound, with 3 leaflets that are also deeply lobed (the blade may at first appear to be palmately compound, but the terminal leaflet is stalked) (usually growing in shaded habitats; stems and leaves with a rather strong, unpleasant odor)
                                             Geranium robertianum (*Plate* 29)
                                                 Herb-Robert; eu; SF
  BB Leaves palmately lobed
    C Sepals (3-4 mm long) not narrowing to slender tips; fruit not hairy, but often transversely wrinkled (pedicels 5-20 mm long)
                                                  Geranium molle
                              Dove's-foot Geranium; eu
    CC Sepals narrowing to slender tips; fruit hairy, but not transversely wrinkled
      D Sepals 3-5 mm long; perennial
        E Petals 6-8 mm long, notched; pedicel, as fruit matures, 10-20 mm long
                Geranium potentilloides [*G. microphyllum* and *G. pilosum*]
                                  Cinquefoil Geranium; au
        EE Petals 4-6 mm long, not notched; pedicel, as fruit matures, 4-16 mm long
                                  Geranium retrorsum
                            New Zealand Geranium; au
      DD Sepals 5-9 mm long; annual
        E Pedicel, as fruit matures, 15-30 mm long (ultimate divisions of upper leaves blunt; petals pale purple)
                                  Geranium bicknellii
              [includes *G. bicknellii* var. *longipes*] (*Plate* 29)
                            Bicknell Geranium; Mo, Ma-n
        EE Pedicel, as fruit matures, not more than 15 mm long
          F Pedicels 8-15 mm long; petals purple; ultimate divisions of upper leaves mostly pointed
                        Geranium dissectum (*Plate* 29)
                            Cutleaf Geranium; eu
          FF Pedicels 2-7 mm long; petals light pink; ultimate divisions of upper leaves mostly blunt
                          Geranium carolinianum
                            Carolina Geranium
AA With 5 anther-bearing stamens
  B Leaf blades at most only shallowly lobed, somewhat heart-shaped or kidney-shaped, owing to a wide notch at the base
    C Sepals and petals 4-5 mm long; leaves scattered along the stem, which may be more than 20 cm tall
                                Erodium malacoides
                        Soft Stork's-bill; me; Al, CC
    CC Sepals 8-10 mm long, petals 10-15 mm long; leaves mostly basal, the stem usually not more than 2 cm long (except for the flower peduncle)
                             Erodium macrophyllum
                           Largeleaf Filaree; Ma
  BB Most leaf blades pinnately compound or deeply lobed in a pinnate pattern, not heart-shaped or kidney-shaped
    C Most leaves pinnately compound
      D Leaflets so deeply divided that the leaves are nearly bipinnate; tip of sepals with a translucent bristle 1-1.5 mm long (prostrate to 50 cm tall, sometimes very compact (widespread)
                        Erodium cicutarium (*Plate* 29)
                             Redstem Filaree; eua
      DD Leaflets toothed and sometimes also shallowly lobed, but rarely so deeply divided that the leaves are nearly bipinnate; tip of sepals with a narrow prolongation about 1 mm long, but this not a bristle
                                Erodium moschatum
                          Whitestem Filaree; eu
    CC Leaves pinnately lobed, but not truly compound
      D Sepals 7-8 mm long in flower, 13-15 mm long in fruit, the tip often reddish, but sometimes green; styles on fully mature fruit 8-12 cm long (widespread)
                            Erodium botrys (*Plate* 29)
                            Broadleaf Filaree; me

DD Sepals 5-6 mm long in flower, 10 mm long in fruit, the tip
     green; styles on fully mature fruit 4-9 cm long
                                        Erodium brachycarpum
                  Southern European Stork's-bill; me; SCr, SCl, Ma

Grossulariaceae (formerly in Saxifragaceae)--Gooseberry Family

Our representatives of the Gooseberry Family are currants and gooseberries, all of which belong to the genus *Ribes*. These are substantial
woody shrubs with alternate and usually palmately lobed leaves. In their
flowers, borne in racemes, there are 5 petals and 5 stamens, and most of
the calyx is fused to the pistil. The fleshy fruit therefore develops below
the 5 calyx lobes. The 2 styles of the pistil correspond to the 2 seed-
producing divisions of the fruit.

Several native species, especially *Ribes sanguineum* var. *glutinosum*, *R.*
malvaceum, *R. speciosum*, and *R. aureum* var. *gracillimum* are excellent
subjects for gardens. They are easily propagated from cuttings, and can
also be grown from seed. The 4 species mentioned are generally available at
sales of native plants, as well as at some commercial nurseries.

A   Branches without spines
    B   Calyx lobes and petals yellow (sometimes tinged with red); lobes of
        leaves usually not toothed   *Ribes aureum* var. *gracillimum* (*Plate* 30)
                                            Golden Currant; Al-s
    BB Calyx lobes and petals usually pink or rose, rarely white; lobes of
       leaves usually toothed
        C   Leaves dull green, glandular and hairy on the upper surface;
            calyx lobes and petals pale pink to bright rose   *Ribes malvaceum*
                                            Chaparral Currant; CC, Ma-s
        CC Leaves bright green, slightly hairy on the upper surface, but not
           glandular; calyx lobes and petals pale pink, occasionally white
           (widespread)   *Ribes sanguineum* var. *glutinosum* (*Plate* 30)
                                            Pinkflower Currant
AA Branches with stout spines at the nodes (these spines are sometimes
   lacking in *Ribes divaricatum* var. *pubiflorum*)
    B   Corolla (and also the calyx) scarlet; stamens protruding well beyond
        the corolla; the 4 calyx lobes equal in length to the 4 petals
                                        Ribes speciosum (*Plate* 30)
                                    Fuchsia-flower Gooseberry; SCl-s
    BB Corolla not scarlet; stamens shorter than the corolla, or only the
       anthers protruding; the 5 calyx lobes longer than the 5 petals
        C   Fruit not bristly
            D   Leaves 2-5 cm wide; calyx lobes green or purplish; petals
                white   *Ribes divaricatum* var. *pubiflorum* [*R. divaricatum*]
                                            Straggly Gooseberry
            DD Leaves 1-2 cm wide; calyx lobes yellow; petals white
                                            Ribes quercetorum
                                            Oak Gooseberry; Al-s
        CC Fruit bristly
            D   Most bristles on fruit with glands at the tip
                E   Calyx lobes green to purplish, petals white or greenish
                    white; leaves up to 5 cm wide; petioles as long as the
                    width of the leaves   *Ribes victoris*
                                    Victor Gooseberry; Ma, Sn, Na, Sl; 4
                EE Calyx and petals reddish; leaves up to 4 cm wide; petioles
                    1-2.5 cm long   *Ribes menziesii* [includes *R. menziesii*
                                    vars. *leptosmum* and *senile*] (*Plate* 30)
                                        Canyon Gooseberry; SLO-n
            DD Most bristles on fruit without glands (petals white or
               whitish)
                E   Calyx lobes purplish red; fruit 14-16 mm wide
                                        Ribes roezlii var. *cruentum*
                                        Sierra Gooseberry; Na, Sn-n
                EE Calyx lobes greenish, whitish, or purplish; fruit 9-10 mm
                    wide   *Ribes californicum* (*Plate* 30)
                                        Hillside Gooseberry; Mo-Me

Gunneraceae (formerly in Haloragaceae)--Gunnera Family

The Gunnera Family consists of some large-leaved waterside plants native to South America. *Gunnera tinctoria* [*G. chilensis*], called Gunnera, is from Chile and has become sparingly naturalized. It is huge for a herbaceous plant. The leaves have somewhat heart-shaped, palmately lobed blades 1-2 m wide, and the fleshy petioles are often more than 1 m long. The small flowers, in dense panicles, have 2 tiny sepals, 2 hoodlike petals, 2 stamens, and a pistil that develops into a small red fruit.

Haloragaceae--Water-milfoil Family

Water-milfoils are perennials rooted in the mud of freshwater lakes and ponds. The leaves of our species are arranged in whorls around the flexible stems and are divided into slender, sometimes almost hairlike lobes. The inconspicuous flowers, with a cuplike, obscurely 4-lobed calyx and sometimes 4 tiny petals, are borne in the axils of the submerged leaves or in an inflorescence that is raised up out of the water. Staminate flowers, with 4 or 8 stamens, are usually above the pistillate flowers, which have 1-4 stigmas (often feathery) and which produce small fruits that split apart into 4 hard, 1-seeded nutlets.

The family includes some plants that are very unlike water-milfoils and that are not included in the key below. One of these, *Haloragis erecta*, Seaberry, a native of New Zealand, is a terrestrial shrub that has become established in a few places. It has 4-angled stems, slightly elongated opposite leaves about 3 or 4 cm long, and clusters of small flowers that originate in the axils of the leaves. Each flower has a 4-lobed calyx, 4 petals, and 8 stamens, and a pistil whose fruit-forming portion is 4 angled. The fruit, as it matures, becomes distinctly 4-winged.

A   All leaves pinnately lobed (flower clusters in the axils of the
       submerged leaves)   *Myriophyllum aquaticum* [*M. brasiliense*]
                                                    Parrot's-feather; sa
AA Most leaves pinnately lobed, but those beneath the flowers, which are in
     an inflorescence raised above the water, bractlike
   B   Submerged leaves with fewer than 26 lobes; lobes usually not opposite
          one another (flower clusters above the water) *Myriophyllum sibiricum*
                                                         Water-milfoil
   BB Submerged leaves usually with more than 28 lobes; lobes opposite one
        another   *Myriophyllum spicatum*
                                           Eurasian Water-milfoil; eua

Hippocastanaceae--Buckeye Family

The Buckeye Family consists of deciduous trees and shrubs that have opposite, palmately compound leaves. The flowers, produced in dense pani-cles, are slightly irregular because neither the 5 lobes of the calyx nor the petals (also usually 5) are of equal size. All of the flowers have several stamens, but only a few, in the upper part of each panicle, have a pistil. The tough-skinned fruit, sometimes spiny, consists of 3 divisions that eventually crack apart to release the large, shiny seeds that resemble true chestnuts. The seeds of buckeyes are, however, very poisonous.

Several species of the principal genus, *Aesculus*, are native to eastern North America, and one of them, *A. glabra*, the Buckeye, is the State Flower of Ohio. The Horse Chestnut, *A. hippocastanum*, is widely cultivated in the United States and Canada, as well as in Europe, to which it is indigenous. The only speces native to our region is *Aesculus californica* (*Plate* 30), the California Buckeye. It is a small tree, sometimes more than 8 m tall, common on hillsides and in ravines. The flowers, with white or pale pink petals, are in showy panicles. The fruits remain on the leafless trees long into the fall and usually produce 1 or 2 seeds, each 2-3 cm wide.

Hippuridaceae (formerly in Haloragaceae)--Mare's-tail Family

The Mare's-tail Family is one of the smallest, there being only a few species. These plants are perennials that grow at the edge of lakes, ponds, and sluggish streams, and they are often partly submerged. They propagate themselves vegetatively by creeping stems rooted in the mud. The narrow leaves, in whorls, stand out stiffly from the upright stems. Our only species, *Hippuris vulgaris* (*Plate* 83), called Mare's-tail, grows up to 30 cm tall, and its leaves are 1-3 cm long. The flowers, which are less than 2 mm high, are sessile in the axils of the leaves. Most of them have a stamen as well as a pistil; the pistil, except for its slender style, is fused to the calyx, which has only rudimentary lobes. Some flowers may have only a stamen or only a pistil, not both. Petals are never present. The fruit, about 2 mm long, is a 1-seeded nutlike structure.

Hydrophyllaceae--Waterleaf Family

Except for one shrub, Yerba-santa (*Eriodictyon californicum*), all of the numerous local representatives of the Waterleaf Family are herbaceous. The flowers have 5 calyx and corolla lobes, 5 stamens (attached to the tube of the corolla and alternating with its lobes), and a pistil that is usually partly or completely partitioned into 2 divisions. The inflorescence is often a tightly-coiled, one-sided cluster.

Nemophila menziesii var. *menziesii*, called Baby-blue-eyes, *Emmenanthe penduliflora* var. *penduliflora*, or Whispering-bells, and some species of *Phacelia* are attractive plants for gardens in which natives are encouraged. *Eriodictyon* is a possibility, but its habit of growth is not compact enough to suit everyone, and it suffers from a fungus that causes its foliage to look as though it had been dusted with soot.

This family is mostly restricted to the western part of North America, and it gives us no bothersome weeds. A few of the phacelias, including *Phacelia nemoralis*, which is fairly common in our area, have stinging hairs. Contact with them may lead to an unpleasant rash.

A   Leaves either smooth-margined or toothed, but not lobed; style of pistil divided to the base (corolla white; in dry gravelly soil)
    B   Shrub up to 200 cm tall (evergreen); leaves up to more than 10 cm long, usually toothed, hairy mostly on the underside (dark green, shiny, the older ones often blackened by a fungus; widespread)
                                        Eriodictyon californicum (*Plate* 31)
                                            Yerba-santa; SB-n
    BB Herb up to 10 cm tall (annual); leaves up to 1.5 cm long, not toothed, and hairy on both surfaces
                        Nama californicum [*Lemmonia californica*]
                                            Nama; La-s
AA Leaves lobed, the lobes sometimes separated nearly to the base and also toothed; style of pistil divided only at the tip, if at all
    B   Leaf blades about as wide as, or wider than, long (1-4 cm wide) with several shallow marginal lobes; nearly all prominent leaves basal, the upper leaves much reduced (corolla about 1 cm long, white, with some yellow; in moist, rocky habitats near the coast)
                              Romanzoffia californica [*R. suksdorfii*]
                                         Mistmaiden; SCr-n
    BB Leaf blades longer than wide, usually deeply lobed; leaves either mainly basal, or well developed on the stems as well
        C   Flowers hanging downward when open; corolla (about 1 cm long) pale yellow or pink (inflorescence a terminal cluster; leaves 3-10 cm long, with 10-20 short lobes)
            D   Corolla pale yellow
                        Emmenanthe penduliflora var. *penduliflora* (*Plate* 31)
                                          Whispering-bells
            DD Corolla pink, drying white   *Emmenanthe penduliflora* var. *rosea*
                                  Rose Whispering-bells; SCl-s
        CC Flowers not hanging downward when open; corolla mostly white, blue, or violet, sometimes tinged with yellow

D   Flowers not in coiled, one-sided inflorescences
  E   Flowers concentrated in dense terminal clusters; leaves
    alternate, often mainly basal, 5-16 cm long, usually with
    light blotches (corolla white to violet; usually in
    shade)   *Hydrophyllum occidentale* (*Plate* 31)
                                     Heliotrope; Mo-n
  EE Flowers either solitary in the leaf axils, or in loose
    racemes (these terminal or in the axils); some leaves
    opposite (sometimes the upper alternate), not mainly
    basal, without white blotches
    F   Upper stems and flower peduncles either glandular or
      with down-curved prickles
      G   Upper stems and flower peduncles glandular, without
        downcurved prickles; primary lobes of leaf blades
        again deeply lobed; flowers in loose racemes
        (corolla white, tinged with yellow; on fire burns,
        but also in shaded areas)
                      Eucrypta chrysanthemifolia
                                      Eucrypta
      GG Upper stems and flower peduncles not glandular, but
        with downcurved prickles; primary lobes of leaf
        blades not lobed again; flowers sometimes solitary
        in the leaf axils, more often in few-flowered
        terminal inflorescences (in shaded habitats)
        H   Corolla 1.5-3 cm wide, blue, lavender, or
          purple, with darker markings; prickles usually
          visible without a hand lens *Pholistoma auritum*
                            Fiesta-flower; La-S
        HH Corolla up to 1 cm wide, white, sometimes with
          a purple spot on each lobe; prickles visible
          only with a hand lens   *Pholistoma membranaceum*
                       White Fiesta-flower; CC-s
    FF Plants not glandular and without downcurved prickles
      G   Corolla usually more than 1 cm wide (sometimes more
        than 3 cm) (in moist areas)
        H   Corolla bright blue, with a light center
          (widespread)
          Nemophila menziesii var. *menziesii* (*Plate* 31)
                                Baby-blue-eyes
        HH Corolla white, or white with blue lines, and
          with very small black spots (coastal)
                     Nemophila menziesii var. *atomaria*
                     White Baby-blue-eyes; SCl-n
      GG Corolla (white to blue) not more than 1 cm wide
        H   Corolla 5-10 mm wide, well exceeding the calyx
          (upper leaves sometimes alternate; in partly
          shaded areas; common)   *Nemophila heterophylla*
                       Variableleaf Nemophila; SB-n
        HH Corolla 2-6 mm wide, only slightly longer than
          the calyx (in moist, often shaded areas)
          I   Corolla 2-4 mm wide, the lobes without dark
            veins or spots; upper leaves sometimes
            alternate, the lobes usually 5
                       Nemophila parviflora
                   Smallflower Nemophila; Mo-n
          II Corolla 3-6 mm wide, the lobes with dark
            veins, spots, or a blotch; leaves all
            opposite, the lobes usually 7
                     Nemophila pedunculata
                     Meadow Nemophila
DD Flowers in coiled, one-sided inflorescences (the coiling,
  after the fashion of the scroll of a violin, becomes less
  apparent after the flowers open)
  E   Larger leaves deeply divided into several lateral lobes on
    both sides, these lobes at least half as large as the
    terminal lobe (lobes sometimes also toothed or divided)
    F   Primary lobes of leaves smooth-margined (flowers pale
      blue or lavender, sometimes white; less than 20 cm
      tall, annual)   *Phacelia breweri*
                     Brewer Phacelia; CC-SB
    FF Primary lobes of leaves deeply divided or toothed

G Corolla (up to 2 cm wide) bright blue, with a paler
   center; calyx lobes not more than 3 times as long
   as wide, and widest below the middle (annual)
                          Phacelia ciliata (*Plate* 31)
                                      Field Phacelia
GG Corolla, if blue, without an obviously paler
   center; calyx lobes more than 3 times as long as
   wide, and not widest below the middle
   H Stamens not projecting beyond the corolla;
      flowers not crowded in the inflorescence
      (corolla about 2 cm wide, bluish purple; in
      sandy soil; annual)   *Phacelia douglasii*
                              Douglas Phacelia; SF-s
   HH Stamens projecting conspicuously beyond the
      corolla; flowers crowded in the inflorescence
      I Sepals decidedly widest above the middle;
         corolla dirty white to pale bluish;
         perennial, usually spreading outward from
         the base
         J Stems below the inflorescence with soft,
            spreading hairs   *Phacelia ramosissima*
                             var. *ramosissima* (*Plate* 31)
                            Branched Phacelia; SCL(SCM)-n
         JJ Stems below the inflorescence with
            coarse, erect hairs
                    Phacelia ramosissima var. *latifolia*
                    [*P. ramosissima* var. *suffrutescens*]
                               Inland Phacelia; SCl-s
      II Sepals not decidedly widest above the
         middle; corolla bluish purple or lavender
         (sometimes pale blue, cream, or nearly
         white in *Phacelia distans*); annual, usually
         upright or bushy
         J Inflorescence usually without small
            leaves where clusters of flowers branch
            off; calyx lobes usually at least 4-5
            times as long as wide; maturing fruit
            decidedly longer than wide, hairy only
            near the apex; corolla bluish purple or
            lavender   *Phacelia tanacetifolia*
                                   Tansy Phacelia; La-s
         JJ Inflorescence typically with small
            leaves where clusters of flowers branch
            off; calyx lobes usually about 3 times
            as long as wide; maturing fruit nearly
            round, hairy to below the middle;
            corolla bluish purple, cream, or nearly
            white   *Phacelia distans*
                                      Wild Heliotrope
EE Larger leaves either not lobed, or divided into only 1-2
   (rarely more) shallow lobes on each side of the basal
   portion, these lobes much smaller than the terminal lobe
   (leaves sometimes also toothed)
   F Leaves toothed (the large teeth sometimes with smaller
      teeth)
      G Stamens not projecting out of the corolla (annual)
         H Coarse teeth of leaves without smaller teeth;
            corolla 7-11 mm long, the lobes lavender to
            purple, the tube yellow; leaves and stems not
            especially hairy   *Phacelia suaveolens*
                               Sweetscent Phacelia; SCl-n
         HH Coarse teeth of leaves with smaller teeth;
            corolla 4-5 mm long, white or pale blue;
            leaves and stems usually very hairy
                                       Phacelia rattanii
                               Rattan Phacelia; SLO-n
      GG Stamens projecting out of the corolla (restricted
         to coastal areas)
         H Corolla 10-12 mm long, lavender; perennial
                          Phacelia bolanderi (*Plate* 83)
                              Bolander Phacelia; Sn-n
         HH Corolla 5-7 mm long, dull white; annual
                                      Phacelia malvifolia
                               Stinging Phacelia; SLO-n

FF Leaves not obviously toothed
    G Leaves without small lateral lobes; up to 20 cm
       tall (corolla white to pale lavender, about 4 mm
       wide; annual; in shade of chaparral, near summit
       of peaks)   *Phacelia phacelioides*
                 Mount Diablo Phacelia; CC(MD)-SCl(MH); 1b
    GG Large leaves often, but not always, with 1-2
       (sometimes more) lobes on each side, these usually
       much smaller than the terminal lobe; generally
       more than 20 cm tall, unless sprawling
       H Corolla 10-18 mm long, purplish blue (annual,
          branching at the base, not often more than 30
          cm tall; mostly in open areas)
                              Phacelia divaricata
                    Divaricate Phacelia; Me-Mo, SB
    HH Corolla not more than 7 mm long, mostly white
       I Leaves rarely with more than 1 lobe on each
          side; with stinging hairs (corolla greenish
          white; biennial or short-lived perennial,
          generally single-stemmed, often more than 1
          m tall; usually in moist, wooded areas)
                        Phacelia nemoralis (*Plate* 83)
                      Bristly Phacelia; SB-n
      II Leaves usually with at least 2 lobes on
         each side; without stinging hairs
         (perennial)
         J Calyx lobes about twice as long as wide,
           overlapping one another, especially as
           fruit matures; larger leaves often with
           more than 2 lobes on each side (corolla
           white)   *Phacelia imbricata*
                          Rock Phacelia
        JJ Calyx lobes usually at least 3 times as
          long as wide, not overlapping one
          another; larger leaves rarely with more
          than 2 lobes on each side
         K Stems (including those of
           inflorescences) glandular-hairy
           (corolla white; on serpentine)
                      Phacelia corymbosa
                Serpentine Phacelia; Sn-n
        KK Stems hairy, but not glandular
          (corolla pale lavender or nearly
          white; stems sometimes upright,
          sometimes sprawling, branching near
          the base)
                  Phacelia californica (*Plate* 31)
                 California Phacelia; SCl-n

Hypericaceae--St. John's-wort Family

All of our species of the St. John's-wort Family have opposite leaves. Their flowers have 5 sepals and 5 yellow petals, and usually there are numerous stamens (sometimes arranged in clusters). The fruit dries as it matures, cracking open to release its small seeds.

Several exotic species of *Hypericum* are widely cultivated. The best known--and perhaps too commonly used in landscaping--is *H. calycinum*, a low, creeping shrub often called Rose-of-Sharon. It is a native of southeastern Europe.

A Plants low, with creeping stems that form mats, the upright flowering
    stems rarely more than 7 cm tall; petals 2-3 mm long; leaves usually not
    more than 1 cm long (in wet places, including sphagnum bogs)
                      Hypericum anagalloides (*Plate* 32)
                      Tinker's-penny; SLO-n
AA Plants upright, not forming mats, the flowering stems more than 12 cm
    tall; petals at least 10 mm long; leaves usually more than 1 cm long
    B Flowering stems not often more than 20 cm tall; leaves folded,
      usually at least 5 times as long as wide, tapering to pointed tips
      (on dry, brushy hillsides)   *Hypericum concinnum* (*Plate* 32)
                      Goldwire; Me-Ma

BB Flowering stems commonly more than 25 cm tall; leaves not folded,
   usually less than 4 times as long as wide, not tapering to pointed
   tips
   C Numerous short branches on each plant not bearing flowers; sepals
      slender, 4-5 mm long; in disturbed areas
                          Hypericum perforatum (*Plate* 83)
                          Klamathweed; eu; SCl-n
   CC Most branches on each plant bearing flowers; sepals oval, 3 mm
      long; in wet meadows and ditches
                          Hypericum formosum var. *scouleri*
                          Scouler St. John's-wort; Mo-n

Juglandaceae--Walnut Family

The two principal genera of the Walnut Family are *Juglans* (walnuts and
butternuts) and *Carya* (hickories and pecans). These are deciduous trees
with alternate, pinnately compound leaves. The staminate flowers are in
drooping catkins and each one is associated with 3 bracts. There are a few
to many stamens, and the calyx, when present, is small and its lobes are
not equally developed. The pistillate flowers are borne singly or in
clusters at the end of branches. They, too, are associated with some bracts
and may have a calyx. The fruit that develops from the pistil has a double
wall. The outer part is a husk that is fleshy at first, but later dries
out; the inner part is a hard nutshell familiar to everyone who has cracked
open a walnut or pecan to get at the large seed.

Only one member of the family is native to our region and it is
considered to be endangered (1b). This is *Juglans californica* var. *hindsii*
[*J. hindsii*], the Northern California Black Walnut, limited to the
foothills of the inner Coast Ranges. It is found around campsites formerly
used by Native Americans. The tree reaches a height of about 25 m. Its
leaves usually have 15-19 leaflets, and its fruit is 3-3.5 cm wide. Farther
south in California, it is replaced by *J. californica* var. *californica*,
which does not grow so tall. The leaves of this variety usually have 11-15
leaflets, and the fruit is commonly 2-3 cm wide.

Lamiaceae (Labiatae)--Mint Family

Mints are usually aromatic, but their aromas are not necessarily minty;
some come close to being unpleasant. The family has, nevertheless, given us
many herbs used in cooking and perfumery. Rosemary, French Lavender, Thyme,
various mints, Marjoram, and Sage are just a few of them. It has also
provided remedies reported to be useful in curing one ailment or another;
Horehound, Selfheal, and Motherwort are among the medicinal species. And
even the sleepiest cats usually become animated when stimulated by a packet
of dry Catnip.

The family has a distinctive complex of characters, including a four-
angled stem and opposite leaves, the pairs being arranged successively at
right angles to each other. The flowers, in whorls in the axils of the
upper leaves, may be so crowded that the inflorescence seems continuous.
Each whorl often has bracts and a pair of leaves directly beneath it. The
corolla is usually 2-lipped, with 2 lobes generally forming the upper lip
and 3 forming the lower lip. The 5-lobed calyx may also show some tendency
toward being 2-lipped. There are 4 stamens, in 2 pairs, but those of the
upper pair may be reduced to filaments that lack anthers. The fruiting part
of the pistil is divided into 4 lobes, each of which matures into a nutlet
that encloses a single seed.

The family is a large one, so we should expect some deviations from the
typical formula. For instance, some species have flowers that are
essentially regular, rather than 2-lipped. Others have such a short upper
lip that there appears, at first, to be only one lip. In still others, the
two lateral lobes of the lower lip may seem to be part of the upper lip.

But a mint is a mint, and one will not need much experience to place an
unfamiliar plant correctly in this family.

Salvia columbariae, called Chia, is an attractive annual, easily grown
from seed. In crowded colonies, the flowers form a sea of purplish blue.

For gardens of native plants, the following perennials are especially
recommended. Most of them can be bought at sales of the California Native
Plant Society, from botanical gardens, or from commercial nurseries.

Monardella villosa ssp.
  villosa, Common Coyotemint
Salvia spathacea, Hummingbird
  Sage
Salvia sonomensis, Sonoma Sage

Salvia mellifera, Black Sage
Satureja douglasii, Yerba-buena
  (excellent ground cover for
  somewhat shaded locations)

A   Bracts (these broad, with very prominent veins) with spines usually at
    least 5 mm long, sometimes 10 mm long (on serpentine)
    B   Upper lip of corolla 2-lobed, about equal to the lower lip; style
        hairy   *Acanthomintha lanceolata*
                                         Santa Clara Thornmint; Al, SCl-Mo; 4
    BB Upper lip of corolla not lobed, decidedly smaller than the lower lip;
        style not hairy (stem usually not branching, with a terminal
        inflorescence; anthers pinkish red)
                            Acanthomintha duttonii [*A. obovata* ssp. *duttonii*]
                                               San Mateo Thornmint; SM; 1b
AA Bracts either not spiny, or the spines less than 3 mm long
    B   Corolla irregular, decidedly 2-lipped
        C   Upper lip of corolla divided to its base (stamens markedly curved
            and extending well beyond the corolla)
            D   Stamens 3-7 mm long   *Trichostema oblongum*
                                                  Mountain Bluecurls; Na-n
            DD Stamens 7-20 mm long
                E   Tube of the corolla bent upward at nearly a 90° angle;
                    leaves with sharp tips; petioles of lower leaves not more
                    than 4 mm long; in open fields (plants with a strong odor
                    of vinegar)   *Trichostema lanceolatum* (*Plate* 84)
                                                              Vinegarweed
                EE Tube of the corolla not bent upward; leaves without sharp
                    tips; petioles of lower leaves usually at least 10 mm
                    long; mostly in damp, gravelly habitats *Trichostema laxum*
                                                  Turpentine-weed; Na, Sn-n
        CC Upper lip of corolla not divided to its base
            D   Calyx not 2-lipped, with 10 teeth, each terminating in a
                hooked spine (stems, and sometimes also the leaves, woolly;
                corolla white; somewhat shrubby)   *Marrubium vulgare*
                                                              Horehound; eu
            DD Calyx regular or 2-lipped, with not more than 6 teeth
                (sometimes without any), these either without spines, or the
                spines not hooked
                E   Calyx regular or 2-lipped, with 3-6 teeth and without a
                    projection on the back of the upper lip
                                              Lamiaceae, Subkey (p. 172)
                EE Calyx 2-lipped, neither lip divided into teeth, although
                    the upper lip has a projection on its back
                    F   Corolla white, yellow-tinged (lower leaves reddish)
                                                  Scutellaria californica
                                              California Skullcap; Al-n
                    FF Corolla blue, violet, or purple
                        G   Corolla 2.5-3 cm long; most leaves at least 5 times
                            as long as wide; upper leaves sessile
                                          Scutellaria siphocampyloides
                                              Grayleaf Skullcap; Al-SB
                        GG Corolla 1-2 cm long; most leaves much less than 5
                            times as long as wide; all leaves with petioles
                            H   Hairs on stems short, tightly curled; stems
                                upright   *Scutellaria antirrhinoides*
                                              Snapdragon Skullcap; Sn-n
                            HH Hairs on stems long, straggly; stems falling
                                down to some extent   *Scutellaria tuberosa*
                                    [includes *S. tuberosa* var. *similis*] (*Plate* 33)
                                              Blue Skullcap; Ma-n

BB Corolla nearly regular, with either 4 or 5 lobes that are almost
   equal in size and shape (the lobes are difficult to see in some
   species, in which the corolla is only 2 mm long)
   C   Flowers in a terminal head, this subtended by some wide bracts;
       corolla 12-20 mm long, with 5 lobes; in dry places (aromatic)
       D   Margin of leaves wavy (in sandy soils)   *Monardella undulata*
                                       Curlyleaf Monardella; Ma-s; 4
       DD Margin of leaves not wavy
          E   Outer bracts below the flower heads not bent down, not
              leaflike
              F   Leaf margins rolled under   *Monardella viridis*
                                       Green Monardella; Na; 4
              FF Leaf margins not rolled under
                 G   Bracts silvery, with darker margins and veins
                                       Monardella douglasii
                                       Douglas Monardella; CC-Mo
                 GG Bracts papery throughout (in sandy areas)
                                       Monardella breweri
                                       Brewer Monardella; Al-s
          EE Outer bracts below the flower heads bent down, leaflike
             F   Plants woolly, especially on the underside of leaves
                 (on serpentine)   *Monardella villosa* ssp. *franciscana*
                                       Serpentine Coyotemint; Ma-s
             FF Plants at most sparsely hairy
                G   Plants sparsely hairy; leaves not thickened,
                    sometimes toothed, not glandular on the underside;
                    stems not purple; bracts not fringed (widespread)
                                       Monardella villosa ssp. *villosa*
                           [includes *M. villosa* ssp. *subserrata*] (*Plate* 32)
                                       Common Coyotemint
                GG Plants not hairy; leaves thickened and toothed,
                    glandular on the underside; stems sometimes
                    purple; bracts sometimes fringed
                           Monardella purpurea [includes *M. subglabra* and
                                       M. villosa ssp. *neglecta*]
                                       Siskiyou Monardella
   CC Flowers either in clusters in the leaf axils, or in elongated
      terminal inflorescences, these without wide bracts; corolla 2.5-
      7 mm long, usually with 4 (but sometimes 5) lobes; moist places
      D   Flowers at the end of leafless stems
          E   Leaves with definite petioles   *Mentha* x *piperita*
                           [includes *M. citrata* and *M. piperita*] (*Plate* 84)
                                       Peppermint; eu
          EE Most leaves sessile
             F   Leaves mostly more than twice as long as wide, not
                 woolly; flowers pale lavender   *Mentha spicata*
                                       Spearmint; eu
             FF Leaves not more than twice as long as wide, woolly,
                 especially on the underside; flowers white
                                       Mentha suaveolens
                                       Sweet Mentha; me
      DD Flowers in clusters in the leaf axils
         E   Corolla 2 mm long, whitish; leaves 3-10 cm long, deeply
             toothed, sessile or the petioles very short; not aromatic
             F   At least some of the leaves with petioles; leaf
                 margins deeply and irregularly toothed
                                       Lycopus americanus (*Plate* 84)
                                       Cutleaf Water-horehound
             FF All leaves sessile; leaf margins evenly toothed
                                       Lycopus asper
                                       Bugleweed; Sl
         EE Corolla 5-8 mm long, usually not white; leaves 2-5 cm
             long, either not toothed or only shallowly toothed, with
             definite petioles; aromatic
             F   Leaves 1-2 cm long, those beneath the flower clusters
                 smaller than the lower stem leaves; leaves not
                 toothed; pedicels only slightly if at all colored;
                 corolla lavender   *Mentha pulegium* (*Plate* 32)
                                       Pennyroyal; eu; SCl, Sn, Ma
             FF Leaves 2-5 cm long, about the same size throughout the
                 plant; lower leaves toothed; pedicels purplish;
                 corolla light purple or whitish   *Mentha arvensis*
                           [includes *M. arvensis* var. *villosa*] (*Plate* 84)
                                       Field Mint

Lamiaceae, Subkey: Corolla irregular, 2-lipped; calyx regular or 2-lipped, with 3-6 teeth
A   Calyx 2-lipped, each lip usually with 2-3 teeth
   B   Flowers with 4 functional stamens; upper lip of calyx fanlike, its 3
      teeth much smaller than the 2 teeth of the lower lip
      C   Flowers in the leaf axils (corolla whitish; leaves 2-6 cm long,
         very fragrant; bushy, up to 80 cm tall)   *Melissa officinalis*
                                                    Beebalm; me
      CC Flowers crowded at the end of stems (in moist areas)
         D   Corolla bluish or violet, less than 10 mm long; calyx up to 5
            mm long; leaves up to 3 cm long (common weed)
                                Prunella vulgaris var. *vulgaris*
                    [includes *P. vulgaris* var. *parviflora*] (*Plate* 33)
                                  European Selfheal; eu
        DD Corolla dark violet, 10-20 mm long; calyx 5-10 mm long;
            leaves up to 5 cm long (in Coast Ranges)
                              Prunella vulgaris var. *lanceolata*
                  [includes *P. vulgaris* var. *atropurpurea*]
                                  Narrowleaf Selfheal
   BB Flowers with 2 functional stamens (the other 2 lack anthers); upper
      lip of calyx not fanlike
      C   Shrubs, 1-2 m tall, extensively woody except for recent growth
         (flowers in dense whorls, the corolla blue to white or lavender;
         widespread)   *Salvia mellifera* (*Plate* 33)
                                        Black Sage; CC-s
      CC Herbs, usually much less than 1 m tall, woody only at the base
         D   Corolla 3-4 cm long, purplish red; leaves 10-20 cm long
            (plants very glandular)   *Salvia spathacea* (*Plate* 33)
                                  Hummingbird Sage; Sl-s
        DD Corolla 1-2.5 cm long, blue, purple, or lavender; leaves less
            than 10 cm long
            E   Leaves spiny (and also white-woolly, at least when young);
               corolla lavender, about 2 cm long, the lower lip ragged
               (in sandy soil)   *Salvia carduacea*
                                    Thistle Sage; CC-s
           EE Leaves not spiny; corolla blue or blue-violet, less than 1
             cm long, the lower lip not ragged
             F   None of the leaves lobed; plants forming mats
               (perennial; leaves woolly on the underside)
                                  Salvia sonomensis
                                  Sonoma Sage; Na-n
            FF Basal leaves pinnately lobed, the lobes sometimes
               divided again; not forming mats
               G   Basal leaves up to 10 cm long, many lobes divided
                   again; annual   *Salvia columbariae* (*Plate* 33)
                                                Chia
               GG Basal leaves up to 6 cm long, the lobes not divided
                  again; perennial   *Salvia verbenacea*
                                  Verbena Sage; eu; Al
AA Calyx regular, although sometimes the lobes are not equal
   B   Shrubs, 1-2 m tall, extensively woody except for recent growth
      (flowers solitary in the leaf axils, the corolla white, pale
      lavender, or pink; calyx becoming enlarged as fruit develops;
      widespread)   *Lepechinia calycina* (*Plate* 32)
                                      Pitcher Sage
   BB Herbs, usually much less than 1 m tall, not woody except perhaps at
      the base
      C   Calyx lobes separate almost to the base, with 2 lobes longer than
         the others
         D   Flowers readily visible among the leaves and bracts; corolla
            1-1.5 cm long
            Pogogyne douglasii [includes *P. douglasii* ssp. *parviflora*]
                                  Douglas Pogogyne; Sn
        DD Flowers almost completely hidden by the leaves and bracts;
            corolla less than 1 cm long
            E   Flowers in the axils of most leaves; corolla without
               spots; stems prostrate (in moist areas)
                          Pogogyne serpylloides (*Plate* 33)
                              Thymelike Pogogyne; SLO-n
           EE Flowers mostly in the axils of leaves near the tip of the
             stem; corolla with dark spots on the lower lip; stems
             mostly erect (in wet places that dry out in spring)
                               Pogogyne zizyphoroides
                                  Sacramento Pogogyne; SCl-n

CC Calyx lobes less than half as long as the calyx, and mostly equal
   to one another (they are slightly unequal in *Glecoma hederacea*)
   D Flowers in the axils of most leaves along the entire length
      of the stem, either solitary or in groups of 2 or 3 (in
      shaded areas)
      E Flowers solitary; corolla 6-8 mm long, white to purple;
         calyx lobes equal in length (leaves with a pleasing
         fragrance)   *Satureja douglasii* (*Plate* 84)
                                                          Yerba-buena
      EE Flowers in groups of 2 or 3; corolla 10-22 mm, bluish or
         purple; calyx lobes slightly unequal in length (in moist
         areas)   *Glecoma hederacea*
                                                      Ground-ivy; eu
   DD Flowers in the terminal portion of the stems, in whorls that
      may be distinct and well spaced or continuous and crowded
      E Whorls of flowers continuous, crowded on stem length for
         about 6 cm
         F Corolla rose or violet; stamens protruding 6-8 mm
            beyond the corolla; up to 2 m tall (in moist habitats)
                                          Agastache urticifolia (*Plate* 84)
                                              Nettleleaf Horsemint; SLO-n
         FF Corolla white with purplish marks; stamens protruding
            beyond the corolla only slightly, if at all; less than
            1 m tall
            G Corolla white with purple veins, the tube 6-7 mm
               long, the upper lip 3-4 mm long (in moist
               habitats)   *Stachys pycnantha*
                                          Shortspike Hedgenettle; Ma, CC-s
            GG Corolla white with purple dots, the tube 8-10 mm
               long, the upper lip 2 mm long   *Nepeta cataria*
                                                            Catnip; eua
      EE Whorls of flowers distinct, well spaced
         F Stems with 1-4 whorls, each with more than 20 flowers
            (corolla 6-7 mm long, white; calyx hairy; in moist
            areas)   *Pycnanthemum californicum*
                                                        Mountain Mint
         FF Stems with about 7 whorls, each with up to 10 flowers
            G Tube of corolla 18-20 mm long, the upper lip 7-9 mm
               long; stamens protruding 4-5 mm from the corolla
               (in moist areas)   *Stachys chamissonis*
                                              Chamisso Hedgenettle; SLO-n
            GG Tube of corolla about 6-16 mm long, the upper lip
               2-6 mm long; stamens protruding 1-3 mm from the
               corolla
               H Leaves narrowed at the base, with long matted
                  hairs (corolla tube 10-15 mm long; in moist
                  areas)   *Stachys ajugoides* var. *ajugoides*
                                              Bugle Hedgenettle; Sn-s
               HH Leaves not obviously narrowed at the base,
                  without matted hairs (although there may be
                  separate long hairs)
                  I Corolla white, the tube 6 mm long (in moist
                     areas)   *Stachys stricta*
                                              Sonoma Hedgenettle; Sn-n
                  II Corolla rose or purple, the tube more than
                     6 mm long
                     J Stems only slightly hairy; largest
                        leaves 1-3 cm long; petioles up to 3 cm
                        long or absent
                        K Upper leaves sessile
                                  Lamium amplexicaule (*Plate* 32)
                                              Clasping Henbit; eua
                        KK Upper leaves with petioles up to 3
                           cm long *Lamium purpureum* (*Plate* 32)
                                                    Red Henbit; eu
                     JJ Stems obviously hairy; largest leaves 4-
                        18 cm long; petioles up to 6 cm long
                        (in moist areas)
                        K Stems glandular as well as hairy;
                           flowers 6 in each whorl
                                      Stachys bullata (*Plate* 34)
                                     California Hedgenettle; SF-s

KK Stems not glandular; flowers not
more than 3 in each whorl
(widespread)   *Stachys ajugoides*
var. *rigida* [includes *S. rigida* sspp.
quercetorum, *rigida*, and *rivularis*]
Rigid Hedgenettle

Lauraceae--Laurel Family

The Laurel Family is a primarily tropical group, but it includes a few species of temperate climates. The European *Laurus nobilis* is famous as a symbol of glory. From its branches were made the laurel wreaths that crowned heroes and poets of centuries past. The "bay leaves" used for seasoning soups and stews also come from this tree. Our only representative of the family is *Umbellularia californica* (*Plate* 34), California Bay Laurel. This tree, often more than 30 m tall, grows from southern Oregon to San Diego County, mostly within about 75 miles of the coast, but is also found in the Sierra Nevada. Its leaves, generally about 3-8 cm long and 1.5-3 cm wide, are like those of *Laurus nobilis* in being alternate, rather tough, and aromatic. They can be used for seasoning, but have a flavor different from those of *L. nobilis*. The flowers, produced in loose clusters, are small, with 6 yellowish green sepals about 6 mm long, a pistil, and 9 stamens. There are no petals. As a rule, not more than 3 flowers in each cluster reach the fruiting stage. The fruit, about 2.5 cm long, has the shape of a plump olive. It is fleshy and becomes purplish as it ripens. There is a single large seed.

The California Bay Laurel is not difficult to grow from seed, but seedlings take a long time to reach respectable size. Nevertheless, this fine tree deserves to be planted in parks and gardens where it can be given the room it needs to form its broad crown. Once it has reached a large size, however, the accumulation of its fallen leaves on the ground will discourage the growth of many other plants.

Lentibulariaceae--Bladderwort Family

Bladderworts are aquatic plants, totally submerged except for their inflorescences. The so-called leaves (they are specialized branches of the stems) are divided into very slender lobes, and the flowers, mostly yellow, have a 2-lipped corolla, with a saclike spur on the lower lip. Small bladders located on the "leaves" (or on other side branches in some species) function as traps for microscopic organisms that wander into them, fail to escape, and eventually are digested.

A   "Leaves" up to about 4 cm long, the 2 primary lobes dividing
dichotomously, but unequally, several times, so that there are numerous
ultimate lobes; inflorescence with 6-20 flowers; lower lip of corolla
12-15 mm long, the spur slightly longer   *Utricularia vulgaris* (*Plate* 85)
Common Bladderwort
AA "Leaves" not more than 1 cm long, usually with only 2 lobes;
inflorescence with 1-3 flowers; lower lip of corolla 6-10 mm long, the
spur shorter   *Utricularia gibba*
Swollenspur Bladderwort; ?

Limnanthaceae--Meadowfoam Family

The Family Limnanthaceae consists of only 2 genera, both restricted to North America, and mostly found in the Pacific states. They are annuals with pinnately compound leaves, and they grow in moist habitats. Only one species, *Limnanthes douglasii*, occurs in our region. The flowers typically have 5 sepals and 5 petals; a distinctive feature of the petals is a U-shaped band of hairs near the base. There are 10 stamens and a pistil that has 5 stigmas and that is partitioned lengthwise into 5 divisions, each of which becomes a nutlet with a single seed.

For a century and a half, our species has brought pleasure to gardeners in Great Britain and Europe, and it is widely grown in America, too. It must have sun and a reasonable amount of soil moisture throughout the growing season, for it is a plant that lives mostly around vernal pools and in similar situations.

A   Petals at least partly yellow
   B   Petals yellow with white tips; leaflets 5-11, smooth-margined to lobed; in Coast Ranges
                                 Limnanthes douglasii ssp. *douglasii* (*Plate* 34)
                                 Douglas Meadowfoam; SB-n
   BB Petals entirely yellow; leaflets 7-13, toothed or lobed; coastal
                                 Limnanthes douglasii ssp. *sulphurea*
                                 Point Reyes Meadowfoam; Ma(PR); 1b
AA Petals white, sometimes aging pink (in Coast Ranges)
   B   Leaves sometimes bipinnate, the leaflets very narrow, with only a few teeth, if any; petals (sometimes aging pink) with pink or cream-colored veins   *Limnanthes douglasii* ssp. *rosea*
                                 Pale Meadowfoam
   BB Leaves pinnate, the leaflets not especially narrow, with teeth or lobes; petals with purple veins   *Limnanthes douglasii* ssp. *nivea*
                                 Snow Meadowfoam; SLO-n

Linaceae--Flax Family

The flowers of our flaxes have 5 separate petals (these often fall away rather early), 5 nearly separate sepals, and 5 stamens. The stamens alternate with the petals and are usually joined together at the base, forming a low collar. The pistil is partitioned lengthwise into 5 divisions, each of which produces 2 flattened seeds. After the fruit has ripened and dried, its wall generally splits apart into 10 valves.

The Common Flax, *Linum usitatissimum*, whose species name means "of maximum usefulness," has been associated with humans for a long time. Linen, made from its long fibers, is believed to be the oldest known textile. In the past, flax fibers were also used for making nets and ropes. The seeds are pressed to obtain linseed oil, which is incorporated into paints, inks, and varnishes. The material left after the oil has been squeezed out is fed to cattle.

Several exotic flaxes, including *L. usitatissimum*, are grown for their attractive flowers. Some of our native species, most of which are annuals, should be interesting if thickly seeded in sunny, open places.

A   Petals blue (rarely white), 8-15 mm long
   B   Petals 8-10 mm long; pedicels 5-18 mm long   *Linum bienne* (*Plate* 34)
                                 Narrowleaf Flax; me; SM-n
   BB Petals 10-15 mm long; pedicels 10-30 mm long
      C   Leaves 1-3.5 cm long, with 3 main veins; inner surface of sepals hairy near the margin   *Linum usitatissimum*
                                   Common Flax; eu
      CC Leaves 1-2 cm long, with 1 main vein; inner surface of sepals not hairy near the margin (widespread)
                 Linum lewisii [*L. perenne* ssp. *lewisii*] (*Plate* 34)
                                 Western Blue Flax
AA Petals yellow, white, pink, or rose, generally not more than 8 mm long
   B   Petals white to pink, sometimes streaked with rose
      C   Petals 2-3 mm long (white to pink, sometimes streaked with rose)
                            Hesperolinon micranthum
                             Smallflower Western Flax
      CC Petals 6-8 mm long
         D   Pedicels 1-8 mm long; petals pink to rose
                            Hesperolinon congestum
                           Marin Western Flax; Ma-SM; 1b
        DD Pedicels 6-20 mm long; petals white to pale pink
                          Hesperolinon spergulinum
                         Slender Western Flax; Sn, Na

BB Petals yellow
    C Petals about 6 or 7 mm long (styles 3)   *Hesperolinon breweri*
                                            Brewer Western Flax; CC(MD); 1b
    CC Petals 3-4 mm long
        D Most leaves opposite, the upper one toothed; plants not
           extensively branched   *Sclerolinon digynum*
                                                   Yellow Flax
        DD Most leaves not opposite, not toothed; plants extensively
           branched
           E Styles 2   *Hesperolinon bicarpellatum*
                          Two-carpellate Western Flax; Na; 1b
           EE Styles 3   *Hesperolinon clevelandii*
                          Cleveland Western Flax; Na, SCl

Loasaceae--Blazing-star Family

All of our species of the Blazing-star Family belong to a single genus,
Mentzelia, and all are natives. The stems and leaves are often bristly-
hairy. Except in one, the flowers are large and showy, with 5 yellow petals
that have a silken sheen. The base of the petals, and also the numerous
stamens, are attached to the calyx, which is 5-lobed. The fruiting portion
of the pistil, usually elongated, is beneath the other flower parts. The
calyx lobes generally persist on the fruit as this ripens, dries out, and
opens at the top to release its seeds.

The one that is especially well suited to cultivation as an annual in a
sunny, well drained location is *Mentzelia lindleyi*.

A Petals mostly 5-8 cm long; biennial, up to 100 cm tall (petals pale
   yellow, at least 3 times as long as wide, tapering to a point)
                                      Mentzelia laevicaulis
                                       Giant Blazing-star
AA Petals either 2-4 cm long or less than 0.5 cm long; annual, usually not
   more than 60 cm tall
    B Petals 20-40 mm long; leaf blades pinnately lobed (petals golden
       yellow, with an orange-red base, rounded, with a tiny projection at
       the tip)   *Mentzelia lindleyi* (*Plate* 34)
                                Lindley Blazing-star; Al-Mo
    BB Petals 3-4 mm long; leaf blades toothed but not deeply lobed
        C Petals pale yellow, without an orange spot at the base
                          Mentzelia micrantha (*Plate* 34)
                               Golden Blazing-star
        CC Petals yellow, with an orange spot at the base *Mentzelia dispersa*
                                  Nevada Stickleaf

Lythraceae--Loosestrife Family

The flowers of loosestrifes are produced singly or in clusters in the
axils of the upper leaves, and the inflorescence as a whole is elongated.
In our species, the calyx is tubular, ribbed lengthwise, and has 5-7 lobes;
alternating with the calyx lobes are some other flattened structures, often
conspicuous. There are 5-7 petals, and an equal number of stamens. The
elongated fruit, tightly enclosed by the calyx but free from it, is
partitioned into 2 divisions.

A Calyx not much longer than wide; leaves strictly opposite and somewhat
   clasping the stem; flowers with 4 petals (flowers, these nearly sessile,
   in groups of 2-5 in the leaf axils; petals 1-2 mm long, purple; in wet
   places)   *Ammannia coccinea*
                                 Longleaf Ammannia; Ma
AA Calyx at least twice as long as wide; leaves alternate or opposite
   (sometimes both patterns occur on the same plant), not clasping the
   stem; flowers usually with 6 petals
    B Petals at least 4 mm long; up to more than 100 cm tall (petals rose-
       purple)
       C Petals mostly 4-6 mm long; largest leaves not often more than 1
          cm wide, the length usually more than 7 times the width
                          Lythrum californicum (*Plate* 35)
                               California Loosestrife; Ma, Sl

   CC Petals mostly 7-10 mm long; largest leaves up to about 2 cm wide,
      the length usually less than 5 times the width (common water-
      garden plant)   *Lythrum salicaria*
                                             Purple Loosestrife; eu
BB Petals not often more than 2 mm long; rarely more than 40 cm tall
   C Flowers sessile; petals up to 2 mm long, rose-purple, sometimes
      white; up to 50 cm tall   *Lythrum hyssopifolium* [*L. hyssopifolia*]
                                                    Grass-poly; eu
   CC Flowers on distinct pedicels; petals about 1 mm long, pale
      purple; usually prostrate, rarely more than 20 cm tall (in
      drying pools)   *Lythrum tribracteatum*
                                       Threebract Loosestrife; me; Sl

Malvaceae--Mallow Family

   The flowers of mallows have 5 separate petals that are usually rolled up
together in the bud stage, and a 5-lobed calyx. There are numerous stamens,
their filaments joined to form a tube around the pistil. The anthers may be
concentrated at the top of the tube or scattered along part of its length.
The pistil is partitioned into several seed-producing divisions, and these
may break apart after the fruit has ripened and dried.
   The best-known garden plants of this family are the Hollyhock, believed
to come from China, and varous species of *Hibiscus*. Europe has sent us
several weedy species belonging to the genus *Malva*, and a few other exotic
mallows have become established in situations where their presence is not
especially objectionable. One of them is the Tree-mallow, *Lavatera arborea*,
native to Europe. It is now rather common in coastal areas. *Lavatera
assurgentiflora*, native to the Channel Islands of southern California,
grows well in relatively frost-free situations, sometimes escaping. Other
possible garden candidates among our native mallows are species of
Malacothamnus and *Sidalcea*. They are available at sales of botanical
gardens and the California Native Plant Society.

A Anthers scattered over much of the tube formed by the united stalks of
   the stamens; leaf blades usually more than 5 cm long (sometimes more
   than 10 cm long) (often more than 1 m tall)
   B Stigmas more or less globular, at the tip of the styles; leaf blades
      heart-shaped, decidedly longer than wide, toothed but not lobed, the
      tip sharply pointed; petals 6-10 cm long (white or pale pink), with
      a dark crimson blotch; along the Sacramento and San Joaquin Rivers
      (annual; in wet places)   *Hibiscus lasiocarpus* [*H. californicus*]
                                         California Hibiscus; CC; 2
   BB Stigmas slender, oriented lengthwise on the styles; leaf blades about
      as long as wide, deeply lobed, the tip blunt or rounded; petals less
      than 5 cm long, and without a dark crimson blotch; coastal
      C Petals 2.5-4.5 cm long, rose, with darker veins (shrub, native to
         Channel Islands, naturalized elsewhere) *Lavatera assurgentiflora*
                                              Island Mallow; 1b
      CC Petals mostly 1.5-2 cm long, either uniformly pink or lilac, or
         pale purple-red with darker veins confined to the base
         D Petals pink or lilac; leaves sparsely hairy, not cottony;
            annual herb   *Lavatera cretica*
                                          Crete Mallow; me; SF, Ma
         DD Petals pale purple-red, with darker veins at the base; leaves
            cottony; perennial shrub (the most common *Lavatera* in our
            region)   *Lavatera arborea* (*Plate* 35)
                                          Tree Mallow; eu; Me-SM
AA Anthers in a single or double ring at the tip of the tube formed by the
   united stalks of the stamens; leaf blades usually less than 5 cm long
   B Stigmas more or less globular, at the tip of the styles
      C Shrubs, woody at least at the base, usually at least 100 cm tall
         D Calyx white-woolly, the hairs almost hiding the calyx lobes
            of flowers that have not opened (petals pink to rose, 1-2 cm
            long)   *Malacothamnus fremontii* (*Plate* 35)
                                          Fremont Mallow; CC
         DD Calyx hairy, but not white-woolly, the hairs not hiding the
            calyx lobes (the lobes distinct before the flowers open)
            Malacothamnus fasciculatus [includes *M. arcuatus* and *hallii*]
                                       Chaparral Mallow; CC-SM, SCl

CC Herbs, not often more than 50 cm tall
    D   Petals up to 8 mm long, dull red (leaves usually deeply
        lobed)   *Modiola caroliniana*
                                                      Wheel Mallow; sa
    DD Petals at least 10 mm long, not dull red
        E   Petals cream or yellowish, 1 cm long; leaves toothed, but
            not deeply lobed; leaves and stems usually grayish or
            whitish, very hairy, most of the larger hairs in clusters
            of at least 10; plants low, the stems reclining
            Malvella leprosa [*Sida leprosa* var. *hederacea*] (*Plate* 85)
                                                        Alkali Mallow
        EE Petals pinkish lavender to purple, 1-3 cm long; leaves
            deeply lobed; leaves and stems not especially grayish or
            whitish, but nevertheless hairy, the larger hairs in
            clusters of fewer than 5; plants mostly upright
                                                     Eremalche parryi
                                                     Parry Mallow; Al-s
BB Stigmas slender, oriented lengthwise on the styles (herbs, usually
    not more than 1 m tall)
    C   Each flower with 3 bracts just beneath the calyx
        D   Anthers in a double ring at the top of the tube formed by the
            united stalks of the stamens *Sidalcea hickmanii* ssp. *viridis*
                                                 Marin Checkerbloom; Ma; 1b
        DD Anthers in a single ring at the top of the tube formed by the
            united stalks of the stamens (common weeds)
            E   Bracts below calyx nearly as long as wide (petals pink to
                blue-violet, up to 12 mm long; up to about 60 cm tall)
                                                    Malva nicaeensis (*Plate* 35)
                                                          Bull Mallow; eua
            EE Bracts below calyx at least twice as long as wide
                F   Petals about twice as long as the calyx (stems
                    reclining; leaf blades shallowly lobed)
                                                     Malva neglecta (*Plate* 35)
                                                        Common Mallow; eua
                FF Petals barely longer than the calyx (petals whitish to
                    pale pink, 4-5 mm long)   *Malva parviflora*
                                                        Cheeseweed; eua
    CC Flowers without 3 bracts beneath the calyx (there may, however,
        be single bracts in the inflorescence)
        D   Leaves (especially the larger ones) only shallowly divided
            into 3-7 lobes (inflorescence a much branched panicle; petals
            white, 7-15 mm long; plants hairy, perennial, up to 1.5 m
            tall; coastal)   *Sidalcea malachroides*
                                                 Mapleleaf Checkerbloom; Mo-n; 1b
        DD Leaves (at least the upper ones) deeply lobed or compound
            E   Upper leaves palmately compound, the leaflets sometimes
                divided again; basal leaves up to 2 cm wide (plants
                sometimes sparsely hairy; usually in moist habitats)
                F   Petals (1-2 cm long) pink, sometimes with a purple
                    center; basal leaves (these often not persisting)
                    deeply lobed; leaflets of upper leaves sometimes lobed
                    (annual, usually slender)   *Sidalcea hartwegii*
                                                 Hartweg Checkerbloom; Me-Na
                FF Petals white to light purple; basal leaves with broad
                    marginal teeth or very shallow lobes, or both;
                    leaflets of upper leaves not divided again (stems
                    sometimes rooting at the nodes)
                    G   Petals 1-2 cm long, white to light purple; annual
                        (in moist areas)
                                        Sidalcea calycosa ssp. *calycosa* (*Plate* 35)
                                                 Annual Checkerbloom; Me, Ma, Na
                    GG Petals 2-2.5 cm long, light purple; perennial
                                                 Sidalcea calycosa ssp. *rhizomata*
                                                 Perennial Checkerbloom; Me, Ma(PR); 1b
            EE Most leaves deeply lobed, but not compound; lobes usually
                divided again, or at least coarsely toothed; basal leaves
                up to 2.5 cm wide
                F   Lobes of uppermost leaves, in whose axils the flowers
                    originate, divided again; stipules of upper leaves
                    about 10 mm long, often lobed; basal leaves toothed;
                    petals 1.5-3 cm long, rose to purple, sometimes with a
                    purple center; annual (leaves hairy)
                                                     Sidalcea diploscypha
                                                     Fringed Checkerbloom; SLO-n

FF Lobes of uppermost leaves, in whose axils the flowers
   originate, usually not divided again; stipules of
   upper leaves not more than 5 mm long, not lobed; basal
   leaves lobed and toothed; petals 1-2.5 cm long, light
   to dark pink (rarely white), commonly with white
   veins; perennial
   G   Base of plant, stipules, and calyx purplish (leaves
       hairy; coastal) *Sidalcea malvaeflora* ssp. *purpurea*
                           Purple Checkerbloom; Me, Sn
   GG Plants not purplish
       H   Stem leaves divided into narrow lobes; leaves
           sparsely hairy, especially on the upper
           surface   *Sidalcea malvaeflora* ssp. *laciniata*
                           Linearlobe Checkerbloom; Sn-SLO
       HH Stem leaves not divided into narrow lobes;
          leaves hairy on both surfaces (widespread)
                                       Sidalcea malvaeflora
                               ssp. *malvaeflora* (*Plate* 35)
                               Common Checkerbloom; Me-s

Menyanthaceae (formerly in Gentianaceae)--Buckbean Family

Menyanthes trifoliata (*Plate* 85), called Buckbean, is the only species
of Menyanthaceae occurring within the area covered by this book. It grows
in bogs and at the margin of lakes and ponds. The old stems, prostrate and
rooted in mud, give rise to shoots that bear the leaves and upright flower
racemes. Each leaf has 3 narrowly oval leaflets, up to about 10 cm long, at
the top of a rather fleshy petiole. The calyx is usually 5-lobed, sometimes
6-lobed. The tube of the corolla extends well above the calyx, and the 5 or
6 corolla lobes, which are white to faintly purplish, have crowded,
scalelike outgrowths on the upper surface. The stamens are attached to the
corolla tube, and alternate with the corolla lobes.

Molluginaceae (formerly in Aizoaceae)--Carpetweed Family

Carpetweeds and their relatives form a relatively small family, mostly
represented in warmer regions. The flowers lack petals, and the calyx
(unlike that of the Aizoaceae, a closely related family) is not fused to
the pistil. The species found in our area have 5 calyx lobes and either 5
or 10 stamens. The fruit is partitioned lengthwise into 3 divisions.

A   Leaves and stems not hairy; leaves about 4 or 5 times as long as wide,
    narrowing gradually to short, often indistinct petioles, decidedly
    whorled (usually 4 or 6 in each whorl); flowers a few at each node, with
    long, slender pedicels; widespread in disturbed areas
                                   Mollugo verticillata (*Plate* 85)
                                                      Carpetweed; sa
AA Leaves and stems at least somewhat hairy (often very hairy); leaves not
   much longer than wide, narrowing rather abruptly to petioles that are
   nearly as long as the blades, not obviously whorled (there are usually 2
   substantial leaves plus 1 or 2 small leaves at each node); flowers
   crowded into globular heads at the nodes; mostly in dried soil bordering
   ponds   *Glinus lotoides*
                                                 Glinus; sa; La-SCl

Myricaceae--Wax-myrtle Family

Wax-myrtles and their relatives are shrubs or small trees with
alternate, aromatic leaves. A few species are deciduous; the rest, like
Myrica californica and *M. pensylvanica*, the one from which bayberry wax is
derived, are evergreen. Erect catkins consisting of pistillate flowers are
borne near the tip of the branches; each pistil, which develops into a
small, 1-seeded fruit, has 2 or more small bracts below it. Lower down on
the branches, there are drooping catkins consisting of staminate flowers.
Each of these is next to a single bract.

Myrica californica (*Plate* 85), the Wax-myrtle, is found near the coast from southern California to Washington. It is commonly up to 4 m tall, but occasional specimens are much larger and nearly treelike. The leaves, mostly 7-10 cm long and 1.5-2 cm wide, are glossy on the upper surface. Staminate flowers, in catkins only 1-2 cm long, have 7-16 stamens; the pistillate catkins are about 8-12 mm long, and the fruits are nearly globular, 6-8 mm in diameter, and fairly hard. They are purplish and covered with a deposit of whitish wax. This shrub will be an extremely attractive component of a garden devoted to indigenous plants, and is sold by some commercial establishments as well as by nonprofit organizations that promote the use of native species. It requires a moderate amount of moisture.

Myrtaceae--Myrtle Family

The Myrtle Family is a large one, with perhaps 75 genera and 3000 species of trees and shrubs. Only a few are native to temperate regions, and none is native to our region. The most conspicuous introduced member of the family is *Eucalyptus globulus*, the Blue Gum, from Tasmania. This large tree is the main source of oil of eucalyptus, and has been widely planted for timber in various parts of the world. In the our area it now occupies such large tracts of land, including portions of some of the Regional Parks, that it is often taken for a California native. Unfortunately, where it is successful, the soil, covered with its aromatic leaves, can support little other vegetation. Furthermore, old groves are fire hazards, owing in part to the flammability of the accumulations of dry leaves. Other species of *Eucalyptus*, not discussed in this book, have also become naturalized, but not so extensively as Blue Gum.

Luma apiculata [*Eugenia apiculata*], called Temu, a South American shrub or small tree up to 2 m tall, has become established in a few places. It is very different from a eucalyptus, for instead of having a dry, woody fruit, it has a fleshy black fruit about 1 cm wide. Its leaves, up to about 2.5 cm long, are opposite and dark green, lacking the bluish tint often characteristic of leaves of eucalyptus.

Nyctaginaceae--Four-O'Clock Family

The several genera of Nyctaginaceae found in California are mostly restricted to the deserts, or at least to the southern part of the state. In this family, the leaves are opposite, and the flowers do not have a corolla; the bell-shaped or tubular calyx resembles a corolla, however. There are bracts below each flower or each flower cluster; these are sometimes united to form a cup, and they may also be colored. Stamens and a pistil are present in each flower. The fruit, frequently with winglike ridges, contains a single seed.

The only representatives native to our region are the sand-verbenas, members of the fleshy-leaved genus *Abronia*. One of the garden four-o'clocks, *Mirabilis jalapa*, is sometimes found as an escape in vacant lots or along roadsides; a native species, *M. laevis*, with rose-purple calyx, occurs only as far north as Monterey County.

A   Flowers solitary, each one situated in a cup consisting of united
     bracts; stigma rounded; leaves not fleshy; escaped garden plant
                                                    Mirabilis jalapa
                                                    Four-o'clock; sa
AA Each flower cluster with several bracts below it; stigma elongated;
     leaves fleshy; on sandy beaches
   B   Calyx yellow; leaf blades about as long as wide
                                          Abronia latifolia (*Plate* 36)
                                                    Yellow Sand-verbena
     BB Calyx pink or rose, sometimes very pale; leaf blades mostly longer
          than wide

C   Corolla tube 9-13 mm long
                    Abronia umbellata ssp. *umbellata* (*Plate* 36)
                                        Coast Sand-verbena; Sn-s
CC Corolla tube 6.5-10 mm long   *Abronia umbellata* ssp. *breviflora*
                                        Pink Sand-verbena; Ma-n; 1b

Nymphaeaceae--Waterlily Family

The Nymphaeaceae consists of aquatic perennials whose leaves have nearly circular blades. These float or are raised slightly above the water. The stems, rooted in mud, are thick and spread laterally, producing new clusters of leaves and flowers at intervals. The number of sepals and petals varies, even within a species. Our one native, *Nuphar luteum* ssp. *polysepalum*, Yellow Pondlily, usually has 7 to 9 sepals, at least 12 petals, and numerous stamens. The fruit is large and consists of several seed-producing divisions located below a concave stigma that has numerous barlike styles arranged in a circle.

A   Flowers about 5 cm wide, usually raised slightly above the water
    surface; sepals deep yellow (occasionally tinged with red), some of them
    4-5 cm long, the petals much smaller and mostly hidden by the stamens
                    Nuphar luteum ssp. *polysepalum* [*N. polysepalum*] (*Plate* 86)
                                        Yellow Pondlily; SLO-n
AA Flowers up to 10 cm wide, floating on the water surface; sepals and
    petals about 4 cm long, of various colors, showy (sometimes naturalized
    in lakes and ponds)
    B   Flowers yellow   *Nymphaea mexicana*
                                        Yellow Waterlily; na
    BB Flowers white or pink   *Nymphaea odorata*
                                        White Waterlily; na

Oleaceae--Olive Family

The Olive, *Olea europaea*, is just one of the economically or horti-culturally important plants in its family, which also includes *Ligustrum* (privets), *Syringa* (lilac), *Jasminum* (jasmines), *Forsythia*, and *Fraxinus* (ashes). Collectively these form a diversified assemblage. Some, for instance, have fleshy fruit; others have a fruit that splits apart when dry; and the ashes have fruit with a membranous, winglike expansion that aids in dispersal by wind.

Two of the 3 local representatives are ashes. Their flowers, in crowded panicles, have a small calyx with 4 toothlike lobes. There are either 2 stamens or a pistil, or both. One of our species has 2 petals, the other has none.

In *Forestiera pubescens*, the Desert-olive, the flowers with 4 stamens are sessile; those with a pistil are on pedicels. There are no petals. This interesting shrub is finding some favor with native-plant enthusiasts, for it can be grown in areas that receive little rainfall.

A   Leaves not compound (mostly 4-5 cm long, 1-1.5 cm wide); flowers in
    crowded clusters, these not drooping (but some of the fruits may droop
    by the time they have ripened); fruit fleshy, blue-black, 5-7 mm long
    (shrub or small tree, rarely more than 3 m tall
                    Forestiera pubescens [*F. neomexicana*] (*Plate* 86)
                                                Desert-olive
AA Leaves pinnately compound; flowers in drooping panicles; fruit dry, 2-5
    cm long, about half of it consisting of a membranous wing
    B   Trees up to more than 20 m tall; leaves mostly 20-25 cm long, the
        leaflets smooth-margined and sessile; staminate and pistillate
        flowers in separate racemes and without petals (along streams and in
        moist places)   *Fraxinus latifolia* (*Plate* 86)
                                                Oregon Ash
    BB Large shrubs or small trees, rarely more than 6 m tall; leaves rarely
        more than 12 cm long, the leaflets toothed and stalked; most flowers
        with stamens and a pistil, and with 2 white petals (on hillsides and
        in canyons)   *Fraxinus dipetala*
                                                California Ash

Onagraceae--Evening-primrose Family

The Onagraceae consists mostly of herbaceous plants, although some garden fuchsias, which are native of Central and South America, become woody. In the flowers, the fruit-forming portion of the pistil is united with the calyx. The other flower parts--typically 4 sepals, 4 petals, and 8 stamens--are attached to a portion of the calyx that usually extends at least slightly above the fruit. (In the key, this is called the flower tube.) There are exceptions, however, to all of these criteria, other than the position of the developing fruit.

Of our native species, one perennial is a particularly outstanding candidate for rockeries, the edges of paths, and some other situations where it can be allowed to form large masses. This is *Epilobium canum*, the California Fuchsia, whose flowers are usually scarlet, occasionally pink or white. The plant blooms at the end of summer and well into the autumn, but it has drawbacks: unless cut back after flowering, it may be unsightly, and it spreads rampantly by underground stems. Another perennial that may be useful in places where its height (up to more than 1 m) will not be a disadvantage is *Oenothera elata* ssp. *hookeri*, the Hooker Evening-primrose. It has large yellow flowers that open at dusk and remain open through the night and next morning, and also on cloudy days. Among the annuals, some species of *Clarkia* are of exceptional beauty and are easily grown in sunny situations. Seeds are available from a few commercial sources, as well as from botanical gardens and societies that promote cultivation of native plants.

A   Plants aquatic, rooting in muddy bottoms of lakes, ponds, or marshes, and partly submerged, unless left behind by receding water; calyx not extended as a flower tube above the fruit-forming part of the pistil, but the sepals persisting on the fruit after the rest of the flower has fallen away
   B   Petals absent; stamens 4; fruit not more than 0.5 cm long
                       Ludwigia palustris [*L. palustris* var. *pacifica*]
                                         Common Water-primrose
   BB Petals 5 or 6, 1-1.5 cm long, yellow; stamens usually 10; fruit about
       2 cm long
       C   Bracts just below each fruit (usually 2, each not more than 1.5
           mm long) broadly triangular; flowering stems usually floating or
           lying on mud; leaves generally widest near the middle
                                         Ludwigia peploides
                                         Yellow Water-primrose
       CC Bracts just below each fruit narrowly triangular (the length
           about twice the width at the base); flowering stems usually
           upright, up to more than 1 m tall; leaves generally (but not
           always) widest above the middle
                       Ludwigia hexapetala [*Jussiaea uruguayensis*] (*Plate* 86)
                                         Sixpetal Water-primrose; sa; Ma
AA Plants not truly aquatic, though some (especially certain species of
    Epilobium) grow in very wet habitats; calyx often extended, as a short
    or long flower tube, above the fruit-forming part of the pistil, but
    this tube falling with the sepals, petals, and stamens
    B   Flowers with only 2 sepals and 2 petals (these white, about 1 mm long
        and 2-lobed); fruit with many hooked hairs (in moist woodlands)
                                         Circaea alpina ssp. *pacifica*
                                         Enchanter's-nightshade; Ma-n
    BB Flowers with 4 sepals and 4 petals; fruit without hooked hairs
        C   Fruit not splitting lengthwise into 4 divisions when mature
            (fruit about 1 cm long, 4-angled [especially above the middle,
            where it is widest] with 1 or a few seeds; petals pink or white,
            5-8 mm long with slender basal portions)   *Gaura drummondii*
                                         Gaura; mx; Al
        CC Fruit splitting lengthwise into 4 divisions when mature
            D   Petals not yellow
                E   Flower tube 2-3 cm long, scarlet, like the sepals and
                    petals (in some forms, rarely encountered in the wild,
                    the flowers are pink or white; widespread)
                       Epilobium canum [*Zauschneria californica*] (*Plate* 38)
                                         California Fuchsia; Sn-s

EE Flower tube (if present) much less than 1 cm long, and
   neither it nor the sepals and petals scarlet
   F Fruit lumpy, rarely with more than 10 seeds, these in
      2 rows (1 row in each of 2 chambers) (plants with very
      slender branches; petals about 3 mm long, changing
      from white to pink; fruit up to 1.5 cm long; seeds
      without a tuft of hairs)   *Gayophytum heterozygum*
                                                     Gayophytum
   FF Fruit not lumpy, usually with many seeds, these either
      in 4 rows (1 row in each of 4 chambers) or 8 rows (2
      rows in each of 4 chambers)
      G Plants at least 1 m tall (upright and rarely
         branched); inflorescence a terminal raceme usually
         with at least 15 flowers; leaves sometimes more
         than 15 cm long; petals pink or lilac-pink (rarely
         white), 1-1.5 cm long, the upper 2 petals slightly
         smaller than the lower 2; seeds with a tuft of
         silky hairs at one end; in disturbed areas,
         including those where there have been fires
                    Epilobium angustifolium ssp. *circumvagum*
                            [*E. angustifolium*] (*Plate* 38)
                                                      Fireweed
      GG Plants rarely so much as 1 m tall; not conforming
         in all respects to the other criteria in choice G
                                    Onagraceae, Subkey (p. 184)
DD Petals yellow
   E Petals generally more than 2 cm long; stigma with 4
      slender lobes; flowers not opening before late afternoon
      F Most leaves along the stem not more than 3 times as
         long as wide (flowers opening in the evening)
            Oenothera glazioviana [*O.* x *erythrosepala*] (*Plate* 38)
                           Biennial Evening-primrose; eu?; Al, SF-n
      FF Most leaves along the stem at least 4 times as long as
         wide
         G Petals generally 2.5-4 cm long
                Oenothera elata ssp. *hookeri* [includes *O. hookeri*
                    sspp. *hookeri* and *montereyensis*] (*Plate* 86)
                                  Hooker Evening-primrose; SLO-n
         GG Petals 2-2.5 cm long
                    Oenothera wolfii [*O. hookeri* ssp. *wolfii*]
                            Wolf Evening-primrose; Ma-n; 1b
   EE Petals not more than 2 cm long (usually considerably
      smaller); stigma disk-shaped or nearly globular, not
      deeply divided into slender lobes; flowers opening in the
      morning
      F Stems and branches very short; leaves all in a basal
         cluster (petals about 1-1.5 cm long; widespread)
                    Camissonia ovata [*Oenothera ovata*] (*Plate* 36)
                                              Suncup; SLO-n
      FF Stems at least 20 cm long, upright or sprawling;
         leaves (these sometimes reduced) along the stems as
         well as in a basal cluster
         G Leaves up to 4 mm wide; fruit cylindrical, not at
            all 4-angled
            H Petals less than 3 mm long; fruit mostly 1.5-
               2.5 cm long, without a beak; coastal
                                          Camissonia strigulosa
                                          Hairy Suncup; SLO-n
            HH Petals 5-8 mm long; fruit 2-4 cm long, with a
               well defined beak; inland
                  Camissonia campestris [*Oenothera campestris*]
                                          Mojave Suncup; CC-s
         GG Leaves mostly 5-20 mm wide; fruit 4-angled
            H Petals mostly 10-15 mm long
                                    Camissonia cheiranthifolia
                        [*Oenothera cheiranthifolia*] (*Plate* 36)
                                          Beach Primrose
            HH Petals mostly 2-3 mm long
               I Stems mostly somewhat prostrate; upper
                  leaves sessile, but not clasping the stem;
                  coastal
                      Camissonia micrantha [*Oenothera micrantha*]
                                          Small Suncup; Ma-s

II Stems mostly upright; upper leaves clasping
   the stems; primarily inland
                Camissonia hirtella [*Oenothera hirtella*]
                         Self-pollinating Suncup

Onagraceae, Subkey: Flowers with 4 sepals and 4 petals (these not yellow);
flower tube (if present) much less than 1 cm long; fruit splitting
lengthwise into 4 divisions when mature; usually less than 1 m tall
A   Petals (these deep pink or reddish pink) deeply divided into 3 lobes
    B   Petals about as long as wide, the middle lobe narrower than the 2
        outer lobes   *Clarkia breweri*
                                  Brewer Clarkia; Al-s; 4
    BB Petals about twice as long as wide, the middle lobe about equal to
        the outer lobes   *Clarkia concinna* (Plate 37)
                                                Redribbons; SCl-n
AA Petals not divided into 3 lobes
    B   Petals white (turning pink in age), 2-3 cm long; leaves deeply
        pinnately lobed (flowers opening in late afternoon; restricted to
        sand dunes near Antioch, Contra Costa County)
                Oenothera deltoides ssp. *howellii* (Plate 38)
                     Antioch Dunes Evening-primrose; CC; 1b
    BB Petals, if white, less than 1.5 cm long; leaves not pinnately lobed
        C   Sepals, in flowers that have opened, not turned back and not
            normally adhering to one another; petals less than 5 mm long
            (except *Epilobium ciliatum* ssp. *watsonii* and *E. densiflorum*);
            petals notched at the tip to form 2 distinct lobes
            D   Petals usually notched for at least one-third their length;
                leaves alternate; seeds without a tuft of hairs at one end
                E   Fruit nearly 4-angled, owing partly to prominent
                    lengthwise ridges between the slight ridges that mark the
                    4 partitions (petals 2-5 mm long, white to pale pink;
                    seeds in 2 rows in each division of the fruit; mostly in
                    drying mud of vernal pools)
                    Epilobium cleistogamum [*Boisduvalia cleistogama*]
                              Self-fertilizing Willowherb; Sl
                EE Fruit nearly cylindrical, there being no lengthwise ridges
                    other than those that mark the 4 partitions
                    F   Petals usually more than 4 mm long (but sometimes only
                        3 mm); seeds in 1 row in each division of the fruit
                        Epilobium densiflorum [*Boisduvalia densiflora*]
                                        Denseflower Willowherb
                    FF Petals mostly 1-3 mm long; seeds in 2 rows in each
                        division of the fruit
                        G   Petals 2-3 mm long (rarely 3.5 mm); leaves 1-1.5 cm
                            long   *Epilobium pygmaeum* [*Boisduvalia glabella*]
                                              Smooth Willowherb
                        GG Petals 1-2 mm long (rarely 3 mm); leaves 1-4 cm
                            long   *Epilobium torreyi* [*Boisduvalia stricta*]
                                           Narrowleaf Willowherb; SCl-n
            DD Petals notched for less than one-third their length; leaves
                either entirely alternate, or partly alternate and partly
                opposite; seeds with a tuft of silky hairs at one end (this
                feature, more reliable than the depth of the notch in the
                petals, can be seen on seeds from fruit that is ripening,
                even if still green)
                E   Stems peeling, especially in the lower portion of the
                    plant (but in *Epilobium minutum* the peeling is sometimes
                    scarcely noticeable); flowers white to pink or rose;
                    annual
                    F   Lower leaves mostly opposite, upper leaves alternate
                        (flower tube about 1 mm long; petals up to 5 mm long,
                        pink; rarely more than 25 cm tall; often in wet
                        places, but sometimes in relatively dry habitats)
                                                Epilobium minutum
                                                 Minute Willowherb
                    FF Leaves alternate (or some of the upper ones in
                        clusters) (in dry habitats)
                        G   Flower tube less than 1 mm long (sometimes longer);
                            corolla white; rarely more than 40 cm tall
                                                Epilobium foliosum
                                              Manyflower Willowherb
                        GG Flower tube usually 2-5 mm long; corolla white to
                            rose; sometimes more than 100 cm tall
                            Epilobium brachycarpum [*E. paniculatum*] (Plate 38)
                                              Panicled Willowherb

EE Stems not peeling; perennial (usually in wet habitats)
   F Leaf veins (other than the midrib) inconspicuous;
     inflorescence often slightly nodding; seeds with small
     bumps, but without ridges; not often so much as 60 cm
     tall (petals white to pink, mostly 2-5 mm long)
                               Epilobium halleanum
                          Hall Willowherb; Ma, SCr
  FF Leaf veins conspicuous; inflorescence upright; seeds
     with lengthwise ridges; often more than 100 cm tall
     G Petals rarely more than 4 mm long, pale pink or
       nearly white   *Epilobium ciliatum* ssp. *ciliatum*
          [includes *E. adenocaulon* vars. *adenocaulon*,
            holosericeum, *occidentale*, and *parishii*]
                          Common Willowherb
     GG Petals usually 5-10 mm long, usually deep pink or
       purple, but sometimes pale pink
               Epilobium ciliatum ssp. *watsonii* [includes
            E. watsonii vars. *franciscanum* and *watsonii*]
                    Watson Willowherb; SLO-n
CC Sepals, in flowers that have opened, turned back, and sometimes
  adhering by the tips to form pairs or a single group of 4;
  petals not deeply notched into 2 lobes (except in *Clarkia
  biloba*), more than 5 mm long (usually at least 10 mm long)
  D Petals (these 1-2.5 cm long, pink or purplish pink) deeply
    divided into 2 lobes   *Clarkia biloba* (*Plate* 36)
                        Bilobe Clarkia; CC
  DD Petals not divided into lobes (but the margins may be
    slightly ragged)
    E Petals widening abruptly from a narrow, stalklike region
      (this may be one-third to one-half the entire length of
      the petal)
      F Narrow, stalklike portion of each petal with a pair of
        teeth or small, rounded lateral lobes
                    Clarkia rhomboidea (*Plate* 37)
                        Tongue Clarkia
      FF Narrow, stalklike portion of each petal without
        lateral teeth or lobes
        G Developing fruit, flower tube, and sepals with some
          long hairs; leaves green; Coast Ranges and coastal
          locations   *Clarkia unguiculata* (*Plate* 38)
                       Elegant Clarkia; Me-s
        GG Developing fruit, flower tube, and sepals without
          long hairs; leaves gray-green; inner Coast Ranges
                    Clarkia tembloriensis
                    Temblor Clarkia; Al-s
    EE Petals without a narrow, stalklike region, widening
      gradually, rather than abruptly, from the base (in
      Clarkia epilobioides, the basal portion is narrow, but it
      is very short, much less than one-fourth the entire
      length of the petal)
      F Petals with a narrow basal portion, white to pale
        cream, without darker markings, but becoming pink as
        they age (petals up to about 1 cm long; in shaded
        areas)   *Clarkia epilobioides* (*Plate* 37)
                    Willowherb Clarkia; SF-s
      FF Petals without a narrow basal portion, pink, lavender-
        pink, purple, or a related color, often with darker
        markings
        G Immature fruit either with 4 lengthwise grooves, or
          without noticeable grooves (rarely with 8 grooves
          in *Clarkia gracilis* ssp. *sonomensis*)
          H Tip of flowering stems, with unopened buds,
            drooping
            I Petals generally 1-2 cm long, uniformly
              pink (but sometimes a little darker at the
              base; widespread)
                  Clarkia gracilis ssp. *gracilis*
                  Summer's-darling; SCl-n
           II Petals 2-4 cm long, mostly pink, but
             usually with a bright red central spot
             Clarkia gracilis ssp. *sonomensis* (*Plate* 37)
                  Slender Clarkia; Ma, Na-n
          HH Tip of flowering stems not drooping

       I   Petals pink or lavender-pink, except for a
          red blotch at the base
       J   Petals mostly 10-30 mm long; flower tube
          4-10 mm long (widespread)
                      Clarkia rubicunda [includes
          C. rubicunda ssp. *blasdalei*] (*Plate* 38)
                                Godetia; SFBR
      JJ Petals 5-13 mm long; flower tube 1-3 mm
          long (on serpentine)
                    Clarkia franciscana (*Plate* 37)
                       Presidio Clarkia; SF; 1b
     II Petals pink, lavender-pink, or white,
        except for red streaks or a red blotch near
        the center, but without a red blotch at the
        base (the base may, however, be darker than
        the rest of the petal)
       J   Fruit mostly longer than the internode
          above the node to which it is attached
                    Clarkia amoena ssp. *amoena*
                       Farewell-to-spring; Sn, Ma
      JJ Fruit mostly shorter than the internode
          above the node to which it is attached
          (in general, the flowers in this
          subspecies are less crowded than in
          ssp. *amoena*, but the 2 subspecies
          intergrade)
               Clarkia amoena ssp. *huntiana* (*Plate* 36)
                       Hunt Clarkia; Ma, Na-n
GG Immature fruit with 8 lengthwise grooves
   H   Plants of seacoast habitats, usually prostrate
      or with only the tip of the stems upright; tip
      of leaf blades blunt (petals 5-11 mm long,
      pale lavender or pink, without a spot and with
      a whitish base; in sandy soil)
                    Clarkia davyi (*Plate* 37)
                       Davy Clarkia; SM-n
HH Plants not necessarily limited to seacoast
   habitats, usually upright; tip of leaf blades
   not blunt
     I   All 4 sepals usually remaining attached to
        one another and thus directed to one side
        (in chaparral or foothill woodland;
        widespread)   *Clarkia affinis*
                       Small Clarkia; Na-s
     II Sepals usually turned back individually or
        remaining attached in pairs
       J   Petals (2-2.5 cm long) very pale
          lavender except for a wedge-shaped
          purple spot that begins near the middle
          and widens toward the tip (in
          chaparral)   *Clarkia imbricata*
                    Vine Hill Clarkia; Sn; 1b
       JJ Petals pale pink or pale lavender to
          deep purple or deep wine red (if pale,
          there may be a red or purple spot near
          the middle, at the base, or at the tip,
          or there may be darker flecks or
          streaks, but none of these markings
          wedge-shaped; an extremely variable and
          widespread species!)
          K   Leaves mostly less than 5 times as
             long as wide; flowers crowded; fruit
             usually longer than the internode
             above the node to which it is
             attached (petals 1-2.5 cm long;
             coastal and inland)
                 Clarkia purpurea ssp. *purpurea*
                     Purple Clarkia; Ma, Sl-s
          KK Leaves mostly more than 6 times as
             long as wide; flowers usually not
             crowded; fruit shorter than the
             internode above the node to which it
             is attached

L   Petals 1.5-2.5 cm long; stamens
    curving away from the style, and
    not reaching the stigma (in
    chaparral or foothill woodland)
                            Clarkia purpurea
                    ssp. *viminea* (*Plate* 37)
                            Large Clarkia
LL Petals less than 1.5 cm long
    (mostly 1 cm long); stamens
    usually reaching the stigma and
    clinging to it (the most common
    subspecies)   *Clarkia purpurea*
                ssp. *quadrivulnera* (*Plate* 37)
                            Winecup Clarkia

Orobanchaceae--Broom-rape Family

Broom-rapes are parasites, attached to the roots of other flowering
plants. They are not green and have no typical leaves, although there are
scalelike structures on the lower portion of the stems. The flowers have a
5-lobed calyx and a 2-lipped corolla, 2 lobes usually forming the upper lip
and 3 forming the lower one. The 4 stamens are attached to the corolla
tube. The fruit, dry at maturity, encloses a single seed-forming chamber,
and this is a main point of difference between the Orobanchaceae and
Scrophulariaceae (the Snapdragon Family); in the latter, the fruit consists
of 2 divisions.

In Europe, there is a species that parasitizes shrubby members of the
Pea Family, including those called brooms. Hence the name Broom-rape.

A   Flowers so crowded on the stem that the inflorescence slightly resembles
    the cone of a pine or fir; filaments of stamens hairy at the base
    B   Bract beneath each flower usually widest at the middle or below the
        middle, and generally pointed at the tip; above-ground portion up to
        about 15 cm tall; parasitic on *Gaultheria shallon*
                                    Boschniakia hookeri
                            Small Groundcone; Ma-n; 2
    BB Bract beneath each flower usually widest near the tip, which is
        generally blunt, almost squared off, or has 1-3 teeth; above-ground
        portion up to about 20 cm tall; parasitic on *Arbutus menziesii* and
        species of *Arctostaphylos*   *Boschniakia strobilacea* (*Plate* 39)
                                    California Groundcone
AA Flowers not so crowded on the stem that the inflorescence resembles the
    cone of a pine or fir; filaments of stamens not hairy at the base
    B   Pedicels of flowers at least 3 cm long
        C   Pedicels 1 to several, arising from an underground stem (this has
            only a few small scales); corolla usually violet or purple,
            sometimes yellowish; parasitic on species of *Saxifraga*, *Sedum*,
            and some Asteraceae
                    Orobanche uniflora [includes *O. uniflora* ssp. *occidentalis*
                            and vars. *purpurea* and *sedi*] (*Plate* 39)
                                    Naked Broom-rape
        CC Pedicels several to many, arising from one or more stems that are
            partly above ground (the above-ground portion, like the
            underground portion, has conspicuous scales); corolla purplish,
            yellowish, or yellowish tinged with purple (yellowish without
            exception in some localities, such as Mount Diablo); parasitic
            on species of *Artemisia*, *Eriogonum*, *Eriodictyon*, and other
            shrubs   *Orobanche fasciculata*
                    [includes *O. fasciculata* var. *franciscana*] (*Plate* 39)
                                    Clustered Broom-rape
    BB Pedicels of flowers not often more than 2 cm long (sometimes much
        shorter) (up to 30 cm tall)
        C   Calyx with 4 lobes; parasitic on tomato and on species of
            Amaranthus and *Xanthium* (plants up to 25 cm tall, branching from
            near the base; corolla lobes blue, the tube yellow)
                                    Orobanche ramosa
                            Hemp Broom-rape; eu; Al
        CC Calyx with 5 lobes; parasitic on various native plants

D   Flowers nearly sessile along a stem that may be 30 cm tall;
corolla yellowish, purplish, or brownish, not obviously
streaked (basal portion of stem often somewhat bulbous;
parasitic on *Adenostoma fasciculatum* and other chaparral
plants)   *Orobanche bulbosa*
                              Chaparral Broom-rape; Sl, Ma-s
DD Flowers on pedicels that are usually 1-2 cm long (sometimes
shorter, however, especially in the upper part of the
inflorescence); corolla yellowish or pinkish and usually with
distinct darker streaks (mostly parasitic on shrubs of the
family Asteraceae: *Grindelia*, *Baccharis*, *Artemisia*, etc.)
E   Corolla pinkish, 28-35 cm long, the lips 10-12 mm long;
anthers hairy; found inland   *Orobanche californica*
               ssp. *jepsonii* [*O. grayana* var. *jepsonii*]
                               Jepson Broom-rape; Sn-SCr
EE Corolla yellowish, 20-25 mm long, the lips 8 mm long;
anthers not hairy; found inland or on the coast
                          Orobanche californica ssp. *californica*
                      [*O. grayana* vars. *nelsonii* and *violacea*]
                               California Broom-rape; Mo-n

Oxalidaceae--Oxalis Family

Of the few species of the Oxalis Family that occur in our region, only 2
are natives. One of these is the Redwood Sorrel, *Oxalis oregana*, which
grows in coniferous forests. It is an excellent ground cover for
cultivation in places that are reasonably moist and at least partly shaded.
Forms with deep rose-pink flowers are especially attractive.

The family is characterized by 3-lobed leaf blades that look like those
of clovers, and by fleshy underground rhizomes that make some of the weedy
types difficult to eradicate. The calyx is 5-lobed; the 5 petals are
separate, however, and in bud they overlap one another and have a slight
twist. There are 10 stamens, the filaments of which are united at the base.
The fruit-forming part of the pistil is typically partitioned into 5
divisions.

Several of our worst garden weeds belong to this family. Some are so
attractive when in flower that many gardeners simply acquiesce and tolerate
them rather than fight them. *Oxalis pes-caprae*, the so-called Bermuda
Buttercup (it is a native of southern Africa), produces little bulblike
structures both below and above ground, and these are effective devices for
propagating the species vegetatively.

A   Petals white, deep pink, purplish red, or reddish violet
B   Flowers solitary (petals white or pale pink, often with delicate pink
streaks); in moist coniferous forests   *Oxalis oregana* (*Plate* 39)
                                       Redwood Sorrel; Mo-n
BB Flowers several in each cluster; weeds in gardens and parks
C   Pedicels and calyx hairy; petals light to deep pink or purplish
pink, with darker streaks   *Oxalis rubra* (*Plate* 39)
                                  Windowbox Oxalis; sa; Ma
CC Pedicels and calyx not hairy; petals purplish red or violet
                          Oxalis latifolia [*O. martiana*]
                               Mexican Oxalis; mx; Ma
AA Petals yellow
B   Petals 15-20 mm long
C   Plants stemless (except for the flower peduncles); leaves all
basal, the leaflets 10-25 mm long   *Oxalis pes-caprae*
                              Bermuda Buttercup; af
CC Plants with stems 15-30 cm tall; leaves not all basal, the
leaflets about 10 mm long   *Oxalis incarnata*
                              Yellow Oxalis; af; SF
BB Petals not more than 12 mm long
C   Stems creeping and rooting at the nodes, somewhat elastic when
pulled on; petals generally not more than 8 mm long; common weed
in gardens   *Oxalis corniculata* (*Plate* 39)
                              Creeping Oxalis; eu
CC Stems upright or falling down, but not rooting at the nodes;
petals 8-12 mm long; coastal   *Oxalis albicans* ssp. *pilosa*
                              Hairy Oxalis

Paeoniaceae--Peony Family

Peonies are similar to some members of the Buttercup Family, and they have often been included in that group. A major point of distinction is that the numerous stamens and several pistils of a peony are attached to a fleshy disk; this is not characteristic of the Buttercup Family. Furthermore, in a peony the anthers on the innermost stamens are the first to produce pollen; in buttercups it is the outermost stamens whose anthers generally produce pollen first. Peonies have 5 sepals that persist until the pistils, which are many-seeded as they are in some Ranunculaceae, have ripened. There are generally 5 or 6 petals.

We have one representative, *Paeonia brownii* (*Plate* 87), the Western Peony. The flowers are more than 3 cm wide and occur singly at the end of stems. The petals, maroon or bronze in the center and yellow or greenish at the edge, are rounded. In our region, the Western Peony occurs above 3000' in the Coast Ranges from Santa Clara County northward.

Papaveraceae (includes Fumariaceae)--Poppy Family

The Poppy Family has been broadened to include plants that were formerly assigned to the Fumariaceae, or Fumewort Family. In true poppies, the 4 to 6 petals are all alike, whereas in fumeworts 2 are different from the other 2. In both assemblages within the family, there are usually 2 sepals, although some of the poppies have 3. The number of stamens ranges from several to many. The fruiting part of the pistil, though initially partitioned lengthwise into 2 or more divisions, ripens as a single unit, producing many seeds. An exception is *Platystemon californicus*, Creamcups, in which the several divisions separate as the fruit begins to mature.

Our two native representatives of the fumewort group, both in the genus *Dicentra*, are perennials. Most of our true poppies are annuals or at most biennials, but *Eschscholzia californica*, the State Flower, has perennial forms, and *Dendromecon rigida*, the Bush Poppy, is a woody shrub. Nearly all of the herbaceous species are attractive when grown from seed in a more or less natural setting.

A One or two petals curving downward, and rounded at the base to form a
   slight pouch or spur; with 6 stamens, these in groups of 3 in line with
   the outer petals
   B One petal forming a spur
      C Corolla (including spur) 7-10 mm long (sometimes 12 mm), the
         lower portion usually deep pink or purplish red, the tip purple;
         leaf lobes sometimes more than 2 mm wide
                                      Fumaria officinalis (*Plate* 40)
                                                        Fumitory; eu
      CC Corolla not more than 5 mm long, the lower portion usually pale
         pink, the tip purple; leaf lobes rarely so much as 1 mm wide
                                                 Fumaria parviflora
                                              Smallflower Fumitory; eu
   BB Each of 2 petals forming a pouch
      C Petals dull rose-purple; mostly less than 40 cm tall; in shady,
         moist woods   *Dicentra formosa* (*Plate* 40)
                                                  Bleeding-heart; Mo-n
      CC Petals yellow, falling away early; sometimes more than 100 cm
         tall; on rather dry ridges   *Dicentra chrysantha* (*Plate* 39)
                                                Golden Eardrops; Me-s
AA All 4 petals (6 in *Argemone munita*) similar, none of them with a slight
   pouch at the base; with numerous stamens
   B Sepals joined to form a conical cap that covers the petals before
      they unfold (the cap is then shed)
      C With a prominent collarlike rim just below the line where the
         sepals are attached; petals usually orange, yellow, or yellow
         with orange at the base, in some forms as long as 5 cm (the
         State Flower of California; widespread)
                                      Eschscholzia californica (*Plate* 40)
                                                        California Poppy

CC Without a distinct rim just below the level where the sepals are
   attached; petals yellow, not more than 2.5 cm long, and usually
   shorter than this
   D   Petals 4 mm long; sepals 3-4 mm long
                                         Eschscholzia rhombipetala
                        Diamond-petal California Poppy; Al, CC; 1a(1950)
   DD Petals 7-25 mm long; sepals 6-18 mm long
      E   Seeds spherical, burlike, owing to many sharp projections
                                              Eschscholzia lobbii
                                                 Frying-pan Poppy
      EE Seeds slightly elongated, nearly smooth, with a
         networklike pattern   *Eschscholzia caespitosa* (*Plate* 40)
                                                     Tufted Poppy
BB Sepals, though shed early, not forming a conical cap over the petals
   C   Leaves opposite or in whorls (sometimes mostly at the base of the
       plant, but nevertheless opposite)
       D   Stamens usually 12, in 2 series; petals white or cream,
           without yellow at the base; most basal leaves broadest above
           the middle; fruit usually at least 8 times as long as wide
           (slightly twisted) (up to 20 cm tall)   *Meconella californica*
                                              California Meconella; SFBR
       DD Stamens many more than 12; petals cream with yellow at the
          base; basal leaves not broadest above the middle; fruit about
          3-4 times as long as wide
          E   Fruit consisting of at least 6 divisions, these separating
              early, and usually constricted into beadlike units;
              leaves often present to at least the middle of the stem
              (widespread)   *Platystemon californicus* (*Plate* 40)
                                                        Creamcups
          EE Fruit consisting of only 3 divisions, these remaining
             firmly attached to one another until the fruit cracks
             open to release seeds; nearly all leaves basal
                     Meconella linearis [*Hesperomecon linearis*]
                                            Narrowleaf Meconella
   CC Leaves alternate
       D   Petals (2-3 cm long) yellow; evergreen shrubs, woody below
                                       Dendromecon rigida (*Plate* 39)
                                                    Bush Poppy; Sn-s
       DD Petals white or various shades of orange or red, but not
          yellow; annual, never woody
          E   Petals 6, up to 5 cm long, white; stems and lobes of
              leaves with yellow spines; stout, somewhat bushy, up 100
              cm tall
                  Argemone munita [includes *A. munita* ssp. *rotundata*]
                                                 Prickly Poppy; CC-s
          EE Petals 4, up to about 2.5 cm long, reddish or orange;
             stems and leaves without spines; slender and rather
             delicate, not more than 60 cm tall
             F   Petals usually orange-red, with a purplish spot at the
                 base; pistil with a distinct style, this with a nearly
                 globular stigma; sap yellowish
                               Stylomecon heterophylla (*Plate* 40)
                                                    Wind Poppy; La-s
             FF Petals brick-red, with a green spot at the base;
                pistil without a distinct style, but capped by a
                lobed, disklike stigma; sap milky *Papaver californicum*
                                                    Fire Poppy; Ma-s

Philadelphaceae (formerly in Saxifragaceae)--Mock-orange Family

In the past, the plants making up the Philadelphaceae have usually been
assigned to either the Saxifragaceae or the Grossulariaceae. They differ
from both of these families, however, in having opposite leaves. In some of
the genera, furthermore, the flowers have many stamens. (Saxifrages do not
have more than 10, and 5 is the usual number for currants and
gooseberries.)

Our only representative is called Yerba-de-selva, *Whipplea modesta*
(*Plate* 87). It forms a low ground cover, and is perhaps underrated as a
subject for lightly shaded places in wild gardens. Unlike most members of
the family, which are substantial shrubs, it is only slightly woody at the

base. It is deciduous, with nearly sessile oval leaves. The small flowers, in upright inflorescences, have 5 or 6 calyx lobes and the same number of white petals 3 mm long. The calyx cup is fused to the lower half of the globular fruit, which is dry at maturity.

Two other California natives are widely cultivated and grow well here. One of them, *Carpenteria californica*, the Tree Anemone, is an evergreen shrub with large flowers. It is indigenous to low mountain areas in Fresno County. *Philadelphus lewisii*, called Mock-orange, is deciduous and uninteresting when dormant, but its flowers have a delightful fragrance. It is found from Lake and Mendocino Counties northward, and also in the Sierra Nevada.

Phytolaccaceae--Pokeweed Family

Pokeweed, *Phytolacca americana*, is a perennial native to the eastern United States and Canada. It is an attractive plant and the source of a useful red dye, long used for coloring candy, wine, cloth, and paper. Pokeweed has therefore been cultivated, or at least encouraged, even in areas where it grows wild. It is now well established in southern Europe, and there are a few places in California, including our region, where it has become a moderately common weed.

Pokeweed reaches a height of more than 2 m. Its stems are ribbed and often reddish, and they branch dichotomously. The narrowly oval leaves are large--up to 25 cm long--and the small flowers, which are greenish or pinkish, are borne in dense, elongated inflorescences. The fruit is interesting: a cluster of 10 fleshy units, thus something like a blackberry. It starts out reddish, but becomes purplish black as it ripens.

Plantaginaceae--Plantain Family

All of our plantains belong to a single genus, *Plantago*, characterized by basal leaves and small flowers concentrated in dense, terminal inflorescences. There are 4 sepals, 4 somewhat papery corolla lobes, and 2 or 4 stamens. The fruit is partitioned into 2 major divisions, each producing 1 or more seeds.

We have an interesting variety of local plantains. Two of them, *Plantago major* and *P. lanceolata*, are aggressive weeds. When in flower they produce much pollen, to which some persons are allergic.

A   Most leaf blades about two-thirds as wide as long (widespread, perennial
       weed)   *Plantago major*
                  [includes *P. major* vars. *pilgeri* and *scopulorum*] (*Plate* 87)
                                                           Common Plantain; eu
AA Leaf blades rarely so much as half as wide as long
    B   Leaf blades pinnately divided into slender lobes; inflorescence
        drooping (but becoming upright after fruiting has begun) (mostly
        near the coast)   *Plantago coronopus*
                                                      Cutleaf Plantain; eu; Mo-n
    BB Leaf blades not pinnately lobed (in *Plantago maritima*, however, there
        are sometimes coarse, irregularly distributed teeth); inflorescence
        not drooping
        C   Leaves obviously succulent (perennial; on sandy beaches, salt
            marshes, and upper levels of rocky shores)
                  Plantago maritima [includes *P. maritima* ssp. *juncoides* and
                                    var. *californica*] (*Plate* 41)
                                                                Sea Plantain
        CC Leaves not obviously succulent
            D   Leaves commonly more than 1.5 cm wide; usually at least 20 cm
                tall when in flower
                E   Inflorescence up to about 4 cm long; leaves usually 2-2.5
                    cm wide (common weed)   *Plantago lanceolata* (*Plate* 87)
                                                            English Plantain; eu

EE Inflorescence often more than 10 cm long; leaves usually
   more than 2.5 cm wide (mostly in coastal areas)
                          Plantago subnuda [P. hirtella var. galeottiana]
                                                    Coast Plantain
DD Leaves rarely so much as 1 cm wide; not often more than 15 cm
   tall when in flower
   E Flowers with 2 stamens; inflorescence often more than 2.5
      cm long (in drying vernal pools and at the border of salt
      marshes)   Plantago elongata
         [includes P. elongata ssp. pentasperma and P. bigelovii]
                                       Linearleaf Plantain; SLO-n
   EE Flowers with 4 stamens; inflorescence not more than 2.5 cm
      long
      F Corolla lobes spreading outward, the developing fruit
         thus exposed; leaves 3-12 cm long; flowering stalk 5-
         25 cm long (widespread)   Plantago erecta
                [P. hookeriana var. californica] (Plate 41)
                                               California Plantain
      FF Corolla lobes forming a beaklike covering over the
         developing fruit; leaves 1-6 cm long; flowering stalk
         1.5-7 cm long   Plantago truncata ssp. firma
                                       Chile Plantain; sa; Sn, Ma

Platanaceae--Sycamore Family

Sycamores and plane trees, all of which belong to a single genus, have
alternate, deciduous leaves. The larger blades are palmately lobed and the
base of the petioles are prominently dilated. The bark flakes off in
substantial pieces, and this confers an attractive mottling on the trunk
and large branches. The small flowers are crowded into globular clusters
attached to drooping racemes. All of the flowers on a particular raceme are
either staminate or pistillate. Staminate flowers have 3-8 small sepals, an
equal number of stamens, and 3-6 petals. Pistillate flowers likewise have
3-8 sepals, an equal number of pistils, and 3-6 petals.

The only species native to our region--to all of California, in fact--is
Platanus racemosa (Plate 87), the Western Sycamore. It grows mostly along
streams in foothills and valleys, and reaches a height of about 25 m. The
leaf blades, usually with 5 pointed lobes, are up to 15 cm wide. The heads
of the short-lived staminate flowers are about 1 cm wide. The pistillate
flowers, also in heads, develop into achenes. Round aggregates of these,
about 2.5 cm wide, are called "buttonballs," and persist into the autumn.

Plumbaginaceae--Leadwort Family

In plants of the Leadwort Family, many of which live in coastal
habitats, including salt marshes, the flowers have a 5-lobed calyx (the
tube of which is usually ribbed lengthwise) and a 5-lobed corolla. The
lobes of the corolla, however, may be separate almost to the base, and thus
appear to be completely free petals. The 5 stamens, attached to the
corolla, are in line with its lobes. The pistil, with 5 styles, develops
into a dry fruit containing a single seed.

A feature of some members of the family is the papery texture of the
calyx or corolla, or both. A few species of Limonium are cultivated,
usually under the obsolete genus name Statice. Their "everlasting" flowers,
when dried, look much as they did when fresh.

A Flowers in a dense head at the end of unbranched stems, each head with
   broad, papery bracts beneath it; corolla lobes (bright pink) separated
   nearly to the base; basal leaves not often more than 3 mm wide;
   flowering stems up to 30 cm tall (usually on rocky shores above the
   high-tide line, but sometimes on backshores of sandy beaches or in other
   open habitats)   Armeria maritima ssp. californica (Plate 41)
                                                    Thrift; SLO-n
AA Flowers in clusters on a branched panicle, each cluster with scalelike
   bracts beneath it; corolla lobes on a long tube; basal leaves usually at
   least 30 mm wide; flowering stems usually more than 30 cm tall

B Basal leaves pinnately lobed (calyx lobes purplish pink, papery;
corolla lobes pale yellow, soon withering; stems conspicuously
winged; in disturbed areas at the coast) *Limonium sinuatum*
Winged Sea-lavender; eu
BB Basal leaves not lobed
  C Corolla lobes grayish violet; basal leaves still green at the
  time of flowering (mostly near the high-tide line in salt
  marshes, common) *Limonium californicum* (*Plate* 41)
  California Sea-lavender
  CC Corolla lobes white; basal leaves withered by the time of
  flowering (in disturbed areas) *Limonium otolepis*
  Asian Sea-lavender; as; SF

Polemoniaceae--Phlox Family

Except for one species of *Phlox*, all our members of this family are
herbaceous plants. Most, in fact, are annuals. The flowers nearly always
have 5 calyx lobes, 5 corolla lobes, 5 stamens (these are attached to the
corolla), and a pistil consisting of 3 seed-producing divisions. Certain
species of *Navarretia*, however, do not follow this arrangement.

All of our species are natives. It is not often that we can say that
about a family that has so many representatives. *Collomia grandiflora*,
species of *Gilia* (particularly *G. tricolor*), and species of *Linanthus*
grow well from seed sown in sunny places.

A Leaves pinnately compound, with 11-21 oval leaflets (corolla up to 2.5
cm wide, salmon, pink, or purple; widespread)
Polemonium carneum (*Plate* 88)
Jacob's-ladder; SM-n
AA Leaves not truly compound (but the blades may be deeply toothed or
divided into several to many lobes, which are often narrow)
  B Leaves smooth-margined or shallowly divided into a few teeth near the
  tip
    C Leaves alternate
      D Corolla often more than 1 cm wide, salmon, pinkish yellow, or
      sometimes white; leaf blades not toothed; stems usually not
      branched; often more than 50 cm tall
      Collomia grandiflora (*Plate* 41)
      Largeflower Collomia
      DD Corolla less than 1 cm wide, yellow at the throat, but with
      purplish lobes; leaf blades toothed (or shallowly divided
      into 3 lobes near the tip); stem usually branched; rarely
      more than 10 cm tall (on serpentine soil)
      Collomia diversifolia
      Serpentine Collomia; Na; 4
    CC Leaves opposite
      D Corolla lobes only 1-2 mm long, usually rose, lavender, or
      white; annual, rarely more than 15 cm tall, with soft leaves;
      common in grassy places
      Phlox gracilis [*Microsteris gracilis*] (*Plate* 42)
      Slender Phlox
      DD Corolla lobes about 10 mm long, bright pink; woody-based
      perennial, sometimes more than 30 cm tall, with tough leaves;
      in rocky habitats *Phlox speciosa* ssp. *occidentalis*
      Showy Phlox; Sn-n
  BB Leaves (at least some of them) deeply lobed, the primary lobes
  sometimes divided again
    C Calyx lobes of unequal length
      D Inflorescence cobwebby-hairy, the hairs intertwined; tip of
      leaf lobes not spinelike (corolla lobes blue, the tube and
      throat yellow) *Eriastrum abramsii*
      Abrams Eriastrum; La-SB
      DD Inflorescence, if hairy, not cobwebby, the hairs not
      intertwined; tip of leaf lobes spinelike
      Polemoniaceae, Subkey (p. 196)
    CC Calyx lobes of equal length, or nearly so
      D Leaves mostly opposite, at least in the lower part of the
      plant, and palmately divided into slender, usually almost
      needlelike lobes

E Corolla tube much longer than the calyx (usually at least
    twice as long)
   F   Corolla with a red spot at the base of each lobe (rest
        of corolla pink, with a white or yellow throat)
                                          Linanthus ciliatus
                                              Whiskerbrush
   FF Corolla without a red spot at the base of each lobe
      G   Corolla lobes generally 5-8 mm long, usually at
            least one-third as long as the corolla tube; more
            than 10 cm tall
         H   Calyx lobes not hairy on the back, but hairy on
                the margin   *Linanthus androsaceus*
                                      Pinklobe Linanthus; Mo-n
         HH Calyx lobes hairy on the back as well as on the
                margin (widespread)
                        Linanthus parviflorus [includes *L. androsaceus*
                            sspp. *croceus* and *luteus*] (*Plate* 42)
                                        Common Linanthus; Me-s
      GG Corolla lobes 3-5 mm long, not often more than one-
            fifth as long as the corolla tube; usually less
            than 10 cm tall
         H   Corolla lobes rose, pink, or white, the tube
                and throat usually yellow (commonly forming
                dense colonies; widespread)
                                    Linanthus bicolor (*Plate* 42)
                                      Bicolor Linanthus; SLO-n
         HH Corolla uniformly bright yellow
                                        Linanthus acicularis
                                      Bristly Linanthus; Al-n; 4
EE Corolla tube less than twice as long as the calyx
    (sometimes not appreciably longer than the calyx)
   F   Flowers nearly sessile, the pedicels less than 5 mm
        long
      G   Flowers solitary (but 2 flowers may be close
            together); calyx not hairy; corolla lobes white,
            tinged with purple on the back, the throat yellow
            (south of San Francisco, the flowers usually open
            in the evening; north of the Bay, the flowers
            usually open in the daytime) *Linanthus dichotomus*
                        [includes *L. dichotomus* ssp. *meridianus*]
                                              Evening-snow; Na
      GG Flowers usually in heads of at least 5 or 6; calyx
            hairy; corolla lobes white, pale pink, or lilac,
            the throat yellow
                            Linanthus grandiflorus (*Plate* 42)
                                    Largeflower Linanthus; Sn-s; 4
   FF Pedicels of flowers at least 5 mm long, usually much
        longer
      G   Corolla without a purple throat, the lobes longer
            than the corolla tube; calyx lobes united by a
            transparent membrane for about half their length
            (corolla 1-2 cm long, white to pale pink or lilac,
            very open, the tube not extending above the calyx
            lobes; often more than 40 cm tall; in sandy soil)
                                        Linanthus liniflorus
                                    Flaxflower Linanthus; CC-SLO
      GG Corolla usually with a purple throat, the lobes
            shorter than the corolla tube; calyx lobes united
            for at least two-thirds their length
         H   Corolla sometimes more than 15 mm long, the
                lobes usually 5 mm long, pink or blue with
                yellow at their bases (on serpentine)
                                        Linanthus ambiguus
                                    Serpentine Linanthus; SM, SCl, SB
         HH Corolla about 10 mm long, the lobes rarely more
                than 4 mm long, white, pale pink, lilac, or
                violet
                        Linanthus bolanderi [includes *L. bakeri*]
                                      Bolander Linanthus; Me-CC
DD Leaves alternate, the lower ones, at least, divided pinnately
    or bipinnately (there may be only 1 or 2 lobes on each side,
    however)
   E   Corolla tube at least 3 times as long as the calyx

F   Lower leaves with several lobes on each side, these
    almost symmetrically arranged and about the same size,
    and sometimes subdivided further (corolla lobes
    pinkish violet, the tube and throat purple)
                                        Gilia tenuiflora
                        Slenderflower Gilia; SCl(SCM)-s
FF Lower leaves with lobes usually of unequal size, not
    often symmetrically arranged, and not subdivided
    further
    G   Corolla lobes pink or violet-pink, the tube red-
        violet or purple; leaves rarely with more than 1
        lobe on each side, these generally much smaller
        than the terminal lobe   *Allophyllum divaricatum*
                                        Straggling Gilia; Mo-n
    GG Corolla lobes violet-blue, the tube the same color;
        leaves often with more than 3 lobes on each side,
        the longest lobes sometimes as long as, or longer
        than, the terminal one   *Allophyllum gilioides*
                                            Purple Gilia
EE Corolla tube not more than twice as long as the calyx
    F Flowers concentrated in dense globular heads (usually
       50-100 flowers in a head)
       G   Corolla 5-8 mm long, the lobes up to 2 mm wide
           H   Inflorescence not hairy or only slightly hairy
               (widespread)
                            Gilia capitata ssp. *capitata* (*Plate* 42)
                                            Globe Gilia; Ma-n
           HH Inflorescence densely hairy at the base
                            Gilia capitata ssp. *tomentosa*
                    Hairy Globe Gilia; Ma, Sn, CC(MD), SCl(MH)
       GG Corolla 7-13 mm long, the lobes 1.5-3.5 mm wide
           H   Corolla dark blue-violet; plants without a
               cluster of leaves at the base
                            Gilia capitata ssp. *staminea*
                                            Pale Gilia; n, CC-SB
           HH Corolla light blue-violet; plants with a
               cluster of leaves at the base (coastal)
                            Gilia capitata ssp. *chamissonis*
                                            Dune Gilia; Ma-SF
    FF Flowers not in dense globular heads (but in *Gilia
        achilleifolia*, there may be up to 50 flowers in a
        crowded inflorescence)
        G   Deepest portion of each cleft between calyx lobes
            distended outward so as to form a slight lip like
            that on a pitcher; flower clusters with a few
            broad leaves just below them (corolla uniformly
            rose, pink, or white; leaf lobes broad, with
            toothed margins; calyx lobes narrowing to slender
            tips)   *Collomia heterophylla* (*Plate* 41)
                                        Variedleaf Collomia; Mo-n
        GG Deepest portion of each cleft between calyx lobes
            not distended outward; flower clusters without
            broad leaves just below them
            H   Corolla uniformly blue, violet-blue, or white,
                without any yellow or orange
                I   Flowers 8-25 in each inflorescence, on
                    pedicels up to 2 mm long (in sandy soil;
                    widespread)
                            Gilia achilleifolia ssp. *achilleifolia*
                                [*G. achilleaefolia*] (*Plate* 42)
                                        California Gilia; Ma, CC-s
                II Flowers 2-7 in each inflorescence, on
                    pedicels 1-30 mm long
                            Gilia achilleifolia ssp. *multicaulis*
                                [*G. achilleaefolia* ssp. *multicaulis*]
                                        Small California Gilia; Ma, CC-s
            HH Corolla not the same color throughout (corolla
                tube or throat, or both, with considerable
                yellow or orange, and with dark purple spots)
                I   Corolla tube 1-1.5 cm long (corolla tube
                    yellow or orange, the throat with 5 pairs
                    of dark purple spots; lobes pale violet)
                                        Gilia tricolor (*Plate* 42)
                                            Bird's-eye Gilia

II Corolla tube less than 1 cm long
    J   Corolla tube and lobes blue-violet, the
        throat yellow with 5 dark purple spots
                            Gilia millefoliata
                San Francisco Gilia; SFB-n
    JJ Corolla tube yellow, lobes blue violet,
        the throat yellow with 5 pairs of dark
        purple spots   *Gilia clivorum*
                      Grassland Gilia; Sl-s

Polemoniaceae, Subkey: Leaves (at least some of them) deeply lobed, the
primary lobes sometimes divided again; calyx lobes of unequal length; tip
of leaf lobes spinelike   *Navarretia*
A   With 4 corolla lobes, calyx lobes, and stamens (unusual for the family!)
    (corolla cream or very pale yellow; in moist areas)
                      Navarretia cotulifolia [*N. cotulaefolia*]
                              Cotula Navarretia; Sn, Me-SB
AA With 5 corolla lobes, calyx lobes, and stamens
    B   Bracts beneath the inflorescence broad, the width at the middle (not
        including the lobes) sometimes more than 6 mm (lobes more or less
        evenly distributed along the margin)   *Navarretia atractyloides*
                                      Hollyleaf Navarretia
    BB Bracts (not including the lobes) not more than 4 mm wide (except in
        Navarretia heterodoxa)
        C   Hairs on main stems turned downward, often pressed against the
            stem (in moist areas, including drying vernal pools)
            D   Stigma deeply divided into 2 or 3 lobes
                E   Stigma divided into 2 lobes; corolla lobes pale blue or
                    white, 4-6 mm long, with 1 main vein
                                Navarretia intertexta (*Plate* 88)
                                      Needleleaf Navarretia
                EE Stigma divided into 3 lobes; corolla lobes pale blue,
                    about 2.5 mm long, with 3 main veins *Navarretia tagetina*
                                      Marigold Navarretia; Na, Sn-n
            DD Stigma rather indistinctly divided into 2 lobes (corolla
                white; mostly growing in drying mud of ponds and lakes,
                sometimes very compact)
                E   Vein on each corolla lobe not branched; corolla 5-7 mm
                    long   *Navarretia leucocephala* ssp. *bakeri* [*N. bakeri*]
                                      Baker Navarretia; Sn-n; 1b
                EE Vein on each corolla lobe branched; corolla 7-9 mm long
                            Navarretia leucocephala ssp. *leucocephala*
                                      Whiteflower Navarretia; SB-n
        CC Hairs on main stems not turned downward
            D   Corolla yellow or cream-yellow (in drying vernal pools)
                    Navarretia nigelliformis [*N. nigellaeformis*]
                                      Adobe Navarretia; CC-SLO
            DD Corolla white, blue, or purple
                E   Corolla purple, with a dark spot at the base of each lobe
                                      Navarretia jepsonii
                                Jepson Navarretia; Na; 4
                EE Corolla without a dark spot at the base of each lobe
                    F   Corolla tube 1 cm long or longer, the lobes 2-2.5 mm
                        long, blue
                        G   Corolla lobes with 3 dark veins
                                    Navarretia pubescens (*Plate* 42)
                                      Downy Navarretia; SLO-n
                        GG Corolla lobes without dark veins
                            H   Corolla tube up to 12 mm long; primary lobes of
                                leaves irregularly spaced, often divided
                                further; corolla blue, usually pale (plant
                                with a strong skunklike odor; widespread)
                                              Navarretia squarrosa
                                                Skunkweed; Mo-n
                            HH Corolla tube 12-16 mm long; lobes of leaves
                                nearly regularly spaced, not divided further;
                                corolla blue or purple   *Navarretia viscidula*
                                              Sticky Navarretia; SM-n
                    FF Corolla tube less than 1 cm long, the lobes less than
                        2 mm long
                        G   Most bracts below the inflorescence with only 1 or
                            2 lobes on each side (corolla tube 3-4 mm long,
                            the lobes about 1 mm long, pink, purple, or white;
                            at elevations above 1000') *Navarretia divaricata*
                                              Mountain Navarretia; Na

GG Most bracts with several lobes on each side
  H Bracts not becoming broader toward the middle,
     the lobes rather evenly distributed along the
     margin; corolla lobes blue, the throat
     lighter, with purple veins   *Navarretia mellita*
                    Honeyscent Navarretia; SLO-n
  HH Bracts becoming obviously broader toward the
     middle (sometimes reaching a width of about 5
     mm, not including the lobes), most of the
     lobes arising from the lower half; corolla
     uniform in color
     I Corolla purple, with the stamens protruding
       out of it (on dry rocky slopes)
                       Navarretia heterodoxa
                 Calistoga Navarretia; Na, Sn-SCl
     II Corolla lavender to white, the stamens not
       protruding out of it (on serpentine)
                        Navarretia rosulata
                 [*N. heterodoxa* ssp. *rosulata*]
                Marin County Navarretia; Ma; 1b

Polygalaceae--Milkwort Family

Milkwort flowers are irregular and superficially resemble those of the Pea Family, but they are really very different. Two of the 5 sepals--those at the sides of a flower--are larger than the others and are usually colored like petals; the uppermost sepal, furthermore, is keeled. In the one genus represented in California, *Polygala*, there are 3 petals, and these are united at the base. The lower one, somewhat boat-shaped, encloses the 8 stamens and the style of the pistil. A further specialization is that the filaments of the stamens are joined to one another to form an incomplete tube, and this in turn is attached to the base of the lateral petals. The fruit-forming part of the pistil is partitioned into 2 halves, each producing a single seed.

Our only species, *Polygala californica* (Plate 43), Milkwort, is a bushy, slender-stemmed perennial, often somewhat woody at the base. It reaches a height of about 35 cm and has oblong or elliptical, short-petioled leaves 1-4 cm long. The flowers, in short racemes, are bright to deep pink. It is found in the outer Coast Ranges from San Luis Obispo County to southern Oregon.

The name *Polygala*, meaning "much milk," is believed to be based on an old idea that some plants of this group, if eaten by cows, stimulate the production of milk. "Milkwort" is also tied to this notion.

Polygonaceae--Buckwheat Family

Our representatives of this family--which does indeed include the 2 species of cultivated buckwheats--are mostly herbs, but a few are shrubs. Although the inflorescences are often conspicuous, the individual flowers are usually small. They have no petals, but the calyx lobes, of which there are commonly 4, 5, or 6, may resemble petals and are sometimes richly colored. The number of stamens ranges from 3 to 9. The fruit, at maturity, is dry, and most of it is occupied by the single seed.

Numerous species of this family contribute to colorful displays of wildflowers, especially in dry, well drained habitats. Some of the eriogonums are excellent subjects for gardens devoted to native plants. Two genera, *Rumex* and *Polygonum*, include weeds whose rating as nuisances ranges from merely objectionable to pernicious.

A Leaves with stipules, these united and forming sheaths that encircle the
    stem (nodes often swollen)
    B Calyx with 6 lobes; leaves often mainly basal; flowers often greenish
      and in racemes

C   Some leaf blades with a pair of basal lobes; mostly less than 30
cm tall, the upright stems arising from perennial underground
rhizomes; leaves very sour; pistillate and staminate flowers on
separate plants (widespread)
Rumex acetosella [*R. angiocarpus*] (*Plate* 44)
Sheep-sorrel; eu
CC Leaf blades without basal lobes; more than 30 cm tall (sometimes
smaller in *Rumex maritimus*), the upright stems arising from a
basal rootstock; leaves not especially sour; most flowers with a
pistil and stamens
D   Most basal leaves at least 15 cm long, the margins sometimes
conspicuously wavy
E   None of the 3 inner calyx lobes with a swelling (alkaline
areas)   *Rumex occidentalis* [includes *R. fenestratus*]
Western Dock; SFBR-n
EE At least 1 of the 3 inner calyx lobes with a swelling
F   All of the 3 inner calyx lobes with a conspicuous,
raised swelling (widespread)
Rumex crispus (*Plates* 44, 89)
Curly Dock; eua
FF Only 1 of the 3 inner calyx lobes with a swelling
Rumex obtusifolius
[includes *R. obtusifolius* ssp. *agrestis*] (*Plate* 89)
Bitter Dock; eu; SCr-n
DD Basal leaves usually less than 15 cm long, the margins not
obviously wavy (sometimes slightly wavy in *Rumex
conglomeratus*)
E   Inner 3 calyx lobes (which closely surround the small,
nutlike fruit) with distinctly toothed margins
F   Inner 3 calyx lobes with only 1-3 slender teeth on
each side, these teeth longer than the width of the
rest of the lobe; annual (found in wet areas)
Rumex maritimus
[includes *R. fueginus* and *R. persicarioides*] (*Plate* 89)
Golden Dock; SM, Ma-n
FF Inner 3 calyx lobes usually with more than 3 teeth on
each side, these shorter than the width of the rest of
the lobe; perennial   *Rumex pulcher*
Fiddle Dock; me
EE Inner 3 calyx lobes with smooth margins (in moist areas)
F   Leaf blades usually more than 5 times as long as wide,
tapering gradually to the petiole
Rumex salicifolius (*Plate* 44)
Willow Dock; Ma-s
FF Leaf blades usually less than 5 times as long as wide,
tapering abruptly to the base (the base therefore
nearly rounded)   *Rumex conglomeratus* (*Plate* 89)
Green Dock; eu
BB Calyx with 5 lobes; leaves usually not mainly basal; flowers usually
not greenish, either in racemes or in the leaf axils
C   Leaf blades triangular or heart-shaped (usually 2-6 cm long)
(annual, with stems trailing on the ground or clambering over
low vegetation; if not flowering, could be mistaken for a
morning-glory)   *Polygonum convolvulus*
Black Bindweed; eu
CC Leaf blades not triangular or heart-shaped
D   Flowers in crowded terminal inflorescences, these leafless
(in wet places, and sometimes partly submerged)
E   Stems usually with a single terminal raceme (sometimes a
pair of racemes); some leaves with petioles at least 1.5
cm long (other leaves on the same plant may have shorter
petioles or be completely sessile)
F   Petioles of lower leaves often as much as, or more
than, two-thirds as long as the blades; upper leaves
sessile, clasping the stems (inflorescence 1-6 cm
long, the flowers white or pink)
Polygonum bistortoides
Western Bistort; Ma-n
FF Petioles of lower leaves much shorter than the blades;
upper leaves with at least short petioles
G   Inflorescence not often more than 3 cm long, the
peduncles not hairy
Polygonum amphibium var. *stipulaceum*
Shore Knotweed

GG Inflorescence usually 3-10 cm long, the peduncles
glandular-hairy   *Polygonum amphibium*
                    var. *emersum* [*P. coccineum*] (*Plate* 44)
                                              Swamp Knotweed
EE Stems usually with 2 to several racemes; leaves either
sessile or with petioles less than 1.5 cm long
  F Calyx with glandular dots
     G Inflorescence nodding at the tip; calyx up to 4 mm
        long, greenish with rose or whitish tips; up to 60
        cm tall   *Polygonum hydropiper* (*Plate* 88)
                                              Marshpepper; eu
     GG Inflorescence not nodding at the tip; calyx up to 3
        mm long, brownish with whitish tips; up to 100 cm
        tall   *Polygonum punctatum*
                                              Water Smartweed
  FF Calyx without glandular dots (up to more than 80 cm
     tall)
     G Leaves 3-10 cm long, mostly sessile, often with a
        darker blotch; inflorescence up to 2.5 cm long
        (calyx pink or purplish)
                          Polygonum persicaria (*Plate* 44)
                                              Lady's-thumb; eu
     GG Leaves 5-20 cm long, mostly with short but distinct
        petioles, without a darker blotch; inflorescence
        up to 7 cm long
        H Racemes usually more than 7 at the end of the
           main stem, this often not branched; racemes
           very dense, the flowers mostly touching one
           another; petioles 1-1.5 cm long; annual
                              Polygonum lapathifolium
                                              Willow-weed
        HH Racemes usually fewer than 7 at the end of the
           main stem or its branches (there are usually
           several branches); racemes rather loose, the
           flowers usually not touching one another;
           petioles usually less than 1 cm long;
           perennial   *Polygonum hydropiperoides*
           [includes *P. hydropiperoides* var. *asperifolium*]
                                              Waterpepper
DD Flowers borne in the leaf axils singly or in small groups (in
Polygonum paronychia, the flowers are concentrated at the tip
of the stem, but they are nevertheless in the leaf axils)
  E Flowers concentrated at the tip of the stem; midrib of
     leaf blades raised prominently on the underside; brownish
     remains of stipules obvious on the stem; on backshores of
     sandy beaches (woody at the base; often forming low
     mats)   *Polygonum paronychia* (*Plate* 44)
                                              Beach Knotweed; Mo-n
  EE Flowers scattered along the stem; midrib of leaf blades
     not raised prominently on the underside; remains of
     stipules not obvious on the stem; not mainly near sandy
     beaches (but sometimes around salt marshes)
     F Flowers sessile and usually solitary in the leaf axils
        (in dry areas)
        G Flowers rather crowded all along the branches of
           the inflorescence; leaves usually about 1 cm long;
           not woody at the base, up to 20 cm tall, annual
                              Polygonum californicum
                                              California Knotweed
        GG Flowers not crowded on the branches of the
           inflorescence, except at the tip; leaves often
           less than 1 cm long; woody at the base, up to 60
           cm tall, perennial   *Polygonum bolanderi*
                                              Bolander Knotweed; Na-n
     FF Flowers on short pedicels and forming small clusters
        in the leaf axils
        G Fruit protruding beyond the calyx lobes (at the
           edge of salt marshes)   *Polygonum marinense*
                                              Marin Knotweed
        GG Fruit not protruding beyond the calyx lobes
           H Calyx 2.5-3 mm long (flowers not hidden among
              the leaves)

```
                    I   Leaves near the flower heads less than 0.5
                        cm long; mostly in dry areas, including
                        those with serpentine soils, but sometimes
                        on backshores of sandy beaches
                            Polygonum douglasii ssp. spergulariiforme
                                            [P. spergulariaeforme]
                                            Fall Knotweed; La-n
                    II  Leaves near the flower heads more than 1 cm
                        long; in dry areas   Polygonum ramosissimum
                                            Yellowflower Knotweed; na?
              HH  Calyx 1.5-2 mm long
                    I   Calyx not hidden among the leaves; upright
                                        Polygonum argyrocoleon
                                        Silversheath Knotweed; as
                    II  Calyx nearly hidden among the leaves;
                        prostrate    Polygonum arenastrum (Plate 89)
                                            Common Knotweed; eu
AA  Leaves without stipules
    B   Flowers not originating within a cuplike involucre (but each flower
        has 1 or several bracts beneath it) (leaves up to 2 cm long)
        C   Bract single, 2-lobed; leaves opposite, the blades fan-shaped,
            with a notch at the tip; calyx (reddish) about 1 mm long,
            usually 6-lobed, the lobes separate nearly to the base; growing
            in the shade of shrubs                 Pterostegia drymarioides
                                                   Pterostegia
        CC  Bracts (these differ little from the leaves, except in being
            small) several, not 2-lobed; leaves mostly in whorls of 4-5,
            slender, the upper ones joined together basally and ending in
            hooked tips; calyx about 4 mm long, usually 5-lobed, the lobes
            separate to about the middle; in sandy, exposed habitats
                              Lastarriaea coriacea [Chorizanthe coriacea]
                                            Lastarriaea; CC-Mo
    BB  Individual flowers, or groups of flowers, within a cuplike, several-
        lobed involucre (sometimes the involucres are in clusters)
        C   Each involucre with a single flower (rarely 2 flowers); midrib of
            each lobe of the involucre prolonged as a spine (in sandy or
            rocky soil)
            D   Upper half of the involucre (except for the midribs of the
                lobes) thin, membranous, and not hairy (leaf blades 1-5 cm
                long; up to 1 m tall) Chorizanthe membranacea (Plates 43, 88)
                                            Pink Spineflower; Me-s
            DD  Only the upper third, or less, of the involucre thin,
                membranous, and not hairy
                E   One spine of the involucre much longer than the others
                    (calyx lobes not bristle-tipped; leaf blades up to 2 cm
                    long; prostrate, with stem tips rising up to 10 cm)
                                            Chorizanthe clevelandii
                                            Cleveland Spineflower
                EE  Spines of the involucre of different lengths, but 1 spine
                    not obviously much longer than the others
                    F   Calyx lobes bristle-tipped (leaf blades up to 5 cm
                        long; calyx 2-3 mm long; prostrate, with stem tips
                        rising up to 50 cm)          Chorizanthe cuspidata
                        [includes C. cuspidata vars. marginata and villosa]
                                    San Francisco Spineflower; Sn-SCr; 1b
                    FF  Calyx lobes not bristle-tipped
                        G   Spines of involucre straight; calyx 4-6 mm long
                            (leaf blades 1-5 cm long; up to 30 cm tall)
                                                     Chorizanthe valida
                                            Sonoma Spineflower; Sn, Ma; 1b
                        GG  Spines of involucre hooked or curved; calyx not
                            more than 4 mm long
                            H   Flowers solitary or in clusters of not more
                                than 5; tube of involucre 3-ribbed; leaf
                                blades up to 1 cm long; calyx 1-2 mm long;
                                prostrate         Chorizanthe polygonoides
                                            Knotweed Spineflower; Ma
                            HH  Flowers in clusters of at least 5; tube of
                                involucre 6-ribbed; leaf blades 1-5 cm long;
                                calyx 3-4 mm long; either upright, or
                                prostrate with stem tips rising
                                                     Chorizanthe robusta
                                     [includes C. pungens var. hartwegii]
                                     Robust Spineflower; SF, Al-Mo; 1b
```

CC Each involucre with several to many flowers; midrib of each lobe
   of the involucre not prolonged as a spine
   D Leaves along the stem as well as in a basal whorl (the upper
      leaves sometimes much reduced, however)
      E Plants not woody, annual (flower cluster single on each
         peduncle; calyx white to rose)
         F Leaves very slender, the petioles short or absent
                                            Eriogonum angulosum
                                     Anglestem Buckwheat; CC-s
         FF Leaves elliptic, the petioles as long as the blades
            (on clay and serpentine soils)   *Eriogonum argillosum*
                                     Clay-loving Buckwheat; SCl-Mo; 4
      EE Plants woody, at least at the base, perennial (leaves
         either sessile or with short petioles)
         F Leaves 15-30 mm long and 5-10 mm wide; flower clusters
            scattered along the branches (calyx white with green
            or rose veins)   *Eriogonum wrightii* var. *trachygonum*
                                           Wright Buckwheat
         FF Leaves 6-15 mm long and 2-6 mm wide; flower clusters
            (these usually in umbels) at the end of branches
            G Leaf blades narrow, 2-3 mm wide, sessile; calyx
               white or pinkish; plants woody throughout
                               Eriogonum fasciculatum (*Plate* 43)
                                        California Buckwheat
            GG Leaf blades elliptic, 5-6 mm wide, with petioles;
               calyx yellow, later reddish; plants woody only at
               the base
                  Eriogonum umbellatum var. *bahiiforme* (*Plate* 43)
                                     Sulfurflower Buckwheat; La-SB
   DD Leaves mainly confined to a basal whorl (but there may be a
      few on the lower portion of the stem)
      E Flowers scattered all along the length of the stems, not
         restricted to the tip of the stems or to the nodes
         F Involucre not more than 1.5 mm long (in sandy soil)
                                            Eriogonum elegans
                                     Elegant Buckwheat; SCl-SLO
         FF Involucre 2-5 mm long
            G Stems and involucres densely hairy; leaf blades
               oblong   *Eriogonum roseum*
                                           Rose Buckwheat
            GG Stems and involucres scarcely, if at all, hairy;
               leaf blades nearly circular
               H Outer and inner lobes of the calyx about the
                  same size and shape, and all lobes slightly
                  hairy near the base (calyx lobes white or
                  rose; uncommon)   *Eriogonum covilleanum*
                                     Coville Buckwheat; Al-SB
               HH Outer lobes of the calyx distinctly larger and
                  broader than the inner lobes, and none of the
                  lobes hairy (calyx lobes white, yellowish,
                  rose, or red) (Note: the following 2 species
                  are difficult to separate.)
                  I At least some branches of the inflorescence
                     in whorls of 3 or more
                                   Eriogonum vimineum (*Plate* 43)
                                     Wicker Buckwheat; Mo-n
                  II Branches of the inflorescence rarely in
                     whorls (sometimes on serpentine)
                                   Eriogonum luteolum var. *luteolum*
                                     Greene Buckwheat; SFBR-n
      EE Flowers almost entirely restricted to the tip of the stems
         and to places where branches originate (but occasionally
         also scattered along the upper portion of the stems)
         F Involucre hairy
            G Flower clusters 1.5-3 cm wide; woody at the base
               (leaves not strictly basal, the blades 1.5-5 cm
               long)   *Eriogonum latifolium* (*Plate* 43)
                                     Coast Buckwheat; SLO-n
            GG Flower clusters not more than 1.5 cm wide; not
               woody
               H Flower clusters less than 5 mm wide; annual, up
                  to 30 cm tall (leaves basal, the blades 2-5 cm
                  long)   *Eriogonum truncatum*
                     Mount Diablo Buckwheat; CC(MD); 1a(1940)

                              HH Flower clusters 10-15 mm wide; perennial, up to
                                 100 cm tall
                                 I   Leaves basal, the blades 2-4 cm long, the
                                     margins mostly flat
                                               Eriogonum nudum var. *oblongifolium*
                                                        Hairy Buckwheat; Na-n
                                 II Leaves on lower stem, the blades 1-3 cm
                                    long, the margins usually wavy
                                               Eriogonum nudum var. *decurrens*
                                            Ben Lomond Buckwheat; CC(MD)-SCr; 1b
                   FF Involucre not hairy
                      G   Flower clusters 10-15 mm wide; leaves 1-7 cm long;
                          up to 100 cm tall
                          H   Leaf blades 3-7 cm long, wavy (calyx white or
                              pink, sometimes yellowish, usually not hairy;
                              outer Coast Ranges and near the coast)
                                               Eriogonum nudum var. *auriculatum*
                                                    Curledleaf Buckwheat; Sn-Mo
                          HH Leaf blades 1-5 cm long, usually not wavy
                              I   Calyx hairy, yellow or white (inner Coast
                                  Ranges)   *Eriogonum nudum* var. *pubiflorum*
                                                    Hairyflower Buckwheat
                              II Calyx not hairy, mostly white to pink,
                                 rarely yellow (Coast Ranges; widespread)
                                     Eriogonum nudum var. *nudum* (*Plates* 43, 88)
                                                    Nakedstem Buckwheat; SFBR-n
                   GG Flower clusters up to 6 mm wide; leaves less than
                      3.5 cm long; not more than 40 cm tall (on shale
                      and serpentine)
                      H   Leaves up to 3.5 cm long, some usually present
                          at lower stem nodes; calyx rose-red
                                                        Eriogonum luteolum
                                         var. *caninum* [*E. caninum*] (*Plate* 43)
                                              Tiburon Buckwheat; Ma, Al, CC
                      HH Leaves up to 2 cm long, strictly basal; calyx
                          white to rose   *Eriogonum covilleanum*
                                                    Coville Buckwheat; Al-SB

Portulacaceae--Purslane Family

The flowers of most purslanes are distinctive in having only 2 sepals.
The number of petals varies from 3 to 16, but in *Claytonia*, *Calandrinia*,
and *Montia*, the three genera that account for more than half of our
species, there are typically 5. A pistil and usually at least 3 stamens are
present in each flower, and the fruit, dry when mature, encloses several to
many seeds.

Gardeners are familiar with *Portulaca grandiflora*, native to Brazil, an
annual grown for the vivid flowers it produces in summer. They also know
Portulaca oleracea, the Common Purslane, a slightly succulent, mat-forming
nuisance that survives being stepped on and that has a knack for colonizing
places where it is not wanted. No one, however, can resist the charm of
some of the native species, especially the richly colored calandrinias,
which grow mostly in sunny grassland habitats, and *Lewisia rediviva*. The
latter, called Bitterroot, is the State Flower of Montana, but we are
privileged to have it near the top of Mount Diablo and in some other rocky
places in the inner Coast Ranges.

A   Much of the calyx joined to the fruiting portion of the pistil (thus a
    part of the fruit develops below the level where the 2 calyx lobes
    become free) (flowers opening only in sunshine; petals widest near the
    tip, yellow; leaves widest above the middle; leaves and stems succulent,
    prostrate; widespread)   *Portulaca oleracea* (*Plate* 90)
                                                Common Purslane; eu
AA Calyx not joined to the pistil
    B   Flowers with 4-8 sepals (leaves numerous, basal, narrow, and fleshy;
        flowers solitary on stems that are rarely more than 3 cm long;
        petals 12-18, about 2 cm long, pink or white; stamens numerous; in
        rock crevices or rocky soil)   *Lewisia rediviva* (*Plate* 45)
                                                Bitterroot; SLM-n

BB Flowers with only 2 sepals
   C Sepals very unequal (one may be twice as long as the other); style of pistil not branched (flowers, with 4 petals, either in dense clusters that resemble umbels, or in elongated, slightly curved inflorescences)
      D Inflorescence a dense umbel-like cluster (or resembling the paw of a cat, hence the common name); style long and slender; petals, as they age, twisted around the style; fruit nearly round (flowering stems prostrate, reddish; petals pink or white; sepals notched, often with a greenish center, but otherwise whitish or pink)   *Calyptridium umbellatum*
                                                                 Common Pussypaws; SCM-n
      DD Inflorescence elongated, slightly curved; style short; petals, as they age, folding into a cup over the developing fruit; fruit egg-shaped to elongated
         E Fruit egg-shaped, broadest near the base, shorter than the sepals; base of sepals with a distinct notch
                                      Calyptridium quadripetalum
                                Fourpetal Pussypaws; Na, Sn; 4
         EE Fruit somewhat elongated, at least 1.5 times as long as the sepals; base of sepals without a notch
                                  Calyptridium parryi var. *hesseae*
                    Santa Cruz Mountains Pussypaws; SCl(SCM)
  CC Sepals equal or only slightly unequal; style of pistil with 3 branches
      D Stamens more than 5 (sometimes only 5 in *Calandrinia breweri*); corolla uniformly deep rose-red, nearly magenta, but occasionally white
         E Leaves rather evenly scattered on the stem; fruit about as long as the sepals; stamens 7-14 (widespread)
                                      Calandrinia ciliata
              [includes *C. ciliata* var. *menziesii*] (*Plate* 45)
                                              Redmaids
         EE Leaves mostly basal; fruit decidedly longer than the sepals (sometimes twice as long); stamens 5-7
                                      Calandrinia breweri
                          Brewer Calandrinia; Sn-s; 4
    DD Stamens 5 or fewer; corolla either uniformly white or pink, or with deep pink or rose lines on a lighter background
      E Stem leaves alternate
         F Petals 7-10 mm long; perennial (petals pink, or white with pink lines; in rocky habitats that are moist; often with horizontal reproductive stems that connect one plant to another)   *Montia parvifolia* (*Plate* 90)
                                        Springbeauty; Mo-n
        FF Petals not more than 5 mm long (and sometimes absent); annual
          G Leaf blades oval or nearly triangular, about as long as wide; inflorescence a leafy panicle; stamens 5; petals white or pale pink, 3-4 mm long (in woods)   *Montia diffusa*
                                    Diffuse Montia; Ma-n
          GG Leaf blades narrow, many times as long as wide; inflorescence a leafless raceme (or sometimes nearly an umbel); stamens 3; petals white, about 5 mm long (in moist areas)
                                Montia linearis (*Plate* 90)
                              Linearleaf Montia; CC-n
      EE Stem leaves below the inflorescence opposite (there are only 2 substantial stem leaves, and they may be joined together on one or both sides [if joined on both sides, they form a nearly circular disk])
         F Inflorescence with numerous bracts (up to 1 cm long) (petals white or pale pink, usually with darker lines)
                Claytonia sibirica [*Montia sibirica*] (*Plates* 45, 90)
                                Candyflower; SCr-n
        FF Inflorescence without bracts, or with only 1 bract just below the lowest flower
         G Stem leaves completely united, forming a nearly circular disk (petals white; blades of basal leaves variable, but typically broad, oval or nearly triangular; common)   *Claytonia perfoliata*
                      [*Montia perfoliata* var. *perfoliata*] (*Plate* 90)
                                  Miner's-lettuce

                    GG Stem leaves united only on one side
                       H   Plants floating on fresh water or lying on
                           muddy banks; petals white
                                   Montia fontana [includes *M. verna*]
                                          Water Montia; Mo-n
                       HH Plants of dry areas, often on serpentine soil;
                           petals white or pink
                          I   Inflorescence with 3-6 flowers, the
                              pedicels rarely so much as 3 mm apart on
                              the stalk; petals rarely more than 4 mm
                              long (widespread)   *Claytonia exigua*
                                      [includes *Montia spathulata* vars.
                              exigua, *spathulata*, *rosulata*, and *tenuifolia*]
                                              Common Claytonia
                          II Inflorescence with 8-15 flowers, the
                              pedicels generally at least 5 mm apart on
                              the stalk; petals 5-7 mm long
                                      Claytonia gypsophiloides
                                      [includes *Montia gypsophiloides*
                              and *M. perfoliata* var. *nubigena*] (*Plate* 45)
                                      Santa Lucia Claytonia; Me-SLO

Primulaceae--Primrose Family

   Nearly all members of the Primrose Family are herbs whose leaves are
simple and either basal, distributed along the stems, or in a single whorl.
The flowers are solitary in some genera, in racemes or umbels in others.
Both the calyx and corolla have a cuplike or tubular lower portion, but the
lobes, of which there are usually 5, are well developed. The stamens, also
generally 5, are attached to the tube of the corolla. (The number of calyx
lobes, corolla lobes, and stamens in *Trientalis latifolia*, however, varies
from 5 to 7, even on the same plant.) The pistil, which develops into a
many-seeded dry fruit, is usually free of the calyx; only in *Samolus
parviflorus* is part of it united with the calyx cup. An interesting feature
of the fruit of some genera is the way they open: by separating into 2
halves crosswise, rather than splitting apart lengthwise.
   This family includes the cultivated species of *Primula*, *Androsace*,
Soldanella, and *Cyclamen*, introduced into horticulture from the Old World.
The few species of *Primula* native to the mountains of the western states
are next-to-impossible to sustain in gardens. Some species of *Dodecatheon*--
called shootingstars--can be grown from seed, provided that they have
abundant moisture during the winter and spring. *Trientalis latifolia*,
called Starflower, is especially easy to establish, but its aggressive
propagation by underground stems may soon make it a weed. Most gardeners
who want a representative of the family in the garden perhaps have one
already. This is *Anagallis arvensis*, the Scarlet Pimpernel, whose flowers
open only when the sky is clear.

A All obvious leaves in a whorl at the top of the stem, just below the
   flowers (corolla about 1.5 cm wide, with 5-7 lobes, pink; in coniferous
   woods)   *Trientalis latifolia* (*Plate* 45)
                                          Starflower; SLO-n
AA Leaves either all basal, all scattered along the stems, or of both types
   B All or most leaves scattered along the stems (in *Samolus parviflorus*,
      there is also a cluster of basal leaves); flowers in terminal
      racemes or mostly solitary in the leaf axils
      C Leaves alternate (except perhaps on the lower part of the stem)
         D Plants with a cluster of basal leaves; flowers white, in
            terminal racemes; calyx and corolla generally with 5 lobes;
            growing along streams and in salt marshes *Samolus parviflorus*
                                          Water Pimpernel; CC, Sl
         DD Plants without a cluster of basal leaves; flowers very small,
            pink, solitary in the axils of the sessile leaves; calyx and
            corolla (if present) generally with 4 lobes; in moist ground,
            but not typically along streams or in salt marshes
                                          Centunculus minimus
                                                  Chaffweed

CC Leaves opposite (flowers solitary in the leaf axils)
  D Corolla conspicuous, usually pinkish orange, sometimes blue,
     8-10 mm wide; stems mostly close to the ground (flowers
     opening only when the sky is clear; widespread)
                                    Anagallis arvensis (*Plate* 45)
                                        Scarlet Pimpernel; eu
  DD Corolla absent, but calyx white or lavender, less than 5 mm
     wide; stems mostly upright (restricted to coastal salt
     marshes)   *Glaux maritima*
                                        Sea-milkwort; SLO-n
 BB Leaves all basal (but in *Androsace elongata* ssp. *acuta*, there is also
    a whorl of leaflike bracts just below the inflorescence); flowers in
    umbels at the end of leafless stems
  C Corolla lobes about 1 mm long, shorter than the calyx lobes,
     spreading outward; umbel with a whorl of leaflike bracts just
     below it; flowering stems up to about 6 cm tall; corolla white
     or pink; annual (in Coast Ranges, including Mt. Diablo)
                               Androsace elongata ssp. *acuta*
                                        California Androsace; 4
  CC Corolla lobes generally more than 10 mm long, much longer than
     the calyx lobes, turned back; umbel without a whorl of leaflike
     bracts just below it; flowering stems usually more than 12 cm
     tall; corolla maroon to white; perennial
   D Anthers mostly pointed at the tip; tube formed by the
      filaments of the stamens without a yellow spot at the base of
      each anther (calyx and corolla lobes 4 or 5; in shaded areas;
      widespread)   *Dodecatheon hendersonii*
               [includes *D. hendersonii* ssp. *cruciatum*] (*Plate* 45)
                                        Mosquito-bills; SB-n
   DD Anthers usually rounded or blunt at the tip (sometimes
      slightly pointed, however); tube formed by the filaments of
      the stamens with a yellow spot near the base of each anther
    E Anthers up to 3 mm long, usually dark; mostly inner Coast
       Ranges (in moist serpentine soil)
                            Dodecatheon clevelandii ssp. *patulum*
                                        Padre's Shootingstar; SF-SB
    EE Anthers 3-4 mm long, usually yellow, but sometimes dark;
       coastal   *Dodecatheon clevelandii* ssp. *sanctarum*
                                        Coastal Shootingstar; SF-s

Ranunculaceae--Buttercup Family

Plants of the Buttercup Family usually are herbaceous and have alternate
leaves. (The woody, opposite-leaved vines of the genus *Clematis* are the
only exceptions to both of these criteria.) The flowers have numerous
stamens, and there are generally several to many closely associated pistils
that develop into single-seeded fruits, or a few separate pistils that
develop into many-seeded fruits. In one of our representatives, *Actaea
rubra*, the Baneberry, the pistil becomes a fleshy fruit. When both sepals
and petals are present, there are usually 5 of each. Petals may be lacking,
however, and the sepals are sometimes extremely small. When there are no
petals, the sepals are often brightly colored. In *Aquilegia* (columbines)
all 5 petals have long, hollow spurs that project backward between the
sepals; in *Delphinium* (larkspurs), in which there are only 4 petals, the
upper 2 petals have spurs that fit into a spur on one of the sepals.

In certain members of the family, staminate and pistillate flowers are
separate. When this is the case, there may be some flowers that have
stamens as well as pistils. It should also be noted that in certain species
of Ranunculaceae found in the region, not all petals develop normally on
all flowers. The flowers are, nevertheless, essentially regular, except in
Delphinium, in which the 2 upper petals are distinctly different from the 2
lower ones.

Clearly this is a remarkably diverse assemblage. But after a little
experience, one will generally be able to connect a previously unfamiliar
plant with the Buttercup Family. The combination of numerous stamens and a
few or many separate pistils is uncommon outside this family.

Some widely cultivated garden plants belonging to the Ranunculaceae are species of *Anemone*, *Aquilegia*, *Delphinium*, *Helleborus* (including Christmas-rose, Lenten-rose), *Nigella* (Love-in-a-mist), and *Ranunculus*. Several species of *Ranunculus* brought from Europe have become established as weeds in North America, and two of them are now common in our area. It should be mentioned that nearly all members of the family are poisonous if eaten. They are thus dangerous to livestock.

A   Flowers decidedly irregular, with 4 petals, the 2 upper ones with small
   spurs that fit into a larger spur on the uppermost sepal
  B   Flowers red or orange-red (in moist, shaded areas)
                            Delphinium nudicaule (*Plate* 46)
                                     Red Larkspur; Mo-n
  BB Flowers not red or orange-red
    C   Leaves divided into numerous slender lobes (some of the leaves
       thus slightly similar to those of parsley); annual (flowers
       blue, purple, pink, or white; occasionally escaping from
       gardens)   *Consolida ambigua*
                                  European Larkspur; eu
    CC Leaves divided into a few (usually not more than 7) broad primary
       lobes, these again lobed or toothed; perennial
       D   Lower leaves not deeply divided, the separation of the
          primary lobes not deeper than half the length of the blade
          (upper petals white; plants scarcely hairy)
          E   Blades of lower leaves nearly fan-shaped, with 3 primary
             lobes, these usually toothed; sepals bright blue (in wet
             or dry places, sometimes on serpentine)
                          Delphinium uliginosum
                      Swamp Larkspur; Na-n; 4
          EE Blades of lower leaves not fan-shaped, with 5 primary
             lobes, these toothed; sepals dark blue or purplish
             (coastal)   *Delphinium bakeri*
                       Baker Larkspur; Sn, Ma; 1b
       DD Lower leaves deeply divided, the separation of the primary
        lobes usually reaching nearly or fully to the base of the
        blade
        E   Sepals primarily green, white, pink, or yellow (in
          Delphinium californicum ssp. *californicum* they are tinged
          with purple, but are otherwise whitish or greenish, and
          in *D. luteum* they may have purple tips, but are otherwise
          yellow)
          F   Leaf blades up to 15 cm wide; flowers not opening
            fully; usually at least 1 m tall, and sometimes
            reaching more than 2 m (mostly in ravines and other
            moist places)
            G   Sepals greenish or whitish, tinged with purple;
               outer Coast Ranges
                    Delphinium californicum ssp. *californicum*
                     California Larkspur; SF-Mo
            GG Sepals dirty yellow; inner Coast Ranges
                    Delphinium californicum ssp. *interius*
                Hospital Canyon Larkspur; CC-SCl; 1b
         FF Leaf blades rarely more than 6 cm wide; flowers
          opening fully; rarely so much as 1 m tall
          G   Sepals and petals yellow, the sepals sometimes with
            purple tips (on bluffs along the coast)
                       Delphinium luteum
                    Yellow Larkspur; Sn; 1b
          GG Sepals and petals whitish or pinkish
                 Delphinium hesperium ssp. *pallescens*
                Pale Western Larkspur; SCl
       EE Sepals primarily (even if pale) blue, purple, lavender, or
        violet
        F   Leaves not hairy, or at least not hairy on the upper
          surface
          G   Leaves hairy on the underside, somewhat fleshy
            (sepals blue-purple, the lateral ones 11-24 mm
            long; upper petals whitish; mostly on rocky
            hillsides and bluffs near the coast)
                        Delphinium decorum
                    Coast Larkspur; SCr-Mo
         GG Leaves not hairy on either surface, not fleshy

    H Sepals bright blue, white, or pink, the lateral
     ones 9-20 mm long; upper petals white, usually
     with blue lines; inland and Coast Ranges (spur
     curving decidedly upward; lower leaves usually
     with 5 [sometimes 3] gradually widening,
     wedge-shaped lobes that generally have 2-3
     nearly equal teeth)   *Delphinium patens*
          Spreading Larkspur; La-s
    HH Sepals light blue, the lateral ones 10-16 mm
     long; upper petals white or cream; mostly in
     saline grassland areas *Delphinium recurvatum*
          Recurved Larkspur; CC-s; 1b
   FF Leaves either hairy on both surfaces, or hairy only on
    the upper surface
    G Lateral sepals 15-20 mm long (deep purple or blue
     purple); inflorescence usually with not more than
     12 flowers (in oak woodland)
         Delphinium variegatum (*Plate* 46)
          Royal Larkspur; SLO-n
    GG Lateral sepals not more than 16 mm long, usually
     less; inflorescence often with more than 12
     flowers
     H Sepals dark or light blue; spur up to 15 mm
      long; upper petals whitish; veins on underside
      of leaves not rust-colored; usually in
      chaparral, sometimes in oak woodland
           Delphinium parryi
       [includes *D. parryi* ssp. *seditiosum*]
          Parry Larkspur; SCl-s
     HH Sepals dark blue or purple; spur up to 18 mm
      long; upper petals blue with white edges;
      veins on underside of leaves generally rust-
      colored; mostly in oak woodland
       Delphinium hesperium ssp. *hesperium*
          Western Larkspur; SCl-n
AA Flowers essentially regular, the petals (or sepals that resemble petals)
 all of the same general shape (in some species, however, one or more
 petals may be missing or not fully developed, or the petals may fall off
 early)
 B Petals absent, but the sepals may be white or colored and thus look
  like petals
  C Woody-stemmed vines (leaves deciduous, pinnately or bipinnately
   compound, opposite; sepals usually 4, white)
   D Flowers not more than 2 cm wide, in dense panicles; leaves
    with 5-7 leaflets (in moist areas; widespread)
          Clematis ligusticifolia
          Virgin's-bower
   DD Flowers up to 5 cm wide, single or in groups of 3; leaves
    with 3 leaflets   *Clematis lasiantha* (*Plate* 46)
          Pipestems
  CC Herbs
   D Stem leaves (these compound, usually with 3 leaflets) in a
    single whorl of 3 below the single flower (flower usually at
    least 1.5 cm wide, the sepals commonly white or pale blue,
    sometimes pink or purplish; plants sometimes with a deeply
    lobed or compound basal leaf; on moist, shaded slopes)
     Anemone oregana [*A. quinquefolia* var. *oregana*] (*Plate* 91)
          Wood Anemone; Mo-Sn
   DD Stem leaves numerous, compound, with several to many
    leaflets, and plants generally with several to many flowers
    E Stamens and pistils in separate flowers, both types borne
     in panicles; leaves usually more than 12 cm long (up to
     more than 40 cm long); up to more than 100 cm tall; in
     moist areas (sepals 2-5 mm long, greenish white to
     purplish; widespread) *Thalictrum fendleri* var. *polycarpum*
         [*T. polycarpum*] (*Plate* 92)
          Meadowrue
    EE Stamens and pistils present in each flower, the flowers
     solitary; leaves not more than 12 cm long; up to 25 cm
     tall; in dry, shaded areas
     F Sepals 7-10 mm long, white to pink; stamens 23-27;
      ultimate leaf lobes 2-9 mm wide *Isopyrum occidentale*
          Western Rue-anemone; Na, SCl-s

```
            FF Sepals up to 5.5 mm long, white; stamens about 10;
               ultimate leaf lobes not more than 4 mm wide
                                             Isopyrum stipitatum
                                      Rue-anemone; CL, Al-n
  BB Both petals and sepals present (but in Myosurus minimus, the petals
     fall early, and in Actaea rubra there are sometimes none)
     C  Leaves compound, divided ternately one or more times; flowers
        numerous, small, in a rather compact raceme, and usually with 4-
        10 small white petals (sometimes none); pistil single,
        developing into a shiny, fleshy, red or white fruit 5-10 mm wide
        (this fruit is poisonous) (in moist, shady habitats)
                 Actaea rubra [includes A. rubra ssp. arguta] (Plate 91)
                                             Baneberry; SLO-n
     CC Plants not conforming to the description in choice C
        D  All 5 sepals with short spurs; pistils crowded on a slender
           receptacle that may be more than 3 cm long (petals small,
           whitish, falling off early; mostly around pools that dry out
           in summer)                      Myosurus minimus (Plate 91)
                                                       Mousetail
        DD Sepals without spurs (but the petals may have spurs); pistils
           not crowded on an especially long receptacle
           E  Petals without spurs; sepals less than 1 cm long, usually
              green, sometimes whitish or yellowish; pistils several to
              many, each 1-seeded; leaves usually not entirely basal
                                          Ranunculaceae, Subkey
           EE All 5 petals with long, hollow spurs; sepals usually more
              than 1 cm long, orange-red, tinted with yellow; pistils
              usually 5, each many-seeded; leaves (compound) basal
              F  Plants sometimes slightly hairy, but not glandular-
                 hairy; basal leaves mostly bipinnately compound; spur
                 usually less than 2 cm long      Aquilegia formosa
                 [includes A. formosa var. truncata] (Plate 46)
                                                       Columbine
              FF Plants glandular-hairy; basal leaves mostly
                 tripinnately compound; spur 1.8-3 cm long (in moist
                 serpentine soil)                 Aquilegia eximia
                                     Serpentine Columbine; Me-s
Ranunculaceae, Subkey: Flowers essentially regular; sepals less than 1 cm
long, usually green, sometimes yellowish; pistils several to many, each 1-
seeded                                                    Ranunculus
A  Petals white; submerged leaves divided into slender lobes, the floating
   leaves (if present) with broad, 3-lobed blades; plants aquatic
   B  Style of pistils, in freshly opened flowers, 2-3 times as long as the
      portion that develops into a fruit; receptacles of flowers not hairy
      (floating leaves present, 3-lobed)            Ranunculus lobbii
                             Lobb Aquatic Buttercup; SCl, Al-Sn, La; 4
   BB Style of pistils, in freshly opened flowers, not longer than the
      portion that develops into a fruit; receptacles of flowers with
      short hairs
      C  Submerged leaves much shorter than the internodes and usually
         sessile; pedicels of flowers curving back after the fruits have
         begun to ripen (without floating leaves)
                      Ranunculus aquatilis var. subrigidus [R. subrigidus]
                                      Uncommon Water Buttercup; SF
      CC Submerged leaves about as long as the internodes, and usually
         with petioles; pedicels of flowers not curving back after the
         fruits have begun to ripen
         D  With floating leaves as well as submerged leaves
                         Ranunculus aquatilis var. hispidulus (Plate 92)
                                             Water Buttercup; Mo-n
         DD Without floating leaves (all leaves submerged)
                          Ranunculus aquatilis var. capillaceus
                                       Inland Water Buttercup
AA Petals yellow (sometimes becoming white as they age); without distinctly
   different types of leaves; plants terrestrial or growing in marshy
   places
   B  Leaves often toothed, but not compound or even deeply lobed (annual
      with reclining stems; flowers with only 1-3 petals; leaf blades
      mostly oval; in marshy places)              Ranunculus pusillus
                                      Low Buttercup; SCl(SCM), Na-n
   BB Leaves compound or deeply lobed
      C  Main body of the fruit with bristles, these sometimes curved or
         hooklike
```

D Flowers usually with 5 petals, these 5-8 mm long; fruit about 5 mm long, with stout, curved bristles; annual or perennial, in wet places   *Ranunculus muricatus* (*Plate* 46)
                                            Prickleseed Buttercup; eu; SM-n
DD Flowers with 1 or 2 petals, these about 1.5 mm long, or without petals; fruit about 2 mm long, covered with slender, hooked bristles; annual, in shaded habitats
                                   Ranunculus hebecarpus (*Plate* 92)
                                            Downy Buttercup
CC Main body of the fruit without bristles (the style that persists on the fruit may be hooked, however) (perennial)
   D Leaf blades narrow, at least three times as long as wide, smooth-margined (stems reclining, rooting at the nodes; petals 5 or 10, up to 6 mm long; in wet places, and often partly submerged)   *Ranunculus flammula*
                    [includes *R. flammula* var. *ovalis*] (*Plate* 91)
                                            Crowfoot; Ma-n
   DD Leaf blades often as wide or wider than long, compound or divided (at least shallowly) into lobes
      E Plants with horizontal stems that root at the nodes (stems hairy; leaflets 3; petals up to more than 1 cm long; growing along or in ditches and in wet lawns)
                                        Ranunculus repens (*Plate* 47)
                                            Creeping Buttercup; eu; Mo-n
      EE Plants upright, the stems not rooting at the nodes
         F Petals (4 or 5) rarely more than 4 mm long, not all of them developing equally (leaves deeply lobed, hairy; body of fruit [excluding beak] up to 2.5 mm long; in moist shaded areas)   *Ranunculus uncinatus*
                       [includes *R. uncinatus* var. *parviflorus*]
                                            Woodland Buttercup; Ma, Sn-n
         FF Petals at least 5 mm long, more or less equal in size
            G Petals usually more than 8 (but sometimes only 5)
               H Lower leaves hairy or not hairy, lobed or compound; petioles up to 20 cm long; petals 7-22, usually twice as long as wide; body of fruit 2-3 mm long (in Coast Ranges and coastal, widespread)   *Ranunculus californicus*
                                        [includes *R. californicus*
                           vars. *cuneatus* and *gratus*] (*Plate* 46)
                                            California Buttercup
               HH Lower leaves hairy, usually deeply lobed, but not compound; petioles up to 12 cm long; petals 5-17, usually less than twice as long as wide; body of fruit 3.5-5.5 mm long
                                               Ranunculus canus
                            [includes *R. canus* var. *laetus*]
                                            Sacramento Valley Buttercup; CC
            GG Petals 5-8 (body of fruit 2-4 mm long)
               H Lower leaves deeply lobed, but not compound, the petioles up to 11 cm long (inner Coast Ranges, widespread)   *Ranunculus occidentalis*
                          [includes *R.occidentalis* var. *eisenii*]
                                            Western Buttercup; Na-n
               HH Lower leaves usually compound, with 3-7 leaflets, the petioles up to 20 cm long
                  I Lower leaves hairy, usually with more than 3 leaflets; not confined to moist areas
                                        Ranunculus orthorhynchus
                              var. *orthorhynchus* [includes
                           R. orthorhynchus var. *platyphyllus*]
                                            Bird's-foot Buttercup; Al, Ma
                  II Lower leaves not hairy, usually with not more than 3 leaflets; in moist areas
                           Ranunculus orthorhynchus var. *bloomeri*
                                            Bloomer Buttercup; SCl-n

Resedaceae--Mignonette Family

The flowers of mignonettes are concentrated in elongated inflorescences, and the several petals, some or all of which are lobed, may not all be of the same size. There is sometimes a single pistil that produces numerous seeds, sometimes a starlike cluster of pistils that become 1-seeded fruits. The many stamens are concentrated on one side. Several species from the Old World are cultivated and may become established as escapes from gardens.

One species of the family, *Oligomeris linifolia*, is native to the Southwest, including southern California and certain of the Channel Islands.

A   Petals 5 or 6, white
  B   Petals (6) of 2 different sizes (flowers very fragrant)
                                                               Reseda odorata
                                      Garden Mignonette; me
  BB Petals (5 or 6) all about the same size   *Reseda alba*
                                     White Mignonette; me
AA Petals 4, yellow or yellowish (plants used for many centuries as a source of a yellow dye)   *Reseda luteola*
                                      Dyer's-rocket; eu

Rhamnaceae--Buckthorn Family

In our region, the Buckthorn Family is represented by shrubs and treelike plants belonging to the genera *Rhamnus* and *Ceanothus*. These have small flowers, but in some species of *Ceanothus*, the crowded inflorescences are extremely showy, and they may also be deliciously fragrant. The calyx generally has 5 lobes, sometimes 4, and unless petals are lacking, there is a corresponding number of these, as well as of stamens. The stamens originate at the edge of the cup formed by the basal part of the calyx. They are in line with the calyx lobes, but alternate with the petals. There is a single pistil. In *Rhamnus*, this is free of the calyx and becomes a fleshy fruit. In *Ceanothus*, however, it is usually joined to the calyx and develops into a dry fruit.

Of the numerous native species of *Ceanothus*, several are widely cultivated. *Ceanothus gloriosus* var. *gloriosus*, the Point Reyes Ceanothus, is perhaps the most useful of the low-growing species; *C. gloriosus* var. *exaltatus*, called Glorybush, and *C. thyrsiflorus*, the Blue-blossom that is so abundant in some chaparral-covered areas, are attractive shrubs in the 1-3 m range. These and some other species can be purchased at nurseries and botanical gardens that sell native plants.

A   Leaves alternate; secondary veins (arranged pinnately) almost as distinct as the single primary vein (midrib); flowers greenish, in clusters in the leaf axils; fruit fleshy, berrylike
  B   Leaf blades 1-4 cm long; flowers lacking petals, but with 4 sepals; fruit 5-6 mm long
    C   Leaf blades up to 1.5 cm long; teeth, if present, not especially prominent and not spine-tipped; petioles 1-4 mm long (stems sometimes thorn-tipped) *Rhamnus crocea* (*Plate* 47)
                                  Spiny Redberry; La-s
    CC Leaf blades 2-4 cm long; teeth usually prominent and spine-tipped; petioles 2-10 mm long
                Rhamnus ilicifolia [*R. crocea* ssp. *ilicifolia*]
                                Hollyleaf Redberry
  BB Leaf blades 3-10 cm long; flowers usually with 5 petals and 5 sepals; fruit 10-12 mm long
    C   Leaves not hairy, or only sparsely hairy, on the underside (widespread) *Rhamnus californica* (*Plate* 47)
                              California Coffeeberry
    CC Leaves hairy on the underside
      D   Leaves not hairy on the upper surface *Rhamnus tomentella* ssp. *tomentella* [*R. californica* ssp. *tomentella*]
                                Hoary Coffeeberry
      DD Leaves hairy on the upper surface *Rhamnus tomentella* ssp. *crassifolia* [*R. californica* ssp. *crassifolia*]
                         Thickleaf Coffeeberry; Na
AA Leaves alternate or opposite; secondary veins usually not as distinct as the 1 or 3 primary veins; flowers blue, lavender, or sometimes white, usually in clusters at the end of branches; fruit a dry capsule
  B   Leaves opposite (with only 1 primary vein, the midrib)

C   Flowers white (rarely pale blue in *Ceanothus cuneatus*)
    D   Leaves usually smooth-margined (sometimes with a few teeth),
        mostly somewhat wedge-shaped, with a slight notch at the tip
        (usually upright, but sometimes prostrate; widespread)
                Ceanothus cuneatus [includes *C. ramulosus*] (*Plate* 47)
                                              Buckbrush
   DD Leaves with 7 or more teeth, not wedge-shaped, without a
      notch
      E   Leaves somewhat folded along the midline, yellow-green on
          the upper surface, the margin inrolled and with spinelike
          teeth; on dry slopes, including those with serpentine
          (flowers with a musky odor)
                                Ceanothus jepsonii var. *albiflorus*
                                      White Muskbrush; Na
      EE Leaves usually flat, dark green on the upper surface, the
          margin not inrolled, with teeth, but these not spinelike;
          restricted to serpentine   *Ceanothus ferrisae*
                              Coyote Ceanothus; Sl; 1b
CC Flowers blue, lavender, purple, or violet
    D   Margin of leaves (these with sharp or spinelike teeth)
        inrolled
      E   Leaves somewhat folded along the midline, yellow-green on
          the upper surface (flowers with a musky odor; on dry
          slopes, including those with serpentine; up to 60 cm
          tall)   *Ceanothus jepsonii* var. *jepsonii*
                                    Muskbrush; Me-Ma
      EE Leaves usually flat, dark green on the upper surface
          F   Plants up to 2 m tall; leaves up to 2.5 cm long, with
            5-8 spinelike teeth   *Ceanothus divergens*
                          Calistoga Ceanothus; Na; 1b
         FF Plants prostrate, with stem tips rising, the stems
            sometimes rooting at the nodes; leaves up to 2 cm
            long, usually with 3-5 sharp teeth
                 Ceanothus confusus [*C. divergens* ssp. *confusus*]
                        Rincon Ridge Ceanothus; Sn; 1b
   DD Margin of leaves not inrolled (except rarely in *Ceanothus*
      prostratus and *C. sonomensis*)
      E   Plants prostrate, sprawling, or with stem tips rising,
          less than 0.5 m tall
          F   Leaves (1-3 cm long) with 3-9 teeth; at elevations of
            more than 3000' (petioles less than 3 mm long)
                              Ceanothus prostratus
                   [includes *C. prostratus* var. *occidentalis*]
                              Mahala-mat; Sn, Na
         FF Leaves usually with many more than 10 teeth; at
            elevations below 1000'
            G   Leaves 2-5 cm long; petioles up to 4 mm long;
               coastal, in sandy soil
                   Ceanothus gloriosus var. *gloriosus* (*Plate* 47)
                         Point Reyes Ceanothus; Me-Ma; 4
            GG Leaves 1-2 cm long; petioles less than 2 mm long;
               in forests   *Ceanothus gloriosus* var. *porrectus*
                       Mount Vision Ceanothus; Ma; 1b
      EE Plants up to more than 1 m tall
          F   Leaves 1.5-4 cm long, usually with many more than 12
            teeth (petioles up to 4 mm long; in chaparral and
            forests)   *Ceanothus gloriosus* var. *exaltatus*
                             Glorybrush; Me-Ma
         FF Leaves less than 2.5 cm long, with fewer than 12 teeth
            (on dry slopes)
            G   Petioles up to 4 mm long (leaves 6-18 mm long, with
               9-11 teeth, these sometimes spinelike)
                              Ceanothus masonii
                          Mason Ceanothus; Ma; 1b
            GG Petioles less than 2 mm long, sometimes absent
              (leaves with spinelike teeth)
              H   Leaves up to 1.5 cm long, mostly sessile, with
                7 teeth, 3 of these at the tip (sometimes on
                serpentine)   *Ceanothus sonomensis* (*Plate* 47)
                           Sonoma Ceanothus; Sn; 1b
             HH Leaves up to 2.5 cm long, on short petioles,
                with about 10 teeth, these evenly spaced
                          Ceanothus purpureus
                    Hollyleaf Ceanothus; Na; 4

BB Leaves alternate
  C Plants with spiny branches (spines sometimes absent in *Ceanothus spinosus*)
    D Leaves with 1 primary vein (the midrib) from the base; bark olive-green (treelike, up to 6 m tall; petioles 4-8 mm long; flowers pale blue to white)   *Ceanothus spinosus*
                                   Greenbark Ceanothus; SLO-n
    DD Leaves with 3 primary veins from the base; bark dark
      E Leaves 1-4 cm long, whitish on the underside, the teeth, if present, well spaced; petioles 2-3 mm long; flowers white or blue   *Ceanothus leucodermis*
                     Chaparral Whitethorn; Al-s
      EE Leaves 2-6 cm long, not whitish on the underside, the teeth, if present, very crowded; petioles up to 12 mm long; flowers white   *Ceanothus incanus*
                     Coast Whitethorn; SCr-n
  CC Plants without spiny branches (although the branches may be very stiff) (leaves smooth-margined or finely toothed)
    D Leaves with only 1 primary vein (the midrib) (*Ceanothus parryi* and *C. foliosus* sometimes have 3 primary veins)
      E Underside of leaves not hairy, or with hairs only on the veins
        F Leaves up to 8 cm long, deciduous; inflorescence up to 15 cm long, the flowers white or blue (rarely pink)   *Ceanothus integerrimus*
           [includes *C. integerrimus* var. *californicus*]
                                  Deerbrush
        FF Leaves less than 2 cm long, evergreen; inflorescence up to 8 cm long, the flowers blue
          G Upper surface of leaves not hairy, the margin wavy   *Ceanothus foliosus* var. *foliosus*
                     Wavyleaf Ceanothus; SCr-n
          GG Upper surface of leaves at least somewhat hairy, the margin not wavy   *Ceanothus foliosus* var. *vineatus*
                     Vine Hill Ceanothus; Sn; 1b
      EE Underside of leaves hairy (flowers blue)
        F Upper surface of leaves not hairy (inflorescence up to 15 cm long; leaves up to 4.5 cm long) *Ceanothus parryi*
                     Parry Ceanothus; Na, Sn
        FF Upper surface of leaves hairy
          G Leaves up to 2 cm long, the margin not inrolled and the upper surface not roughened and glandular; petioles less than 3 mm long; inflorescence up to 8 cm long   *Ceanothus foliosus* var. *medius*
                     La Cuesta Ceanothus; SCL
          GG Leaves up to 5 cm long, the margin inrolled and the upper surface roughened and glandular; petioles up to 6 mm long; inflorescence up to 5 cm long   *Ceanothus papillosus*
                   Wartleaf Ceanothus; SM-SLO
    DD Leaves usually with 3 primary veins (but sometimes only the midrib is prominent)
      E Underside of leaves hairy
        F Leaves with crowded glandular teeth; margin of leaves not inrolled, and the upper surface hairy; upright, sometimes treelike
          Ceanothus oliganthus var. *sorediatus* [*C. sorediatus*]
                              Jimbrush
        FF Leaves with small teeth, but these not glandular; margin of leaves inrolled, and the upper surface not hairy; prostrate or upright   *Ceanothus griseus* (*Plate* 47)
                     Carmel Ceanothus; Sn
      EE Underside of leaves not hairy, or with hairs only on the veins
        F Leaves (up to 8 cm long) either not toothed or with only a few teeth at the tip, these not glandular; deciduous (flowers white or blue [rarely pink]; inflorescence up to 15 cm long) *Ceanothus integerrimus*
           [includes *C. integerrimus* var. *californicus*]
                                  Deerbrush
        FF Leaves with fine glandular teeth along the margin; evergreen

G   Flowers white, the inflorescence up to 12 cm long;
    leaves 2.5-8 cm long, with a sweet, tobaccolike
    odor; upright, sometimes treelike
                            Ceanothus velutinus var. *hookeri*
                                      Tobacco-brush; Ma-n
GG Flowers mostly deep blue (rarely white), the
    inflorescence up to 8 cm long; leaves 1-5 cm long,
    without a sweet odor; prostrate or upright,
    sometimes treelike (common) *Ceanothus thyrsiflorus*
    [includes *C. thyrsiflorus* var. *repens*] (*Plate* 47)
                                      Blue-blossom

Rosaceae--Rose Family

The Rose Family is a large and diverse group of flowering plants,
difficult to define concisely. The arrangement of pistils and structure of
fruit are particularly variable features. Nevertheless, after you have
learned to recognize some of the common genera, such as *Rubus* (blackberries
and raspberries), *Fragaria* (strawberries), *Rosa* (roses), *Prunus* (cherries),
and *Potentilla* (cinquefoils), you will probably be able to place relatives
of these plants in the same family.

The characteristics given here apply to members of the Rose Family that
grow wild in the region. The leaves are alternate and often have stipules.
The calyx has 5 lobes (but sometimes there is an accessory outgrowth
between each 2 lobes), and there are either 5 petals or none. The stamens
usually number at least 15; like the petals, they are attached to the calyx
cup. There may be a single pistil or several to many of them. When there is
just 1, it may be fused to the calyx cup, so the mature fruit, as in an
apple or pear, shows the calyx lobes at its free end. When there are many,
each pistil may become a 1-seeded dry fruit (as in a cinquefoil), or it may
become a 1-seeded fleshy fruit attached to a conical receptacle (as in a
raspberry or blackberry). In a strawberry, the receptacle is fleshy, but
the individual fruits embedded in it are of the dry, 1-seeded type.

The Rose Family is extremely important, for it includes many trees and
shrubs that yield edible fruit (and seeds, in the case of almonds), as well
as many that are cultivated for their ornamental value.

Native species that are good subjects for wild gardens are available at
plant sales and at some nurseries. Valuable herbaceous types are various
species of *Fragaria* (strawberries) and *Potentilla* (cinquefoils). Among the
shrubs, the following are especially useful.

Amelanchier alnifolia,
  Serviceberry
Heteromeles arbutifolia, Toyon
Holodiscus discolor, Creambush
Oemleria cerasiformis, Osoberry
Physocarpus capitatus, Ninebark

Prunus ilicifolia, Hollyleaf
  Cherry
Rosa californica, California
  Wild Rose
Rosa gymnocarpa, Wood Rose
Rubus spectabilis, Salmonberry

A   Plants neither woody (even at the base) nor spiny-stemmed
    B   Flowers without petals (but the sepals may resemble petals), and with
        a single pistil
        C   Leaves mostly basal, up to 10 cm long, pinnately compound, with
            11-17 leaflets, these pinnately divided into 3-7 lobes; flowers
            numerous, in nearly globular or elongated inflorescences at the
            end of branches; up to 25 cm tall, perennial (in sandy or rocky
            areas)   *Acaena pinnatifida* var. *californica* [*A. californica*]
                                      California Acaena; Sn-s
        CC Leaves scattered along the stems, less than 1 cm long, the blades
            lobed, but on the whole fan-shaped; flowers in small clusters in
            the leaf axils; rarely more than 5 cm tall, annual (leaves pale
            green, with stipules that clasp the stem; usually in colonies)
                        Aphanes occidentalis [*Alchemilla occidentalis*] (*Plate* 48)
                                      Western Dewcup
    BB Flowers with petals, and with more than 1 pistil (leaves compound)
        C   Leaves always with 3 leaflets; fruit a strawberry, in which the
            pistils are embedded in the fleshy receptacle

D Leaflets leathery, densely hairy on the underside; on
   seacoast bluffs and backshores of sandy beaches
                                            Fragaria chiloensis
              [includes *F. chiloensis* ssp. *pacifica*] (*Plate* 48)
                                    Beach Strawberry; SLO-n
DD Leaflets not leathery, only slightly hairy on the underside;
   inland habitats, especially open woods
              Fragaria vesca [includes *F. vesca* ssp. *californica*]
                                               Wood Strawberry
CC Larger leaves sometimes with 3 leaflets, but usually with more;
   each pistil developing into a separate fruit that is dry when
   mature
   D Petals white; stamens 10, of 2 sizes, or at least 2 forms,
      these alternating; leaves pinnately compound
      E Flowers rarely with more than 30 pistils
         F Lower leaves with 7-25 leaflets, these toothed or
            shallowly lobed; flowers with 24-30 pistils (coastal)
                                            Horkelia marinensis
                               Point Reyes Horkelia; Me-SM; 1b
         FF Lower leaves with 21-41 leaflets, these palmately
            divided into several slender lobes; flowers with 9-26
            pistils (in sandy soils)   *Horkelia tenuiloba*
                                  Thinlobe Horkelia; Ma, Sn; 1b
      EE Flowers with more than 50 pistils
         F Bracts beneath the individual flowers shorter than the
            sepals (in coastal, sandy areas) (Note: the
            distinctions between the 2 subspecies of *H. cuneata*
            are not sharp.)
            G Leaves and stems with soft, long hairs, somewhat
               silky to the touch; stems and leaves not glandular
                                     Horkelia cuneata ssp. *sericea*
                                       Kellogg Horkelia; SF-s; 1b
            GG Leaves and stems sometimes hairy, but not silky to
               the touch; stems and leaves usually glandular
                                     Horkelia cuneata ssp. *cuneata*
                                      Wedgeleaf Horkelia; SF-s
         FF Bracts at least as long as the sepals (there are 5
            bracts alternating with the sepals, but attached
            slightly below them)
            G Lower leaves with 7-11 leaflets, those on both
               sides of the rachis deeply toothed, but rarely
               truly lobed (the terminal leaflet, however, is
               often lobed) (flowers with 80-220 pistils)
               Horkelia californica ssp. *frondosa* [*H. frondosa*]
                                        Leafy Horkelia; Me-Sn
            GG Lower leaves generally with 11-21 leaflets, those
               on both sides of the rachis deeply lobed, the
               lobes toothed
               H Bracts usually 3-toothed; flowers with 80-200
                  pistils; coastal
                  Horkelia californica ssp. *californica* (*Plate* 49)
                                      California Horkelia; SCr-n
               HH Bracts rarely toothed; flowers with 50-100
                  pistils; inland (in moist areas)
                  Horkelia californica ssp. *dissita* [*H. elata*]
                                       Lobed Horkelia; Al-n
   DD Petals yellow or cream-colored; stamens 10 or more, all
      alike; leaves pinnately or palmately compound
      E Lower leaves palmately compound, with 5-7 leaflets, these
         deeply toothed (petals pale yellow; leaflets up to more
         than 10 cm long, slightly hairy)
                                     Potentilla recta (*Plate* 49)
                                 Pale Cinquefoil; eua; SF, SCl
      EE Lower leaves pinnately compound (if there are only 3
         leaflets, the terminal leaflet has a long stalk)
         F Flowers borne singly (petals about 1-1.5 cm long,
            bright yellow; lower leaves green above, usually
            whitish-hairy below, and with at least 15 substantial
            leaflets, as well as some very small leaflets;
            abundant at the margin of coastal salt marshes and in
            other wet places)   *Potentilla anserina* ssp. *pacifica*
                            [*P. egedei* var. *grandis*] (*Plate* 49)
                                          Pacific Cinquefoil
         FF Flowers mostly in clusters

G   Petals cream-colored to pale yellow; stems with
glandular hairs (lower leaves mostly with 7-9
leaflets, these much longer than wide; up to 80 cm
tall; common)   *Potentilla glandulosa* (*Plate* 49)
Sticky Cinquefoil
GG Petals bright yellow; stems not glandular (usually
in wet habitats)
H   Petals less than 4 mm long; basal leaves with
3-5 leaflets, these with large teeth
Potentilla rivalis
[includes *P. rivalis* var. *millegrana*]
River Cinquefoil
HH Petals 6-10 mm long; basal leaves usually with
at least 9 leaflets, these deeply lobed
Potentilla hickmanii
Hickman Cinquefoil; Sn-Mo; 1b
AA Plants woody (trees, shrubs, and vines), sometimes spiny-stemmed
B   Some leaves compound (stems, as a rule, spiny)
C   Leaves pinnately compound, with 5 or more leaflets; fruit single,
originating mostly from the tube of the calyx, and therefore
below the sepals; shrubs
D   Sepals soon falling away from the ripening fruit (spines
numerous, straight, slender, not broadened at the base; fruit
bright red; in shaded areas)   *Rosa gymnocarpa* (*Plate* 50)
Wood Rose; Mo-n
DD Sepals persisting on the ripe fruit
E   Sepals toothed or pinnately divided (spines 7-13 mm long,
broadened at the base, often curved; flowers 3-5 cm
wide)   *Rosa eglanteria*
Sweetbrier; eu
EE Sepals not toothed or pinnately divided
F   Calyx tube covered with gland-tipped bristles;
generally less than 30 cm tall
Rosa spithamea [includes *R. spithamea* var. *sonomensis*]
Ground Rose; Me-SLO
FF Calyx tube without gland-tipped bristles (but it may
be hairy, and the pedicels may be glandular); usually
more than 50 cm tall, and often more than 1 m tall
G   Spines slender, straight; pedicels glandular-hairy;
backs of sepals not hairy   *Rosa pinetorum*
Pine Rose; Mo-n
GG Spines stout, broadened at the base, curved;
pedicels not often glandular-hairy (usually close
to streams; common)   *Rosa californica* (*Plate* 50)
California Rose
CC Most leaves either palmately compound or with only 3 leaflets;
fruit an aggregate of smaller fruits, these attached to one
another and to a conical receptacle raised above the calyx, as
in a raspberry or blackberry; shrubs or vines
D   Aggregate of ripe fruits separating freely from the
receptacle (raspberries); shrubs
E   Stems usually covered with a whitish deposit; most leaves
with 3 leaflets (occasionally 5); petals white, less than
1 cm long, shorter than the sepals, and often of unequal
length; aggregate of fruits dark purple (nearly black) or
red, with a whitish coating   *Rubus leucodermis*
Blackcap Raspberry; SCr-n
EE Stems not covered with a whitish deposit; leaves with 3
leaflets; petals purplish red, 1.5-2 cm long, longer than
the sepals; aggregate of fruits yellow, orange, pinkish
orange, or red   *Rubus spectabilis*
[includes *R. spectabilis* var. *franciscanus*] (*Plate* 49)
Salmonberry; SCl-Sn
DD Aggregate of ripe fruits not separating freely from the
receptacle (blackberries); vines
E   Leaflets deeply lobed (spines 5-6 mm long; widely
naturalized)   *Rubus laciniatus*
Cutleaf Blackberry; eu
EE Leaflets not deeply lobed
F   Slender-stemmed vines; flowers in small cymelike
clusters; spines not thickened at the base (common)
Rubus ursinus [includes *R. vitifolius*]
vars. *eastwoodianus* and *vitifolius*] (*Plate* 93)
California Blackberry

    FF Coarse-stemmed vines; flowers in large panicles;
     spines (where present) stout, thickened at the base
     (escape from cultivation; mostly in vacant lots and
     other disturbed areas)
     G Almost all stems bearing spines (widespread)
              Rubus discolor [*R. procerus*]
              Himalayan Blackberry; eu
      GG Stems without spines *Rubus ulmifolius* var. *inermis*
              Thornless Blackberry; eu
  BB Leaves not compound, though they may be deeply lobed
   C Stems spiny
    D Leaves evergreen, at least twice as long as wide, woolly on
     the underside, the margins finely toothed; fruit yellow-
     orange (escaping from gardens) *Pyracantha angustifolia*
              Firethorn; as; Ma
    DD Leaves deciduous, less than twice as long as wide, not woolly
     on the underside, toothed or with shallow lobes; fruit
     blackish (sometimes in moist areas) *Crataegus suksdorfii*
              Black Hawthorn; Ma-n
   CC Stems not spiny (*Prunus subcordata* may have spiny branchlets,
    however)
    D Leaves not palmately lobed (but they may be toothed or
     pinnately lobed, and in *Malus fusca*, many of the leaves have
     a single lobe or an especially large tooth on one or both
     sides near the base)   Rosaceae, Subkey
    DD Most leaves palmately lobed, the lobes separated for about
     one-third to one-half the distance to the midrib (petals
     white)
     E Leaf blades up to about 12 cm wide, usually with 5
      distinct lobes; petioles 2-12 cm long; inflorescence
      open, with only a few flowers (rarely more than 7);
      petals 15-25 mm long; aggregate of fruits a bright red
      raspberry (widespread)   *Rubus parviflorus*
       [includes *R. parviflorus* var. *velutinus*] (*Plate* 93)
              Thimbleberry
     EE Leaf blades not often more than 6 cm wide, usually with 3
      lobes; petioles not more than 2 cm long; inflorescence
      compact, with numerous flowers; petals about 4 mm long;
      fruits dry when mature, 1-5 (usually 3) on each
      receptacle (in moist areas; widespread)
           Physocarpus capitatus (*Plate* 49)
              Ninebark

Rosaceae, Subkey: Shrubs; stems not spiny; leaves smooth-margined, toothed,
or pinnately lobed
A Leaves narrow and rigid, up to about 1 cm long (petals white, about 1.5
 mm long; widespread)   *Adenostoma fasciculatum* (*Plate* 48)
              Chamise
AA Leaves neither narrow nor rigid, much more than 1 cm long
 B Leaves smooth-margined
  C Leaves deciduous, 5-10 cm long, not woolly on the underside;
   staminate and pistillate flowers on separate plants; fruit
   orange, sometimes becoming blue-black before falling (mostly in
   canyons)
    Oemleria cerasiformis [*Osmaronia cerasiformis*] (*Plate* 92)
              Osoberry
  CC Leaves evergreen, up to 4 cm long, woolly on the underside; all
   flowers with stamens and a pistil; fruit red (escapes from
   gardens)
   D Petals pink; leaves up to 3 cm long *Cotoneaster franchetii*
         Pinkflower Cotoneaster; as; SFBR
   DD Petals white; leaves up to 4 cm long *Cotoneaster pannosa*
         Silverleaf Cotoneaster; as; Ma
 BB Leaves with fine to coarse teeth, and sometimes also lobed
  C Teeth at margin of leaves obviously spine-tipped
           Prunus ilicifolia (*Plate* 49)
          Hollyleaf Cherry, Na-s
  CC Teeth at margin of leaves either not spine-tipped, or the spines
   very small
   D Flowers without petals; fruit (dry) with a feathery stigma
    that is usually at least 5 cm long (leaf blades 1.5-3 cm long
    and nearly as wide, with uniform small teeth around the
    margin; widespread)   *Cercocarpus betuloides* (*Plate* 48)
           Mountain-mahogany
   DD Flowers with petals; fruit (fleshy) without a feathery stigma

E Leaves somewhat leathery, evergreen (shrubs up to more
   than 4 m tall; fruit about 7 mm wide, red when mature in
   late autumn and winter; corolla white, 8 mm wide; flowers
   numerous on each peduncle (widespread)
                          Heteromeles arbutifolia (*Plate* 48)
                                                           Toyon
EE Leaves not at all leathery, deciduous
   F Leaves distinctly toothed in the upper half, but
      either only indistinctly toothed or not at all toothed
      in the lower half; petals about 4 times as long as
      wide (petals white, up to 1.5 cm long; flowers in
      short racemes; fruit about 1 cm wide, dull purple)
                          Amelanchier alnifolia (*Plate* 48)
                                                     Serviceberry
   FF Leaves distinctly and rather evenly toothed through-
      out, and usually also shallowly lobed; petals not more
      than twice as long as wide
      G Flowers (these in dense panicles up to more than 15
         cm long) less than 5 mm wide; fruit dry; most
         leaves distinctly lobed, at least in the lower
         half or two-thirds (widespread)
                          Holodiscus discolor [includes
                      H. discolor var. *franciscanus*] (*Plate* 48)
                                                        Creambush
      GG Flowers more than 5 mm wide; fruit fleshy; leaves
         not distinctly lobed (but in *Malus fusca*, there
         may be an especially large tooth or a lobe on one
         or both sides near the base)
         H Fruit like a small apple (about 1 cm wide,
            developing below the sepals (thus the sepals
            are at the tip of the fruit, instead of at the
            base), and with more than 1 small seed; leaf
            blades often with a large tooth or lobe on one
            or both sides near the base (in moist, shaded
            areas)   *Malus fusca* (*Plate* 92)
                                      Oregon Crabapple; Sn, Na-n
         HH Fruit like a cherry or a plum (but leathery,
            instead of fleshy, in *Prunus dulcis*),
            developing above the sepals, and with a single
            large seed; leaf blades without large teeth or
            lobes near the base
            I Leaves not hairy
               J Leaves 7-10 cm long; flowers (2.5-4 cm
                  wide) solitary or in pairs, sessile;
                  fruit leathery, about 3 cm long
                          Prunus dulcis [*P. amygdalus*]
                                                    Almond; as
               JJ Leaves 2-5 cm long; flowers in clusters
                  of 3-10, on pedicels 3-12 mm long;
                  fruit fleshy, 6-8 mm long (bright red
                  or purplish red)
                          Prunus emarginata (*Plate* 49)
                                                   Bitter Cherry
            II Leaves hairy (leaves with distinct
               petioles; flowers either in elongated
               racemes or in clusters of 2-4; fruit red)
               J Leaves up to about 10 cm long, usually
                  widest below the middle; flowers
                  numerous, in elongated racemes; fruit
                  about 1 cm long, distinctly longer than
                  wide; branchlets not spiny (in moist
                  areas)   *Prunus virginiana*
                                    var. *demissa* (*Plate* 93)
                                          Western Chokecherry
               JJ Leaves up to about 5 cm long, usually
                  widest near the middle; flowers in
                  clusters of 2-4; fruit 1.5-2.5 cm long,
                  and nearly as wide; branchlets
                  sometimes spiny
                          Prunus subcordata (*Plate* 93)
                                             Sierra Plum; Mo-n

Rubiaceae--Madder Family

Coffee, Cinchona (the source of quinine), and the Gardenia are perhaps the best known plants of the large and diverse Madder Family. Most of our representatives are completely herbaceous, but some are slightly woody at the base. Only one is a large shrub. This is *Cephalanthus occidentalis* var. *californicus*, the California Button-willow, which occurs from Napa County northward as well as in the Central Valley and in the foothills of the ranges that border it.

The leaves of Rubiaceae are opposite or whorled, and the flowers of our native species, with few exceptions, have a 4-lobed tubular corolla and 4 stamens. The fruit-forming part of the pistil, fused with the calyx, is below the level where the petals and stamens originate. Sepals are usually indistinct or absent. In most of our species, the fruit is very distinctly 2-lobed, the lobes eventually separating, when dry, into 1-seeded nutlets. The genus *Galium* is an easy one to remember, for its fruits, thanks to hooked bristles, cling tenaciously to socks and other clothing; the stems, furthermore, are 4-angled, at least when young.

A   Leaves opposite, in pairs or in threes; flowers (white) in dense, globular clusters; shrub, up to more than 5 m tall (in moist areas)
Cephalanthus occidentalis var. *californicus* (*Plate* 93)
California Button-willow; Na-n
AA Leaves usually in whorls of at least 4; flowers few in each inflorescence (sometimes solitary), not in globular clusters; mostly herbs (but sometimes the old stems become woody), and generally with stems less than 0.5 m long, except in vinelike or sprawling species
B   Corolla pale pink or pale blue, the tube longer than the 4 lobes, which are directed upward as much as outward (calyx with 4-6 well developed sepals; widespread in lawns and orchards)
Sherardia arvensis (*Plate* 50)
Field Madder; me
BB Corolla white, greenish, yellowish, or purplish, the 3-4 lobes much longer than the tube, and spreading outward (stems 4-angled when young; fruit dry at maturity, consisting of 2 somewhat separate, 1-seeded halves)
C   Leaves 5-8 in each whorl
D   Flowers usually at least 6 in each cluster
E   Stems not rough to the touch (whatever hairs are present are soft) (leaves 1-2.5 cm long; fruit smooth; up to 12 cm tall, perennial)   *Galium mollugo*
Hedge Bedstraw; eu; SM
EE Stems rough to the touch, owing to the presence of stiff bristles directed toward the base of the plant
F   Leaves up to 4 cm long; fruit (and fruit-forming portion of the pistil) bristly, roughened by short, sharp hairs or scalelike outgrowths; sprawling, the stems up to 90 cm long, perennial (in wet places)
Galium mexicanum var. *asperulum*
Rough Bedstraw; SCl-n
FF Leaves not more than 1.2 cm long; fruit smooth or with a granular surface, not bristly; up to 30 cm tall, annual (tip of leaves with a sharp point and with marginal bristles directed toward the tip)
Galium divaricatum
Lamarck Bedstraw; me; Sn, Na
DD Flowers usually 3 (but sometimes up to 5) in each cluster
E   Tip of leaves blunt, not ending in a bristle; leaves not more than 2 cm long; fruit smooth (angles of stem bristly; perennial, up to 10 cm long, forming mats; mostly in wet places at elevations above 1500')
Galium trifidum var. *pusillum*
Trifid Bedstraw
EE Tip of leaves ending in a bristle or at least a sharp point; leaves up to 7 cm long; fruit with bristles, these usually hooked at the tip

    F Bristles on the surface of leaf blades mainly
concentrated on the midrib; leaves at least 4 times as
long as wide; annual, sprawling or upright (common)
Galium aparine (*Plate* 50)
Goosegrass; eu?

    FF Bristles on the surface of leaf blades not
concentrated on the midrib; leaves 2 or 3 times as
long as wide; perennial, generally sprawling, the stem
tips rising   *Galium triflorum*
Sweetscent Bedstraw

CC Leaves 4 in a whorl (but sometimes there may be fewer or more
than 4)

  D Leaves tough, sharply pointed at the tip (up to 11 mm long)
(perennial, woody only at the base if at all, forming dense
mats; flowering stems up to 15 cm tall; corolla yellowish; on
dry slopes)   *Galium andrewsii* (*Plate* 50)
Phloxleaf Bedstraw; La-s

  DD Leaves not tough or sharply pointed at the tip

    E Plants not at all woody (leaves in whorls of 4-6)

      F Leaves 4-19 mm long; angles of stem bristly; fruit
smooth; perennial, up to 10 cm tall forming mats
(mostly in wet places) *Galium trifidum* var. *pusillum*
Trifid Bedstraw

      FF Leaves up to 3 mm long; angles of stem not bristly;
fruit bristly, especially near the tip; annual, not
more than 6.5 cm tall (widespread)   *Galium murale*
Tiny Bedstraw; eu; SFBR

    EE Plants woody, at least at the base

      F Angles of stem bristly (corolla yellowish green, about
1 cm wide; perennial, the stems often more than 1 m
long, clambering over and through other plants)

        G Larger leaves (1-1.5 cm long) not more than 4 times
as long as wide; Coast Ranges
Galium porrigens var. *porrigens*
[*G. nuttallii* var. *ovalifolium*] (*Plate* 50)
Climbing Bedstraw

        GG Larger leaves more than 5 times as long as wide;
inner Coast Ranges   *Galium porrigens* var. *tenue*
[*G. nuttallii* var. *tenue*]
Narrowleaf Climbing Bedstraw

      FF Angles of stem not bristly (but they may have hairs)
(leaves up to 2.5 cm long; plants tufted or climbing)

        G Surface of leaves generally not hairy except at or
near the margin (in some specimens, however, there
are hairs scattered over much of the leaf
surface); corolla purplish red or yellowish; stems
up to 25 cm long; in dry, open areas
Galium bolanderi [*G. pubens*]
Bolander Bedstraw; Sl-n

        GG Surface of leaves with short, stiff hairs; corolla
yellowish; stems up to 90 cm long; in moist,
shaded areas   *Galium californicum* (*Plate* 50)
California Bedstraw

Rutaceae--Rue Family

    Only one species of the large Rue Family is native to the region. This
is *Ptelea crenulata* (*Plate* 51), the Hoptree, found mostly in canyons of the
inner Coast Ranges and foothills of the Sierra Nevada. It is a small tree,
up to about 5 m high, with alternate, deciduous leaves which, when rubbed
or bruised, yield an aroma somewhat similar to that of the foliage and
fruit of citrus trees, which belong to the same family. The leaves
typically have 3 approximately oval leaflets about 3-5 cm long. The small
flowers, clustered in cymes, are greenish white. There are 4 or 5 sepals
and an equal number of petals and stamens. The single pistil develops into
a flattened fruit about 1.5 cm wide. Much of this, except for the central
portion that is occupied by the 2 seeds, consists of a thin, nearly
membranous wing.

Salicaceae--Willow Family

There are only 2 genera in the Willow Family, and both are represented in our area by native trees and shrubs usually found in moist areas. The leaves are alternate and deciduous, and the flowers are concentrated in staminate and pistillate catkins, which are on separate plants. There are no petals or sepals, but each flower has a little scalelike bract below it. Furthermore, in *Populus* (cottonwoods), there is an obliquely cup-shaped disk under each flower; in *Salix* (willows), 1 or 2 rather conspicuous glands are located in approximately the same place. Staminate flowers have 1 to many stamens, whereas the pistillate flowers have a single pistil with 2-4 stigmas that correspond to the number of seed-producing divisions. The numerous seeds that develop in the fruit have long hairs that facilitate dispersal by wind.

A Leaf blades less than twice as long as wide, broadest near the base; trees, sometimes more than 25 m tall, and usually with one main trunk
  B Leaves smooth-margined
        Populus balsamifera ssp. *trichocarpa* [*P. trichocarpa*] (*Plate* 94)
                                    Black Cottonwood
  BB Leaves toothed (teeth absent from the tip of the leaf, however)
                                  Populus fremontii
                                Fremont Cottonwood
AA Leaf blades generally at least twice as long as wide, not conspicuously broader near the base than near the middle; shrubs or small trees, rarely so much as 15 m tall, usually much-branched from near the base
  B Leaves with yellow glands on the teeth of the blade and uppermost portion of the petiole (leaves up to more than 15 cm long, but not often more than 2 cm wide, the upper surface shiny; shrub or tree up to 10 m tall)   *Salix lucida* ssp. *lasiandra*
    [includes *S. lasiandra* vars. *lasiandra* and *lancifolia*] (*Plate* 94)
                                Shining Willow
  BB Leaves without yellow glands (there may be other glands, however)
    C Leaves usually less than 6 mm wide (leaves 5-12 cm long, toothed or smooth-margined, hairy, about the same color on both surfaces; shrub up to 7 m tall)   *Salix exigua*
      [includes *S. hindsiana* vars. *hindsiana* and *leucodendroides*]
                              Narrowleaf Willow
    CC Leaves usually at least 6 mm wide
      D Leaves less than 1.5 cm wide (leaves sessile or with petioles 1-2 mm long, toothed or smooth-margined, hairy or not; shrub up to 4 m tall)   *Salix melanopsis*
                          Dusky Willow; Sn-n
      DD Largest leaves usually more than 1.5 cm wide
        E Leaf blades rarely so much as 3 times as long as wide, with margins often rather obviously rolled under, at least in the lower portion
          F Leaves sometimes hairy on the underside, but not silky-hairy; large shrub or substantial tree up to 10 m tall, not necessarily restricted to wet habitats   *Salix scouleriana* (*Plate* 94)
                          Scouler Willow; Mo-n
          FF Leaves silky-hairy on the underside; shrub or tree less than 7 m tall; restricted to wet habitats, such as lake margins and streambank
             Salix sitchensis [includes *S. coulteri*] (*Plate* 94)
                        Sitka Willow; SLO-n
      EE Leaf blades usually at least 4 times as long as wide, the margins rolled under only slightly or not at all
        F Mature leaves densely hairy on the underside, sparsely hairy on the upper surface, usually smooth-margined (or with only a few teeth); shrubs usually not more than 1 m tall (inflorescences appearing before the leaves; near serpentine)   *Salix breweri*
                        Brewer Willow; La-SB
        FF Mature leaves not hairy (younger leaves hairy, however), smooth-margined or with fine teeth; shrubs or trees up to more than 10 m tall

G   Leaf blades usually 4-12 cm long, smooth-margined
    or with fine teeth, slightly rolled under;
    inflorescences up to 7 cm long, appearing before
    the leaves; tree or shrub up to 10 m tall
                                          Salix lasiolepis
              [includes *S. lasiolepis* var. *bigelovii*]
                                            Arroyo Willow
GG Leaf blades usually 7-15 cm long, with fine teeth,
    not rolled under; inflorescences up to 11 cm long,
    appearing with or after the leaves; tree up to 15
    m tall   *Salix laevigata*
        [includes *S. laevigata* var. *araquipa*] (*Plate* 94)
                                             Red Willow

Saururaceae--Lizard's-tail Family

The sole representative of the family Saururaceae in California is
Anemopsis californica (*Plate* 51), called Yerba-mansa. In the past, it was
thought to be useful for treating diseases of the skin and blood. The
flowers are concentrated into a thick, short inflorescence, below which is
a ring of large, white, often red-tinged bracts. A small bract is
associated with nearly every flower. The larger leaves, with blades notched
at the base, are sometimes more than 20 cm long.

Yerba-mansa has creeping underground stems that take root and give rise
to new plants. In favorable habitats, especially somewhat wet, alkaline
soils, it forms dense stands. It is an attractive subject for cultivation
in moist areas near ponds.

Saxifragaceae--Saxifrage Family

With some exceptions, plants that belong to the Saxifrage Family are
perennials that die back each autumn. The principal leaves are concentrated
at the base of the plant, with the upper leaves much reduced or absent. In
the individual flowers, much of the calyx forms a cup, but there are 5
lobes. The fruiting portion of the pistil is sometimes free of the calyx
cup, sometimes partly fused to it. Either way, when the fruit is ripe it is
dry and splits open lengthwise. As a rule, there are 5 petals, and these
are usually attached to the calyx. Some saxifrages have 5 stamens, others
have 10. These, like the petals, are often attached to the calyx cup.

Two common exotic components of local gardens are in this family. One of
them is *Heuchera sanguinea*, Coralbells, which grows wild in New Mexico; the
other is *Bergenia crassifolia*, from Siberia. A California native sold by
florists and supermarkets for indoor cultivation is *Tolmiea menziesii*,
called Piggy-back-plant or by some other name that alludes to its charming
habit of sprouting new plants on its leaves. It is perfectly hardy in this
region and is a good subject for a shady, reasonably moist outdoor habitat.
Some other shade-loving natives that deserve to be grown are *Tiarella
trifoliata* var. *unifoliata*, Sugarscoop, and *Tellima grandiflora*,
Fringecups. The former makes a very attractive ground cover. *Heuchera
micrantha*, Smallflower Alumroot, though not as showy as Coralbells, may be
a useful addition to exposed rockeries devoted to native plants.

A   Flowers solitary at the top of the stem, with white petals 1-1.5 cm
    long, these not lobed (all leaves basal, the blades broad, with smooth
    margins; in wet meadows)
                Parnassia californica [*P. palustris* var. *californica*]
                                    Grass-of-Parnassus; SB-n
AA Flowers mostly in inflorescences, and if with white petals, these either
    less than 1 cm long or divided into lobes
    B   Petals 4 (these purplish brown, almost threadlike); stamens 3 (some
        leaves sprouting plants that may eventually take root
                                    Tolmiea menziesii (*Plate* 51)
                                      Piggy-back-plant; SFBR-n
    BB Petals 5; stamens 5 or 10
        C   Stamens 5 (usually in moist, shaded areas)

D   Leaves scattered along the flowering stems, as well as basal
    (leaf blades up to 8 cm wide, divided into toothed lobes;
    inflorescence usually more than 30 cm long; petals white, 3-4
    mm long, widest above the middle; restricted to shady, wet
    banks and cliffsides)   *Boykinia occidentalis* [*B. elata*]
                                                              Brookfoam
DD Nearly all leaves basal (only 1-2 along each flower stem)
    E   Petals yellow-green, divided into 4-7 slender lobes
        (collectively, the petals form what looks like a
        snowflake)   *Mitella ovalis*
                                                         Mitrewort; Ma
    EE Petals white or pink, not divided into slender lobes
        F   Style of pistil 2-4 mm long, decidedly extended out of
            the flower; inflorescence usually open, the flowers
            not crowded; hairs on calyx not obvious without a hand
            lens; widespread   *Heuchera micrantha*
                    [includes *H. micrantha* var. *pacifica*] (*Plate* 95)
                                            Smallflower Alumroot; SLO-n
        FF Style of pistil less than 2 mm long, not extended out
            of the flower; inflorescence usually dense, the
            flowers crowded; hairs on calyx dense and obvious
            without a hand lens; coastal   *Heuchera pilosissima*
                                               Seaside Alumroot; SLO-n
CC Stamens 10
    D   Pistil with 2 styles
        E   Petals either less than 1.5 mm wide, or broader and
            divided into several lobes
            F   Petals white, slender, not divided; flowers in a
                panicle
                    Tiarella trifoliata var. *unifoliata* [*T. unifoliata*]
                                                  Sugarscoop; SFBR-n
            FF Petals greenish-white to nearly crimson, at least 2 mm
                wide, divided into 5-7 lobes (petals sometimes falling
                early); flowers in a slender raceme
                                       Tellima grandiflora (*Plate* 51)
                                                   Fringecups; SLO-n
        EE Petals at least 2 mm wide, not divided into lobes (white,
            sometimes with darker spots)
            F   Leaf blades not rounded, longer than wide, not more
                than 3 cm wide (hairy on the upper surface, with
                prominent teeth) (filaments wider below than above;
                inflorescence with not more than 20 flowers, these
                often on one side; widespread)   *Saxifraga californica*
                                                  California Saxifrage
            FF Leaf blades rounded, sometimes wider than long, up to
                10 cm wide
                G   Two petals longer than the other 3; leaves up to 10
                    cm wide, with white veins on the upper surface and
                    red veins on the underside; new plants produced
                    from prostrate stems that root at the nodes
                                   Saxifraga stolonifera [*S. sarmentosa*]
                                           Strawberry-geranium; as; Sn
                GG Petals of more or less equal length; leaves up to 7
                    cm wide, with veins of the same color on both
                    surfaces; new plants not produced from prostrate
                    stems (filaments wider above than below)
                                                  Saxifraga mertensiana
                                                   Wood Saxifrage; Sn-n
    DD Pistil with 3 styles
        E   Basal leaves either palmately compound, or divided more
            than halfway to the base of the blade
                                    Lithophragma parviflorum (*Plate* 51)
                                                   Prairie Starflower
        EE Basal leaves not compound, divided less than halfway to
            the base of the blade (some of the stem leaves, however,
            may be divided more than halfway to the base of the
            blade)
            F   Petals divided into lobes at the tip (with 1-3
                alternate leaves on the flowering stems; widespread)
                G   Calyx cup decidedly conical at the base, with at
                    least half of the fruit-forming part of the pistil
                    located in this conical portion; seeds smooth (in
                    moist areas)   *Lithophragma affine* (*Plate* 51)
                                                   Woodland-star

GG Calyx cup squared-off at the base, or only slightly
   conical (in which case much less than half of the
   fruit-forming part of the pistil is in this
   portion of the cup); seeds spiny (widespread)
                  Lithophragma heterophyllum
                              Hill Starflower
FF Petals not divided into lobes at the tip (there may be
   some teeth on the margin, however) (seeds spiny)
   G Flowering stems with 2 opposite leaves; basal
      leaves not more than 2.5 cm wide; calyx cup
      conical at the base; the pedicel twice as long as
      the cup   *Lithophragma cymbalaria*
                  Mission Starflower; SCl-s
   GG Flowering stems with 1-3 alternate leaves; basal
      leaves up to 4 cm wide; calyx cup squared-off at
      the base, or only slightly conical, the pedicel
      not longer than the cup
      H Basal leaves with more than 7 rounded teeth;
         petals 5-12 mm long, usually without teeth at
         the base (common)   *Lithophragma heterophyllum*
                      Hill Starflower
      HH Basal leaves with 3-5 definite lobes; petals 4-
         7 mm long with fine teeth at the base
                    Lithophragma bolanderi
             Bolander Starflower; Me-SFBR

Scrophulariaceae--Snapdragon Family

The Snapdragon Family has numerous representatives here, and some are exceptionally attractive when in flower. Most of our species are herbaceous, but there are a few shrubs. The corolla is usually distinctly 2-lipped, the upper lip being divided into 2 lobes, the lower lip into 3 lobes. But this pattern is not adhered to uniformly. In *Synthyris* and *Veronica*, for instance, the corolla has 4 lobes, the uppermost one usually being the largest. In *Verbascum*, there are 5 nearly equal lobes, so the corolla appears, at first glance, to be regular. The calyx is 5-lobed in most genera, but 4-lobed in some. The arrangement of stamens is also variable. Most commonly there are 2 pairs. Those of 1 pair, however, may lack anthers, or may be absent altogether. In certain genera, there are 5 stamens, at least 4 of which have anthers. The fruiting portion of the pistil is partitioned lengthwise into 2 seed-producing halves. The fruit splits open after it has ripened and dried.

Some of the paintbrushes and owl's-clovers (*Castilleja* and *Triphysaria*) are partial parasites. They have green leaves and can therefore make part of their own food, but their roots are joined to those of various other plants, which supply them with some of the organic nutrients they require.

Many California natives are in cultivation. Some of the best perennials are *Galvezia speciosa*, from the Channel Islands, and various species of *Mimulus* and *Penstemon*. One that is especially useful in dry situations is *Mimulus aurantiacus*, the Bush Monkeyflower. For wet places, including stream banks and edges of pools, *Mimulus cardinalis*, *M. guttatus*, and *M. moschatus* are excellent subjects. Among the annuals that will provide much color is *Collinsia heterophylla*, called Chinesehouses. It will probably do best if partly shaded and not allowed to become too dry.

A Upper lip of corolla modified to form a narrow hood or beaklike
   structure that usually encloses the stamens and at least much of the
   pistil
   B Leaves opposite (the bracts in whose axils the flowers originate may
      be alternate, however)
      C Corolla yellow; inflorescence somewhat loose; calyx lobes equal
         (coastal)   *Parentucellia viscosa*
                   Yellow Parentucellia; eu; Sn
      CC Corolla pink and white; inflorescence dense; calyx lobes unequal
                    Bellardia trixago (*Plate* 52)
                    Bellardia; me; SFBR
   BB Leaves alternate or mostly basal

C   Pollen sacs of each stamen equal and attached at the same level
    to the filament; most leaves at least 10 cm long, pinnately
    lobed, the lobes toothed (corolla purplish red)
    D   Corolla about 2.5 cm long, its lower lip only about one-
        eighth as long as the upper lip; leaves shorter than the
        flowering stems (common)   *Pedicularis densiflora* (*Plate* 54)
                                                        Indian-warrior
    DD Corolla about 1.8 cm long, its lower lip about half as long
        as the upper lip; leaves often as long as the flowering
        stems, or longer   *Pedicularis dudleyi*
                                            Dudley Lousewort; SM; 1b
CC Pollen sacs of each stamen either unequal, or one of them
    attached below the other (in some species of *Triphysaria*, only
    one pollen sac of each stamen is well developed); leaves rarely
    more than 6 cm long, and if pinnately lobed, the lobes smooth-
    margined
    D   Calyx with a definite tube, and usually with 4 nearly equal
        lobes; inflorescence usually with more than 10 flowers
                            Scrophulariaceae, Subkey 1 (p. 229)
    DD Calyx a boat-shaped structure, sometimes notched at the tip,
        above the corolla (caution: below each flower there are 2
        bracts; both of these may be distinctly lobed, but the inner
        one, which faces the lower side of the corolla, often
        resembles the calyx, and could be mistaken for part of the
        calyx); inflorescence with only a few flowers (rarely so many
        as 15)
        E   Leaves fairly broad, mostly less than 8 times as long as
            wide (inflorescence 2-15 cm long; in salt marshes)
            F   None of the leaves lobed; inner bract with 2 teeth
                separated by a notch   *Cordylanthus maritimus*
                                            Saltmarsh Bird's-beak; 1b
            FF Some of the upper leaves with 1 or 2 pairs of short
                lateral lobes; inner bract divided into 3-7 lobes
                G   Plants bristly, and usually branching near the
                    base; corolla pouch and tube only slightly hairy
                    (inland saline habitats)
                        Cordylanthus mollis ssp. *hispidus* [*C. hispidus*]
                                        Hispid Bird's-beak; Sl; 1b
                GG Plants usually soft-hairy, usually branching near
                    the middle; corolla pouch and tube densely hairy
                    (coastal salt marshes)
                            Cordylanthus mollis ssp. *mollis*
                                        Soft Bird's-beak; SF; 1b
        EE Leaves slender, the lower ones mostly more than 8 times as
            long as wide (the upper leaves not so slender, and
            sometimes lobed)
            F   Plants bristly, not glandular; inflorescences usually
                with 5-15 flowers (sometimes only 3 or 4) (corolla
                yellowish with a maroon mark, the pouch white)
                                            Cordylanthus rigidus
                                        Stiff Bird's-beak; SCl-s
            FF Plants not bristly, but usually with some gland-tipped
                hairs; inflorescences with 1-4 flowers
                G   Corolla white, the inside with purple streaks;
                    outer bracts 3-lobed nearly to the base, the lobes
                    slightly broadened near the tip (on serpentine)
                                            Cordylanthus nidularius
                                    Mount Diablo Bird's-beak; CC(MD); 1b
                GG Corolla whitish, but usually with a yellow tip, and
                    some maroon markings; outer bracts either not
                    lobed or lobed for not more than two-thirds their
                    length
                    H   Corolla only slightly marked with maroon (tip
                        of outer bract either not divided or divided
                        into 3 rounded teeth) (Note: this species is
                        not very distinct from *C. tenuis*, below.)
                                            Cordylanthus pilosus
                                            Hairy Bird's-beak; SFBR
                    HH Corolla conspicuously marked with maroon (on
                        serpentine)
                        I   Outer bracts divided, for about half their
                            length, into 3 slender lobes
                                Cordylanthus tenuis ssp. *capillaris*
                                        Pennell Bird's-beak; Sn; 1b

II Outer bracts not lobed
                      Cordylanthus tenuis ssp. *brunneus*
                      Serpentine Bird's-beak; Na, Sn; 4
AA Upper lip of corolla not modified to form a hood or beaklike structure
    that encloses the stamens and at least much of the pistil (it may
    consist of 2 lobes or be undivided; in certain species, moreover, the
    corolla, with either 4 or 5 lobes, may appear to be nearly regular)
    B   Leaves alternate
        C   Flowers with 5 anther-bearing stamens (corolla yellow, with 5
            nearly equal lobes; biennial, flowering the second year)
            D   Plants woolly, owing to abundant soft, whitish hairs;
                pedicels 1-2 mm long (widespread)   *Verbascum thapsus*
                                                Common Mullein; eua
            DD Plants not woolly (but they have some glandular hairs,
                especially above); pedicels 10-15 mm long
                                    Verbascum blattaria (*Plate* 95)
                                                Moth Mullein; eua
        CC Flowers with fewer than 5 anther-bearing stamens
            D   Corolla tube with a slender spur on the lower side near the
                base
                E   Corolla blue or violet, less than 1 cm long (excluding the
                    spur); mostly in sandy habitats   *Linaria canadensis*
                            [includes *L. canadensis* var. *texana*] (*Plate* 95)
                                                Blue Toadflax; SCl-n
                EE Corolla primarily yellow, with orange inside, 1.5-2 cm
                    long (excluding the spur); along roadsides, and in other
                    disturbed habitats   *Linaria vulgaris*
                                        Butter-and-eggs; me; Sn, Ma
            DD Corolla tube with a broad sac on the lower side near the base
                E   Corolla 3-5 cm long; plants robust (widely cultivated,
                    occasionally escaping)   *Antirrhinum majus*
                                            Common Snapdragon; me
                EE Corolla rarely more than 1.5 cm long (except in
                    Antirrhinum multiflorum); plants mostly delicate, and
                    often clinging to other vegetation by slender branches or
                    pedicels
                    F   Stems twining and clinging by slender pedicels, these
                        3-9 cm long (corolla blue, about 1.5 cm long)
                                        Antirrhinum kelloggii (*Plate* 52)
                                                Lax Snapdragon; Ma-s
                    FF Stems sometimes clinging by slender branches, but not
                        twining by pedicels, which are more than 1 cm long
                        G   Plants with slender, somewhat twisted branches that
                            usually cling to other vegetation (corolla
                            lavender-violet or light purple; pedicels 1-4 mm
                            long)
                            H   Corolla 11-17 mm long, without dark veins;
                                upper calyx lobes 8-12 mm long, the others 7-9
                                mm long, longer than the fruit (coastal)
                                        Antirrhinum vexillo-calyculatum
                                            ssp. *vexillo-calyculatum*
                                                Wiry Snapdragon; Sn-SB
                            HH Corolla 8-12 mm long, with some dark veins;
                                upper calyx lobes 4-7 mm long, the others 3-5
                                mm long, shorter than the fruit
                                    Antirrhinum vexillo-calyculatum ssp. *breweri*
                                            [*A. breweri*] (*Plate* 52)
                                                Brewer Snapdragon; Sn-n
                        GG Plants supporting themselves without the help of
                            twisted branches that cling to other vegetation
                            H   Flowers solitary in the leaf axils; corolla 9-
                                11 mm long, white with violet veins; pedicels
                                1-2 mm long; sac on lower side of corolla
                                deeper than wide   *Antirrhinum cornutum*
                                                Spurred Snapdragon; Na-n
                            HH Flowers in terminal racemes; corolla 13-18 mm
                                long, pink to red; pedicels 2-10 mm long; sac
                                on lower side of corolla about as deep as wide
                                (or wider than deep)
                                I   Plants hairy; calyx lobes unequal; sac on
                                    lower side of corolla wider than deep;
                                    annual, usually less than 1 m tall (often
                                    on fire burns)   *Antirrhinum multiflorum*
                                                Manyflower Snapdragon; SCl-s

II Plants not hairy; calyx lobes equal; sac on
lower side of corolla about as wide as
deep; perennial, often more than 1 m tall
(sometimes on serpentine) *Antirrhinum virga*
Tall Snapdragon; Sn; 4

BB Leaves (at least the lower ones) opposite or mostly basal (corolla
tube without a sac or spur near the base)
C   With 5 stamens, one of which lacks an anther (in *Scrophularia
californica*, this stamen is reduced to a scale; look
carefully!); mostly shrubby (except *Scrophularia californica*)
D   Corolla about 1 cm long, the 2 upper lobes larger than the 2
lateral lobes and the lower lobe, which is turned downward
(stem square; corolla brownish-red; leaves coarsely toothed;
generally in moist areas; widespread)
Scrophularia californica (*Plate* 54)
Beeplant

DD Corolla usually at least 1.5 cm long, either with 5 nearly
equal lobes or distinctly 2-lipped, the upper lip with 2
lobes, the lower lip with 3 lobes
E   Corolla scarlet, and consisting of a rather, slender tube
and 5 nearly equal lobes   *Penstemon centranthifolius*
Scarlet-bugler; La-s

EE Corolla not scarlet (but it may be some shade of red), and
distinctly 2-lipped
F   Corolla mostly yellow (but with a brownish upper lip
and with some purple lines), less than 1.5 cm long
Keckiella lemmonii
Lemmon Penstemon; Sl-n

FF Corolla not yellow, usually at least 1.5 cm long
G   Corolla mostly whitish, tinged with pink or
lavender, and with darker lines
H   Corolla usually 2-3 cm long, tinged with violet
or blue-violet
Penstemon grinnellii var. *scrophularioides*
Grinnell Penstemon; SCl(MH)

HH Corolla usually 1.5-2 cm long, tinged with pink
I   Calyx hairy
Keckiella breviflora var. *breviflora*
Gaping Penstemon; Al-s

II Calyx not hairy
Keckiella breviflora var. *glabrisepala*
Hairless Penstemon; Me-Na

GG Corolla mostly lavender, violet, purple, rose, or
red
H   Leaves rarely more than 5 mm wide, and
generally at least 8 times as long as wide
I   Plants not obviously hairy, except perhaps
in the lower portion (widespread)
Penstemon heterophyllus
var. *heterophyllus* (*Plate* 54)
Foothill Penstemon

II Plants with short hairs almost throughout
Penstemon heterophyllus var. *purdyi*
Purdy Penstemon; SB-n

HH Larger leaves mostly at least 10 mm wide, and
usually not more than 5 times as long as wide
I   Larger leaves commonly 10 cm long and 2 cm
wide; up to about 100 cm tall (corolla
lavender to violet-purple or red-purple)
Penstemon rattanii
var. *kleei* [*P. rattanii* ssp. *kleei*]
Santa Cruz Mountains Beardtongue; SCl(SCM); 1b

II Larger leaves up to 4 cm long and less than
2 cm wide; mostly less than 50 cm tall
J   Inflorescence a raceme; corolla dark
rose-purple; leaves with more than 5
teeth
Penstemon newberryi var. *sonomensis*
Sonoma Beardtongue; Na, Sn

JJ Inflorescence a nearly flat-topped
corymb; corolla pink to red; leaves
either smooth-margined or with 3-5
teeth   *Keckiella corymbosa*
Redwood Penstemon; Mo-n

CC With 2 or 4 stamens, these with functional anthers (there is no
   sterile filament); mostly herbaceous plants, except for *Mimulus
   aurantiacus* and species of *Hebe*, which are shrubby
   D   Flowers with 4 stamens
      E   Corolla not more than 3 mm long; leaves either entirely
         basal or limited to the lower portion of the stem, the
         petioles much longer than the blades (there is no obvious
         blade in *Limosella subulata*) (less than 10 cm tall, often
         tufted or forming mats, on muddy shores)
         F   Leaves thickened, the blades not distinct from the
            petioles (may be a native; rare)   *Limosella subulata*
                                               Delta Mudwort; na?; CC
         FF Leaves flat, the blades distinct from the petioles
            G   Lobes of corolla rounded, white or violet-tinged;
               leaf blades rarely more than 2 mm wide; style of
               pistil 0.5-1 mm long   *Limosella acaulis*
                                               Southern Mudwort; Ma-s
            GG Lobes of corolla pointed, white or pale pink; leaf
               blades up to 8 mm wide; style of pistil less than
               0.5 mm long   *Limosella aquatica*
                                               Northern Mudwort; eua; SM-n
      EE Corolla more than 7 mm long (except in *Tonella tenella*);
         most leaves distributed along the stem (but there may
         also be a few to many basal leaves), the petioles usually
         not longer than the blades
         F   Middle lobe of lower lip of the corolla (when there is
            a distinct middle lobe) not forming a sac; corolla not
            curving downward
            G   Corolla about 2.5 mm long, without an obvious tube
               (corolla white, turning to violet on the lobes; in
               shaded areas)   *Tonella tenella*
                                               Smallflower Tonella; SCl-n
            GG Corolla generally more than 7 mm long, and with an
               obvious tube
               H   Sepals distinctly separate (corolla 4-6 cm
                  long, usually pink-purple, the lower side
                  paler and dark-spotted; often more than 1 m
                  tall, usually with a single main stem and a
                  terminal, many-flowered raceme; widely
                  established exotic; poisonous if eaten)
                                          Digitalis purpurea (Plate 95)
                                                        Foxglove; eu
               HH Sepals united for much of their length
                                          Scrophulariaceae, Subkey 2 (p. 231)
         FF Middle lobe of the lower lip of the corolla with a sac
            that encloses the stamens; corolla as a whole curving
            downward (annuals)
            G   Flowers, even the upper ones, mostly in pairs in
               the leaf axils, the pedicels 1.5-2 cm long
               (corolla mostly purple)
               H   Corolla 8-12 mm long   *Collinsia sparsiflora*
                                          var. *sparsiflora* (Plate 53)
                                                  Blue-eyed-Mary; CC-n
               HH Corolla 12-20 mm long
                                          Collinsia sparsiflora var. *arvensis*
                                          Largeflower Blue-eyed-Mary; Na, Sn, CC
            GG Flowers (at least the upper ones) generally crowded
               in whorls of at least 3, the pedicels less than
               1.5 cm long (up to 2 cm long, however, in
               Collinsia multicolor)
               H   Lateral lobes of lower lip of corolla hairy on
                  the inner surface (plants imparting a brownish
                  stain to the skin when handled; corolla yellow
                  to greenish white, with some spots)
                                                  Collinsia tinctoria
                                          Sticky Chinesehouses; Sn-n
               HH Lateral lobes of lower lip of corolla not hairy
                  on the inner surface
                  I   Upper lip of corolla less than half as long
                     as the lower lip; flowers in a single
                     cluster (upper lip of corolla bluish, the
                     lower whitish; coastal, usually on sandy
                     backshores)   *Collinsia corymbosa*
                                          Roundhead Chinesehouses; SF-n

II Upper lip of corolla more than half as long
   as the lower lip; flowers not in a single
   cluster
   J Pedicel longer than the calyx (lower
      flowers single or in pairs, the upper
      ones in whorls of 3 or more; upper lip
      of corolla whitish with spots, the
      lower bluish, keel purple; calyx lobes
      pointed at the tip; in moist, shaded
      areas)   *Collinsia multicolor*
                     San Francisco Collinsia; SF-Mo; 4
   JJ Pedicel equal to or shorter than the
      calyx
      K Corolla dark purple; pedicel about
         equal to the calyx; filaments of the
         upper stamens hairy only near the
         base (calyx lobes blunt at the tip;
         sometimes on serpentine)
                                  Collinsia greenei
                            Green Collinsia; Sn-n
      KK Corolla not dark purple; pedicel
         shorter than the calyx; filaments of
         upper pair of stamens hairy for much
         of their length
         L Upper lip of corolla whitish, the
            lower lip violet; calyx lobes
            pointed at the tip; filaments of
            upper pair of stamens with
            slender outgrowths that project
            upward into the spur of the
            corolla tube; common at the
            margin of woods
            Collinsia heterophylla (*Plate* 53)
                                  Chinesehouses
         LL Corolla uniformly rose to white;
            calyx lobes blunt at the tip;
            filaments of upper pair of
            stamens without outgrowths; in
            open, sandy areas
                          Collinsia bartsiifolia
                           [*C. bartsiaefolia*]
                     White Chinesehouses; La-s
DD Flowers with 2 stamens (corolla with 4 lobes, the uppermost
   lobe usually the largest)
   E Leaves tough and leathery, smooth-margined; shrubby, up to
      more than 1 m tall (naturalized on bluffs and other
      coastal locations)
      F Corolla purplish blue, about 8 mm wide; leaves up to
         10 cm long   *Hebe speciosa*
                                 Showy Hebe; au; SF
      FF Corolla lilac, about 10 mm wide; leaves about 3 cm
         long (believed to be a hybrid)   *Hebe x franciscana*
                                             Hebe; ?; SF
   EE Leaves not tough and leathery, often with toothed margins;
      herbaceous plants, not at all shrubby
      F Leaves almost entirely basal, the petioles generally
         at least 5 cm long, the blades broadly heart-shaped,
         usually more than 2 cm wide (blades coarsely toothed;
         leaves on flowering stems, if present, extremely
         reduced and alternate; corolla purplish blue; in
         moist, shaded areas   *Synthyris reniformis*
                       [includes *S. reniformis* var. *cordata*]
                                         Snowqueen; Ma-n
      FF Leaves scattered along the stems (at least the lower
         ones opposite), sessile or with short petioles, the
         blades not heart-shaped (except sometimes in *Veronica
         filiformis*) and rarely so much as 2 cm wide
         G Flowers in racemes borne in the leaf axils (there
            are typically 2 or more racemes on each main stem)
            H Leaves with short but distinct petioles
               I Leaves with scarcely noticeable teeth;
                  corolla 7-10 mm wide, violet-blue to lilac;
                  wet habitats *Veronica americana* (*Plate* 96)
                                       American Brooklime

II Leaves with coarse teeth; corolla 10-12 mm
wide, bright blue with white dots; a weed
in lawns   *Veronica chamaedrys* (*Plate* 96)
Germander Speedwell; eu; SF
HH Leaves sessile (corolla lilac to lavender-blue;
in wet places)
I Leaves slender, commonly at least 8 times
as long as wide; corolla 5-7 mm wide
Veronica scutellata (*Plate* 96)
Marsh Speedwell; Ma-n
II Leaves broad, not often so much as 4 times
as long as wide; corolla 4-5 mm wide
Veronica anagallis-aquatica (*Plate* 96)
Water Speedwell; eu
GG Flowers either solitary in the leaf axils, or in
crowded racemes at the tip of the stems
H Lower leaves narrow, several times as long as
wide; corolla white (2-2.5 mm wide) (mostly in
moist areas) *Veronica peregrina* ssp. *xalapensis*
Purslane Speedwell
HH Lower leaves broad, not so much as twice as
long as wide; corolla blue
I Corolla 2-3 mm wide; pedicels 1-2 mm long
Veronica arvensis
Common Speedwell; eua
II Corolla generally at least 7 mm wide;
pedicels much more than 2 mm long (flowers
in the axils of widely spaced leaves)
J Flowers in racemes; corolla 7-11 mm
wide; larger leaf blades more than 1 cm
long, noticeably longer than wide, not
heart-shaped   *Veronica persica*
Persian Speedwell; as
JJ Flowers solitary; corolla up to 8 mm
wide; leaf blades rarely more than 1 cm
long, about as long as wide, often
somewhat heart-shaped
Veronica filiformis
Asian Speedwell; as; SF

Scrophulariaceae, Subkey 1: Upper lip of corolla modified to form a narrow
hood or beaklike structure; leaves alternate or mostly basal; pollen sacs
of each stamen either unequal or one of them attached below the other;
calyx with a definite tube, and usually with 4 nearly equal lobes;
inflorescence usually with more than 10 flowers   *Castilleja, Triphysaria*
A Upper lip of corolla much longer than the lower lip
B Foliage white-woolly (tip of bracts usually orange-red; widespread)
Castilleja foliolosa (*Plate* 52)
Woolly Paintbrush
BB Foliage not white-woolly (but it may be hairy)
C Bracts and leaves slender, not divided (tip of bracts red; lower
lip of corolla yellow or purplish, upper lip yellowish; annual)
Castilleja minor ssp. *spiralis*
[*C. stenantha* sspp. *spiralis* and *stenantha*]
Largeflower Paintbrush; La-s
CC Most or all bracts (and often the leaves) divided into lobes
D Upper calyx lobes not separated so deeply as the lower lobes,
and the tips usually curving upward (tips of bracts red or
orange-red)
Castilleja subinclusa ssp. *franciscana* [*C. franciscana*]
Franciscan Paintbrush; Sn-SM
DD Upper calyx lobes separated about as deeply as the lower
lobes, and the tips straight
E Stems and leaves obviously glandular-hairy below the
inflorescence (tips of bracts red or yellow)
Castilleja applegatei ssp. *martinii* [*C. roseana*]
Wavyleaf Paintbrush; La-s
EE Stems and leaves not obviously glandular-hairy below the
inflorescence (in *Castilleja wightii*, however, there are
usually some gland-tipped hairs on the stem)
F Calyx lobes 2-3 mm long, blunt at the tip (tips of
bracts red to yellow; corolla 21-25 mm long, the upper
lip 13-15 mm long; coastal)
Castilleja wightii (*Plate* 53)
Seaside Paintbrush; Me-SM

        FF Calyx lobes 3-6 mm long, pointed at the tips
          G Corolla 18-20 mm long, the upper lip 9-10 mm long;
            tips of bracts usually yellowish (on serpentine)
                Castilleja affinis ssp. *neglecta* [*C. neglecta*]
                      Tiburon Indian-paintbrush; Ma; 1b
          GG Corolla 25-35 mm long, the upper lip of corolla
            more than 16 mm long; tips of bracts usually red
            or purplish (sometimes orange-red or yellow)
            H Leaves 3-9 cm long, often divided into 2-6
              lobes, these not more than 5 mm wide; hairs,
              when present, rather stiff (widespread)
                    Castilleja affinis ssp. *affinis*
                      [includes *C. inflata*] (Plate 52)
                  Common Indian-paintbrush; Ma, Sn, Na-s
            HH Leaves mostly 2-5 cm long, not divided into
              lobes; hairs, if present, not stiff
              I Stems often branching; hairs on leaves and
                bracts light gray; tips of bracts purplish
                red; up to 100 cm tall (naturalized in
                swampy areas, but now very rare)
                      Castilleja chrymactis
                    Alaska Paintbrush; na; Ma(PR)
              II Stems usually not branching; hairs on
                leaves and bracts, if present, not light
                gray; tips of bracts usually red or orange-
                red, occasionally yellow; not more than 50
                cm tall
                  Castilleja miniata [includes *C. uliginosa*]
                      Red Paintbrush; Sn
AA Upper lip of corolla scarcely if at all longer than the lower lip (but
  broader than the lower lip)
  B Stamens with only 1 pollen-producing sac, this at the tip
    C Corolla 4-6 mm long, purplish brown; plants branching from the
      base, so there is not a distinct main stem
            Triphysaria pusilla [*Orthocarpus pusillus*] (Plate 96)
                    Dwarf Owl's-clover; SLO-n
    CC Corolla 10-25 mm long, not purplish brown; plants branching well
    above the base, so there is a distinct main stem
      D Stamens protruding beyond the upper lip of corolla; corolla
        (10-14 mm long) white or cream
            Triphysaria floribunda [*Orthocarpus floribundus*]
                  San Francisco Owl's-clover; Ma-SM; 1b
      DD Stamens not protruding beyond the upper lip of corolla;
      corolla not white or cream
        E Upper lip of corolla yellowish; plants not especially
          hairy, except perhaps on the inflorescence
              Triphysaria versicolor ssp. *faucibarbata*
                  [*Orthocarpus faucibarbatus*]
                  Smooth Owl's-clover; Me-Ma
        EE Upper lip of corolla purple; plants decidedly hairy
          F Lower lip of corolla yellow, purplish at the base, 3-4
          mm deep (common)   *Triphysaria eriantha* ssp. *eriantha*
                  [*Orthocarpus erianthus*] (Plate 54)
                  Yellow Johnnytuck; Me-SLO
          FF Lower lip of corolla rose or white, 4-5 mm deep
                Triphysaria eriantha ssp. *rosea*
                  [*Orthocarpus erianthus* var. *roseus*]
                  Rose Johnnytuck; Me-SLO
  BB Stamens with 2 pollen-producing sacs, 1 at the tip, the other lower
    C Bracts entirely green; upper lip of corolla not protruding much
      beyond the lower lip (corolla yellow or white, usually with 2
      purple spots)
      D Bracts not divided into lobes; corolla tube 1-2 mm wide (up
        to 25 cm tall; in moist areas, including vernal pools)
            Castilleja campestris [*Orthocarpus campestris*]
                  Field Owl's-clover; Sn-n
      DD Bracts divided into lobes; corolla tube more than 2 mm wide
      just below the level where the upper and lower lip diverge
      (in grassland)
        E Corolla yellow (widespread)
            Castilleja rubicundula ssp. *lithospermoides*
           [*Orthocarpus lithospermoides* var. *lithospermoides*]
                  Yellow Creamsacs; SCl-n

EE Corolla white, turning pink with age
Castilleja rubicundula ssp. *rubicundula*
[*Orthocarpus lithospermoides* var. *bicolor*]
White Creamsacs; Na
CC Bracts with purple or yellow tips; upper lip of corolla
definitely protruding beyond the lower lip
  D Upper lip of corolla densely hairy, with a hooklike tip;
stems mainly purple or reddish (corolla purple, the lower lip
often white or yellow at the tip, and with purple dots)
    E Inflorescence purplish at the tip, but greenish for more
than half its length, and without a banded appearance;
bracts 10-20 mm long and 1-2 mm wide at the tip; on
grassy hills (common)   *Castilleja exserta* ssp. *exserta*
[*Orthocarpus purpurascens* var. *purpurascens*] (*Plate* 52)
Purple Owl's-clover; Me-s
    EE Inflorescence purplish for more than half its length, and
with a banded appearance; bracts about 10 mm long, 2-3 mm
wide at the tip; on coastal dunes
Castilleja exserta ssp. *latifolia*
[*Orthocarpus purpurascens* var. *latifolius*]
Banded Owl's-clover; SLO-n
  DD Upper lip of corolla only slightly hairy, nearly straight;
stems not mainly purple or reddish
    E Corolla slender, not noticeably broader at the tip; stigma
not protruding beyond the corolla (corolla whitish or
purplish, with purple dots)
Castilleja attenuata [*Orthocarpus attenuatus*] (*Plate* 52)
Valley-tassels
    EE Corolla becoming broader toward the tip; stigma protruding
beyond the corolla
      F Corolla mostly purplish (the lower lip often
yellowish); leaves usually not more than 3 mm wide,
and usually divided into lobes that are as long as the
basal portion; upright; in grassy fields (widespread)
Castilleja densiflora [*Orthocarpus densiflorus*]
Common Owl's-clover; Me-s
      FF Corolla mostly pale yellow, with some purple dots;
leaves often more than 3 mm wide, and either smooth-
margined, or with lobes that are shorter than the
basal portion; sprawling somewhat; in salt marshes
Castilleja ambigua [*Orthocarpus castillejoides*]
Johnny-nip; Mo-n

Scrophulariaceae, Subkey 2: Leaves (at least the lower ones) opposite, well
distributed along the stem; stamens 4; corolla generally more than 7 mm
long; sepals united for much of their length   *Mimulus*
A Corolla scarlet (up to 80 cm tall; along streams)
Mimulus cardinalis (*Plate* 53)
Scarlet Monkeyflower
AA Corolla not scarlet (but it may be purplish red)
  B Corolla usually dull orange; shrubby, partly woody (up to more than 1
m tall) (widespread on dry hillsides) *Mimulus aurantiacus* (*Plate* 53)
Bush Monkeyflower
  BB Corolla not orange; not at all shrubby or woody (corolla mostly
yellow, purplish red, or pink)
    C Corolla mostly yellow, often with red spots or a red blotch, but
without considerable pink or purplish red (growing in wet, often
partly shaded places)
      D Stems and leaves slimy (and also short-hairy); corolla (about
2 cm long) almost entirely yellow, without prominent red
spots (but it may have some very small brown dots) (up to 30
cm tall)   *Mimulus moschatus*
Muskflower; SCr-n
      DD Stems and leaves not slimy (except sometimes in *Mimulus
floribundus*); corolla usually with prominent red or reddish
brown spots on the lips or inside the tube
        E Calyx of mature flowers flattened from side to side,
inflated; corolla usually at least 2 cm long (up to 1 m
tall, but usually much shorter; widespread)
Mimulus guttatus [include *M. guttatus* sspp.
arvensis, *micranthus*, and *M. nasutus*] (*Plate* 53)
Common Monkeyflower
        EE Calyx of mature flowers cylindrical, not obviously
inflated; corolla rarely so much as 1.5 cm long

   F Midribs of calyx lobes raised up into ridges; leaves
    typically with distinct petioles, the blades usually
    less than twice as long as wide, toothed; up to 50 cm
    tall   *Mimulus floribundus*
             Longflowering Monkeyflower
   FF Midribs of calyx lobes not raised up into ridges;
    leaves sessile, the blades usually more than twice as
    long as wide, smooth-margined; up to 35 cm tall
                Mimulus pilosus
             Downy Monkeyflower
 CC Corolla mostly purplish red or pink, often with yellow inside
  (and in *Mimulus tricolor* and *M. angustatus*, the tube is yellow)
  D Tube of corolla yellow (corolla 30-45 mm long, the lips
   purple and spotted; not more than 12 cm tall; mostly around
   drying pools)
   E Tube of corolla twice as long as the calyx; inside of
    corolla white with a purple-spotted yellow blotch; up to
    12 cm tall   *Mimulus tricolor* (*Plate* 95)
           Tricolor Monkeyflower; Sn-n
   EE Tube of corolla 3-6 times as long as the calyx; inside of
    corolla purple with darker spots; not more than 2 cm
    tall   *Mimulus angustatus*
        Narrowleaf Monkeyflower; Me-Na
  DD Tube of corolla not yellow
   E Pedicel (1.5-2.5 cm long) much longer than the calyx;
    corolla (red-purple) usually falling away after withering
    (up to 8 cm tall)   *Mimulus androsaceus*
         Androsace Monkeyflower; SCl
   EE Pedicel usually shorter than the calyx; corolla usually
    adhering to the pistil after withering
    F Corolla 3-4.5 cm long (fruit obviously lopsided,
     thick-walled, breaking open late or not at all)
     G Lower lip of the corolla less than one-third as
      long as the upper lip; up to 6 cm tall (corolla
      purple, the throat often streaked with gold,
      sometimes closed; on serpentine) *Mimulus douglasii*
           Purple Monkeyflower; SB-n
     GG Lower lip of the corolla more than one-third as
      long as the upper lip (and sometimes fully as long
      as the upper lip); up to 30 cm tall (corolla
      mostly red-purple, but much of the throat yellow
      or gold; in moist areas) *Mimulus kelloggii*
           Kellogg Monkeyflower; Na-n
    FF Corolla not more than 2.5 cm long
     G Corolla lobes essentially equal; stigmas nearly
      equal (corolla 13-20 mm long, usually red-purple,
      except on the throat, which is commonly whitish
      with dark spots; up to 20 cm tall; in sandy
      areas)   *Mimulus layneae*
           Layne Monkeyflower; Na-n
     GG Corolla lobes obviously unequal; stigmas unequal
      H Corolla 8-10 mm long; calyx less than 10 mm
       long, usually longer than the bracts of the
       lower flowers (corolla red-purple or rose;
       fruit symmetrical, thin-walled, usually
       breaking open soon after ripening; up to 15 cm
       tall)   *Mimulus rattanii*
          Rattan Monkeyflower; Ma
      HH Corolla 15-25 mm long; calyx mostly more than
       10 mm long, shorter than the bracts of the
       lower flowers
       I Plants up to 60 cm tall, with a tobaccolike
        odor; fruit symmetrical, thin-walled,
        usually breaking open soon after ripening;
        in dry, open places (corolla red-purple
        except for white folds on the lower side of
        the throat)   *Mimulus bolanderi*
         Bolander Monkeyflower; Me-s
       II Plants up to 8 cm tall, without a
        tobaccolike odor; fruit obviously lopsided,
        thick-walled, breaking open late or not at
        all; in damp places (corolla red-purple)
             Mimulus congdonii
         Congdon Monkeyflower; Me-s

Simaroubaceae (Simarubaceae)--Quassia Family

Our only representative of this family is *Ailanthus altissima*, called Tree-of-heaven. Imported from China, it has become so successfully established in some portions of California that it may seem to be native. This tree, which grows to a height of about 20 m, has alternate, deciduous leaves. These are pinnately compound and may be more than 75 cm long. Some of the numerous individual leaflets, in fact, may reach a length of 15 cm. The small flowers, in terminal panicles, have 5 greenish yellow petals and a 5-lobed calyx. Staminate flowers, with 10 stamens, are mixed with flowers that have a pistil as well as 2 or 3 stamens. Each pistil consists of a few divisions that eventually separate, forming 1-seeded, winged structures.

Solanaceae--Nightshade Family

From one's seat at the dinner table, the Nightshade Family appears to be a very important one, for it includes the Potato, Tomato, Eggplant, and many varieties of Sweet Pepper and Chili Pepper (but not Black Pepper). It also includes Tobacco and plants of medicinal value, as well as some whose fruit or leaves are poisonous. *Petunia* and *Salpiglossis* are the best-known genera of garden plants that belong to this family. Several of the species common in our area are well established weeds.

Both the calyx and corolla are 5-lobed, and there are usually 5 stamens; these are attached to the corolla tube. The fruit, partitioned into 2 or 4 seed-producing divisions, is sometimes dry when mature, sometimes a fleshy berry.

A   Corolla with a definite and often long tube
    B   Shrub or small tree up to more than 5 m tall (leaves slightly bluish; flowers in panicles; corolla 3-4 cm long, mostly tubular, but with short lobes, greenish yellow)　　　*Nicotiana glauca* (*Plate* 54)
                    Tree-tobacco; sa

    BB Herbs not more than 2 m tall
       C   Corolla tube becoming slightly narrower toward the opening; fruit a small, fleshy berry (flowers white; perennial, the stems usually trailing or climbing, up to 1.5 m long)
                 Salpichroa originifolia [*S. rhomboidea*]
                   Lily-of-the-valley Vine; sa
      CC Corolla tube becoming broader toward the opening; fruit dry at maturity, cracking apart into 2 or 4 divisions
         D   Corolla less than 1 cm long, the tube white, the lobes pale blue or purplish and slightly unequal; fruit cracking apart into 2 divisions at maturity (leaf blades about 1 cm long; stems generally falling down)　　　*Petunia parviflora*
                     Wild Petunia
        DD Corolla more than 2 cm long, white (often tinged with bluish purple) or greenish white, the lobes equal; fruit cracking apart into 4 divisions at maturity
           E   Corolla white (often tinged with bluish purple), 6-8 cm long, the upper portion broadly funnel-shaped; fruit usually with prominent soft spines; leaf blades up to 20 cm long　　　*Datura stramonium* (*Plate* 97)
                     Jimsonweed; sa
           EE Corolla greenish white, 2.5-4 cm long, mostly tubular, but with lobes 1-1.5 cm long; fruit without spines; leaf blades rarely more than 10 cm long (calyx with 5 dark stripes)　　　*Nicotiana acuminata* var. *multiflora*
                     Manyflower Tobacco; sa
AA Corolla with an extremely short tube (so short that there may seem to be none)
    B   Plants with spines on the stem, and sometimes also on the leaves and calyx lobes
       C   Corolla yellow; fruit completely enclosed by the calyx lobes (these with numerous slender spines); annual (up to 70 cm tall)
                    Solanum rostratum
                    Buffalo-berry; na
      CC Corolla not yellow; fruit not completely enclosed by the calyx lobes; shrubs

    D   Corolla white, with a purple, star-shaped mark; leaves and
        calyx lobes with stout spines; fruit 30-40 mm wide, yellow;
        up to more than 100 cm tall   *Solanum marginatum*
                                    Whitemargin Nightshade; af; SF-Mo
    DD Corolla light purplish blue; leaves and calyx lobes without
        spines (the spines are mostly on the main stem); fruit 6-8 mm
        wide, orange; not often more than 20 cm tall
                                               Solanum lanceolatum
                                       Lanceleaf Nightshade; sa; CC
BB Plants without spines on the stem
    C   Corolla decidedly blue, violet, or purple (pale purplish in
        Solanum aviculare) (woody shrubs or vines)
        D   Corolla deeply lobed, the lobes much longer than wide (fruit
            red when ripe; some leaves 3-lobed; perennial vine climbing
            over shrubs; usually in wet places)
                                       Solanum dulcamara (*Plate* 97)
                                                   Bittersweet; eua
        DD Corolla shallowly lobed, the lobes much wider than long
            E   Leaf blades mostly more than 10 cm long, and usually
                pinnately divided into a few lobes; corolla pale purplish
                (about 3 cm wide) (shrub up to more than 2 m tall;
                coastal)   *Solanum aviculare*
                                          Australian Nightshade; au; SF-n
            EE Leaf blades rarely more than 8 cm long, sometimes with
                basal lobes, but not pinnately divided; corolla deep blue
                or violet
                F   Hairs on stem not branched or forked
                         Solanum xanti [includes *S. xanti* var. *intermedium*]
                                           Purple Nightshade; Me-s
                FF Some hairs on upper stems branched or forked
                    G   Back of corolla lobes without star-shaped hairs;
                        flowers deep blue or violet; fruit whitish with a
                        green base; upright shrub (widespread)
                                              Solanum umbelliferum
                         [includes *S. umbelliferum* var. *incanum*] (*Plate* 54)
                                             Blue Nightshade; Me-s
                    GG Back of corolla lobes with star-shaped hairs;
                        flowers lavender-violet; fruit yellow; climbing
                        shrub   *Solanum furcatum*
                                          Forked Nightshade; sa; SM-n
    CC Corolla yellow or mostly white (if white, sometimes faintly
        tinged with purple or lavender, but not decidedly blue, violet,
        or purple)
        D   Corolla yellow, with a purple center, shallowly lobed; mature
            calyx enlarged, forming a loose bladderlike covering around
            the entire fruit (annual)
                            Physalis philadelphica [*P. ixocarpa*]
                                           Tomatillo; mx; Ma, Sl-s
        DD Corolla mostly white, sometimes tinged with purple or
            lavender, deeply lobed; mature calyx, if slightly enlarged,
            not completely covering the fruit
            E   Stems and calyx very hairy; mature calyx slightly enlarged
                and partly covering the fruit; fruit yellow when ripe
                                     Solanum sarrachoides (*Plate* 54)
                                             Hairy Nightshade; sa
            EE Stems and calyx only slightly, if at all, hairy; mature
                calyx not enlarged; fruit black or dark purple when ripe
                F   Inflorescence usually branched into 2 umbel-like
                    flower clusters (corolla 10-18 mm wide; perennial)
                                                   Solanum furcatum
                                          Forked Nightshade; sa; SM-n
                FF Inflorescence usually consisting of a single umbel or
                    raceme
                    G   Corolla 3-6 mm wide; calyx lobes turned back when
                        the fruit is ripe (fruit glossy; annual or
                        perennial, up to 80 cm tall)
                                      Solanum americanum [*S. nodiflorum*]
                                         Smallflower Nightshade; sa
                    GG Corolla 10 mm wide; calyx lobes not turned back
                        H   Corolla with greenish spots below the base of
                            the lobes; anthers 2.5-4 mm long; perennial,
                            up to 2 m tall   *Solanum douglasii*
                                             Douglas Nightshade; SM-s

HH Corolla with yellow spots below the base of the
lobes; anthers less than 2.5 mm long; annual,
about 0.5 m tall   *Solanum nigrum* (*Plate* 97)
Black Nightshade; eu

Sterculiaceae--Cacao Family

Nearly all members of the Sterculiaceae--of which there are hundreds--
are found in the tropics. Most famous of them is the Cacao Tree, whose
seeds yield chocolate; it is a native of South America. Our only indigenous
species, the Flannelbush, *Fremontodendron californicum* [includes *F.
californicum* sspp. *crassifolium* and *napense*] (*Plate* 55), is interesting in
the botanical sense and a valuable shrub for gardens, for it is attractive,
easily grown, and drought-resistant.

The family as a whole is characterized by flowers that usually have a 5-
lobed calyx, 5 functional stamens whose filaments are united near the base
(as well as attached to the calyx), and either 5 petals or none. The
funtional stamens may be accompanied by filaments that lack anthers. The
fruit is dry at maturity.

In *Fremontodendron*, there are no petals, but the calyx lobes are broad,
and bright yellow when fresh. The fruit is partitioned into 5 few-seeded
divisions that separate from one another at maturity. The stems and leaves
are covered with star-shaped hairs which readily detach and are irritating
to gardeners.

Tamaricaceae--Tamarisk Family

Several species of *Tamarix*, including *T. ramosissima*, called Tamarisk,
native to Asia, are now well established in various parts of California,
generally along streams. So aggressive are they in many areas of the
southwestern United States that the competing native flora has been
eliminated.

They are shrubs or small trees with long, slender branches. The leaves,
rarely more than 3.5 mm long, are in the form of needles or scales. It is
the habit of these plants to shed, once a year, many of their young twigs,
as well as their leaves. The flowers, concentrated in slender racemes, are
usually only 1-2 mm wide. They have 4 or 5 white or pink petals and an
equal number of sepals and stamens. The single pistil matures into a small
dry fruit, and each of the several seeds within it has a tuft of long
hairs.

Thymelaeaceae--Daphne Family

The Thymelaeaceae should be appreciated by gardeners as the family that
has given them a few European species of *Daphne*. The group is best
represented, however, in the tropics. The family as a whole is
characterized by flowers in which petals are lacking or much-reduced, but
whose 4-5 calyx lobes are colored like petals. There are usually 8 or 10
stamens, joined to the tubular portion of the calyx, and a single pistil.
The pistil sometimes develops into a fleshy fruit (as it does in *Daphne*),
sometimes into a dry, nutlike fruit.

Dirca occidentalis (*Plate* 55), called Western Leatherwood, our only
native species, is included on List 1b of the California Native Plant
Society's *Inventory of Rare and Endangered Vascular Plants of California*
(1994) It is a modest but charming deciduous shrub, up to about 2 m high,
with branches that are exceptionally pliable when they are young. The
leaves, usually appearing after the flowers, are more or less oval, 3-6 cm
long, bright green on the upper surface, and paler and slightly hairy on
the underside. The flowers are sessile and usually turned downward. The
yellow calyx, with a tube that becomes gradually broader away from its

basal portion, has 4 lobes. There are generally 8 stamens. The pistil
ripens into an yellowish green fruit that is approximately egg-shaped and
about 1 cm long.

Tropaeolaceae--Nasturtium Family

The Garden Nasturtium, *Tropaeolum majus*, is a South American plant that
has become naturalized in our area. It may be an annual or perennial, and
sometimes becomes almost vinelike. The leaf petioles are inserted close to
the center of nearly circular blades. The flowers, up to more than 6 cm
wide, usually have orange petals, but the color varies from nearly white to
deep red. Two of the 5 petals are slightly different from the others. The
calyx, with 5 lobes, has a long spur. There are 8 stamens, and these are
not all the same size. The fruit-forming part of the pistil is 3-lobed; at
maturity, each division encloses a single large seed.

Urticaceae--Nettle Family

All nettles in our region are annual or perennnial herbs, mostly with
opposite leaves. The flowers are concentrated in clusters that arise from
the leaf axils. Each flower has either 4 stamens or a pistil that produces
a single seed. The calyx of staminate flowers is 4-lobed, that of
pistillate flowers is either 4-lobed or 2-lobed. Petals are absent.
Stinging hairs are present on the stems and leaves of most species, and
contact with them will probably result in persistent prickly sensations!

A   Plants with alternate leaves and without stinging hairs (leaves oval or
    somewhat elongated, up to 3 cm long; flowers in crowded clusters in the
    leaf axils; up to 50 cm tall)   *Parietaria judaica*
                                                         Pellitory; eua; SF
AA Plants with opposite leaves and with stinging hairs
    B   Flowers in nearly globular clusters (leaves rarely so much as 5 cm
        long; up to 50 cm tall, weak-stemmed; in shaded areas)
                                                     Hesperocnide tenella
                                                     Western Nettle; Na-s
    BB Flowers in racemes or panicles
        C   Leaf blades usually less than 4 cm long; inflorescence up to 2.5
            cm long, with pistillate and staminate flowers mixed in the same
            cluster; usually less than 60 cm tall, annual   *Urtica urens*
                                                         Dwarf Nettle; eu
        CC Leaf blades usually more than 4 cm long; inflorescence up to 7 cm
            long, with pistillate and staminate flowers in separate
            clusters; often more than 100 cm tall, perennial (in moist
            areas)
            D   Stems with many whitish, nonstinging hairs; most leaves at
                least twice as long as wide (widespread)
                Urtica dioica ssp. *holosericea* [*U. holosericea*] (*Plate* 55)
                                                         Hoary Nettle
            DD Stems without numerous whitish, nonstinging hairs; most
                leaves less than twice as long as wide (coastal)
                        Urtica dioica ssp. *gracilis* [*U. californica*]
                                        American Stinging Nettle; Sn-SM

Valerianaceae--Valerian Family

All of our native species of the Valerian Family are small herbaceous
annuals of the genus *Plectritis*. They usually grow crowded together, thus
forming a carpet, but with the exception of some of the more richly colored
forms of *P. congesta*, they are not likely to become popular as garden
plants. They have opposite leaves, 4-angled stems, and crowded
inflorescences. The corolla, with 5 lobes, is slightly to markedly 2-
lipped, and there is usually a saclike spur originating from the lower side
of the long tube. The calyx is scarcely evident, there being no definite
lobes. Three stamens are attached to the inside of the corolla tube. The

fruit, which develops below the place where the corolla tube originates, contains a single seed and is dry when mature.

It should be noted that not all members of the family conform to the description given above. For instance, the flowers of a common garden plant, *Centranthus ruber,* called Red Valerian, have a nearly regular corolla, although the tube has a spur. The stems, moreover, are not 4-angled. This plant often escapes and becomes so well established in some places, especially along the coast, that it may seem to be native.

A Corolla regular or very slightly 2-lipped, ranging in color from white to pale pink (spur at least as long as the corolla tube itself, stout; corolla 2-3.5 mm long; in moist, shaded areas)
        Plectritis macrocera [includes *P. macrocera* var. *grayii*] (*Plate* 55)
                                            Longhorn Plectritis
AA Corolla decidedly 2-lipped, ranging in color from pale pink to red
  B Spur about as long as the corolla tube itself, tapering gradually nearly to a point; corolla usually deep pink, with 1 or 2 darker spots at the base of the middle lobe of the lower lip
    C Corolla 5-9 mm long (widespread)   *Plectritis ciliosa* ssp. *ciliosa*
                                  Longspur Plectritis; Me-Mo
    CC Corolla 1-4 mm long   *Plectritis ciliosa* ssp. *insignis*
                                          Pink Plectritis
  BB Spur absent or less than one-third as long as the corolla tube itself, and rather bluntly rounded; corolla pale pink to deep pink, without spots
    C Corolla 4.5-9.5 mm long   *Plectritis congesta* (*Plate* 55)
                                        Seablush; SLO-n
    CC Corolla up to 3.5 mm long (spur often absent)
                                 Plectritis brachystemon
        [includes *P. congesta* var. *major* and *P. magna* var. *nitida*]
                            Pale Plectritis; Mo-n

Verbenaceae--Verbena Family

The leaves of verbenas and their relatives are opposite or whorled, and the flowers are usually in elongated inflorescences or dense heads. The stems are often 4-angled. The calyx is tubular, 5-lobed in *Verbena*, 2- or 4-lobed in *Phyla*. The corolla is 5-lobed (or apparently 4-lobed, if 2 of the lobes are not well separated), and it may be decidedly 2-lipped. There are 4 stamens, in 2 pairs, and a single pistil. The fruit breaks apart into 4 or 2 nutlets, each containing a single seed. In some genera, not represented in our flora, there are deviations from the general formula given above.

Some verbenas are attractive garden plants, although some species of *Lantana* are pests in some warm temperate and tropical areas of the world. In Europe, *Verbena officinalis*, called European Vervain, is much used for making tea, and also as an herbal remedy.

A Inflorescence up to 1 cm long, on a peduncle 1.5-9 cm long; corolla pink or white; calyx with 2-4 lobes; leaves usually toothed in the upper half (the blades up to 3 cm long); prostrate, forming tight mats (used in erosion control or in place of grass lawns)
                  Phyla nodiflora [includes *Lippia nodiflora* var. *rosea*]
                                  Lemon Verbena; sa
AA Inflorescence often much more than 1 cm long, the peduncle usually shorter than the inflorescence; corolla usually blue or purple; calyx with 5 lobes; leaves coarsely toothed or lobed; upright
  B Leaves usually 1-3 cm long (sometimes up to 6 cm long); bracts beneath individual flowers 4-8 mm long, longer than the flowers themselves   *Verbena bracteata* (*Plate* 97)
                                      Bract Verbena
  BB Leaves 4-15 cm long; bracts beneath individual flowers 3-4.5 mm long, shorter than the flowers themselves
    C Leaves 7-15 cm long, those on the upper stem usually not lobed; inflorescences usually not more than 4 cm long (but up to 12 cm in fruit), dense, often more than 8 on each flowering stem
                                Verbena bonariensis
                      Clusterflower Verbena; sa; Ma

CC Leaves 4-10 cm long, those on the upper stem usually lobed;
inflorescences generally at least 5 cm long (up to 30 cm),
interrupted below, and usually only 1-3 on each flowering stem
  D   Inflorescences 10-30 cm long when fruits are maturing; leaves
somewhat grayish due to fine hairs, the upper surface not
rough to the touch   *Verbena lasiostachys* var. *lasiostachys*
Western Verbena
  DD Inflorescences 3-10 cm long when fruits are maturing; leaves
decidedly green, the upper surface slightly rough to the
touch
Verbena lasiostachys var. *scabrida* [*V. robusta*] (*Plate* 55)
Robust Verbena; Ma-s

Violaceae--Violet Family

Violets have distinctly irregular flowers, the 2 upper petals, 2 lateral
petals, and the lower petal being different. The lower petal, furthermore,
has a saclike spur near its base. The sepals, which are separate from one
another, are also unequal. There are 5 stamens, the 2 lower ones having
winglike lobes that go down into the spur of the lower petal. The fruit is
not partitioned lengthwise into chambers, but its wall consists of 3
valves, and these split apart when the fruit is ripe.

Gardeners who grow native plants will find two species easy. These are
Viola glabella, which is deciduous, and *V. sempervirens*, which is
evergreen. Both are yellow-flowered and both do well in shaded situations
that remain at least slightly moist during the summer.

A   Petals not mostly yellow
  B   Petals 15-25 mm long (petals blue-violet or, more commonly, a
combination of yellow, white, and blue; in disturbed areas) (The
cultivated pansy is derived from this species.)   *Viola tricolor*
Wild Pansy; eu; SF
  BB Petals 7-13 mm long
    C   All 5 petals violet (sometimes pale) and similar in coloration
(widespread)   *Viola adunca* (*Plate* 56)
Western Dog Violet; Mo-n
    CC Petals either primarily cream-colored, or not all alike in their
coloration
      D   Petals mostly cream-colored, often tinged with violet; in
disturbed areas   *Viola arvensis*
Field Violet; eu
      DD Petals white with yellow at the base, the 2 lateral petals
with a dark purple spot near the base, and the backs of the 2
upper petals deep reddish violet; common in redwood forests,
but also found in other habitats   *Viola ocellata* (*Plate* 56)
Western Heart's-ease; Mo-n
AA Petals mostly yellow (but usually with dark lines, and sometimes darker
on the back)
  B   Leaves either compound or deeply divided into lobes
    C   Most leaves pinnately compound, the 3-5 leaflets again deeply
lobed   *Viola douglasii*
Douglas Violet; SLO-n
    CC Leaves palmately compound or lobed
      D   Leaves compound, the 3 leaflets again deeply lobed (in shaded
areas)   *Viola sheltonii*
Shelton Violet; SCl-n
      DD Leaves incompletely divided into several primary lobes, only
certain of these shallowly lobed again
Viola lobata ssp. *lobata*
Lobed Pine Violet; Na-n
  BB Leaves not compound or even deeply lobed
    C   All or most stem leaves located just below the flowers (there may
also be a few basal leaves) (in shaded areas)
      D   Back of 2 upper petals brownish; stipules with prominent
teeth   *Viola lobata* ssp. *integrifolia*
Pine Violet; Na-n
      DD Back of 2 upper petals not brownish; stipules without
prominent teeth   *Viola glabella* (*Plate* 56)
Stream Violet; Mo-n
    CC Leaves scattered along the stem (and sometimes also basal)

D Stems creeping, rooting at the nodes, forming mats; evergreen
   (leaves scarcely hairy, often with purple spots; in shaded,
   moist woods near the coast)   *Viola sempervirens* (*Plate* 97)
                                              Evergreen Violet; Mo-n
DD Stems usually upright or with stem tips rising, not rooting
   at the nodes; deciduous
   E Leaves without a purplish tinge; basal leaves lacking;
      petals orange-yellow (widespread)
                                        Viola pedunculata (*Plate* 56)
                                              Johnny-jump-up; Sn-s
   EE Leaves often with a purplish tinge; basal leaves 1-5;
      petals yellow   *Viola purpurea*
                                                    Mountain Violet

Viscaceae (formerly in Loranthaceae)--Mistletoe Family

All mistletoes of our region are perennial and parasitize trees and
shrubs. The leaves, which are opposite, may be well developed, leathery
structures, or they may be reduced to small scales. The flowers lack petals
and are of 2 types. In those that produce pollen, there are 1 or 2 stamens
and usually 3 or 4 calyx lobes; in those that produce fruit, the cuplike
basal portion of the calyx is fused to the fruiting portion of the pistil,
and there are generally 2 or 3 sepals. The fruit is a sticky or
mucilaginous berry containing a single seed. In the genus *Arceuthobium*, the
fruit sometimes expels the seed explosively.

A Plants greenish yellow, yellow, orange-yellow, or brownish; stems angled
   at least when young; leaves scalelike; fruit slightly flattened,
   purplish (with a whitish coating), explosive; parasitic on coniferous
   trees
   B On *Pseudotsuga menziesii*; usually flowering April-June
                                           Arceuthobium douglasii
                                          Douglas-fir Dwarf-mistletoe
   BB On conifers other than *Pseudotsuga menziesii*, including species of
      Pinus, *Abies*, and *Tsuga*; flowering August-September (Note: the
      following are so similar to each other that some taxonomists refer
      all of them to one species, *Arceuthobium campylopodum*.)

      Arceuthobium abietinum [*A. campylopodum* forma *abietinum*], Fir
         Dwarf-mistletoe; on *Abies grandis*, *A. concolor*
      Arceuthobium californicum, Sugar Pine Dwarf-mistletoe; on *Pinus
         lambertiana*
      Arceuthobium campylopodum (*Plate* 56), Western Dwarf-mistletoe;
         on *Pinus sabiniana*, *P. coulteri*
      Arceuthobium tsugense [*A. campylopodum* forma *tsugensis*], Hemlock
         Dwarf-mistletoe; on *Tsuga heterophylla*

AA Plants greenish yellow or green (usually dull); stems cylindrical;
   leaves either well developed or scalelike; fruit globular, white, straw-
   colored, pink, or reddish, not explosive; some species on coniferous
   trees, others on non-coniferous woody trees and shrubs
   B Leaves reduced to scales (fruit reddish)
      C Parasitic on *Calocedrus decurrens*
                           Phoradendron libocedri [*P. juniperinum* var. *libocedri*]
                                           Incense-cedar Mistletoe
      CC Parasitic on *Juniperus californica*   *Phoradendron juniperinum*
                                                    Juniper Mistletoe
   BB Leaves well developed
      C Leaves 5-8 cm long, greenish yellow (known only from apple trees
         and maples in a few localities in Sonoma County; believed to
         have been introduced by Luther Burbank)   *Viscum album*
                                           European Mistletoe; eua; Sn
      CC Leaves mostly less than 4 cm long, green (usually dull)
         D Clusters of pistillate flowers with 2 flowers at each node;
            fruit usually straw-colored
            E Leaves mostly 1-1.5 cm long, sessile; parasitic on
               Juniperus californica, *Cupressus sargentii*, and perhaps
               other species of *Cupressus*
                              Phoradendron densum [*P. bolleanum* var. *densum*]
                                                    Dense Mistletoe

```
EE Leaves mostly 1.5-3 cm long, with short petioles;
   parasitic on Abies concolor
      Phoradendron pauciflorum [P. bolleanum var. pauciflorum]
                                              Fir Mistletoe
DD Clusters of pistillate flowers with more than 5 flowers at
   each node; fruit white, sometimes tinged with pink
   E  Leaves densely covered with short hairs; parasitic on
      various species of Quercus (oaks), sometimes on other
      trees and shrubs, including Umbellularia californica,
      Adenostoma fasciculatum, and species of Arctostaphylos
      (widespread)           Phoradendron villosum (Plate 56)
                                              Oak Mistletoe
   EE Leaves hairy only when young, if at all; parasitic on
      Fraxinus latifolia, Platanus racemosa, Populus fremontii,
      species of Salix, and cultivated fruit trees
                              Phoradendron macrophyllum
                         [P. tomentosum ssp. macrophyllum]
                                              Bigleaf Mistletoe
```

Vitaceae--Grape Family

Vitis californica (*Plate* 56), the California Wild Grape, widespread in open canyons, resembles rampant cultivated grapes in the way it grows. The fruit, generally 6-10 mm wide and purplish when ripe, is too sour to be popular. This vine is very attractive, however, especially when its leaves turn red in the autumn. It is easily grown in sunny or partly shaded locations, and can be propagated by cuttings.

The family, which consists of woody climbers with branching tendrils, also includes the Virginia Creeper, *Parthenocissus quinquefolia*, and Japanese Ivy, *P. tricuspidata*. The flowers are in clusters whose stalks arise opposite the petioles of the leaves. In *Vitis*, the shallow calyx has a smooth margin, and there are 5 petals--these fall away early--and 5 stamens. In other genera, the calyx has 4-5 toothlike lobes, and a corresponding number of petals and stamens; the fruit, as in *Vitis*, is juicy.

Zygophyllaceae--Caltrop Family

The Zygophyllaceae is a mostly tropical family. The only species in our area is *Tribulus terrestris* (*Plate* 56), called Puncture Vine, which hugs the ground in vacant lots, along railroad tracks, and in similar situations. It was introduced from the Mediterranean region of Europe. The opposite leaves are pinnately compound, and the flowers have 5 sepals, 5 yellow petals, and 10 stamens. The fruit eventually splits apart into 5 hard nutlets, each with 2 long spines and 2 short spines. This plant is now widely established in California, especially in the southern part of the state. The fruit and component nutlets are a menace to animals, and also to humans who enjoy walking barefooted.

MONOCOTYLEDONOUS FAMILIES

Alismataceae--Water-plantain Family

Water-plantains and their relatives are mostly found growing in shallow water at the edges of ponds, lakes, and sluggish streams. The leaves are entirely basal, and in our species they are large and broad-bladed. The flowers are numerous, in whorls on branched stems that often rise up above the leaves. Each flower has 3 sepals, 3 delicate white petals, and 6 to many stamens; the fruiting part of the pistil, dry at maturity, is joined tightly to the single seed within.

A Leaf blades usually arrowhead-shaped, with pointed basal lobes; stamens
  9 to many (petals white)
    B Petals 10-20 mm long; leaf blades up to more than 25 cm long; bracts
    below whorls of flowers often somewhat boat-shaped, blunt at the
    tip, rarely so much as 1.5 cm long   *Sagittaria latifolia* (*Plate* 98)
                                        Wapato

    BB Petals 6-10 mm long; leaf blades not often more than 12 cm long;
    bracts below whorls of flowers flat, pointed at the tip, commonly
    more than 1.5 cm long (sometimes more than 2.5 cm long)
                                   Sagittaria cuneata
                                   Arrowhead; Ma
AA Leaf blades oval or somewhat elongated; stamens 6
    B Petals 8-10 mm long, yellow or white, the margin fringed; fruit with
    a sharp beak; leaf blades 3-8 cm long
                Damasonium californicum [*Machaerocarpus californicus*]
                               Fringed Water-plantain; Sn
    BB Petals 3-6 mm long, pink or white, the margin not fringed; fruit
    without a beak; leaf blades up to 20 cm long
      C Leaf blades usually not more than twice as long as wide; petals
      rounded at the tip, pink or white
               Alisma plantago-aquatica [*A. triviale*] (*Plate* 98)
                             Common Water-plantain
      CC Leaf blades 3 or 4 times as long as wide; petals pointed at the
      tip, rose   *Alisma lanceolatum*
                   Lanceleaf Water-plantain; eua; Sn, Ma

Aponogetonaceae--Cape-pondweed Family

The Cape-pondweed Family consists of a single genus of freshwater plants. The leaves have a characteristic pattern of venation: the several parallel primary veins are linked by many cross-veins. The flowers, in a simple or branched inflorescence held above the water, have 1-3 (usually 2) perianth segments that look like petals, and 6 or more stamens arranged in 2 circles around the 3-6 pistils.

Aponogeton distachyon [*A. distachyus*], Cape-pondweed, native to South Africa, is cultivated in garden pools. It has become naturalized in a few places in California. The leaf blades float, and the flowers, crowded onto a 2-branched inflorescence, have 1 or 2 elliptical, white perianth segments about 1.5 cm long, stamens with purple anthers, and 3 pistils.

Araceae--Arum Family

The calla-lilies, from Africa, and species of *Anthurium*, from tropical America, are familiar examples of the Araceae. The flowers are small and crowded onto a clublike inflorescence (the spadix). Originating below the spadix, and sometimes surrounding it almost completely, is a bract (the spathe). This is often richly colored. All flowers may have a few stamens and a pistil, or the staminate flowers may be concentrated in the upper part of the spadix, the pistillate ones in the lower portion. Small scalelike structures, if present, are the only perianth parts. In many members of this large family, the inflorescences produce unpleasant odors.

These attract insects that normally lay their eggs on dung or dead animals.
The insects thus inadvertently pollinate the flowers.

A   Bract cream or white; leaf blades 15-45 cm long, arrowhead-shaped
                                    Zantedeschia aethiopica
                                        Calla-lily; af
AA Bract yellow; leaf blades 30-150 cm long, not arrowhead-shaped (in
    swampy habitats near the coast)   *Lysichiton americanum* (Plate 57)
                                    Yellow Skunk-cabbage; SM(SCM)-n

Commelinaceae--Spiderwort Family

Tradescantia fluminensis, called Spiderwort, has become established in
some damp habitats within our region. It is a native of South America, much
cultivated in gardens and as a pot plant. Its creeping stems, rooting
freely at the nodes, bear oval leaves about 4 cm long. The leaves are often
tinged with violet on the underside, and their bases form sheaths around
the stem. The flowers, about 1 cm wide, are in clusters, each of which has
a pair of elongated bracts beneath it. The 3 petals are white, the 3 sepals
green. There are 6 stamens and one pistil, which matures as a dry capsule
containing several seeds.

Cyperaceae--Sedge Family

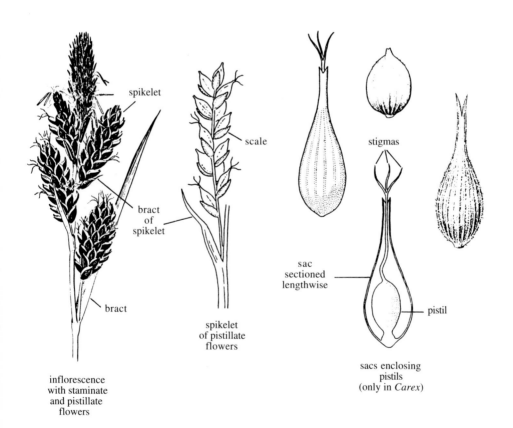

spikelet

scale

stigmas

bract
of
spikelet

sac
sectioned
lengthwise

pistil

spikelet
of pistillate
flowers

bract

sacs enclosing
pistils
(only in *Carex*)

inflorescence
with staminate
and pistillate
flowers

Most sedges grow in aquatic habitats, or at least in places that are
wet. They are somewhat grasslike, but the stems are usually solid and 3-
angled. The leaves are slender, but the basal portion of each one generally
has a substantial sheath that clasps the stem; the edges of the sheath, as

is the case in certain grasses, are fused. In the inflorescence, which may
have one or more bracts below it, the flowers, called florets, are
clustered in spikelets. A spikelet may also have a bract, and still another
kind of bract, termed a scale, is associated with each floret. The scales
persist to the fruiting stage, and they sometimes have an awn at the tip;
furthermore, the lowest scale of a spikelet may be enlarged. A floret
ordinarily has 3 stamens (rarely 2) and a pistil with 2-4 feathery stigmas.
There are sometimes bristles that are believed to represent perianth parts.
In one of our genera, *Carex*, each pistillate floret and the fruit that
develops from it lies within a sac.

A Each floret or fruit not enclosed within a sac; all florets with a
   pistil and stamens
   B Stems triangular; spikelets flattened, the florets and scales in 2
      rows (the 2 rows may not be obvious in *Kyllinga brevifolia*, since
      the inflorescence is globular)
      C Inflorescence consisting of a globular head (4-8 mm long) of
         sessile spikelets; spikelets 3 mm long, with a single fertile
         floret (bracts 1-4 cm long)   *Kyllinga brevifolia*
                                                    Kyllinga; sa; SF
      CC Inflorescence either consisting of several clusters of spikelets
         on pedicels (or, if the inflorescence is a single globular head
         of sessile spikelets, it is usually at least 10 mm long);
         spikelets 3-10 mm long, each with several florets
         D Stalk of inflorescence with ridges (separate the spikelets to
           see the stalk)
           E Bracts (2 or 3) below the inflorescence 2-4 cm long;
              leaves 2 or 3 on each stem; not more than 20 cm tall
              (spikelets 4-10 mm long; annual)
                              Cyperus squarrosus [*C. aristatus*]
                                         Awned Cyperus
           EE Bracts at least 8 cm long; leaves at least 5 on each stem;
              up to more than 50 cm tall
              F Bracts 2-6, each 8-12 cm long; spikelets 6-30 mm long;
                up to 50 cm tall, perennial   *Cyperus esculentus*
                                       Yellow Nutgrass
              FF Bracts 4-10, each about 20 cm long; spikelets 3-10 mm
                long; up to 100 cm tall, annual *Cyperus erythrorhizos*
                                         Redroot Cyperus
         DD Stalk of inflorescence without ridges
           E Bracts below the inflorescence more than 4, each often
              over 20 cm long; leaves (sometimes lacking blades) more
              than 5 on each stem; up to 90 cm tall (perennial)
              F Spikelets 10-20 mm long; leaf blades present
                            Cyperus eragrostis (*Plate* 57)
                                         Tall Cyperus
              FF Spikelets 5-10 mm long; leaf blades absent
                              Cyperus involucratus
                                African Cyperus; af
         EE Bracts 2-4, each not more than 18 cm long; leaves 2-4 on
           each stem; up to 50 cm tall (spikelets 4-10 mm long)
            F Leaves 2 on each stem, shorter than the stem; stigmas
              2; perennial
                    Cyperus niger [includes *C. niger* var. *capitatus*]
                                  Brown Umbrella Sedge
            FF Leaves 2-4 on each stem, as long as the stem; stigmas
              3; annual
              G Scales with pointed tips that curve outward
                                Cyperus acuminatus
                                Shortpoint Cyperus
              GG Scales with rounded, straight tips
                            Cyperus difformis
                      Flat Sedge; eu; Sn, Na, Ma, SF
   BB Stems cylindrical or triangular; spikelets not flattened, the florets
     and scales in several rows and arranged spirally
     C Plants not more than 4 cm tall, annual; florets without bristles
       (bracts below the inflorescence 1-3, these not more than 1 cm
       long; scales with elongated, narrow tips)
                  Lipocarpha occidentalis [*Hemicarpha occidentalis*]
                                  Western Lipocarpha

CC Plants at least 7 cm tall (usually much taller); florets with
   bristles (lift the scales to see the bristles)
   D Florets with numerous white bristles, these extending far
      beyond the scales (bract below the inflorescence single, 1-2
      cm long; up to 60 cm tall; mostly in sphagnum bogs)
                                       Eriophorum gracile
                         Slender Cottongrass; SF, Sn-n
   DD Florets with not more than 10 bristles, these not extending
      far, if at all, beyond the scales
      E Spikelets solitary at the tip of stems; inflorescence
         without a bract below it; scales without awns (stems
         cylindrical; leaves reduced to sheaths; plants forming
         clumps)
         F Spikelets usually more than 7 mm long (up to 25 mm),
            with at least 10 florets (sometimes many more)
            G Leaf sheaths less than 2 cm long; annual (stigmas
               2; up to 50 cm tall; in marshes and ponds)
                                       Eleocharis obtusa
                             Blunt Spikerush; Ma-n
            GG Leaf sheaths at least 2 cm long; perennial
               H Spikelets up to 25 mm long, the scales with
                  green midribs; stigmas 2 (up to 1 m tall; in
                  marshes   *Eleocharis macrostachya*
                                   Common Spikerush
               HH Spikelets up to 15 mm long, the scales without
                  green midribs; stigmas 3
                  I Scales light-colored; up to 150 cm tall (in
                     saline or alkaline habitats)
                                       Eleocharis rostellata
                             Beaked Spikerush; Ma, Sn
                  II Scales dark-colored; up to 40 cm tall (in
                     moist areas)   *Eleocharis montevidensis*
                     [includes *E. montevidensis* var. *parishii*]
                                   Montevidio Spikerush
         FF Spikelets not more than 7 mm long, usually with fewer
            than 10 florets (stigmas 3)
            G Leaf sheaths 2-3 cm long; up to 40 cm tall, the
               stems not delicate; in moist meadows (spikelets 4-
               7 mm long)   *Eleocharis pauciflora* [includes
                  E. pauciflora var. *suksdorfiana*] (*Plate* 99)
                                   Fewflower Spikerush; Ma-n
            GG Leaf sheaths not more than 1 cm long; not more than
               20 cm tall, the stems delicate; found in wet
               habitats, including salt marshes
               H Spikelets up to 7 mm long, the scales dark
                  brown; up to 20 cm tall *Eleocharis acicularis*
                                   Needle Spikerush; Sn, Ma, Me
               HH Spikelets not more than 4 mm long, the scales
                  straw-colored, greenish, or yellowish
                  (sometimes reddish brown); up to 8 cm tall
                  I Scales straw-colored   *Eleocharis radicans*
                                   Spongystem Spikerush
                  II Scales greenish or yellowish (sometimes
                     reddish brown)   *Eleocharis parvula*
                     [includes *E. parvula* var. *coloradoensis*]
                                   Small Spikerush; Na, Ma; 4
      EE Spikelets in clusters either at the tip of stems or in the
         leaf axils; inflorescence with at least one bract below
         it; scales usually with short (and sometimes stout) awns
         F Clusters of spikelets 1-5, in the leaf axils (those
            lower on the stem usually on peduncles); leaves
            usually less than 3 mm wide (spikelets 4-6 mm long; in
            bogs)
            G Scales whitish; up to 70 cm tall
                                       Rhynchospora alba (*Plate* 99)
                                   White Beakrush; Sn; 4
            GG Scales brownish; up to 100 cm tall
               H Spikelets globular   *Rhynchospora globularis*
                                   Roundhead Beakrush; Sn; 2
               HH Spikelets not globular *Rhynchospora californica*
                                   California Beakrush; Ma(PR); Sn; 1b
                                   and *Rhynchospora capitellata*
                                   [*R. glomerata* var. *capitellata*]
                                   Brown Beakrush; Sn

FF Clusters of spikelets usually more than 5, at the tip
   of stems; leaves usually more than 3 mm wide (except
   in *Scirpus cernuus* and *S. koilolepis*)
   G Bracts below the inflorescence 2-5, these up to 30
      cm long; stems leafy (up to 1.5 m tall)
      H Stems cylindrical; spikelets 3-6 mm long, the
         scales without awns (in freshwater habitats)
                                    Scirpus microcarpus
                                       Smallfruit Bulrush
      H Stems sharply triangular; spikelets 10-25 mm
         long, the scales with stout awns
         I Inflorescence open, spreading, most of the
            clusters of spikelets on peduncles more
            than 2.5 cm long; leaves 8-16 mm wide;
            stigmas 3; in freshwater habitats
                                    Scirpus fluviatilis
                                    River Bulrush; Na, SM
         II Inflorescence usually dense, compact, the
            clusters of spikelets usually sessile just
            above the bracts; leaves usually 4-6 mm
            wide (occasionally up to 15 mm wide);
            stigmas 2; in salt marshes and freshwater
            habitats   *Scirpus robustus*
                                        Robust Bulrush
   GG Bract solitary, not more than 10 cm long; stems
      either triangular or cylindrical, the leaves
      mainly near the base
      H Inflorescence less than 1 cm long (with 1-3
         spikelets); spikelets not more than 5 mm long;
         each scale folded around the floret, usually
         with a green midrib (awn, if present, less
         than 1 mm long); annual, not more than 20 cm
         tall (in saltwater and freshwater marshes)
         I Inflorescence with 1-3 spikelets, the bract
            below it 10-25 mm long, much longer than
            the inflorescence (scales conspicuously
            keeled)   *Scirpus koilolepis*
                                    Keeled Bulrush; Me-s
         II Inflorescence with 1 spikelet, the bract 2-
            5 mm long, scarcely longer than the
            spikelet   *Scirpus cernuus*
                   [includes *S. cernuus* var. *californicus*]
                                        Low Bulrush
      HH Inflorescence 1-5 cm (or more) long; spikelets
         8-18 mm long; scales flattened and keeled,
         usually flecked with brown or red; perennial,
         at least 50 cm tall (often over 100 cm tall)
         I Inflorescence usually spreading, 3-5 cm
            wide, the bract (1-3 cm long) shorter than
            the inflorescence; scales usually flecked
            with red; up to 4 m tall; in freshwater
            habitats and in ponds or seeps at the edge
            of salt marshes (leaf blades, if present,
            up to 8 cm long; stems up to 2 cm wide)
            Scirpus acutus var. *occidentalis* (*Plate* 99)
                                       Hardstem Bulrush
         II Inflorescence not spreading, not more than
            2 cm wide, the bract longer than the
            inflorescence; scales brownish; up to 2 m
            tall; in saltwater and freshwater habitats
            J Inflorescence (its bract 3-10 cm long)
               with 1-7 spikelets; lower scales of
               spikelets with awns more than 4 mm
               long; stems usually not more than 4 mm
               wide; up to 1 m tall
                     Scirpus pungens [*S. americanus* vars.
                        longispicatus and *monophyllus*]
                                    Common Threesquare
            JJ Inflorescence (its bract not more than 3
               cm long) with 5-12 spikelets; lower
               scales of spikelets with awns less than
               1 mm long; stems up to 10 mm wide; up
               to 2 m tall   *Scirpus americanus*
                                       American Bulrush

AA Each pistillate floret or fruit enclosed within a sac (which often has a characteristic beak); florets either pistillate or staminate (individual spikelets are typically either staminate or pistillate, and both types are usually present in the inflorescence) (often both the scale of the lowest floret of a spikelet, and the bract of the lowest spikelet in the inflorescence, are enlarged)

B   Spikelets sessile
  C   Inflorescence consisting of 1 or 2 spikelets, with a total of not more than 10 florets on each stem; leaves less than 2 mm wide; stigmas 3
    D   Sac around the pistillate floret or fruit 3-5 mm long; awn of scale of lowest floret of each spikelet not more than 5 mm long, the other scales of the florets either without awns or with awns shorter than the sacs; inflorescence less than 4 mm wide; in wet areas   *Carex leptalea*
                                                          Flaccid Sedge; Ma-n; 2
    DD Sac 5-7 mm long; awn of scale of lowest floret of each spikelet 10-40 mm long, the scales of the other florets usually with shorter awns; inflorescence up to 7 mm wide; in dry forests (stems cylindrical)   *Carex multicaulis*
                                                       Manystem Sedge
  CC Inflorescence (usually at least 5 mm wide) consisting of 2 or more spikelets, each with more than 10 florets; leaves 2-5 mm wide; stigmas 2
    D   Lowest bract of the inflorescence as long as, or longer than, the rest of the inflorescence
      E   Plants of moist or wet areas; inflorescence (1-2 cm long) usually more than 10 mm wide; lowest bract leaflike (leaves sometimes longer than the stem; up to 60 cm tall)   *Carex athrostachya*
                                        Slenderbeak Sedge; Ma-n
      EE Plants of dry, open areas; inflorescence less than 10 mm wide; bract not leaflike
        F   Each pistillate floret or fruit with a scale longer than the sac; scales with awns; beak of each sac with conspicuous teeth; inflorescence 2-5 cm long; leaves up to 3 mm wide; up to 80 cm tall, sometimes not forming clumps   *Carex tumulicola*
                                        Foothill Sedge; Mo-n
        FF Floret or fruit with a scale slightly shorter than the sac; scales mostly without awns; beak of each sac with very small teeth, if any; inflorescence up to 2.5 cm long; leaves up to 4 mm wide; up to 45 cm tall, forming clumps   *Carex subfusca* [*C. teneraeformis*]
                                                Rusty Sedge
    DD Lowest bract shorter than the rest of the inflorescence
      E   Lowest bract at least as long as one spikelet (sometimes nearly as long as the rest of the inflorescence)
        F   Weed in lawns and other disturbed areas (inflorescence 1-2 cm long; floret scale shorter than the sac; up to 65 cm tall)   *Carex leavenworthii*
                                    Leavenworth Sedge; na; Al
        FF Native plants in bogs, marshes, or other wet habitats
          G   Each pistillate floret or fruit with a scale longer than the sac (except for the protruding stigmas) (beak of each sac less than one-third as long as the scale; inflorescence usually less than 2 cm long; up to 75 cm tall, usually not forming clumps)   *Carex praegracilis*
                                              Field Sedge
         GG Floret or fruit with a scale shorter than, or about as long as, the body of the sac (excluding the beak)
           H   Scales with awns (inflorescence usually less than 1 cm wide and not more than 5 cm long; leaves 3-5 mm wide; beak of floret sac about half as long as the scale, with conspicuous teeth; up to 90 cm tall)   *Carex bolanderi*
                                        Bolander Sedge; Mo-n
          HH Scales without awns

I   Beak (often reddish) with very small teeth,
if any (the teeth, when present, less than
one-third as long as the scale of the
floret); leaves (2-4 mmm wide) shorter than
the stems (up to 80 cm tall)
Carex ovalis [*C. tracyi*]
Tracy Sedge; Ma-n
II Beak with conspicuous teeth; leaves
sometimes longer than the stems
J   Beak of each floret sac about as long as
the scale; leaves 2-3 mm wide;
inflorescence up to more than 7 mm
wide; plants up to 60 cm tall
Carex echinata ssp. *phyllomanica*
[*C. phyllomanica*]
Coastal Sedge; SCr-n
JJ Beak about half as long as the scale;
leaves 3-5 mm wide; inflorescence not
more than 6 mm wide; plants up to 80 cm
tall   *Carex deweyana* ssp. *leptopoda*
[*C. leptopoda*]
Shortscale Sedge; Scr-n
EE Lowest bract, if present, less than the length of one
spikelet (beak of each sac without conspicuous teeth; in
bogs, marshes, or other habitats that are at least
seasonally wet)
F   Each pistillate floret or fruit with a scale longer
than the sac; scales with awns (inflorescence 1-2.5 cm
long and usually less than 10 mm wide; leaves
sometimes longer than the stems; up to 50 cm tall
Carex simulata
Shortbeak Sedge
FF Floret or fruit with a scale shorter than, or about as
long as, the sac; scales without awns
G   Inflorescence usually over 1 cm wide, so dense that
individual spikelets often obscured
H   Beak of each floret sac about half as long as
the scale; up to 120 cm tall; moist areas at
elevations of up to 5200'   *Carex subbracteata*
Smallbract Sedge
HH Beak less than one-fourth as long as the scale;
up to 80 cm tall; in coastal marshes or boggy
areas at elevations of up to 1000'
Carex harfordii [includes *C. montereyensis*]
Harford Sedge; SLO-n
GG Inflorescence usually not more than 1 cm wide, not
so dense as to obscure individual spikelets
H   Inflorescence up to 8 cm long and 1 cm wide,
usually with 10 or more spikelets; each
spikelet with about 20 florets (beak of each
floret sac less than one-fourth as long as the
scale; up to 120 cm tall)   *Carex feta*
Greensheath Sedge; SCl-n
HH Inflorescence not more than 3.5 cm long and
less than 1 cm wide, with fewer than 10
spikelets; each spikelet with about 10 florets
I   Beak of each floret sac about as long as
the scale; leaves sometimes longer than the
stems; up to 60 cm tall   *Carex echinata*
ssp. *phyllomanica* [*C. phyllomanica*]
Coastal Sedge; SCr-n
II Beak half as long as the scale; leaves
shorter than the stems; up to 120 cm tall
(in vernal pools)   *Carex gracilior*
Slender Sedge
BB At least some spikelets on stalks
C   Each sac around the pistillate floret or fruit covered with fine
hairs (look carefully with a hand lens) (lowest bract of the
inflorescence leaflike; stigmas 3)
D   Leaves hairy (especially along the edge), up to 12 mm wide
(in moist habitats)   *Carex gynodynama*
Hairy Sedge; Mo-n

DD Leaves not hairy, less than 5 mm wide (stems may be reddish
brown at the base; pistillate spikelets often hidden in the
basal leaves)
  E Lowest spikelet of the inflorescence usually more than 2
cm long (5-7 mm wide), compact, with more than 15
florets; lowest bract up to more than 20 cm long; sacs of
florets (reddish brown) densely hairy; up to 100 cm tall,
in wet or dry habitats, sometimes a garden weed (scales
purplish brown, with a lighter center, pointed, narrower
than the sacs; beak of each sac with teeth about 1 mm
long)   *Carex lanuginosa*
                                          Woolly Sedge
  EE Lowest spikelet of the inflorescence usually less than 1.5
cm long, not compact, few-flowered; lowest bract up to 10
cm long (usually much shorter); sacs not densely hairy;
up to 35 cm tall (usually shorter), in dry habitats
     F Uppermost spikelet of the inflorescence often less
than 1 cm long, with fewer than 10 florets; edges of
bract reddish brown at the base (where it clasps the
stem); beak of each floret sac with very small teeth;
up to 15 cm tall   *Carex brevicaulis*
                              Shortstem Sedge; Sl0-n
    FF Uppermost spikelet of the inflorescence 1-2 cm long,
with more than 10 florets; edges of bract not reddish
brown; beak with conspicuous teeth; usually more than
15 cm tall   *Carex globosa*
                             Roundfruit Sedge
CC Sac not hairy
  D Lowest bract of the inflorescence leaflike, longer than the
lowest spikelet and often longer than the rest of the
inflorescence   Cyperaceae, Subkey
  DD Lowest bract, if present, hairlike or bristlelike, but not
leaflike, seldom much longer than the lowest spikelet (but in
Carex stipata, the bract is sometimes up to 5 cm long and
leaflike at the base) (stigmas 2; in moist or wet habitats)
     E Scales of florets with conspicuous awns (inflorescence not
interrupted, crowded; leaves 4-7 mm wide; up to 70 cm
tall)   *Carex dudleyi*
                              Dudley Sedge; Mo-n
    EE Scales with pointed tips, but without awns
      F Inflorescence interrupted (scale of floret sometimes
longer than the sac)
        G Leaves 3-6 mm wide; inflorescence up to 8 cm long;
sacs brownish black; up to 120 cm tall
                               Carex cusickii
                          Cusick Sedge; SLO-n
        GG Leaves 1-3 mm wide; inflorescence up to 5 cm long;
sacs brownish; up to 70 cm tall   *Carex diandra*
                              Panicled Sedge
     FF Inflorescence not interrupted (leaves 3-8 mm wide)
        G Inflorescence (very compact) up to 5 cm long;
ligule of leaf conspicuous; floret sacs widest
near the middle; up to 70 cm tall   *Carex densa*
            [includes *C. breviligulata* and *C. vicaria*]
                             Dense Sedge
        GG Inflorescence up to 10 cm long; ligule of leaf not
conspicuous; sacs widest at the base; up to 90 cm
tall   *Carex stipata*
                          Awlfruit Sedge; Sn-s

Cyperaceae, Subkey: Each pistillate floret or fruit enclosed within a sac,
this not hairy; at least some spikelets on stalks; lowest bract of the
inflorescence leaflike, longer than the lowest spikelet and often longer
than the rest of the inflorescence   *Carex* (in part)
A Leaves with very small bumps on the upper surface, and sometimes with a
rectangular pattern on the lower surface (the pattern is evidence of
partitions within the leaves), some leaves usually more than 1 cm wide
(stigmas 3; in moist or wet habitats)
  B Lowest spikelet of the inflorescence up to 7 mm wide (5-16 cm long);
sac of each floret 3 mm long, the beak 1 mm long (curved, and with
very small teeth) (leaves 8-20 mm wide)   *Carex amplifolia*
                             Ampleleaf Sedge; SM-n
  BB Lowest spikelet of the inflorescence up to 14 mm wide; sac 5-6 mm
long, the beak 1-4 mm long

C   Sac of each floret 1 mm wide, the tapered beak 2-4 mm long
    (including teeth 1-2 mm long); lowest spikelet of the
    inflorescence up to 7 cm long; leaves 6-16 mm wide
    (inflorescence bristly)   *Carex comosa*
                                    Bristly Sedge; La-SCr; 2
CC Sac 2-3 mm wide, the beak less than 2 mm long (and with very
    small teeth); lowest spikelet of the inflorescence up to 10 cm
    long; leaves 2-12 mm wide   *Carex utriculata* [*C. rostrata*]
                                        Beaked Sedge; SCr-n
AA Leaves without bumps or a rectangular pattern, often less than 1 cm wide
   B   Leaves not more than 4 mm wide (lowest spikelet of the inflorescence
       2-5 cm long; in moist or wet habitats)
       C   Each pistillate floret or fruit with a scale longer than the sac;
           terminal spikelet of the inflorescence pistillate at the tip and
           staminate below (bract shorter than, or as long as, the
           inflorescence; stigmas 3; up to 1 m tall)   *Carex buxbaumii*
                                        Buxbaum Sedge; Ma-n
       CC Floret or fruit with a scale shorter than the sac; terminal
           spikelet of the inflorescence either entirely staminate, or at
           least staminate at the tip
           D   Lowest bract longer than the rest of the inflorescence; up to
               130 cm tall (terminal spikelet of the inflorescence staminate
               at the tip and pistillate below; stigmas 3; sometimes found
               on serpentine)   *Carex serratodens*
                                                      Bifid Sedge
           DD Lowest bract shorter than the rest of the inflorescence; up
               to 80 cm tall
               E   Scale, with whitish margins, about as wide as the sac;
                   beak of sac one-fourth as long as the scale, with teeth;
                   stigmas 3; may be on serpentine   *Carex mendocinensis*
                                                      Mendocino Sedge; Ma-n
               EE Scale, with dark brown margins, much narrower than the
                   sac; beak less than one-fourth as long as the scale, and
                   without teeth; stigmas 2; not found on serpentine
                                            Carex nudata (*Plate* 98)
                                                      Torrent Sedge
   BB Some leaves more than 4 mm wide
      C   Leaves hairy (bract below the inflorescence usually longer than
          the inflorescence; leaves 3-12 mm wide; each pistillate floret
          or fruit with a scale shorter than the sac, the beak toothed;
          stigmas 3; up to 90 cm tall; in moist or wet habitats)
                                                      Carex gynodynama
                                                      Hairy Sedge; Mo-n
      CC Leaves not hairy
          D   Leaves longer than the stems (bract below the inflorescence
              longer than the inflorescence; leaves 3-7 mm wide; floret or
              fruit with a scale about half as long as the sac, the beak 2-
              3 mm long, with conspicuous teeth; stigmas 3; up to 1 m tall;
              in moist or wet habitats)
                                    Carex vesicaria var. *major* [*C. exsiccata*]
                                        Inflated Sedge; SCl-n
          DD Leaves shorter than the stems (usually much shorter)
              E   Plants not more than 15 cm tall (bract longer than the
                  inflorescence; leaves 2-5 mm wide; floret sacs without
                  beaks; stigmas 2; in moist or wet habitats)
                                    Carex saliniformis [*C. salinaeformis*]
                                        Deceiving Sedge; SCr-n
              EE Plants more than 50 cm tall
                  F   Lowest bract equal to, or longer than, the rest of the
                      inflorescence (beak of floret sac less than one-third
                      as long as the scale; up to 1 m tall)
                      G   Bract less than 10 cm long (leaves up to 9 mm wide)
                          H   Awn of floret scale about as long as the rest
                              of the scale; lowest spikelet up to 8 cm long;
                              terminal spikelet up to 6 cm long; in vernal
                              pools   *Carex barbarae*
                                                      Santa Barbara Sedge
                          HH Awn, if present, much shorter than the rest of
                              the scale; lowest spikelet up to 6 cm long;
                              terminal spikelet up to 4 cm long; in dry or
                              wet habitats, sometimes a garden weed (scales
                              dark, narrower than the sacs)
                                                      Carex nebrascensis
                                                      Nebraska Sedge

GG Bract at least 10 cm long (often more than 19 cm
   long) (floret sac not more than 4 mm long, the
   scale often longer; in moist or wet habitats)
   H Floret scales without awns; terminal spikelet
      8-14 cm long; lowest spikelet up to 20 cm long
      (bract up to 20 cm long; leaves 6-12 mm wide)
                                        Carex schottii
                                   Schott Sedge; SCl-s
   HH Scales with awns; terminal spikelet not more
      than 8 cm long; lowest spikelet not more than
      14 cm long
      I Leaves not more than 5 mm wide; lowest
         spikelet up to 14 cm long (terminal
         spikelet up to 6 cm long; bract up to 50 cm
         long)   *Carex obnupta* (*Plate* 98)
                                   Slough Sedge; SlO-n
      II Some leaves up to at least than 9 mm wide;
         lowest spikelet 8 cm long
         J Terminal spikelet not more than 4 cm
            long; bract up to 25 cm long (awn of
            floret scale sometimes longer than the
            rest of the scale)   *Carex lyngbyei*
                                   Lyngbye Sedge; Ma-n
         JJ Terminal spikelet up to 8 cm long; bract
            up to 50 cm long   *Carex aquatilis*
                                var. *dives* [*C. sitchensis*]
                                    Sitka Sedge; SCr-n
FF Lowest bract shorter than the rest of the
   inflorescence (in moist or wet habitats)
   G Leaves (about 5 mm wide) not more than 15 cm long,
      strictly at the base of the plant; bract 2-3 cm
      long (spikelets 1-2 cm long; stigmas 3; up to 90
      cm tall)   *Carex luzulina*
                              Luzula-like Sedge; Ma-n
   GG Leaves more than 15 cm long, or if shorter not
      confined to the base of the plant; bract more than
      3 cm long
      H Leaves not more than 5 mm wide (bract not more
         than 10 cm long; floret sacs less than 5 mm
         long)
         I Floret or fruit with a greenish scale, this
            about as wide as the sac, and usually with
            an awn; spikelets not more than 2 cm long;
            stigmas 3; up to 60 cm tall   *Carex albida*
                                    White Sedge; SN; 1b
         II Scale dark brown, much narrower than the
            sac, and without an awn; spikelets up to 5
            cm long; stigmas 2; up to 100 cm tall
                                          Carex senta
                                    Rough Sedge; Sl-s
      HH Some leaves more than 5 mm wide (floret scales
         usually with short awns)
         I Bract 10-15 cm long; lowest spikelet not
            more than 4 cm long, terminal spikelet not
            more than 3 cm; floret sacs 5-6 mm long;
            stigmas 3 (in continually moist habitats)
                                     Carex hendersonii
                                   Henderson Sedge; Sn-n
         II Bract less than 10 cm long; lowest spikelet
            up to 8 cm long, terminal spikelet up to 6
            cm; sacs less than 5 mm long; stigmas 2
            J Awn of floret scale about as long as the
               rest of the scale; lowest spikelet up
               to 8 cm long; terminal spikelet up to 6
               cm long; in vernal pools *Carex barbarae*
                                    Santa Barbara Sedge
            JJ Awn, if present, much shorter than the
               rest of the scale; lowest spikelet up
               to 6 cm long; terminal spikelet up to 4
               cm long; in dry or wet habitats,
               sometimes a garden weed (scales dark,
               narrower than the sacs)
                                     Carex nebrascensis
                                     Nebraska Sedge

Hydrocharitaceae (includes Najadaceae)--Waterweed Family

Our representatives of the Hydrocharitaceae are submerged aquatics. Their leaves, opposite or arranged in whorls of 3 or more, are sessile, rather narrow, and neither compound nor lobed. Two of the species of *Najas* have very flexible stems and leaves, and could be confused with *Zannichellia palustris* (Zannichelliaceae). In *Zannichellia*, however, all the leaves are distinctly opposite, whereas in *Najas* many of the pairs are accompanied by a second pair, and such clusters of 4 leaves may seem to be whorls.

A   Leaves and stems very flexible (except in *Najas marina*); flowers without perianth segments; staminate flowers with a single stamen; surface of fruit with a networklike pattern (this may not be visible, however, at low magnification)
  B   Leaves stiff, up to 3 mm wide, with marginal teeth up to nearly 1 mm long; stems often with spinelike outgrowths; plants with either staminate or pistillate flowers, but not both
                        Najas marina (*Plate* 100)
                        Hollyleaf Waternymph; La-s
  BB Leaves very flexible, not more than 2 mm wide, either smooth-margined or with extremely fine teeth; stems without spinelike outgrowths; plants with both staminate and pistillate flowers
    C   Surface of fruit dull, the networklike pattern of the surface visible with a hand lens; anthers usually with only 1 chamber (often in brackish water)   *Najas guadalupensis*
                        Common Waternymph
    CC Surface of fruit shiny, the networklike pattern of the surface usually not visible with a hand lens; anthers usually with 4 chambers   *Najas flexilis*
                        Slender Waternymph
AA Leaves and stems not especially flexible; flowers usually with 3 sepals (or 3 calyx lobes) and 3 petals (but the petals may be scarcely visible); staminate flowers with at least 3 stamens; surface of fruit without a networklike pattern (Note: in species of *Egeria* and *Elodea*, the plants in a particular population may never, or only rarely, produce flowers.)
  B   Underside of leaves, along the midrib, roughened by conical bumps; horizontal stems (from which the upright stems arise) thickened to form tuberous structures (leaves mostly 1-2 cm long, usually in whorls of 4-8; a serious pest)   *Hydrilla verticillata*
                        Hydrilla; eua
  BB Underside of leaves, along the midrib, not roughened by conical bumps; horizontal stems not thickened to form tuberous structures
    C   Leaves commonly at least 20 mm long, mostly in whorls of 3-6 (except on the lower portion of the stem, where they are usually opposite); petals of staminate flowers 8-10 mm long, those of pistillate flowers 6-7 mm long (cultivated in aquaria and garden pools, occasionally escaping; all populations in California seem to consist of staminate plants)   *Egeria densa* [*Elodea densa*]
                        Brazilian Waterweed; sa
    CC Leaves rarely more than 15 mm long, mostly opposite or in whorls of 3; petals not often more than 5 mm long
      D   Leaves 9-15 mm long and up to 3 mm wide, blunt or abruptly pointed at the tip; petals of pistillate flowers about 2.5 mm long, those of staminate flowers about 5 mm long
                        Elodea canadensis (*Plate* 100)
                        Common Waterweed
      DD Leaves 6-13 mm long and up to 2 mm wide, tapering rather gradually to a pointed tip; petals absent or less than 1 mm long   *Elodea nuttallii*
                        Nuttall Waterweed

Iridaceae--Iris Family

Irises and their relatives are perennials with rather narrow leaves. The flowers have 3 sepals, 3 petals, and 3 stamens, these being in line with the sepals. The petals and sepals are sometimes similar, sometimes different. The fruit-forming part of the pistil is below the petals and sepals. Flowers arise from between a pair of substantial bracts.

This family gives us not only cultivated irises, but also crocuses,
tigridias. gladioli, and some other valuable garden plants. Some of them,
including the Common Freesia (*Freesia refracta*) of South Africa, have
escaped and become established in places where they have been dumped or to
which they have escaped by some other route.

Our native species, if planted in appropriate situations, are excellent
subjects for wild gardens. Some of them may be purchased from nurseries.
Most of them can also be grown from seed, although they are not likely to
flower before the third or fourth year after germination. *Iris douglasiana*
is particularly useful under oaks and *Sisyrinchium bellum* readily reseeds
itself in a variety of habitats. *Sisyrinchium californicum* requires
moisture; wet soil just beyond the edge of a pool should suit it nicely.

A   Only 1 flower between each pair of bracts; flowers orange-red (a garden
    hybrid)   *Crocosmia x crocosmiiflora* [*C. crocosmiflora*]
                                                                 Montbretia; af
AA More than 1 flower between each pair of bracts; flowers not orange-red
   B   Petals and sepals similar, and not more than 1.5 cm long; flowering
       stems with 2 winglike margins
       C   Petals and sepals purplish blue; styles not divided; widespread
           in grassy places   *Sisyrinchium bellum* (*Plate* 57)
                                                                 Blue-eyed-grass
       CC Petals and sepals yellow; styles deeply divided; mostly
          restricted to wet situations (coastal)
                                              Sisyrinchium californicum (*Plate* 57)
                                                           Golden-eyed-grass; Mo-n
   BB Petals and sepals decidedly different, and generally longer than 2.5
      cm; flowering stems cylindrical
      C   Flowers mostly bright yellow; up to more than 100 cm tall, rooted
          in mud at the edge of lakes, ponds, and streams (cultivated in
          water gardens and sometimes escaping)   *Iris pseudacorus*
                                                                 Yellowflag; eu; Sn
      CC Flowers not bright yellow (but sometimes pale yellow or cream);
         rarely more than 40 cm tall, and strictly terrestrial
         D   Flowering stems with numerous leaves that clasp for most of
             their length, and that overlap (these leaves are often tinged
             purplish red) (petals and sepals cream or pale yellow, with
             darker veins and sometimes with a lavender tinge, especially
             on the sepals; mostly in woods)   *Iris purdyi*
                                                                 Purdy Iris; Sn-n
         DD Flowering stems with leaves that clasp for not more than half
            their length, and that do not overlap
            E   Perianth tube about the same length as, or shorter than,
                the fruiting portion of the pistil
                F   Perianth tube about as long as the fruiting portion of
                    the pistil; tip of fruiting portion of pistil (just
                    below the perianth tube) with several small, blunt
                    lobes; petals and sepals predominately dark lilac or
                    purple, but sometimes pale, in which case the darker
                    streaks are especially prominent; leaves commonly at
                    least 1 cm wide (on grassy hillsides near the coast)
                                                    Iris douglasiana (*Plate* 57)
                                                                 Douglas Iris
                FF Perianth tube much shorter than the fruiting portion
                   of the pistil; tip of fruiting portion of pistil
                   without small, blunt lobes; petals and sepals mostly
                   lavender blue (on the petals, most of the color is
                   concentrated in streaks, and there is often a yellow
                   streak along the midline); leaves not often so much as
                   1 cm wide (coastal)   *Iris longipetala*
                                                                 Coast Iris; Me-Mo
            EE Perianth tube much longer than the fruiting portion of the
               pistil, and flaring appreciably at the top, just below
               the petals and sepals
               F   Perianth tube flaring gradually at the top, this
                   portion longer than wide; petals and sepals cream,
                   tinged with yellow, with faint darker veins (in
                   slightly shaded situations)   *Iris fernaldii* (*Plate* 57)
                                                                 Fernald Iris; Sn, Sl-SCr

FF Perianth tube flaring abruptly at the top, this
   portion approximately bowl-shaped, about as wide as
   long; petals and sepals cream, yellowish, or purplish
   blue, with darker veins (grassy slopes and oak woods)
                                        Iris macrosiphon
                                        Ground Iris; SCl-n

Juncaceae--Rush Family

In their general appearance, and in the way their leaf sheaths envelop
the stems, most rushes resemble grasses. The leaves are often reduced,
however, to the point that only the sheaths remain. Furthermore, when leaf
blades are present, these are more often cylindrical than flattened.

The flowers, usually concentrated in terminal inflorescences, differ
from those of grasses in that they have definite perianth segments, 3 of
which correspond to sepals, 3 others to petals. There are 2, 3 or 6
stamens, and a pistil with 3 slender styles. The fruit, often with a little
beak at the tip, contains numerous seeds in a single chamber or in 3
chambers.

The inflorescence typically has 2 bracts below it. The lower one is
longer than the upper one, and may appear to be a continuation of the stem.
There are usually additional bracts associated with the individual flowers.
In identification of rushes, the appearance and disposition of bracts, leaf
blades, and sheaths are important. It may also be necessary to have mature
fruit.

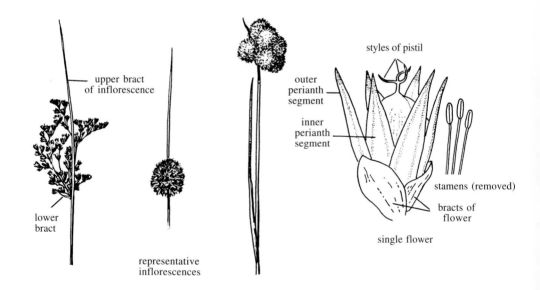

upper bract
of inflorescence

lower
bract

representative
inflorescences

styles of pistil

outer
perianth
segment

inner
perianth
segment

stamens (removed)

bracts of
flower

single flower

A Leaf blades flattened and very flexible (much like those of most
   grasses), with long hairs, and often shredding at the base when young;
   edges of leaf sheaths fused together; stems hollow; mostly in woodland
   habitats   *Luzula comosa* [includes *L. subsessilis*] (*Plates* 57, 101)
                                                              Wood Rush
AA Leaf blades, if present, generally stiff (even if flattened, as they are
   in *Juncus falcatus* and a few other species), without hairs, and not
   shredding at the base; edges of leaf sheaths not fused; stems not
   hollow; mostly in wet places, but sometimes on backshores of sandy
   beaches

B   Stems branching, with flowers (usually single) in the leaf axils (in
     Juncus bufonius var. *congestus*, most of the flowers are very close
     together at the tip of the branches, and in *J. ambiguus* they are
     also often concentrated in this way) (annuals, rarely so much as 20
     cm tall; bracts beneath single flowers or clusters of flowers
     membranous, up to 2 mm long)
     C   Inner 3 perianth segments (corresponding to petals) rounded at
          the tip; tip of fruit bluntly rounded (flowers often
          concentrated near the tip of branches; mostly in saline habitats
          at the coast)   *Juncus ambiguus* [*J. bufonius* var. *halophilus*]
                                                              Coastal Rush
     CC   Inner 3 perianth segments, like the outer ones, sharply pointed
          at the tip; tip of fruit conical
          D   Outer perianth segments (corresponding to sepals) 3-4 mm long
               (flowers scattered rather evenly along the branches; in mud
               of drying pools and on stream banks)
                                        Juncus bufonius var. *occidentalis*
                                                         Western Toad Rush
          DD   Outer perianth segments 4-7 mm long
               E   Flowers scattered rather evenly along the branches
                    (sometimes a weed in gardens, but common also in many
                    other moist habitats)
                                     Juncus bufonius var. *bufonius* (*Plate* 100)
                                                                 Toad Rush
               EE   Most flowers concentrated near the tip of the branches,
                    which are often slightly coiled (in saline habitats)
                                          Juncus bufonius var. *congestus*
                                                       Congested Toad Rush
BB   All stems originating from the base of the plant, and not branching
     below the inflorescence
     C   Flowers usually not more than 3 at the tip of each stem; stamens
          2 or 3; not often more than 5 cm tall (annual)
          D   Flowers (reddish), commonly 3 at the tip of each stem;
               usually 3-5 cm tall   *Juncus kelloggii* (*Plate* 101)
                                                              Kellogg Rush
          DD   Flowers usually single on each stem; usually less than 3 cm
               tall (plants with few leaves)
               E   Fruit protruding 1 mm beyond the flower; perianth segments
                    with a red midrib; at elevations above 2000' (plants with
                    few flowers and stems)   *Juncus hemiendytus* (*Plate* 101)
                                                     Self-pollinating Rush; Na-n
               EE   Fruit not protruding so much as 1 mm beyond the flower;
                    perianth segments with a green midrib; from lowland
                    habitats to elevations of 4500' (plants forming clumps)
                                                           Juncus uncialis
                                                          Dwarf Rush; SLO-n
     CC   Flowers more than 5 (and usually many) in each inflorescence;
          stamens usually 6, but sometimes 3; more than 10 cm tall
          D   Lower bract below the inflorescence cylindrical (use hand
               lens), appearing to be a continuation of the stem, and longer
               than the inflorescence, so that the inflorescence appears
               lateral; leaves often reduced to only the sheaths
               (inflorescence either spreading or compact, the flowers
               usually numerous, but sometimes solitary) (Note: the lower
               bracts of *Juncus occidentalis* and *J. tenuis*, keyed under
               choice DD, may appear to be continuations of the stems, but
               both of these rushes have flattened bracts, and both have
               some distinct leaf blades.)
               E   Flowers 4-8 mm long (sometimes only 3.5 mm long in *Juncus*
                    mexicanus and *J. balticus*); fruit with a sharp beak
                    (stamens 6)
                    F   Flowers usually 6-7 mm long, the perianth segments
                         reddish brown near the margin (lower bract 10-25 cm
                         long; stems cylindrical; leaf blades absent, the
                         sheaths dark to light brown; up to more than 90 cm
                         tall, often in rows because they arise from horizontal
                         underground stems; at the edges of salt marshes or on
                         backshores of sandy beaches)
                                                 Juncus lesueurii (*Plate* 101)
                                                           Salt Rush; SLO-n
                    FF   Flowers 4-5 mm long (rarely 6 mm long in *Juncus*
                         mexicanus), the perianth segments with a transparent,
                         whitish margin)

G   Stems flattened and twisted; lower bract up to 15
     cm long; perianth segments green, striped on both
     sides with reddish brown; up to 60 cm tall (leaf
     sheaths yellowish to brown, the blades rarely
     present)   *Juncus mexicanus*
                                         Mexican Rush; La-s
GG Stems cylindrical or somewhat flattened, not
     twisted; lower bract up to 20 cm long; perianth
     segments brownish or greenish, with a green
     midrib; up to more than 100 cm tall (leaf sheaths
     sometimes with a bristle up to 1 cm long; plants
     often in rows because they arise from horizontal
     underground stems)   *Juncus balticus*
                                              Baltic Rush
EE Flowers not more than 3.5 mm long; fruit without a sharp
    beak (but it may be pointed) (leaves without blades, the
    sheaths usually dull to dark brown)
    F   Stems 2-5 mm wide, up to 130 cm tall; inflorescence
         usually at least 4 cm long (lower bract rarely as long
         as one-third the length of the stem; leaf sheaths 5-15
         cm long, dull brown to very dark brown; perianth
         segments stiff, pale brown to straw-colored, the edge
         light tan; stamens 3; plants forming clumps; in inland
         boggy areas)
                        Juncus effusus var. *pacificus* (*Plate* 101)
                                           Pacific Bog Rush
    FF Stems usually less than 2 mm wide, mostly less than 90
        cm tall; inflorescence usually less than 4 cm long
        (plants forming clumps)
        G   Perianth segments upright at maturity, the margin
             dark brown; fruit somewhat 3-angled; inflorescence
             up to 4 cm wide; lower bract usually less than
             one-third the length of the stem; leaf sheaths 4-
             15 cm long, reddish brown at the base; stamens 3;
             coastal boggy areas   *Juncus effusus* var. *brunneus*
                                             Brown Bog Rush
        GG Perianth segments (with a green midrib) spreading
             apart at maturity, the margin mostly reddish
             brown; fruit globular (with a small beak);
             inflorescence up to 6 cm wide; bract usually more
             than one-third the length of the stem; leaf
             sheaths 2-10 cm long, dark brown at the base;
             stamens 6; in moist but not boggy areas (stems
             sometimes bluish green)   *Juncus patens*
                                             Spreading Rush
DD Lower bract below the inflorescence usually flattened and
    either shorter than the inflorescence or directed away from
    the axis of the stem, so that the inflorescence appears to be
    terminal; leaf blades usually present
    E   Inflorescence loose, with single flowers scattered along
         its branches, and with bracts below each flower (leaves
         basal, 5-20 cm long, the blade more than half as long as
         the stem; flowers green to reddish, about 4 mm long;
         stamens 6; up to 60 cm tall)   *Juncus tenuis*
                                               Slender Rush
    EE Inflorescence consisting of one or more dense clusters of
        flowers, with bracts below each cluster and sometimes
        also below each flower
        F   Lower bract below the inflorescence (less than 1 mm
             wide and up to 10 cm long) protruding beyond the
             inflorescence at a slight angle, superficially
             stemlike (it is flattened, however) (flowers more than
             4 mm long; stamens 6; leaves less than 1 mm wide,
             mainly basal, less than half as long as the stem; up
             to 60 cm tall, forming clumps)
                       Juncus occidentalis [*J. tenuis* var. *congestus*]
                                              Western Rush
        FF Lower bract either not protruding beyond the
            inflorescence or, if longer than the inflorescence,
            directed laterally at a wide angle, and not stemlike
            G   Leaf blades either cylindrical or flattened, but
                 not folded and not clasping the stem, sometimes
                 with ridges across the blade (making them appear
                 jointed); stems cylindrical or partly flattened

H   Inflorescence 5-15 cm wide, often with 2 or
    more whorls of spreading branches; flower
    clusters 4-6 mm wide, more than 10 in each
    inflorescence; leaves shorter than the stems
    (flower clusters brownish, with fewer than 15
    flowers; leaf blades 1-3 mm wide, appearing to
    be jointed, with a white membrane 4-6 mm long
    where each one joins its sheath; fruit
    protruding beyond the perianth segments;
    stamens 6; up to 1 m tall   *Juncus dubius*
                                         Mariposa Rush
HH Inflorescence less than 5 cm wide, without
    whorls of spreading branches; flower clusters
    more than 6 mm wide (sometimes only 5 mm wide
    in *Juncus covillei*), fewer than 10 in each
    inflorescence; leaves sometimes as long as the
    stems
    I   Each flower cluster with at least 20
        flowers (these dark brown); leaf blades (1-
        3 mm wide) appearing to be jointed, and
        with a white membrane 4-6 mm long where
        each one joins its sheath; stamens 3; up to
        100 cm tall   *Juncus bolanderi* (*Plate* 100)
                                  Bolander Rush; SCl-n
    II Each flower cluster with not more than 15
        flowers; leaf blades (flattened, 2-4 mm
        wide) not appearing jointed, and without a
        white membrane; stamens 6; up to 30 cm tall
        J   Flower clusters 5-10 mm wide, 1-6 in
            each inflorescence; flowers light
            brown, 2-5 mm long; fruit protruding up
            to 1 mm beyond the perianth segments;
            leaf blades not sickle-shaped (plants
            forming clumps)   *Juncus covillei*
                                     Coville Rush; Ma-n
        JJ Flower clusters 10-15 mm wide, 1-3 in
            each inflorescence; flowers dark brown,
            4-6 mm long; fruit (sometimes globular)
            either not protruding beyond the
            perianth segments, or just barely; leaf
            blades slightly sickle-shaped (and
            stiff) (stems sometimes spreading
            horizontally and rooting)
                         Juncus falcatus (*Plate* 100)
                                     Sickleleaf Rush
GG Leaf blades flattened, folded, and clasping the
    stem (they are similar to the leaf blades of an
    iris), the ridges, if present, usually not
    continuing completely across the blade (except
    sometimes in *Juncus phaeocephalus*); stems
    flattened (Note: the following species are
    difficult to separate.)
    H   Fruit not protruding beyond the perianth
        segments; flowers 4-6 mm long (leaf blades
        usually not more than 5 mm wide; flowers dark
        brown or greenish brown; beak less than one-
        fourth as long as the body of the fruit;
        stamens 6)
        I   Inflorescence 2-5 cm wide, its branches few
            and ascending; flower clusters 10-15 mm
            wide, 1-5 in each inflorescence, each
            cluster often with more than 50 flowers; up
            to 50 cm tall
                Juncus phaeocephalus var. *phaeocephalus*
                                     Brownhead Rush
        II Inflorescence up to 12 cm wide, its
            branches many and spreading; flower
            clusters 5-10 mm wide, up to 25 in each
            inflorescence, each cluster with not more
            than 15 flowers; up to 90 cm tall
                Juncus phaeocephalus var. *paniculatus*
                                 Panicled Brownhead Rush
    HH Fruit protruding (sometimes just barely) beyond
        the perianth segments; flowers 3-4 mm long

I   Beak about one-fourth as long as the body
    of the fruit, the tip tapering gradually
    (inflorescence up to 12 cm wide, its
    branches many and spreading; flower
    clusters 5-10 mm wide, up to 25 in each
    inflorescence, each cluster with not more
    than 15 flowers; leaf blades usually not
    more than 5 mm wide; flowers light to dark
    brown; stamens 6; up to 60 cm tall)
                                  Juncus oxymeris
                                  Pointed Rush
II Beak less than one-fourth as long as the
    body of the fruit, the tip not tapering
    gradually
    J   Leaf blades usually not more than 5 mm
        wide; flowers greenish brown or dark
        brown; stamens 3; not more than 50 cm
        tall   *Juncus ensifolius*
                                  Threestem Rush; Na-n
    JJ Leaf blades often over 5 mm wide (up to
        12 mm wide); flowers straw-colored to
        dark brown; stamens 6; up to 90 cm
        tall   *Juncus xiphioides*
                                  Irisleaf Rush

Juncaginaceae (includes Lilaeaceae)--Arrow-grass Family

Members of the Arrow-grass Family are semi-aquatic, or at least grow in
wet places. In our region, one or more species of *Triglochin* will be found
in almost any salt marsh; *Lilaea scilloides*, Flowering Quillwort, is
occasionally encountered in freshwater habitats, particularly vernal pools.
The leaves are all basal, somewhat grasslike in general appearance and in
having sheaths at the base. In species of *Triglochin*, the flowers are in
crowded, elongated inflorescences at the top of unbranched stems that
usually rise above the leaves. They have 3 or 6 sepals and stamens, and the
fruiting part of the pistil is partitioned lengthwise into 3 or 6
divisions. The staminate flowers are characterized by a distinct bract; the
pistillate flowers rarely have this structure. In *Lilaea*, each
inflorescence consists of flowers that have either a pistil or a stamen.
There are also pistillate flowers of a different type at the base of the
plant. These are peculiar in that the style of the pistil is often more
than 5 cm long.

A   Terminal inflorescences up to 2 cm long, the flowers either with a
    single stamen or a pistil; other flowers at the base of the plant
    consisting of a pistil with a long style (flowering stems 6-20 cm long,
    equal to, or shorter than, the leaves, these 1-5 mm wide, cylindrical)
                                          Lilaea scilloides
                                          Flowering Quillwort
AA Terminal inflorescences 3-40 cm long, the flowers all with stamens (3 or
    6) and a pistil; flowers absent at the base of the plant
    B   Flowering stems usually 1 or 2 (up to 20 cm long), equal to, or
        shorter than, the leaves; sepals 3; fruiting portion of the pistil
        consisting of 3 divisions; fruit about as wide as long, almost 2 mm
        long, each division 3-ribbed (leaves flattened, 1-2 mm wide; racemes
        up to 12 cm long)   *Triglochin striata*
                                          Threerib Arrow-grass
    BB Flowering stems usually 2 or more, longer than the leaves; sepals 6;
        fruiting portion of the pistil consisting of 6 divisions; fruit
        decidedly longer than wide, usually 3-4 mm long, the divisions not
        obviously ribbed
        C   Flowering stems up to 70 cm tall, with racemes up to 40 cm long;
            leaves flattened, 1.5-5 mm wide, the ligules usually pointed,
            rarely slightly notched   *Triglochin maritima* (*Plate* 102)
                                             Seaside Arrow-grass
        CC Flowering stems usually not more than 40 cm tall, with racemes up
            to 15 cm long; leaves cylindrical, about 1 mm wide, the ligules
            obviously bilobed   *Triglochin concinna* (*Plate* 102)
                                             Saltmarsh Arrow-grass

Lemnaceae-Duckweed Family

It may be hard to believe that the little duckweeds that float at the surface of ponds and streams are flowering plants. But they do indeed have tiny flowers. There are no leaves. Most of our species, which belong to the genus *Lemna*, consist of flattened green stems from which 1 or more simple, unbranched roots originate. The flowers, without petals or sepals, are in pouches along the edge of the stem; 1 or 2 of them consist of a single stamen, another consists of a pistil. In the genera *Wolffia* and *Wolffiella*, roots are lacking, and the flower pouch is on the upper surface of a flattened or nearly spherical stem. It should be noted that flowers of duckweeds are rarely observed, and perhaps are never formed in some populations. The plants reproduce mostly by budding, and this explains why they are often attached to one another.

A   Plants without any roots
    B   Plants nearly spherical, less than 1.5 mm long, and raised above the surface of the water   *Wolffia columbiana*
                                                                                 Wolffia
    BB Plants flattened, oblong, 2.5-6 mm long, the width about half the length, and not raised above the surface of the water   *Wolffiella lingulata*
                                                          Mudmidget; Ma, Sn, SM-n
AA Plants with at least 1 root
    B   Plants with 2 or more roots
        C   Plants usually 5-8 mm long (but sometimes slightly smaller or larger), with 5-12 roots and 5-15 veins   *Spirodela polyrrhiza* (*Plate* 102)
                                                          Common Duckmeat
        CC Plants 2.5-5 mm long, with 2-5 roots and 3-5 veins   *Spirodela punctata* [*S. oligorrhiza*]
                                                          Small Duckmeat; Al
    BB Plants with only 1 root
        C   Upper surface of the stem very flat and smooth and with only one vein or none
            D   Plants 2.5-5 mm long, narrowly elliptical, usually about 3 times as long as wide, adhering in groups of up to 10   *Lemna valdiviana*
                                                          Uncommon Duckweed; La-s
            DD Plants 1-2.5 mm long, oval, less than twice as long as wide, generally solitary or in pairs   *Lemna minuscula* [*L. minima*]
                                                          Least Duckweed
        CC Upper surface of the stem bumpy and with 3-5 veins (but some or all of these may not be readily visible)
            D   Plants less than 3 mm long, the tip (the end opposite that which may still be attached to another plant) often with a slight projection; veins not distinct; upper part of roots with winglike membranes   *Lemna aequinoctialis*
                                                            Valley Duckweed
            DD Plants mostly more than 3 mm long, the tip evenly rounded; veins rather distinct; upper part of roots without winglike membranes
                E   Plant usually symmetrical, the length usually almost twice the width; lower surface nearly flat; upper surface dark green   *Lemna minor* (*Plate* 102)
                                                         Common Duckweed
                EE Plant usually (not always) lopsided, the length not more than one and a half times the width; lower surface usually inflated (but sometimes nearly flat); upper surface generally light green, but mottled   *Lemna gibba*
                                                         Swollen Duckweed

Liliaceae (includes part of Amaryllidaceae)--Lily Family

The Lily Family is well represented in the region. A few of our genera have sometimes been placed in the Amaryllis Family (Amaryllidaceae). All species are perennial, but most are herbaceous and die back, after flowering and fruiting, to underground bulbs, corms, or rhizomes. There are usually 3 sepals and 3 petals; when these components of the flower are much

alike, they are all referred to as perianth segments. There are typically 6
stamens. The fruit, usually dry and cracking apart at maturity, but fleshy
in certain types, is partitioned into 3 divisions. There are a few
exceptions to the formula stated above. For instance, in *Maianthemum* there
are only 4 perianth segments and 4 stamens, and the fruit consists of 2
divisions; in *Smilax*, which is a woody vine, some flowers are pistillate,
others staminate.

Hundreds of exotic species of the Lily Family are in cultivation. They
include hyacinths, tulips, and onions, as well as true lilies. Many natives
of our western states have also been grown, not only in North America but
in other parts of the world. Some of them are easily started from seed, but
seedlings cannot be expected to bloom until about the fourth year. Both
seeds and flowering-size bulbs raised from seed are often available at
sales of the California Native Plant Society and botanical gardens, and
also at commercial nurseries that specialize in natives. It is important to
choose these plants carefully with respect to the type of situation in
which they are to be grown. Species of *Allium*, *Brodiaea*, *Camassia*, *Lilium*,
and *Calochortus* generally prefer open, sunny places; species of *Trillium*,
Erythronium, *Smilacina*, and *Maianthemum* usually do best in shaded habitats,
and also require moisture over a rather long growing period.

A With 3 broad leaves, these in a whorl at the top of the stem, where
    there is a single flower (in moist, shaded areas)
    B Flower on a distinct peduncle at least 2 cm long; petals white,
        becoming dull rose as they age; leaves uniformly green
                                    Trillium ovatum (*Plate* 103)
                                        Western Trillium; Mo-n
    BB Flower sessile; petals white, whitish, yellowish, pink, or dark
        purple; leaves mottled
        C Petals whitish, yellowish, or dark purple; fruiting portion of
            pistil purplish; top of the filaments, between the anther sacs,
            purplish
            Trillium chloropetalum [includes *T. chloropetalum* var. *giganteum*]
                                        Giant Trillium; Mo-n
        CC Petals white to pink; fruiting portion of pistil usually green,
            but sometimes slightly tinged with purple; tip of filaments,
            between the anther sacs, green   *Trillium albidum* (*Plate* 61)
                                        Sweet Trillium; SFBR-n
AA Leaves not necessarily in whorls, but if so arranged, not restricted to
    the top of the stem, and not especially broad
    B Leaves not entirely or almost entirely basal, at least some
        substantial leaves scattered along the stem or in whorls (stem
        leaves sometimes absent, however, in certain species of *Calochortus*)
        C Stems twining, woody (at least in old growth), persisting from
            year to year, and with tendrils (leaves nearly heart-shaped;
            petals small, greenish or yellowish; pistillate and staminate
            flowers on separate plants; fruit fleshy, black; in moist
            areas)   *Smilax californica*
                                        Greenbrier; Na-n
        CC Stems neither twining nor woody, not persisting from year to year
            (except in *Xerophyllum tenax*), and without tendrils
            D Perianth segments not more than 7 mm long (inflorescence a
                terminal raceme or panicle 2-12 cm long)
                E Perianth segments 4 (2.5 mm long); leaf blades heart-
                    shaped, on long petioles (coastal, in moist, shaded
                    areas)   *Maianthemum dilatatum* (*Plate* 103)
                                        False Lily-of-the-valley; SM, Ma-n
                EE Perianth segments 6; leaf blades not heart-shaped, nearly
                    or quite sessile (mostly in shaded habitats)
                    F Flowers more than 20, in a panicle; perianth segments
                        1-2 mm long; up to 90 cm tall   *Smilacina racemosa*
                        [includes *S. racemosa* var. *amplexicaulis*] (*Plate* 60)
                                        Fat Solomon
                    FF Flowers 5-15, in a loose raceme; perianth segments 4-6
                        mm long; rarely more than 40 cm tall
                                            Smilacina stellata
                        [includes *S. stellata* var. *sessilifolia*] (*Plate* 103)
                                        Slim Solomon
            DD Perianth segments more than 10 mm long

E   Plants not bushy (branching sparingly if at all)
Liliaceae, Subkey 1 (p. 262)
EE Plants branching (leaves 3-15 cm long, sessile; flowers 1-
7 at the tip of stems; perianth usually white or creamy
white, sometimes faintly greenish; in moist coniferous
woods)
F   Stamens about as long as, or longer than, the perianth
segments; stigma not divided   *Disporum hookeri*
Fairybell; Mo-n
FF Stamens shorter than the perianth segments; stigma
divided into 3 lobes   *Disporum smithii* (*Plate* 102)
Largeflower Fairybell; SCr-n
BB Leaves entirely or almost entirely basal, in some species withering
before flowering time
C   Flowers in an umbel (or umbel-like racemes), this sometimes with
smaller umbels or individual flowers below it (the flowering
stem in *Scoliopus* is extremely short, but the pedicels are at
least 10 cm long) (leaves entirely basal)
D   Perianth segments united for some distance above the base, so
that a definite tube is formed
E   With only 2 bracts below the umbel (flowers fragrant;
perianth segments dull white, greenish at the base, and
with a reddish or brownish midrib on the outer surface;
an escape from gardens)   *Nothoscordum inodorum*
False Garlic; sa; Ma
EE Usually with at least 3 bracts below the umbel (and not
conforming to other characteristics described in choice
E)   Liliaceae, Subkey 2 (p. 264)
DD Perianth segments separate to the base, or nearly so, not
forming a tube
E   Plants with only 2 leaves, these 5-10 cm wide (10-20 cm
long), often with dark blotches (flowers with mottled
green and purple petals 1.5 cm long; pedicels at least 10
cm long, arising from a very short, scarcely noticeable
peduncle; in moist, shaded areas in the Coast Ranges)
Scoliopus bigelovii
Fetid Adder's-tongue; SCr-n
EE Plants usually with more than 2 leaves (if there are only
2, they are not more than 1.5 cm wide), usually without
blotches
F   Leaves 5-12 cm wide; flowering stems up to more than
50 cm tall, sometimes with smaller umbels or single
flowers below the main umbel; in redwood forests
(perianth rose-purple, 10-18 mm long)
Clintonia andrewsiana (*Plate* 59)
Red Bead Lily; Mo-n
FF Leaves (these may have withered by flowering time) not
more than 1.5 cm wide; flowering stems not often so
much as 50 cm tall, and rarely with flowers below the
main umbel; mostly in exposed sunny habitats
G   Umbel, before flowering, enclosed by 3 or more
separate bracts; individual flowers with very
small bracts; leaves, if bruised, without the odor
of onion (perianth segments 3-6 mm long, greenish
white, with a brownish midvein; sometimes on
serpentine)   *Muilla maritima* (*Plate* 60)
Common Muilla
GG Umbel, before flowering, enclosed by a sheath that
later splits apart into 2-4 bracts; individual
flowers without bracts; leaves, if bruised, with
the odor of onion   Liliaceae, Subkey 3 (p. 265)
CC Flowers in racemes or panicles
D   Leaves only 2, mottled, commonly more than 2 cm wide (flowers
1 to several in a raceme)
E   Perianth segments 3.5-4 cm long, white with a yellow base,
the base without a band of contrasting color; anthers
yellow; in damp woods above 1500'   *Erythronium helenae*
Napa Fawn Lily; Na, Sn; 4
EE Perianth segments 2.5-3.5 cm long, white to cream, with a
greenish yellow base, this base with a band of yellow-
orange or brown; anthers white; in dry areas, from the
lowlands to elevations of up to more than 1500'
Erythronium californicum (*Plate* 60)
California Fawn Lily; Sn-n

DD Leaves several, not mottled and not often more than 2 cm wide
  E Flowering stems much-branched (flowers scattered, up to
      about 3 cm wide, opening abruptly in late afternoon;
      perianth segments white, with a green or purplish
      midvein; leaves 20-70 cm long and 6-25 mm wide, with wavy
      margins)
      F Flowering stems up to more than 100 cm tall, branching
        above, but not from near the base (widespread)
                  Chlorogalum pomeridianum var. *pomeridianum*
                                Common Soap-plant
      FF Flowering stems prostrate or up to 40 cm tall,
        branching from near the base (coastal)
            Chlorogalum pomeridianum var. *divaricatum* (*Plate* 59)
                             Soap-plant; Sn-Mo
  EE Stems rarely if ever branched (but there may be several to
    many pedicels arising from the otherwise unbranched
    peduncle)
    F With at least 50 leaves that persist from one year to
      the next; flowering stems (up to more than 150 cm
      tall) with hundreds of cream-colored flowers (leaves
      2-6 mm wide and 30-100 cm long)
                        Xerophyllum tenax (*Plate* 103)
                                Beargrass; Mo-n
    FF Leaves fewer than 15, not persisting from year to
      year; flowering stems usually with fewer than 50
      flowers
      G Leaves entirely basal (less than 1 cm wide)
        H Perianth segments 3-6 mm long, white,
          yellowish, or greenish; leaves 5-20 cm long
                          Tofieldia occidentalis
                    [*T. glutinosa* ssp. *occidentalis*]
                        Western Tofieldia; Sn-n
        HH Perianth segments 20-35 mm long, blue or
          purplish blue (occasionally white); leaves 15-
          60 cm long (usually in habitats that are moist
          in winter). (Note: *Camassia quamash* and *C.*
          leichtlinii ssp. *suksdorfii* are usually
          considered to be distinct, because in the
          former the perianth segments, one of which is
          spaced farther apart than the others, turn
          back as they wither, whereas in the latter
          they are evenly spaced and twist around the
          developing fruit.)   *Camassia quamash*
                [includes *C. quamash* ssp. *linearis*
         and *C. leichtlinii* ssp. *suksdorfii*] (*Plate* 59)
                        Common Camas; Ma, Na-n
    GG Leaves mainly basal, but some stem leaves present,
      these reduced (perianth segments greenish yellow,
      cream, or white)
      H Style 1; perianth segments united at the base,
        forming a slender tube about as long as the
        free lobes; flowering stems up to 50 cm tall
        (leaves 10-30 cm long, less than 1 cm wide; in
        the Coast Ranges)   *Odontostomum hartwegii*
                      Hartweg Odontostomum; Na
      HH Styles 3; perianth segments more or less free,
        not forming a tube; flowering stems often more
        than 50 cm tall
        I Leaves 4-10 mm wide (10-40 cm long);
          stamens (in mature flowers) as long as, or
          longer than, the perianth segments (these
          4-6 mm long) (in moist or dry, grassy
          places)   *Zigadenus venenosus* (*Plate* 61)
                            Death Camas
        II Leaves usually more than 10 mm wide,
          stamens shorter than the perianth segments
          J Perianth segments 5-7 mm long; leaves
            often longer than the stems; on
            serpentine soils and usually where wet
               Zigadenus micranthus var. *fontanus*
                        [*Z. fontanus*]
                    Serpentine Star Lily; Ma

JJ Perianth segments 8-15 mm long; leaves
   shorter than the stems; widespread on
   dry slopes
                    Zigadenus fremontii (*Plate* 61)
                              Common Star Lily

Liliaceae, Subkey 1: Leaves not basal or almost entirely basal, usually at
least some substantial leaves scattered along the stem or in whorls;
perianth segments more than 1 cm long
A Perianth segments all similar, not clearly differentiated into petals
   and sepals
   B Styles not divided into 3 slender lobes (the stigma, however, is 3-
      lobed); perianth segments 4-11 cm long
      C Flowers upright; perianth (4-7 cm long) white with purple spots
         when first open, then becoming reddish (in chaparral or
         forests)   *Lilium rubescens*
                              Redwood Lily; SCr-n; 4
      CC Flowers nodding; perianth not primarily white when first open,
         and usually with considerable orange or yellow (in moist areas)
         D Perianth 5-11 cm long, mostly yellow, with tinges of red and
            with maroon spots
                    Lilium pardalinum ssp. *pardalinum* (*Plate* 60)
                                          Leopard Lily
         DD Perianth 5-7 cm long, mostly deep orange, but green at the
            center, crimson on the outside, and usually dotted with
            maroon   *Lilium pardalinum* ssp. *pitkinense* [*L. pitkinense*]
                              Pitkin Marsh Lily; Sn; 1b
   BB Styles divided into 3 slender lobes; perianth segments not more than
      4 cm long
      C Leaves present on the upper part of the stem, but not on the part
         close to the ground (flowers nodding)
         D Perianth segments orange-red or scarlet, with some yellow
            markings on the inner surface (the segments 1.5-4 cm long,
            conspicuously curled up at the tip) (in shaded areas)
                                          Fritillaria recurva
                    [includes *F. recurva* var. *coccinea*] (*Plate* 60)
                              Scarlet Fritillary; Na, Sl-n
         DD Perianth segments not orange-red
            E Perianth segments 1-4 cm long, mostly purplish brown, with
               greenish yellow markings, but sometimes greenish yellow
               with purple mottling almost throughout, the tips straight
               (in shaded areas)
                    Fritillaria affinis [*F. lanceolata*] (*Plate* 60)
                                          Checker Lily
            EE Perianth segments 1-2 cm long, mostly greenish yellow, but
               often marked with reddish purple, the tips slightly bent
               back   *Fritillaria eastwoodiae* [*F. phaeanthera*]
                              Butte County Fritillary; Na; 1b
      CC Leaves present on the lower part of the stem
         D Perianth segments mottled, at least on one surface (usually
            on serpentine)
            E Flowers nodding; perianth segments up to 3 cm long,
               whitish, with purple spots and lines, the tips sometimes
               bent back; leaves not sickle-shaped   *Fritillaria purdyi*
                              Purdy Fritillary; Na-n; 4
            EE Flowers upright; perianth segments up to about 2 cm long,
               greenish on the outer surface, yellow and rust-brown
               inside, the tips straight; leaves sickle-shape
                                          Fritillaria falcata
                              Talus Fritillary; SCl; 1b
         DD Perianth segments not mottled, but they may have green lines
            (tips mostly straight) (flowers nodding)
            E Perianth segments (up to 4 cm long) dark brown, purplish,
               or greenish purple on both surfaces   *Fritillaria biflora*
                              Chocolate Fritillary; Na, SM-s
            EE Perianth segments not dark brown, purplish, or greenish
               purple on both surfaces (at least the outer surface will
               be light)
               F Perianth segments up to 16 mm long, white, with green
                  lines; flowers slightly fragrant *Fritillaria liliacea*
                              Fragrant Fritillary; Sn-Mo; 1b
               FF Perianth segments 18-35 mm long, greenish white on the
                  outer surface, purplish brown inside; flowers with an
                  unpleasant odor   *Fritillaria agrestis*
                              Stinkbells; Me-SLO; 4

AA Petals and sepals distinctly different
  B Flowers bell-shaped, generally nodding
    C Petals white (2-2.5 cm long)   *Calochortus albus*
                               White Globe Lily; SF-s
    CC Petals yellow
      D Inner surface of petals without hairs, or with only a few
         hairs near the basal gland (petals 16-20 mm long, deep
         yellow; north Coast Ranges)   *Calochortus amabilis*
                     Golden Globe Lily; Ma, Na, Sn-n
      DD Much of the inner surface of petals hairy (petals pale
         yellow)
        E Petals 25-33 mm long, about as long as the sepals,
           sparsely hairy on the inner surface; stems usually
           branched; in woodlands or chaparral
                Calochortus pulchellus (*Plate* 59)
                Mount Diablo Fairy-lantern; CC(MD); 1b
        EE Petals 35-45 mm long, longer than the sepals, with long
           hairs on the inner surface; stems mostly not branching;
           in open areas, on serpentine   *Calochortus raichei*
                     Cedars Fairy-lantern; Sn; 1b
  BB Flowers broadly bowl-shaped, generally facing up (in open areas)
    C Petals mainly yellow or yellow-green, sometimes with red-brown or
       purple on the inner surface
      D Petals fan-shaped at the tip, scarcely hairy, except for
         short hairs at the base of the inner surface; petals deep
         yellow, sometimes with lines or a red-brown mark inside
                  Calochortus luteus (*Plate* 58)
                    Yellow Mariposa Lily; Me-s
      DD Petals pointed at the tip, with long hairs over much of the
         inner surface, or at least along the inner margin and at the
         base; petals greenish yellow, with purplish markings inside
         (on serpentine)   *Calochortus tiburonensis* (*Plate* 59)
                   Tiburon Mariposa Lily; Ma; 1b
    CC Petals white, cream, lavender, or lilac (occasionally yellow),
       often with red, yellow, or purple on the inner surface
      D Petals (12-25 mm long) with long hairs over much of the inner
         surface (petals white, pinkish, or purplish, without markings
         at the base)   *Calochortus tolmiei*
                       Pussy-ears; SCr-n
      DD Petals with hairs only at the base of the inner surface
        E Petals 12-18 mm long (petals white or pink, the inside
          often with a purple mark at the base; stems usually
          branched, up to 25 cm tall; often on serpentine soils)
                Calochortus umbellatus (*Plate* 59)
                Oakland Startulip; La-SCl; 4
        EE Petals at least 20 mm long (smaller in *Calochortus*
         uniflorus)
         F Gland at the base of the inner surface of petals
           bordered by a fringed membrane (petals 2-4 cm long,
           whitish with green stripes, the inside often with
           purple spots at the base; basal leaves 2-4 mm wide;
           stems rarely branched, up to 50 cm tall; in dry areas,
           at elevations above 4000')   *Calochortus invenustus*
                   Plain Mariposa Lily; SCl(MH)-s
        FF Gland at base of the inner surface of petals not
         bordered by a membrane
         G Petals 15-28 mm long (lilac, the inside sometimes
           with a purple spot near the base); basal leaves 5-
           20 mm wide; stems rarely branched, less than 5 cm
           tall; in moist areas   *Calochortus uniflorus*
                   Largeflower Startulip; Me-Mo
         GG Petals usually more than 25 mm long (up to 50 mm
           long); basal leaves not more than 6 mm wide; stems
           usually branched, up to 60 cm tall; in dry areas
          H Petals deep lilac, the inner surface sometimes
            with a purple spot; hairs at the base of the
            inner surface collectively resembling a growth
            of fungus; longer hairs scattered on the lower
            half of the inner surface (in Coast Ranges)
                    Calochortus splendens
                  Splendid Mariposa Lily; La-s

HH Petals white to lavender or pale lilac
(occasionally yellow), the inner surface with
a red, purple, or brown mark, this often
surrounded by yellow; hairs absent from the
base of the inner surface, but present about 5
mm from the base, this concentration not
funguslike; hairs not present on the rest of
the inner surface
I   Concentration of hairs on the inner surface
of petals rectangular (inner surface with a
dark red mark in the center and usually
with a paler red blotch above)
Calochortus venustus (*Plate* 59)
Butterfly Mariposa Lily; SF-s
II Concentration of hairs on the inner surface
of petals not retangular
J   Concentration of hairs on the inner
surface of petals shaped like a
crescent or an inverted V; inner
surface with a brown or purplish
central mark, this surrounded by bright
yellow   *Calochortus superbus*
Superb Mariposa Lily
JJ Concentration of hairs on the inner
surface of petals shaped like two
crescents or inverted Vs; inner surface
with a red-brown mark, this surrounded
by pale yellow (at elevations above
1500')   *Calochortus vestae*
Clay Mariposa Lily; Na, Sn-n

Liliaceae, Subkey 2: Leaves usually basal; flowers in an umbel (or umbel-
like raceme), this usually with at least 3 bracts below it; perianth
segments united at the base, thus forming at least a short tube
A Flowering stems up to more than 100 cm long, often twining through
shrubbery or lying on the ground, but the stem contorted even if upright
(perianth rose, pinkish, or purplish, narrowed in the middle, the tube
5-7 mm long; functional stamens 3; inland, in dry areas)
Dichelostemma volubile [*Brodiaea volubilis*]
Snake Lily; Sl-n
AA Flowering stems rarely more than 50 cm long, upright, not contorted
B   Perianth some shade of yellow (the segments with a dark midvein on
the outer surface; functional stamens 6, 3 longer than the others;
in dry areas)
C   Perianth deep yellow, the tube 7-10 mm long; filaments deeply
divided into 2 lobes, the anthers in the clefts (coastal)
Triteleia ixioides [*Brodiaea lutea*]
Golden Triteleia; SM-SLO
CC Perianth deep to pale yellow, the tube 4-5 mm long; filaments not
divided into 2 lobes
Triteleia lugens [*Brodiaea lugens*] (*Plate* 61)
Uncommon Triteleia; Sn-Sl
BB Perianth purple, violet, lavender, blue, or white
C   Umbel (or umbel-like raceme) dense, congested, the pedicels not
more than 1.5 cm long (usually much shorter)
D   With 6 functional stamens (these are of 2 different sizes,
but all have anthers); perianth tube not narrowed just below
the lobes; peduncle usually smooth (perianth segments usually
blue, bluish purple, or pinkish purple, but sometimes white;
widespread)
Dichelostemma capitatum [*Brodiaea pulchella*] (*Plate* 102)
Bluedicks
DD With 3 functional stamens (if 3 sterile stamens are present,
these are small and lack anthers); perianth tube noticeably
narrowed just below the lobes; peduncle often slightly rough
E   Inflorescence a much-condensed raceme; 3 small sterile
stamens present (perianth segments pinkish purple or
bluish purple; widespread)   *Dichelostemma congestum*
[*Brodiaea congesta*] (*Plates* 60, 102)
Ookow; SCl-n
EE Inflorescence more like an umbel than a raceme; sterile
stamens absent (perianth segments pinkish purple or
violet)   *Dichelostemma multiflorum* [*Brodiaea multiflora*]
Manyflower Bluedicks; SM, SCl-n
CC Umbel not dense, most of the pedicels at least 2 cm long

D Perianth white or very pale blue or lilac
    E With 3 functional stamens and 3 sterile stamens; perianth
       segments without a green midvein (often on serpentine)
       (perianth 3-4 cm long)
                             Brodiaea californica var. *leptandra*
                                 California Brodiaea; Sn, Na
    EE With 6 functional stamens; perianth segments with a green
       midvein
       F Perianth 1-1.5 cm long; pedicels up to 5 cm long; all
          stamens the same length and attached at the same level
                Triteleia hyacinthina [*Brodiaea hyacinthina*]
                                  White Triteleia; Mo-n
       FF Perianth 1.5-2.5 cm long; pedicels up to more than 10
          cm long; stamens of different lengths and attached at
          different levels (often on serpentine)
                    Triteleia peduncularis [*Brodiaea peduncularis*]
                            Longray Triteleia; Mo-n
DD Perianth bright to dark blue, violet, or purple (occasionally
   white in *Triteleia laxa*)
    E With 6 functional stamens (these attached at 2 levels)
       (perianth blue, blue-purple, or white, 2-3.5 cm long;
       pedicels up to 9 cm long; widespread)
                    Triteleia laxa [*Brodiaea laxa*] (*Plate* 61)
                                Ithuriel's-spear
    EE With 3 functional stamens and 3 sterile stamens (these
       without anthers)
       F Lobes of the perianth not more than 1.5 cm long
          (usually much less); flowering stems not more than 7
          cm tall (sterile stamens 4-8 mm long, about equal to
          the functional stamens)
          G Pedicels 1-5 cm long; filaments of functional
             stamens with 2 slender outgrowths arising just
             below the anthers (Coast Ranges)
                              Brodiaea stellaris
                        Starflower Brodiaea; Sn-n
          GG Pedicels 3-15 cm long; filaments of functional
             stamens without slender outgrowths arising just
             below the anthers   *Brodiaea terrestris*
                       [*B. coronaria* var. *macropoda*] (*Plate* 58)
                            Dwarf Brodiaea; SLO-n
       FF Lobes of the perianth at least 1.5 cm long (usually
          longer); flowering stems more than 10 cm tall
          G Sterile stamens flat, 6-9 mm long, usually equal to
             the functional stamens (pedicels 5-10 cm long;
             filaments of functional stamens without slender
             outgrowths arising just below the anthers)
                            Brodiaea elegans (*Plate* 58)
                         Elegant Brodiaea; Mo-n
          GG Sterile stamens slightly inrolled, 8-15 mm long,
             longer than the functional stamens
             H Pedicels 4-10 cm long; filaments of functional
                stamens with 2 slender outgrowths arising just
                below the anthers   *Brodiaea appendiculata*
                        Grassland Brodiaea; Na-SCl
            HH Pedicels 1-5 cm long; filaments of functional
               stamens without slender outgrowths arising
              just below the anthers (mostly inland)
                            Brodiaea coronaria
                            Harvest Brodiaea

Liliaceae, Subkey 3: Leaves basal, not more than 1.5 cm wide (but these may
have withered by flowering time); flowers in umbels, these with 2-4 united
bracts beneath them; perianth segments not forming a tube   *Allium*
A Leaves flattened, often more than 5 mm wide, sometimes sickle-shaped;
   flowering stems flattened or 3-angled (sometimes cylindrical in *Allium
   cratericola*) (in dry areas)
    B Perianth white; leaves 2-3, not sickle-shaped; flowering stems 3-
       angled (2 of the edges somewhat winged) (escaping from gardens)
                                     Allium neapolitanum
                    Naples Onion; me; Na, Sn, Ma, SF
    BB Perianth rose to purple; leaves 1-2, generally sickle-shaped;
       flowering stems usually flattened (often on serpentine)
       C With 2 leaves; perianth rose to purple; stems winged; found
          mostly at elevations above 3000'   *Allium falcifolium* (*Plate* 58)
                                  Sickleleaf Onion; SCr-n

CC Usually with only 1 leaf (this sometimes straight); perianth
purple, but sometimes pale; stems not winged; usually found at
elevations below 2500'   *Allium cratericola*
                                                Crater Onion; Na-n
AA Leaves, whether flattened or nearly cylindrical, generally less than 5
mm wide, not sickle-shaped; flowering stems cylindrical
B Plants generally in scattered, dense colonies, owing to production of
bulbs along underground stems; in moist areas (leaves 2-3; pedicels
up to 2.5 cm long; perianth pink or lilac, occasionally white)
                                                Allium unifolium (*Plate* 58)
                                                Clay Onion; Mo-n
BB Plants not in dense colonies (even when abundant, the plants are
usually separate, having originated from seed, rather than from
underground stems (even if they have underground stems, they do not
produce large colonies); in dry areas
C With 6 prominent outgrowths at the tip of the fruiting portion of
the pistil
D With 2-4 leaves (perianth 5-9 mm long)
E Perianth segments nearly 3 times as long as wide, white or
tinged with pink; leaves 2-4, cylindrical; pedicels 4-16
mm long (sometimes on serpentine)   *Allium amplectens*
                                         Paper Onion; Al, SCl
EE Perianth segments not more than twice as long as wide,
rose to purple (rarely white) with a darker base; leaves
2, more or less flattened; pedicels 10-20 mm long
                                         Allium campanulatum
                                         Sierra Onion; Mo-n
DD With only 1 leaf (this cylindrical)
E Perianth 6-12 mm long, dark red-purple; stamens shorter
than the perianth (often on serpentine) *Allium fimbriatum*
                                         Fringed Onion; Na-s
EE Perianth 5-8 mm long, lavender or white; stamens equal to,
or slightly longer than, the perianth   *Allium howellii*
                                         Howell Onion; SCl
CC Without outgrowths at the tip of the fruiting portion of the
pistil (stamens shorter than the perianth)
D Perianth segments with a prominent green or red midvein
(otherwise white to pale pink); leaves 2 (perianth 4-9 mm
long; pedicels 5-12 mm long; sometimes on serpentine
                                         Allium lacunosum (*Plate* 58)
                                         Wild Onion; Ma-s
DD Perianth segments without a prominent green or red midvein;
leaves often more than 2
E Margin of upper portion of inner perianth segments
roughened by very small, nearly hemispherical outgrowths
(use hand lens) (leaves 2-3; perianth 8-15 mm long)
F Slender tip of outer perianth segments curving
backward (perianth white to purplish rose; pedicels 6-
25 mm long; inflorescence with up to 40 flowers)
                                         Allium acuminatum (*Plate* 58)
                                         Hooker Onion; CC(MD)-n
FF Slender tip of outer perianth segments not distinctly
curved
G Perianth red-purple (sometimes white), the segments
about 3 times as long as wide, widest below the
middle; inflorescence with up to 20 flowers;
pedicels up to 20 mm long (sometimes on
serpentine)   *Allium bolanderi*
                                      Bolander Onion; SCl
GG Perianth rose-purple, the segments about twice as
long as wide, widest near the middle;
inflorescence with up to 40 flowers; pedicels up
to 35 cm long (in sandy soil)   *Allium crispum*
                                      Crinkled Onion; CC-s
EE Margin of upper portion of inner perianth segments nearly
smooth
F Perianth pink to rose (8-11 mm long) (leaves 2-3;
pedicels 7-15 mm long; filaments about 3 times as wide
at the base as over most of their length; widespread)
                                         Allium serra (*Plate* 58)
                                         Serrated Onion; La-SCl
FF Perianth red-purple

G   Leaves 2-3; perianth 10-15 mm long; pedicels 10-40
    mm long; stigmas often 3-lobed   *Allium peninsulare*
                          Peninsular Onion; SFBR-s
GG Leaves 3-6; perianth 9-12 mm long; pedicels 5-20 mm
    long; stigmas not lobed (coastal)
                                Allium dichlamydeum
                             Coastal Onion; Me-Mo

Orchidaceae--Orchid Family

The Orchid Family, consisting entirely of perennials, is one of the
largest groups of flowering plants, with over 15,000 species. All of ours,
unlike most of those in the tropics, are rooted in soil rather than in bark
or in moss growing on trees. The flowers are markedly 2-lipped, the lowest
of the 3 petals (the lip petal) being different from the other 2. The 3
sepals are similar to one another. There is a single functional stamen in
all of our species, except those of the genus *Cypripedium*, which have 2
functional stamens. The filaments of the stamens are united to the style of
the pistil. The fruiting part of the pistil, partitioned lengthwise into 3
divisions, is below the level where the other flower parts originate. Each
fruit produces numerous small seeds.

Orchids, like many other plants, have a symbiotic relationship with
fungi that penetrate their roots. Some species, in fact, have no
chlorophyll, and depend entirely on their fungal associates. The complex of
orchid and fungus requires very special conditions, and these, at least in
the case of terrestrial orchids, are not likely to be met in gardens.
Attempts to cultivate terrestrial orchids will almost certainly result in
failure, and it is therefore a mistake to disturb these plants in nature,
unless they happen to be in the path of a bulldozer. Even then it would be
best for experts in botanical gardens to try to transplant them to new
situations. It should be noted, however, that *Epipactis helleborine*, a
European species, is now a benign weed in certain portions of the Pacific
Northwest, and has become established in some places within the San
Francisco Bay Region.

A   Plants saprophytic, lacking chlorophyll, therefore not green; leaves
    reduced to scales (found in dark coniferous woods)
    B   Plants white, except for a slight yellow tinge on the petals
                    Cephalanthera austiniae [*Eburophyton austinae*] (*Plate* 61)
                                          Phantom Orchid; Mo-n
    BB Plants mostly brownish (sometimes yellowish in *Corallorhiza*
        maculata), the petals with dark spots or streaks
        C   Lip petal white, with purple spots, the other petals and the
            sepals mostly brownish purple (in certain areas, some plants may
            be yellowish)   *Corallorhiza maculata* (*Plate* 62)
                                          Spotted Coralroot
        CC Lip petal and other petals, as well as the sepals, with purplish
            or reddish brown streaks   *Corallorhiza striata* (*Plate* 62)
                                          Striped Coralroot; SCl(SCM)-n
AA Plants with green foliage; leaves ample, not reduced to scales
    B   Lip petal inflated, the portion that projects forward closed to form
        a sac; flowers 1 or a few (except in *Cypripedium californicum*)
        C   Plants with a single basal leaf and a single flower (predominant
            color of petals and sepals rose-purple, although other colors
            are evident; in damp woods)   *Calypso bulbosa* (*Plate* 61)
                                          Fairy-slipper; SM, Ma-n
        CC Plants either with 2 opposite leaves or several alternate leaves,
            and usually with more than 1 flower
            D   Leaves 2, opposite; flowers (several) in a cluster not far
                above the leaves; up to about 25 cm tall (petals and sepals
                up to about 2.5 cm long, mostly greenish brown or greenish
                yellow, and with brown streaks or margins; in rocky areas)
                                          Cypripedium fasciculatum
                                  Clustered Lady's-slipper; SCr-n; 4
            DD Leaves several, alternate; flowers solitary in the leaf
                axils; up to about 50 cm tall

E   Flowers 1-3; lip petal white, streaked with purple; other
    petals (2 of which are twisted) up to 5 cm long, mostly
    dark brown; in moist woods   *Cypripedium montanum*
                          Mountain Lady's-slipper; SCl(SCM)-n; 4
EE Flowers up to more than 10; lip petal white or pale rose,
    streaked or spotted with purple; other petals (none of
    these twisted) less than 2.5 cm long, mostly greenish or
    brownish yellow; in wet rocky areas
                          Cypripedium californicum (*Plate* 62)
                          California Lady's-slipper; Ma-n; 4
BB Lip petal not inflated, the portion that projects forward not closed
   to form a sac (but the basal portion of this petal may be deeply
   concave or drawn out into a long spur); flowers generally numerous
   (except in *Epipactis gigantea*)
   C   Flowers with distinct pedicels
       D   Sepals and petals commonly more than 1.2 cm long, the lip
           petal up to 2 cm long and 3-lobed (flowers rarely more than
           10; mostly around springs, seepage areas, small streams, and
           ponds)   *Epipactis gigantea* (*Plate* 62)
                                                          Stream Orchid
       DD Sepals and petals not more than 1.2 cm long, the lip petal
           not lobed   *Epipactis helleborine* (*Plate* 62)
                                                  Helleborine; eu; SFBR
   CC Flowers without distinct pedicels
       D   Lip petal not drawn out into a spur (in the Coast Ranges)
           E   Leaves dark green with white veins, persisting throughout
               the year; inflorescence not obviously twisted; in dry
               areas, mostly in woods   *Goodyera oblongifolia* (*Plate* 104)
                                                    Rattlesnake-plantain; Ma-n
           EE Leaves uniformly light green, generally withering by the
               time of flowering; inflorescence obviously twisted; in
               moist or wet areas
               F   Perianth usually white (sometimes cream); lip petal
                   rounded at the tip; in moist meadows
                                       Spiranthes romanzoffiana (*Plate* 104)
                                                  Hooded Lady's-tresses
               FF Perianth usually yellowish (sometimes cream); lip
                   petal usually with a pointed tip; usually in bogs and
                   marshes   *Spiranthes porrifolia*
                                                  Western Lady's-tresses; Mo-n
       DD Lip petal drawn out into a spur
           E   Inflorescence with leaves along much of its length; in wet
               areas or near streams (flowers pure white, the spur of
               the lip petal conspicuously curved; lip petal abruptly
               broadened near the base)   *Platanthera leucostachys*
                                  [*Habenaria dilatata* var. *leucostachys*]
                                          Whiteflower Bog Orchid; SLO-n
           EE Inflorescence without leaves (all leaves basal, and
               usually withering before flowering time); mostly in dry
               areas
               F   Spur of the lip petal twice as long as this petal, and
                   also much longer than the fruiting part of the pistil;
                   inflorescence up to 60 cm tall   *Piperia elegans*
                   [includes *Habenaria elegans* vars. *elegans* and *maritima*]
                                                  Elegant Rein Orchid
               FF Spur of the lip petal only slightly longer than the
                   petal, and shorter than the fruiting part of the
                   pistil; inflorescence up to 30 cm tall (in chaparral,
                   oak woodland, grassland)   *Piperia unalascensis*
                                  [*Habenaria unalascensis*] (*Plate* 62)
                                                  Alaska Rein Orchid

Poaceae (Gramineae)--Grass Family

by Richard G. Beidleman

With the exception of bamboos, which are woody and sometimes treelike,
most grasses are easily recognized as such. In our region, this family,
with about 200 species, is as well represented as the Sunflower Family.
Unfortunately, about 40% of our grasses were introduced from other parts of
North America, Europe (especially the Mediterranean area), Asia, Central

and South America, Africa, and Australia. Beginning with the early Spanish
colonists, in the eighteenth century, the seeds of exotic grasses arrived
with hay, ballast, mattress fillers, and clothing, as well as in the hair
of farm animals. Some species, moreover, were introduced intentionally,
either as crop plants or for landscaping and erosion control.

The importance of grasses--especially wheat, barley, oats, rice, corn,
and rye--in providing food for humans and livestock should be understood by
everyone. Many species, furthermore, are cultivated for their ornamental
foliage, interesting habit of growth, erosion-controlling attributes, and
suitability for lawns. So successfully have introduced grasses taken over
wild areas, however, that native species are no longer found in many
places. Accounting in part for the decline in the number of native species
are overgrazing by livestock, conversion of natural areas into cropland,
and urbanization, with its varied impacts. About 20% of our native grasses
are annuals; the rest are perennials. Some of our perennial species are
being propagated for use in landscaping. They are not only attractive, but
survive summer drought and fit in with other native plants.

As common as grasses are, and as simple as they may appear to be, they
are difficult to identify. The parts that are generally used in
identification tend to be small--so small in some cases that even a 10x
hand lens is not adequate. Furthermore, certain structures of grasses are
unlike those of other plants. Thus they require a special set of terms. In
the key presented here, technical terms have been kept to a minimum. Each
major unit of the inflorescence is called a spikelet. At the base of it are
two glumes, and above these are one or more florets. A floret typically
consists of a pistil and/or stamens (usually 3), enclosed within two
bracts. The lower bract, known as the lemma, is almost always larger and
better developed than the upper one (the palea) and at least partially
encloses the palea. The pistil becomes a 1-seeded dry fruit, and in the key
this is referred to as the grain. Some florets may be sterile, lacking
structures present in fertile florets.

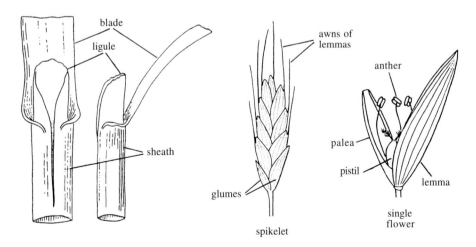

The lemma and glumes are especially important in identification of
grasses. It is necessary to examine them carefully to see if they have
hairlike or bristlelike projections at the tip called awns, riblike
lengthwise thickenings (veins), or other differentiations.

In the field, when you first encounter a grass, be sure to observe its
overall growth form, and especially the appearance of parts that may be
disturbed upon collection. For instance, the stems and leaves are often
grouped together in a dense basal clump; other types are simple individual

plants, or form a sod, or spread outward by means of prostrate stems. You will have to look closely at the leaf blades and leaf sheaths. The blades may be flat, folded, long or short, or have some special attributes. On the upper side of the base of the blade, where this attaches to the sheath, there is usually a collarlike structure, or a circle of long hairs, called the ligule, and there may also be a pair of small lobes. The edges of the sheath may overlap or be partly or completely fused. The hairiness of some parts of the plant may be important. The general appearance of the inflorescence is also used extensively in the key. Some inflorescences are very open, with spreading branches and branchlets, while others are short-branched and compact, at least early in their development.

Many species exhibit considerable variation and may appear in more than one section of the key. Furthermore, certain grasses freely interbreed, and even experts are often unable to agree on identification of a particular specimen. So even if you just come close, you are to be congratulated!

A Stems at least 1 cm wide, tough and inflexible; leaves up to more than 5 cm wide (inflorescence present much of the year, plumelike, often more than 50 cm long; up to more than 3 m tall)
   B Leaves evenly spaced on the upright stem; inflorescence up to 12 cm wide and 60 cm long (generally in moist places, but sometimes on dry river bottoms)   *Arundo donax* (*Plate* 62)
                                                          Giantreed; eu
   BB Leaves mainly basal, forming a dense clump more than 2 m high; inflorescence up to 15 cm wide and more than 100 cm long
      C Leaf sheaths densely hairy; leaf blades dark green, not curled toward the tip; up to 7 m tall (noxious weed)   *Cortaderia jubata*
                                                  Hairy Pampas Grass; sa
      CC Leaf sheaths only sparsely hairy, if at all; leaf blades pale green or blue-green, curled toward the tip; up to 4 m tall
                                                      Cortaderia selloana
                                                  Smooth Pampas Grass; sa
AA Stems usually less than 0.5 cm wide, flexible; leaves not more than 2 cm wide (except in *Sorghum* and species of *Spartina*, in Division IV)
   B Spikelets partly concealed in each inflated leaf sheath, the blades falling off early; leaves 3-6 cm long, stiff, spreading outward, usually several on each stem; plants branching extensively at the base, forming mats   *Crypsis vaginiflora* [*C. niliaca*]
                                                      Prickle Grass; eu
   BB Plants not conforming in all respects to the description in choice B
      C Spikelets and/or florets either situated above stout bristles that arise from the base of the spikelet (*see Hordeum, Plate* 108) (the bristles may be modified glumes), or enclosed within a spiny bur (inflorescence sometimes headlike)
                                          Poaceae, Division I (p. 271)
      CC Spikelets and/or florets neither enclosed within burs nor situated above stout bristles
         D Spikelets partly sunken into or pressed against concave sections of the stem (*see Lolium perenne, Plate* 108) (as a spikelet matures, however, it may bend outward) (inflorescence not branched)   Poaceae, Division II (p. 273)
         DD Spikelets not sunken into or pressed against the stem
            E Inflorescence headlike (*see Ammophila, Plate* 62) (usually very dense and compact, but sometimes loose and interrupted), either without branches or with the spikelets so concentrated that the branches and branchlets, if present, are difficult to distinguish
                                          Poaceae, Division III (p. 275)
            EE Inflorescence not headlike (usually open and spreading), generally with obvious branches
               F Inflorescence with obvious branches, these often divided into branchlets or long pedicels that resemble branchlets (*see Aira, Plate* 104)
                                          Poaceae, Division IV (p. 282)
               FF Inflorescence not branched (spikelets usually on very short pedicels) (lemma with 7 parallel veins and a terminal awn, this arising between 2 membranous teeth; prominent teeth present on the upper bract of the floret [the palea]; glumes papery, without awns; up to 1.5 m tall)

G   Lemma about 5 mm long, with awn more than 6 mm
    long; plants forming clumps (in moist areas)
                              Pleuropogon californicus
                          California Semaphore Grass; Al-n
GG Lemma about 8 mm long, with awn up to 3 mm long;
    plants not forming clumps, upright
                              Pleuropogon hooverianus
                      North Coast Semaphore Grass; Me-Ma; 1b

Poaceae, Division I: Spikelets and/or florets either situated above stout
bristles that arise from the base of the spikelet (the bristles may be
modified glumes), or enclosed within a spiny bur (inflorescence sometimes
headlike)
A   Spikelets enclosed within a spiny bur (in disturbed areas, especially
    sandy places; annual, up to 50 cm tall, the stems sometimes falling)
    B   Bur with a dense ring of slender bristles around its base (with
        larger spines on the inner lobes of the bur)   *Cenchrus echinatus*
                                              Southern Sandbur; na; Sl
    BB Bur without a ring of slender bristles around its base, but with many
        stout spines throughout
        C   Bur 4-6 mm long (including spines), with about 12 spines, these
            mostly broader at the base   *Cenchrus incertus*
                                              Coast Sandbur; na
        CC Bur 6-12 mm long (including spines), with more than 14 spines,
            these mostly narrow throughout *Cenchrus longispinus* (*Plate* 106)
                                              Mat Sandbur; na; Sl, Mo
AA Spikelets (or florets) not enclosed within a bur, but situated above or
    among what appear to be 2 or more bristles (if there are only 2, they
    may be modified glumes; in some cases, both glumes and florets appear
    bristlelike)
    B   Spikelets of two very different types in the same inflorescence: the
        fertile spikelets, containing several florets (5 mm long) that often
        fall off early, are hidden among the bristly sterile spikelets that
        have empty florets (inflorescence globular, one-sided, with awns up
        to 2 cm long)   *Cynosurus echinatus* (*Plate* 63)
                                              Hedgehog Dogtail; eu
    BB All spikelets in the inflorescence similar
        C   Inflorescence with 2-4 spikelets, these completely hidden within
            the leaf sheath; stems prostrate and rooting (basal bristles
            long and slender, without plumelike hairs; leaves and stems
            hairy)   *Pennisetum clandestinum*
                                              Kikuyu Grass; af; SF, Al
        CC Inflorescence with numerous spikelets in a loose to compact
            cylindrical head at the end of the stem; stems upright
            D   Bristles numerous (2-4 cm long), the basal portion densely
                covered with delicate, plumelike hairs (inflorescence up to
                35 cm long, often lavender; plants forming dense clumps)
                                              Pennisetum setaceum
                                              Fountain Grass; af; Al
            DD Bristles 2 or more, the basal portion without plumelike hairs
                (but there may be small, unbranched hairs)
                E   Spikelets 1 or 2 at each node of the inflorescence (Note:
                    the spikelets in this choice are sometimes difficult to
                    see, but those of *Hordeum*, keyed under choice EE, are
                    easily seen.)
                    F   Spikelets without awns; fertile floret 1 (the lower
                        floret sterile); glumes not bristlelike (and shorter
                        than the florets) (bristles 3-15; spikelets 3 mm long,
                        separating from the bristles at maturity)
                        G   Bristles 3-8 mm long, usually 5-15 below each
                            spikelet (the bristles are most easily counted at
                            the base of the inflorescence); leaf blades with
                            long, delicate white hairs only at the base;
                            inflorescence upright, 2-8 cm long; lemma of
                            fertile floret ridged crosswise, the sterile lemma
                            5-veined   *Setaria pumila* [*S. glauca*]
                                              Bristly Foxtail; eu
                        GG Bristles more than 9 mm long, usually 3 (sometimes
                            4 or 5) below each spikelet; leaf blades usually
                            uniformly covered with short hairs; inflorescence
                            drooping, 6-20 cm long; lemma of fertile floret
                            not ridged crosswise, the sterile lemma 7-veined
                                              Setaria faberi
                                              Foxtail; as; Ma, Sl

FF Spikelets with awns (but these sometimes very short);
   fertile florets 2 or more; glumes bristlelike
   G Awns less than 0.5 cm long; spikelets more or less
      flattened, with one of the broad sides facing the
      stem
      H Inflorescence usually less than 6 cm long and
         up to 1.5 cm wide; spikelet 1 at each node; up
         to 60 cm tall, the leaves often extending to
         or beyond the tip of the inflorescence
         (coastal)   *Leymus pacificus* [*Elymus pacificus*]
                                    Pacific Wild Rye; Me-Mo
      HH Inflorescence usually more than 9 cm long and
         less than 0.8 cm wide; spikelets usually 2 at
         each node; up to 130 cm tall, the leaves
         generally not extending beyond the mature
         inflorescence (in moist, alkaline areas)
                      Leymus triticoides [*Elymus triticoides*]
                                    Wet-meadow Wild Rye
   GG Awns up to 10 cm long; spikelets not obviously
      flattened (usually 2 at each node, but sometimes
      either 1 or 2 at each node in *Taeniatherum caput-
      medusae*)
      H Glumes and lemma sometimes with more than 1
         awn; glumes (including the awns) 2.5-8.5 cm
         long; florets falling with the glumes at
         maturity (glumes with evident veins; lemma awn
         up to 10 cm long; leaf blades more than 1.5 mm
         wide; stems 1-3 mm wide, not wirelike)
         I Glumes divided into 3 or more bristlelike
            parts; inflorescence dense, usually not
            hidden in the leaf sheath
                                    Elymus multisetus
                           [*Sitanion jubatum*] (Plate 64)
                                    Big Squirreltail
         II Glumes undivided, or divided into only 2
            parts; inflorescence sparse, often partly
            hidden in the leaf sheath
                      Elymus elymoides [*Sitanion hystrix*]
                                    Squirreltail
      HH Glumes and lemma each with 1 awn; glumes
         (including the awns) not more than 4 cm long;
         florets falling from the glumes at maturity
         I Bristlelike glumes (including the awns)
            1.5-4 cm long, not so wide as 1 mm at the
            base, without obvious veins; lemma awn up
            to 10 cm long (tending to spread outward);
            leaf blades less than 1 mm wide, mostly
            basal; stems less than 1 mm wide, wirelike;
            inflorescence nearly as wide as long
                      Taeniatherum caput-medusae
                           [*Elymus caput-medusae*]
                                    Medusahead; eu
         II Bristlelike glumes (including the awns)
            less than 2.5 cm long, widened to 1 mm at
            the base, with obvious veins; lemma awn up
            to 4 cm long; leaf blades up to 15 mm wide,
            not mostly basal; stems up to 4 mm wide;
            inflorescence much longer than wide
                                    Elymus canadensis
                           Canadian Wild Rye; na; Sn-n
EE Spikelets 3 at each node of the inflorescence (*Elymus
   canadensis*, keyed in choice E, above, sometimes has 3 or
   4 spikelets at each node.) (each group of 3 spikelets
   readily detaching from the stem; lateral spikelets often
   present only as 1-3 bristles; awns usually less than 8 cm
   long; inflorescence sometimes partly enclosed in the leaf
   sheath)
   F Mature inflorescence almost as wide as long, nodding;
      awns 4-8 cm long (delicate and eventually spreading)
      (base of leaf blades without overlapping membranous
      lobes; lateral florets reduced, much shorter than the
      central floret)   *Hordeum jubatum*
                                    Foxtail Barley

FF Inflorescence much longer than wide, not nodding; awns
   generally less than 5 cm long
   G   Some bristles fringed with delicate hairs; base of
       leaf blades with 2 overlapping membranous lobes;
       lateral florets not reduced, sometimes longer than
       the central floret (Note: the following 2 species
       are difficult to separate.)
       H   Florets about 1.5 mm wide; lemma awn of the
           central floret 10-25 mm long; inflorescence
           with 5-7 groups of 3 spikelets in each cm
                     Hordeum murinum ssp. *glaucum* [*H. glaucum*]
                                       Glaucous Barley; eu
       HH Central florets about 2-3 mm wide; lemma awn of
           the central floret 25-45 mm long;
           inflorescence with 3-5 groups of 3 spikelets
           in each cm (widespread)   *Hordeum murinum*
                   ssp. *leporinum* [*H. leporinum*] (*Plate* 64)
                                          Hare Barley; eu
   GG Bristles not fringed with hairs (but there may be
       small, stiff barbs); base of leaf blades without
       overlapping membranous lobes; lateral florets
       often reduced, not longer than the central floret
       H   Leaf blades (3-8 mm wide) not covered with fine
           hairs (but scattered hairs sometimes present);
           lateral spikelets on curved pedicels (central
           spikelet sessile; inflorescence sometimes
           purplish; lemma awn up to 10 mm long; plants
           forming dense clumps, in moist areas)
                Hordeum brachyantherum ssp. *brachyantherum*
                                          Meadow Barley
       HH Leaf blades covered with very fine hairs;
           lateral spikelets on straight pedicels
           I   Inflorescence generally less than 1 cm
               wide; spikelets purplish; lemma awn of
               central and lateral florets 4-8 mm long;
               plants forming dense clumps (leaf blades 2-
               3 mm wide)
                    Hordeum brachyantherum ssp. *californicum*
                                       [*H. californicum*]
                                       California Barley
           II Inflorescence up to 2 cm wide; spikelets
               not purplish; lemma awn of central floret
               8-15 mm long, lateral florets either
               without awns or the awns less than 4 mm
               long; plants not forming dense clumps
               (usually in moist, alkaline areas)
               J   Inflorescence up to 1.5 cm wide and 6 cm
                   long; spikelet bristles (these are
                   slender glumes) straight, not inflated
                   at the base; lemma of central floret 7-
                   8 mm long, its awn 8-10 mm long;
                   lateral florets about 6 mm long,
                   without awns   *Hordeum depressum*
                                          Low Barley
               JJ Inflorescence up to 2 cm wide and 4 cm
                   long; spikelet bristles slightly curved
                   upwards, somewhat inflated at the base;
                   lemma of central floret 6 mm long, its
                   awn 10-15 mm long; lateral florets much
                   reduced compared to central floret,
                   awns (if present) up to 4 mm long
                        Hordeum marinum ssp. *gussoneanum*
                           [*H. geniculatum*] (*Plate* 108)
                                   Mediterranean Barley; eu

Poaceae, Division II: Spikelets partly sunken into or pressed against
concave sections of the stem (as a spikelet matures it may bend outward
from the concavity)
A   Spikelets entirely without awns (look at several spikelets)
    B   Spikelets sunken into one side of a flattened, fibrous stem (the
        spikelets bending out, however, as they mature) (lower glume papery,
        1 mm long; upper glume about 4 mm long, almost as long as the
        florets; leaf blades up to 7 mm wide, rounded at the tip; plants
        creeping; annual; a common weed in lawns)   *Stenotaphrum secundatum*
                                          St. Augustine Grass; sa?

BB Spikelets alternately sunken into both sides of a nearly cylindrical
   stem (introduced into coastal areas)
   C   Inflorescence curved to markedly sickle-shaped, the tip sometimes
       directed downward; spikelets with 1-2 florets; annual
       D   Inflorescence 6-10 cm long, markedly sickle-shaped; basal
           leaves very congested, curled; mature spikelets so sunken
           into the stem that they are barely visible (glumes sharp-
           pointed)   *Parapholis incurva*
                                                    Sickle Grass; eu
       DD Inflorescence 10-20 cm long (but less than 2 mm wide),
           slightly curved; basal leaves scarcely congested or curled;
           mature spikelets projecting out from the stem like spines
                           Hainardia cylindrica [*Monerma cylindrica*]
                                                    Thintail; eu; La-s
   CC Inflorescence not obviously curved; spikelets with more than 2
       florets; perennial
       D   Spikelets with a broad surface facing the stem; both glumes
           present; inflorescence 4-6 cm long (florets and glumes
           pointed)   *Elytrigia juncea* [*Agropyron junceum*]
                                                    Wheat Grass; eu; SF
       DD Spikelets with a narrow edge facing the stem (often tilting
           away from the stem); upper glume usually absent;
           inflorescence 10-30 cm long (widespread)
                                       Lolium perenne (Plates 64, 108)
                                                    Perennial Rye Grass; eu
AA Spikelets with awns at least 1 mm long (look at several spikelets)
   B   Awns mostly more than 1 cm long
       C   Awns up to about 8 cm long, 3 on each glume and lemma
           (inflorescence with fewer than 10 spikelets, these inflated at
           maturity, only 1 at each node; glumes spiny)
                                                    Aegilops triuncialis
                                                    Barbed Goat Grass; me; Ma
       CC Awns usually less than 2 cm long, restricted to the lemmas,
           although the glumes sometimes sharp-pointed
           D   Spikelets with a narrow edge facing the stem; glumes almost
               as long as, or longer than, the aggregate of florets in the
               spikelet (excluding the awns); one glume often missing
               (spikelets usually with 5-7 florets; up to 90 cm tall)
                                                    Lolium temulentum
                                                    Darnel; me
           DD Spikelets with a broad surface facing the stem; glumes
               distinctly shorter than the whole spikelet, 2 present on each
               spikelet
               E   Inflorescence 5-20 cm long, usually with more than 30
                   spikelets, 2 at each node, all along the stem; spikelets
                   less than 1.5 cm long, with 2-6 florets; up to 120 cm
                   tall, forming dense clumps (widespread)
                                             Elymus glaucus (Plate 107)
                                                    Blue Wild Rye
               EE Inflorescence not more than 7 cm long, with 1-5 spikelets,
                   1 at each node, concentrated at the top of the stem;
                   spikelets 2-3.5 cm long, with more than 8 florets; up to
                   30 cm tall, not forming dense clumps
                                                    Brachypodium distachyon
                                                    Purple Falsebrome; me
   BB Awns not more than 1 cm long
       C   Spikelets scarcely visible, sunken into the narrow stem (the stem
           about 1 mm wide), with 1 floret; inflorescence 4-11 cm long; up
           to 30 cm tall (awns 2-5 mm long; annual)   *Scribneria bolanderi*
                                                    Scribneria; SLO-n
       CC Spikelets readily visible, with more than 1 floret; inflorescence
           7-30 cm long; up to more than 80 cm tall
           D   Spikelets with a broad surface facing the stem, always with 2
               glumes (perennial)
                       Elytrigia repens [*Agropyron repens*] (Plate 107)
                                                    Quack Grass; eua; SF
           DD Spikelets with a narrow edge facing the stem, usually with
               only 1 glume (the lower one)
               E   Glume(s) almost as long as, or even longer than, the rest
                   of the spikelet (excluding the lemma awns) (florets
                   usually 5-7; base of leaf blade without 2 overlapping
                   membranous lobes; annual)   *Lolium temulentum*
                                                    Darnel; me

EE Glume(s) much shorter than the whole spikelet (excluding
   the lemma awns) (widespread)
      F Lower lemmas of spikelets without awns, the upper
         lemmas with awns up to 10 mm long; stems cylindrical;
         spikelets with 8-20 florets; annual or biennial
                              Lolium multiflorum (*Plate* 64)
                                        Italian Rye Grass; eu
      FF Lower and upper lemma awns about 1 mm long; stems
         slightly flattened; spikelets with 6-10 florets;
         perennial   *Lolium perenne* (*Plates* 64, 108)
                                     Perennial Rye Grass; eu

Poaceae, Division III: Inflorescence headlike (usually very dense and
compact, but sometimes loose and interrupted), either without branches, or
the spikelets so concentrated that the branches and branchlets, if present,
are difficult to distinguish
A Inflorescence very dense and compact, usually not interrupted to the
   extent that the stem is visible; individual spikelets (and branches, if
   present), difficult to see
   B Awns, if present, not more than 1 mm long
      C Spikelets usually green-striped (they may also be purplish)
         (sometimes difficult to see and sometimes absent in *Phleum
         pratense* and *Agrostis densiflora*, keyed under this choice)
         D Inflorescence widest above the middle (up to 9 cm long and 1-
            2 cm wide); glumes notched, the notch separating 2 unequal
            teeth; spikelets detaching from the mature inflorescence in
            clusters of 6 or more, leaving naked sections of stem (plants
            forming dense clumps, the inflorescence often partly or
            wholly enclosed by the sheath of a short leaf)
                                          Phalaris paradoxa
                                      Hood Canary Grass; me
         DD Inflorescence not widest above the middle; glumes neither
            notched at the tip nor toothed; spikelets not detaching in
            clusters of 6 or more
            E Inflorescence narrowly cylindrical (less than 1 cm wide
               and up to 20 cm long); both glumes ending in a stout awn
               about 1 mm long (the awns of the 2 glumes form a U-shaped
               configuration) (spikelets flattened, sometimes purplish,
               edged with green)   *Phleum pratense*
                                       Cultivated Timothy; eua
            EE Inflorescence not narrowly cylindrical; glumes not awn-
               tipped
               F Spikelets not conspicuously flattened, less than 1 mm
                  wide when not yet open; inflorescence more than 3
                  times as long as wide, the stem often visible through
                  it; glumes not flattened or keeled, 2-4 mm long, with
                  one faint green stripe (plants forming dense clumps;
                  coastal)   *Agrostis densiflora* [includes *A. clivicola*
                           vars. *clivicola* and *punta-reyesensis*]
                                     California Bent Grass; SCr-n
               FF Spikelets flattened, 1.5-4 mm wide when not yet open;
                  inflorescence not more than 3 times as long as wide
                  (except in *Phalaris angusta*), the stem not visible
                  through it; glumes flattened and keeled, 4-8 mm long,
                  with distinct green striping (up to more than 90 cm
                  tall)
                  G Inflorescence 6-15 cm long; glumes with a narrow
                     keel (inflorescence 8-15 mm wide; leaves up to
                     more than 1 cm wide; annual; in wet areas)
                                             Phalaris angusta
                                Narrow Canary Grass; Sn, Sl-s
                  GG Inflorescence generally less than 5 cm long; glumes
                     with a broad keel (spikelets with 1 or 2 hairy,
                     sickle-shaped sterile florets [1 mm long] at the
                     base of the fertile floret)
                     H Glumes 4-6 mm long; spikelets with 1 sterile
                        floret (annual)   *Phalaris minor*
                                  Littleseed Canary Grass; me
                     HH Glumes 6-8 mm long; spikelets with 2 sterile
                        florets (in moist to wet areas)
                        I Spikelets, when not yet open, 1-2 mm wide,
                           usually tinged with purple; sterile florets
                           whitish; perennial, forming dense clumps,
                           up to 150 cm tall   *Phalaris californica*
                                  California Canary Grass; SLO-n

II Spikelets, when not yet open, 3-4 mm wide,
not tinged with purple; sterile florets
brownish; annual, not forming dense clumps,
up to 90 cm tall   *Phalaris brachystachys*
Shortspike Canary Grass; me; CC, SLO
CC Spikelets not green-striped (occasionally in *Koeleria macrantha*
and *Phalaris aquatica*, keyed under this choice)
D   Spikelets with one fertile floret (there may also be
inconspicuous sterile florets) (inflorescence more than 4.5
times as long as wide, often cylindrical)
E   Spikelets with minute awns barely projecting beyond the
spikelet; inflorescence (compact) up to 7 cm long and
less than 0.6 cm wide (leaf blades less than 0.4 cm wide,
up to 10 cm long; in wet places, at elevations above
3000')   *Alopecurus aequalis*
[includes *A. aequalis* var. *sonomensis*]
Shortawn Foxtail
EE Spikelets without awns; inflorescence up to 30 cm long and
at least 1 cm wide
F   Spikelets 2 mm long, arranged in whorled clusters;
inflorescence (1-4 cm wide) loosely and irregularly
branched (not cylindrical); not more than 75 cm tall
(florets falling with the glumes at maturity; in moist
areas)   *Agrostis viridis* [*A. semiverticillata*]
Whorled Bent Grass; eu
FF Spikelets 3-15 mm long, not in whorled clusters;
inflorescence cylindrical; up to 200 cm tall
G   Spikelets 2-4 mm long; glumes much less than 1 mm
wide, with small spiny hairs, especially on veins
(inflorescence dense, but sometimes interrupted;
not more than 85 cm tall, forming dense clumps;
coastal)
Agrostis densiflora [includes *A. clivicola*
vars. *clivicola* and *punta-reyesensis*]
California Bent Grass; SCr-n
GG Spikelets 4-15 mm long; glumes 1-2 mm wide, not
hairy
H   Inflorescence up to 15 cm long and 1.5 cm wide,
widest at the base and tapering toward the
tip; spikelets 4-6 mm long; leaf blades flat,
up to 15 mm wide (with a sterile floret 1 mm
long at the base of the fertile floret; in
moist areas)   *Phalaris aquatica*
Harding Grass; me; Ma
HH Inflorescence up to 30 cm long and 3 cm wide,
widest in the middle and tapering at both
ends; spikelets 12-15 mm long; leaf blades 10
mm wide when flat (the margins become rolled
up as they age) (plants forming clumps; on
coastal sand dunes)
Ammophila arenaria (*Plate* 62)
European Beach Grass; eu
DD Spikelets with more than 1 fertile floret
E   Spikelets and/or florets of 2 distinctly different types
in the same inflorescence (pry open the spikelet)
F   Spikelets all similar; inflorescence up to 15 cm long
and more than 1.5 cm wide; plants forming dense clumps
(florets of 2 types in each spikelet: the lower one
with a shiny grain, the upper one purplish gray and
with a hooked awn on its lemma)   *Holcus lanatus*
Common Velvet Grass; eu
FF Spikelets of 2 distinct types: one without fertile
florets, the other with large fertile florets (the 2
spikelet types mixed all along the stem);
inflorescence 4-8 cm long and up to 1 cm wide; plants
not forming dense clumps
Cynosurus cristatus (*Plate* 63)
Crested Dogtail; eu; Na, SF
EE Spikelets and/or florets in the inflorescence essentially
all alike (florets falling from the glumes at maturity;
plants forming clumps)

F   Spikelets flattened; florets 5-10, stiff and curved,
    tightly ascending, the lemma of each one ending in a
    spinelike tip on the outer edge, and with more than 10
    prominent veins; leaf blades 2-4 cm long; usually less
    than 8 cm tall (rarely up to 12 cm), often prostrate,
    with stem tips rising (in vernal pools)
                    Tuctoria mucronata [*Orcuttia mucronata*]
                                Crampton Tuctoria; 1b
FF Spikelets only slightly, if at all, flattened; florets
    2-6, not as described in choice F; some leaf blades
    more than 5 cm long; at least 10 cm tall
    G   Inflorescence up to 6 cm long, somewhat oblong to
        pineapple-shaped, less than 4 times as long as
        wide; spikelets 6-10 mm long; florets usually with
        delicate webbed hairs at the base; leaf blades up
        to 10 cm long, the tip shaped like the bow of a
        boat (especially when folded); on coastal dunes
                                            Poa douglasii
                                Sand-dune Blue Grass; Mo-n
    GG Inflorescence up to 15 cm long, somewhat
        cylindrical, more than 4 times as long as wide
        (sometimes interrupted); spikelets 3-5 mm long;
        florets without webbed hairs; leaf blades up to 20
        cm long, the tip not shaped like the bow of a
        boat; in dry areas   *Koeleria macrantha* (*Plate* 108)
                                                June Grass
BB Awns present, more than 1 mm long
   C   Inflorescence more or less globular, not more than 5 times as
       long as wide (usually less than 4 times as long as wide)
       D   Upper portion of paired, flattened glumes, when mature,
           forming a U- or V-shaped configuration; glumes hairy
           (spikelets with 1 floret)
           E   Inflorescence about twice as long as wide, not softly
               fuzzy in appearance; awns on glumes stout, 2 mm long;
               lemmas without awns; florets falling from the glumes at
               maturity (in wet areas, at elevations of at least 3000')
                                                Phleum alpinum
                                    Mountain Timothy; SF, La-n
           EE Inflorescence 3 or 4 times as long as wide, bushy in
               appearance and soft and fuzzy to the touch; awns on
               glumes delicate, 5-8 mm long; lemma awn less than 2 mm
               long; florets falling with the glumes at maturity
                            Polypogon monspeliensis (*Plate* 64)
                                            Annual Beard Grass; me
       DD Upper portion of paired glumes not forming a U- or V-shaped
           configuration; glumes not hairy (awns attached to the backs
           of the florets; annual)
           E   Awns 10-20 mm long (dark); inflorescence up to 3 cm long
               and about as wide (soft and fuzzy, with grayish, feathery
               hairs); spikelets (excluding the awns) 7-10 mm long, with
               1 floret; plants not clumped   *Lagurus ovatus*
                                            Hare's-tail; me; SFBR
           EE Awns 2-4 mm long; inflorescence usually less than 3 cm
               long and 0.5 cm wide; spikelets 2-3 mm long, with 2
               florets; plants forming small clumps   *Aira praecox*
                                            Early Hair Grass; me; Ma-n
   CC Inflorescence cylindrical, usually at least 6 times as long as
       wide (sometimes less in *Alopecurus saccatus*, keyed here)
       D   Awns up to 2.5 mm long, but sometimes forming only a slight
           projection (measure from tip to base of awn) (inflorescence
           up to 10 cm long and 2.5 cm wide; forming dense clumps)
           E   Glumes of fertile spikelets with short, winglike lateral
               projections and an awnlike tip; inflorescence often
               partly enclosed in the sheath of a short leaf; spikelets
               detaching in clusters of 6 or more from the mature
               inflorescence, leaving naked sections of stem (usually
               only one of the spikelets in each cluster is fertile)
                                            Phalaris paradoxa
                                    Hood Canary Grass; me
           EE Glumes without winglike lateral projections, but with awns
               about 2 mm long; inflorescence not enclosed in a leaf
               sheath; spikelets not detaching in clusters from the
               mature inflorescence (in moist areas)   *Phleum alpinum*
                                    Mountain Timothy; SF, La-n

DD Awns 2.5-10 mm long
   E   Spikelets (excluding the awns) 5-12 mm long (lemma awn 3-5
      mm long; florets falling from the glumes at maturity)
      F   Spikelets 6-7 mm long, with 1 floret; inflorescence
         (up to 9 cm long and 1 cm wide) not interrupted, with
         only a very faint sweet smell, if any (spikelets
         greenish yellow above, shiny and cream-colored at the
         swollen base)   *Gastridium ventricosum*
                                             Nit Grass; eu
      FF Spikelets 5-12 mm long, with 2 sterile florets, the
         lemmas of these with awns, and 1 awnless, fertile
         floret; inflorescence often interrupted, usually
         sweet-smelling
         G   Inflorescence up to 8 cm long and 1.5 cm wide;
            spikelets longer than 7 mm; leaf blades more than
            3 mm wide; perennial, forming dense clumps
                       Anthoxanthum odoratum (*Plate* 63)
                                Sweet Vernal Grass; eu
         GG Inflorescence up to 3 cm long and 1 cm wide;
            spikelets 5-7 mm long; leaf blades not more than 2
            mm wide; annual, not forming dense clumps
                             Anthoxanthum aristatum
                          Vernal Grass; eu; Ma, Sn
  EE Spikelets usually less than 4 mm long (sometimes 5 mm in
    Alopecurus saccatus)
      F   Awns bent (lemma minutely toothed at the tip, its awn
         attached to the middle of the back or lower; annual,
         forming small clumps)
         G   Spikelets with 1 floret, the awn of its lemma 4-6
            mm long; florets falling with the glumes at
            maturity; inflorescence up to 6 cm long and 1 cm
            wide; glumes very hairy (in moist areas)
                  Alopecurus saccatus [includes *A. howellii*]
                                 Pacific Foxtail
         GG Spikelets with 2 florets, the awns of the lemmas 2-
            4 mm long; florets falling from the glumes at
            maturity; inflorescence up to 3 cm long and less
            than 0.5 cm wide; glumes not hairy except slightly
            at the base (in sandy soils)   *Aira praecox*
                           Early Hair Grass; me; Ma-n
      FF Awns not bent
         G   Florets falling from the glumes at maturity; glumes
            without awns, but with sharp tips and small spiny
            hairs, especially on veins; inflorescence up to 10
            cm long and 1.5 cm wide, dense but occasionally
            interrupted (lemma awn up to 3 mm long; plants
            forming dense clumps; coastal)
               Agrostis densiflora [includes *A. clivicola* vars.
                       clivicola and *punta-reyesensis*]
                     California Bent Grass; SCr-n
         GG Florets falling with the glumes at maturity; glumes
            with awns, and at least somewhat hairy;
            inflorescence up to 15 cm long and 3 cm wide, not
            interrupted (bushy in appearance and soft and
            fuzzy to the touch) (paired, flattened glumes, at
            maturity, forming a U- or V-shaped configuration)
            H   Glumes not densely hairy, with awns up to 8 mm
               long, originating just below the tip (look
               carefully; at first glance, the awns may
               appear to originate from the tip); leaf blades
               3-8 mm wide; lemma awn less than 2 mm long;
               inflorescence up to more than 15 cm long
                   Polypogon monspeliensis (*Plate* 64)
                           Annual Beard Grass; me
            HH Glumes densely hairy on the edge, the awns up
               to 10 mm long, originating about one-third the
               distance from the tip; leaf blades usually
               less than 3 mm wide; lemma without an awn;
               inflorescence usually less than 9 cm long
                       Polypogon maritimus
               Mediterranean Beard Grass; me; La, Na
AA Inflorescence a loose head, sometimes interrupted to the extent that the
  stem is visible; spikelets (and branches, if present) often visible

B Spikelets with 1 obvious floret (this fertile, but there may also be
   inconspicuous sterile florets)
C Spikelets without awns
   D Spikelets less than 3 mm long; inflorescence up to 15 cm
      long; up to 75 cm tall, not coarse, the stems sometimes
      falling down and rooting
      E Leaf blades flat, up to 8 mm wide; inflorescence with
         whorled, lobelike clusters of spikelets; florets (lemma 1
         mm long) usually falling with the glumes at maturity;
         stems usually prostrate and rooting at the nodes (in
         moist areas)   *Agrostis viridis* [*A. semiverticillata*]
                                        Whorled Bent Grass; eu
      EE Leaf blades mostly wirelike, but if flat, then generally
         less than 5 mm wide; inflorescence (purple and tan)
         without whorled clusters of spikelets; florets (lemma 2-
         2.5 mm long) falling from the glumes at maturity; stems
         upright   *Agrostis pallens* [includes *A. diegoensis*]
                                             Dune Bent Grass
   DD Spikelets more than 7 mm long (up to 20 mm long);
      inflorescence up to 30 cm long; up to more than 120 cm tall,
      coarse (stems up to 1 cm wide at the base)
      E Inflorescence less than 1.5 cm wide, the spikelets curved
         and tightly overlapping; leaf blades 5-17 mm wide; glumes
         with conspicuous veins; plants not forming dense clumps
         (but sometimes rooting at nodes, and thus forming
         extensive colonies); in salt marshes, becoming submerged
         during high tides   *Spartina foliosa*
                                            California Cord Grass
      EE Inflorescence up to 3 cm wide, the spikelets straight and
         not overlapping; leaf blades 2-5 mm wide; glumes without
         conspicuous veins (but veins evident on the lemma);
         plants forming dense clumps; on coastal sand dunes
                                    Ammophila arenaria (*Plate* 62)
                                      European Beach Grass; eu
CC Spikelets with awns (these sometimes only on the lemmas)
   D Florets with a tuft of whitish hairs at the base (pry out a
      floret and examine it with a hand lens) (spikelets more than
      5 mm long [except in *Muhlenbergia andina*]; forming clumps)
      E Awns easily visible beyond the tip or side of each
         spikelet
         F Awns not bent; spikelets 3-4 mm long (excluding the
            awns); basal hairs of florets sometimes as long as the
            florets; glumes sometimes with short awns
            (inflorescence often purplish; lemma awns 3-8 mm
            long)   *Muhlenbergia andina*
                                          Foxtail Muhly; SCl-n
         FF Awns bent; spikelets 5-10 mm long (excluding the
            awns); basal hairs of florets not as long as the
            florets; glumes without awns (tip of lemma with 4
            minute teeth)
            G Spikelets 10 mm long; awn extending 8-10 mm beyond
               the tip of each spikelet (leaf blades 1-3 mm wide,
               margin rolled up; coastal) *Calamagrostis foliosa*
                                        Leafy Reed Grass; Sn-n; 4
            GG Spikelets 5-6 mm long; awn extending 2 mm beyond
               the side of each spikelet (look at several)
               (inflorescence straw-colored or purple tinged; on
               serpentine)   *Calamagrostis ophitidis*
                                 Serpentine Reed Grass; La, Ma, Sn; 4
      EE Awns hidden within each spikelet or barely protruding
         (lemma awn bent; glumes without awns; inflorescence
         sometimes purplish)
         F Inflorescence 15-30 cm long and up to 3 cm wide; leaf
            blades 6-12 mm wide; up to 1.5 m tall (without white
            hairs where each leaf blade joins the sheath; in moist
            areas)   *Calamagrostis nutkaensis*
                                          Pacific Reed Grass; SLO-n
         FF Inflorescence 7-15 cm long and not more than 1.5 cm
            wide; leaf blades 2-6 mm wide; up to 1 m tall
            G White hairs present where each leaf blade joins the
               sheath; leaf blades 2-4 mm wide; spikelets about 4
               mm long; inflorescence usually less than 1 cm wide
                                          Calamagrostis rubescens
                                              Pine Grass; SLO-n

GG Without white hairs where each leaf blade joins the
sheath; leaf blades up to 6 mm wide; spikelets
about 6 mm long; inflorescence usually 1-1.5 cm
wide   *Calamagrostis koelerioides*
Tufted Pine Grass; SLO-n
DD Florets without a tuft of whitish hairs at the base
E   Glumes with awns 2-8 mm long (inflorescence often
interrupted or lobed, usually 8-20 cm long and up to 3.5
cm wide, greenish yellow to purplish; leaf blades 3-10 mm
wide)
F   Glumes 3 mm long, the tips tapering gradually into
awns 2-3 mm long; lemma awn 1-2 mm long; in salt
marshes   *Polypogon elongatus*
Longspike Beardgrass; sa; CC, SLO
FF Glumes less than 2 mm long, the tips ending abruptly,
with awns 5-8 mm long; lemma awn 2-5 mm long; in moist
areas (the spikelets often in grapelike clusters)
Polypogon australis
Chilean Beard Grass; sa
EE Glumes without awns (but they may have sharp tips)
F   Spikelets with 2 sterile florets (the lemma of these
with awns) on either side of the awnless fertile
floret (inflorescence often interrupted, sweet-
smelling)
G   Inflorescence up to 8 cm long and 1.5 cm wide;
spikelets longer than 7 mm; leaf blades more than
3 mm wide; perennial, forming dense clumps
Anthoxanthum odoratum (*Plate* 63)
Sweet Vernal Grass; eu
GG Inflorescence up to 3 cm long and 1 cm wide;
spikelets 5-7 mm long; leaf blades not more than 2
mm wide; annual, not forming dense clumps
Anthoxanthum aristatum
Vernal Grass; eu; Ma, Sn
FF Spikelets without sterile florets, the fertile floret
with an awn (plants forming clumps)
G   Awn of the lemma originating near its base (the awn
6 mm long, bent, protruding about 2 mm beyond the
side of the spikelet) (inflorescence 7-13 cm long,
up to 1.5 cm wide)   *Calamagrostis ophitidis*
Serpentine Reed Grass; La, Ma, Sn; 4
GG Awn of the lemma originating near the middle or
near the tip
H   Awn 4-5 mm long; lower glume longer than the
upper glume; inflorescence up to 8 cm long and
1 cm wide, not interrupted; leaf blades 1-4 cm
long (spikelets greenish above, shiny and
cream-colored at the swollen base)
Gastridium ventricosum
Nit Grass; eu
HH Awn usually less than 3 mm long; glumes equal;
inflorescence up to 10 cm long and 1.5 cm
wide, occasionally interrupted; leaf blades 2-
12 cm long (coastal)
Agrostis densiflora [includes *A. clivicola* vars.
clivicola and *punta-reyesensis*]
California Bent Grass; SCr-n
BB Spikelets with more than 1 obvious floret (at least one of the
florets fertile, but sometimes the sterile florets are also obvious)
C   Awns more than 10 mm long (lemma awn 10-35 mm long)
D   Glumes absent; leaf blades 10-20 mm wide; plants not forming
clumps (spikelets 3-4 at each node; in shaded areas)
Elymus californicus [*Hystrix californica*]
California Bottlebrush Grass; Sn-SCr; 4
DD Glumes present; leaf blades 4-15 mm wide; plants forming
dense clumps
E   Inflorescence up to 16 cm long, upright; spikelets 2 at
each node; glumes 7-19 mm long, with awns about 1 mm long
(widespread)   *Elymus glaucus* (*Plate* 107)
Blue Wild Rye
EE Inflorescence up to 25 cm long, nodding; spikelets 2-4 at
each node; glumes 10-25 mm long, with awns about 10 mm
long   *Elymus canadensis*
Canadian Wild Rye; na; Sn-n

CC Awns, if present, less than 3 mm long
   D   Spikelets with awns less than 3 mm long
      E   Florets of two different types in each spikelet: the lower
         floret with a shiny grain, the upper one with a hooked
         awn (spikelets 3-6 mm long; florets falling with the
         glumes at maturity; inflorescence up to 15 cm long and
         more than 1.5 cm wide; plants often forming dense
         clumps)   *Holcus lanatus*
                                Common Velvet Grass; eu
      EE Florets similar within each spikelet (there may be
        inconspicuous sterile florets, however)
         F   Inflorescence 1-3 cm long (less than 0.5 cm wide);
            annual (spikelets 3-4 mm long [excluding the awns];
            lemma awn originating below the middle of the back,
            bent; floret 2-4 mm long)   *Aira praecox*
                                Early Hair Grass; me; Ma-n
        FF Inflorescence more than 5 cm long; perennial (coastal)
           G   Spikelets 15-20 mm long; inflorescence 20-30 cm
              long; leaf blades up to 12 mm wide; plants not
              forming dense clumps, but spreading extensively by
              rhizomes
                Leymus x *vancouverensis* [*Elymus* x *vancouverensis*]
                           Vancouver Wild Rye; Ma-n
           GG Spikelets 6-8 mm long; inflorescence up to 20 cm
              long (sometimes purplish); leaf blades 1-2 mm
              wide; plants forming dense clumps (lemma with 2-4
              tiny teeth at the tip; a pedestal, covered with
              white hairs, between the 2 florets)
                   Deschampsia cespitosa ssp. *holciformis*
                         [*D. caespitosa* ssp. *holciformis*]
                         Pacific Hair Grass; SLO-n
   DD Spikelets without awns
      E   Tip of the leaf blades (especially when folded) curved
         upward, resembling the bow of a boat; florets sometimes
         with cottony hairs at the base (in sandy coastal
         habitats)
         F   Spikelets 3-5 mm long, with 3-4 florets; inflorescence
            somewhat open, 1-4 cm long, up to 1 cm wide (florets
            sometimes with sparse cottony hairs at the base)
                              Poa confinis
                          Beach Blue Grass; Ma-n
        FF Spikelets 6-10 mm long, with 3-8 florets;
         inflorescence dense, 2-9 cm long and 1-2 cm wide
         G   Florets 5-7.5 mm long, often with sparse cottony
            hairs at the base; leaf blades 1-2.5 mm wide;
            inflorescence 2-5 cm long   *Poa douglasii*
                        Sand-dune Blue Grass; Mo-n
           GG Florets 3-4.5 mm long, without cottony hairs; leaf
            blades 1-5 mm wide; inflorescence 3-9 cm long
                           Poa unilateralis
                   Ocean-bluff Blue Grass; Mo-n
      EE Tip of leaf blades not curved upward; florets without
        cottony hairs
        F   Spikelets 3-5 mm long, sometimes green-striped and
            purplish (inflorescence up to 18 cm long, sometimes
            interrupted; leaves wirelike; forming dense clumps, up
            to 60 cm tall)   *Koeleria macrantha* (*Plate* 108)
                                      June Grass
        FF Spikelets 6-30 mm long, not green-striped
         G   Leaves mostly less than 10 cm long, tapering to
            sharp tips, the blades 1-4 mm wide and the sheaths
            sometimes overlapping; inflorescence up to 7 cm
            long; spikelets about 10 mm long (in salt marshes
            and moist alkaline habitats)
               Distichlis spicata [includes *D. spicata* vars.
                   nana and *stolonifera*] (*Plate* 106)
                               Salt Grass
          GG Leaves much more than 10 cm long, not tapering to
            sharp tips, the blades 5-15 mm wide and the
            sheaths not overlapping; inflorescence 8-30 cm
            long; spikelets 15-30 mm long (stems up to 1 cm
            wide; on backshores of sandy beaches)
             Leymus mollis [*Elymus mollis*] (*Plates* 63, 108)
                            Dune Grass; SLO-n

282   (M) Poaceae (Division IV)--Grass Family

Poaceae, Division IV: Inflorescence with obvious branches, these often divided into branchlets or long pedicels that resemble branchlets
A Branches of the inflorescence not divided into branchlets or long pedicels that resemble branchlets (pedicels, if present, very short and obscure)
  B Branches of the inflorescence in conspicuous whorls (at least 3 branches in each whorl) (spikelets sessile or on very short pedicels, in 2-3 rows on one side of each branch)
    C Inflorescence with several whorls, each whorl with 2-7 branches (the branches 4-14 cm long); awns present (2-6 mm long); spikelets protruding from the branches; plants upright (forming dense clumps) *Chloris verticillata*
Finger Grass; na; Al
    CC Inflorescence with one whorl of 4 or more branches at the tip (in *Digitaria*, there are occasionally 1-3 additional whorls); awns lacking; spikelets pressed against the branches; plants often prostrate and rooting
      D Branches of the inflorescence (usually 4-5) up to 5 cm long; spikelets 2 mm long, in 2 rows, tightly overlapping; ligule composed of white hairs (common) *Cynodon dactylon* (*Plate* 105)
Bermuda Grass; af
      DD Branches of the inflorescence 5-15 cm long (these sometimes flattened); spikelets 3 mm long, in 2-3 rows, not tightly overlapping; ligule membranous (the leaf may be hairy elsewhere, however) *Digitaria sanguinalis* (*Plate* 106)
Crab Grass; eu
  BB Branches of the inflorescence not in conspicuous whorls
    C Spikelets solitary at the ends of branches, these delicate, much thinner than the main stem (awns up to 4 cm long)
      D Teeth at tip of mature lemma soft and membranous, less than 2 mm long, whitish; branches (pedicels) generally somewhat thicker than the awns; spikelets 3 mm or more wide *Avena fatua* (*Plate* 104)
Wild Oat; eu
      DD Teeth at tip of mature lemma stiff and awnlike, about 4 mm long, reddish brown when mature; branches (pedicels) no thicker than the awns; spikelets less than 2 mm wide
Avena barbata
Slender Wild Oat; me
    CC Spikelets numerous on the branches, these not delicate, about as thick as the main stem
      D Spikelets 1 or 2 at each node, sessile
        E Spikelets up to 2.5 cm long, alternating on both sides of each branch (oriented with a narrow edge facing the branch), each with numerous florets; awns up to 10 mm long (widespread) *Lolium multiflorum* (*Plate* 64)
Italian Rye Grass; eu
        EE Spikelets (flattened) less than 1 cm long, in dense groups on one side of each branch, each with 1 floret; awns lacking or less than 1 mm long (in salt marshes)
          F Leaf blades not more than 4 mm wide; inflorescence with 2-6 branches; stems up to 4 mm wide
Spartina patens
Salt-meadow Cord Grass; na; Sl
          FF Leaf blades usually at least 5 mm wide; inflorescence often with more than 10 branches; stems up to 10 mm wide (sometimes more at the base)
            G Inflorescence dense, the branches pressed against the stem; spikelets curved *Spartina foliosa*
California Cord Grass
            GG Inflorescence somewhat spreading; spikelets mostly straight (established in San Francisco Bay)
Spartina alterniflora
Salt-water Cord Grass; na; SF
      DD Spikelets 1-several at each node
        E Inflorescence with 2 branches (these 3-6 cm long) at the top of the stem (occasionally there is another branch below these) (spikelets without awns, flattened, in 2 rows; leaf blades 2-7 mm wide; plants sometimes creeping and rooting; in coastal, moist habitats)
Paspalum distichum
Knot Grass

  EE Inflorescence usually with 4 or more branches, these
    alternating along the stem (sometimes several branches at
    each node)
    F Awns 5-25 mm long (spikelets, in 3-4 rows on each
      branch, with 1 fertile floret, the enclosed grain hard
      and shiny; glumes and lemma with stiff, bristlelike
      hairs on the margins; leaf blades 6-20 mm wide)
          Echinochloa crus-galli [*E. crusgalli*] (*Plate* 64)
                   Barnyard Grass; eua
    FF Awns, if present, less than 4 mm long
      G Spikelets with up to 12 florets (these 2 mm long,
        with awns 1 mm long) (base of inflorescence
        usually enclosed within the leaf sheath, the leaf
        blade often extending to, or beyond, the tip of
        the inflorescence; spikelets often overlapping; in
        moist, alkaline areas)   *Leptochloa fascicularis*
               Bearded Sprangletop; Mo, Sn, SF
      GG Spikelets with 1 floret (inflorescence with
        delicate hairs)
        H Ligule present, membranous; inflorescence
          branches 4-10 cm long; first glume absent (but
          larger lemma resembles a glume); grains not
          shiny; forming clumps   *Paspalum dilatatum*
                     Dallis Grass; sa
        HH Ligule absent; inflorescence branches less than
          3 cm long; both glumes present (first one
          small); grains shiny; not forming clumps (in
          moist areas) *Echinochloa colona* [*E. colonum*]
                 Jungle Rice; eua; SCl
AA Branches of the inflorescence usually divided into branchlets or long
 pedicels that resemble branchlets (bending the branches out may make the
 branchlets easier to see)
 B Spikelets 2-3 at each node, 1 spikelet sessile and the other 1 or 2
  (sterile) on pedicels (but in *Andropogon*, only the pedicel remains
  along with the sessile fertile spikelet) (fertile spikelets with 2
  florets; florets falling with the glumes at maturity; up to 2 m
  tall)
  C Inflorescence up to about 20 cm long, with dense clusters of
    spikelets, each partly hidden in an elongated sheath and with
    silky hairs sometimes more than 10 mm long; spikelets 2, the
    sterile spikelet reduced to a hairy pedicel; leaf blades up to
    0.5 cm wide (awns up to 20 mm long; perennial, in moist areas)
                   Andropogon virginicus
            Broomsedge Bluestem; na; Sn
  CC Inflorescence up to 40 cm long, if dense, not hidden in a sheath
    and without long silky hairs; spikelets 3, neither the fertile
    nor the sterile spikelets obviously reduced; leaf blades 0.5-10
    cm wide
    D Leaf blades less than 2 cm wide; spikelets elliptical, less
      than 2 mm wide, not tightly clustered (usually colorful);
      perennial   *Sorghum halepense*
                  Johnson Grass; me
    DD Leaf blades 3-10 cm wide; spikelets globular, 3 mm wide,
      tightly clustered; annual   *Sorghum bicolor*
                     Sorghum; af
 BB Spikelets, if more than one at each node, either all sessile or all
  on pedicels
  C Spikelets with only 1 obvious floret, this fertile, but there may
    also be inconspicuous sterile florets (in *Poa bulbosa* the
    apparent floret is actually a bulblet)
          Poaceae, Division IV, Subkey 1 (p. 288)
  CC Spikelets with more than 1 obvious floret (at least one of these
    fertile, but sometimes the sterile florets are also conspicuous;
    pry open a spikelet to see)
    D Awn of lemma at least 1 mm long (sometimes missing in *Bromus*
      secalinus, keyed in Subkey 2)
      E Awn attached at or very near the tip of the lemma (in some
        species it originates between 2 small teeth)
        F Glumes without awns, of unequal length and usually
          shorter than the whole spikelet; florets without long
          white hairs at the base; lemma awn either not
          originating between 2 teeth, or the teeth not awnlike

G   Spikelets (excluding the awns) usually more than
    1.5 cm long; edges of the leaf sheath (at least
    the lower portion) fused together; leaves up to
    more than 10 mm wide, the margin usually flat
    (sometimes rolled up, but never wirelike); ligule
    prominent (sometimes torn and jagged)
                    Poaceae, Division IV, Subkey 2 (p. 293)
GG Spikelets (excluding the awns) usually less than
    1.5 cm long; edges of the leaf sheath not fused
    together (but they overlap); leaves usually less
    than 2 mm wide, the margin usually rolled up and
    wirelike; ligule absent or inconspicuous
                    Poaceae, Division IV, Subkey 3 (p. 295)
FF Glumes of about equal length, and usually longer than
    the aggregate of florets in the spikelet (excluding
    the awns); florets with long whitish hairs at the
    base; awn bent, attached between 2 teeth, these awn-
    like (glumes papery, with 3 or more prominent veins;
    ligule consisting of long hairs; forming dense clumps)
    G   Inflorescence with 4-12 spikelets, the spikelet
        branches ascending; spikelets 4 mm wide; lemma
        with stiff whitish hairs on the back and
        elsewhere, the teeth up to 7 mm long, and the
        largest awn 8-10 mm long   *Danthonia pilosa*
                                    Hairy Oat Grass; au
    GG Inflorescence with 2-5 spikelets, the spikelet
        branches spreading outward; spikelets more than 10
        mm wide; lemma not hairy on the back, the teeth
        less than 5 mm long, and the largest awn usually
        more than 10 mm long   *Danthonia californica*
                                    California Oat Grass; Mo-n
EE Awn (which may bend with age) attached to the back of the
    lemma below its uppermost quarter (florets 2 or 3, the
    lemmas usually with 2-5 very small teeth at the tip)
    F   Awn less than 4 mm long
        G   Spikelets (excluding the awns) less than 4 mm long;
            florets without a small hairy bristle among them,
            the tip of the lemma with 2 teeth (but these
            difficult to see even with a hand lens) (lower
            floret sometimes lacking an awn; awns bent, barely
            projecting beyond the spikelet)
            H   Inflorescence 5 mm wide (7-35 mm long), oblong
                in shape, the lower branches not readily
                visible   *Aira praecox*
                                    Early Hair Grass; me; Ma-n
            HH Inflorescence up to more than 50 mm wide,
                spreading, the lower branches readily visible
                I   Spikelets 1-2 mm long (excluding the awns);
                    awn of the lower floret much shorter than
                    the awn of the upper floret (or absent),
                    only the longer awn extending beyond the
                    spikelet   *Aira elegantissima* [*A. elegans*]
                                    Elegant Hair Grass; me; Ma, Sn
                II Spikelets 2-3 mm long (excluding the awns);
                    awns equal, both extending beyond the
                    spikelet (in sandy coastal areas)
                                    Aira caryophyllea (*Plate* 104)
                            Silver European Hair Grass; eu; Ma-n
        GG Spikelets (excluding the awns) at least 4 mm long;
            florets usually with a small hairy bristle among
            them, the tip of the lemma usually with 4 teeth
            (sometimes 2, 3, or 5), (florets with long upright
            hairs at the base; awn attached between the middle
            and the base of the lemma; plants forming dense
            clumps)
            H   Inflorescence at least 1.5 cm wide (up to 9 cm)
                and up to 20 cm long; awn of lemma originating
                near its base, straight to slightly bent, and
                extending slightly beyond the spikelet; blades
                of basal leaves mostly wirelike, but some as
                much as 3 mm wide (in wet, coastal areas)
                        Deschampsia cespitosa ssp. *holciformis*
                            [*D. caespitosa* ssp. *holciformis*]
                                    Pacific Hair Grass; SLO-n

HH Inflorescence about 0.5 cm wide and up to 30 cm
   long (the branches pressed against the stem);
   awn of lemma originating near its middle,
   straight, and extending about 2 mm beyond the
   spikelet; blades of basal leaves threadlike
   (in wet areas)   *Deschampsia elongata*
                                      Slender Hair Grass

FF Awn at least 4 mm long
   G Awn up to 4 cm long; spikelets (excluding the awns)
      up to 25 mm long; florets without a small hairy
      bristle among them; tip of lemma with 2 teeth
      (glumes longer than the lowest floret and often
      longer than the entire group of florets; up to 75
      cm tall; widespread)
      H Teeth at tip of mature lemma soft and
         membranous, less than 2 mm long, whitish;
         branches (pedicels) generally somewhat thicker
         than the awns; spikelets 3 mm or more wide
                                      Avena fatua (*Plate* 104)
                                             Wild Oat; eu

      HH Teeth at tip of mature lemma stiff and awnlike,
         about 4 mm long, reddish brown when mature;
         branches (pedicels) no thicker than the awns;
         spikelets less than 2 mm wide   *Avena barbata*
                                      Slender Wild Oat; me
   GG Awn not more than 1 cm long; spikelets (excluding
      the awns) not more than 12 mm long; florets with a
      small hairy bristle among them; tip of lemma with
      2-5 teeth
      H Awn 7-10 mm long, curved or bent, attached at
         the beginning of the uppermost quarter of the
         lemma; spikelets 6-12 mm long; glumes much
         shorter than the whole spikelet; tip of lemma
         with 2 teeth (these needlelike) (in moist,
         shaded areas)
         Trisetum canescens [*T. cernuum* var. *canescens*]
                                      Tall Trisetum; SCr-n

      HH Awn 4-7 mm long, usually bent, attached about
         halfway from the tip of the lemma to its base;
         spikelets 4-8 mm long; glumes almost as long
         as, sometimes even longer than, the aggregate
         of florets in the spikelet; tip of lemma
         usually with 4 teeth (sometimes 2, 3, or 5)
         (inflorescence up to 9 cm wide, the branches
         spreading; glumes 5-6 mm long; florets with
         long upright hairs at the base)
         I Awns (these originating near the bases of
            the lemmas) extending beyond the spikelet
            for about 3-4 mm; leaves few, 1 mm wide and
            less than 10 cm long; branchlets of the
            inflorescence stiff and ascending, with few
            spikelets (these at the tip of the
            branchlets); up to 60 cm tall (moist areas)
                         Deschampsia danthonioides (*Plate* 106)
                                      Annual Hair Grass

         II Awns barely extending beyond the spikelet;
            leaves numerous, 1.5-3 mm wide and more
            than 30 cm long; branchlets of the
            inflorescence nodding, with numerous
            spikelets; up to 120 cm tall (plants
            forming dense clumps; in wet coastal areas)
                         Deschampsia cespitosa ssp. *cespitosa*
                                 [includes *D. caespitosa* sspp.
                                    beringensis and *caespitosa*]
                                      Tufted Hair Grass; Ma-n
DD Awn, if present on lemma, less than 1 mm long
   E Each spikelet with 2 (or 3) decidedly different floret
      types (in *Ehrharta erecta*, one is often inconspicuous)
      (inflorescence up to 15 cm long, spreading at maturity;
      spikelets usually 3 mm long)

F   Each spikelet with 3 floret types (one often hidden):
a sterile floret with cross grooves on its lemma, a
sterile floret with a smooth lemma, and a fertile
floret with a smooth lemma; florets falling from the
glumes at maturity; plants not clumped
Ehrharta erecta
Ehrharta; af; Al

FF Each spikelet with 2 floret types: the lower one
producing a hard, shiny grain, the upper one strictly
staminate, its lemma bearing a small hooked awn;
florets falling with the glumes at maturity; plants
often forming dense clumps (inflorescence grayish or
purplish)   *Holcus lanatus*
Common Velvet Grass; eu

EE Each spikelet with 2 or more similar florets
F   Florets (3) side-by-side in each spikelet (these tan
to purple) (florets, at maturity, falling as a group
from the flattened glumes; plants sweet-smelling; leaf
blades up to 17 mm wide; in shaded areas)
Hierochloë occidentalis
Sweet Grass; Mo-n

FF Florets forming an alternating series in each spikelet
G   Spikelets inflated and papery (sometimes almost as
wide as long), resembling a rattlesnake rattle
H   Inflorescence upright, with numerous spikelets;
mature spikelets 3 mm long, 4-5 mm wide
(widespread)   *Briza minor*
Little Quaking Grass; me

HH Inflorescence drooping, with relatively few
spikelets; mature spikelets 12-25 mm long, 10-
13 mm wide   *Briza maxima* (Plate 63)
Big Quaking Grass; me; SLO-n

GG Spikelets not inflated and papery, and not
resembling a rattlesnake rattle
H   With a tuft of white hairs where the leaf blade
joins the sheath; florets (often crowded) 3-
veined (look at several); spikelets with 6-35
florets
I   Plants low, densely matted and creeping,
rooting at the nodes; spikelets (saw-
toothed in appearance) with 10-35 florets,
these tightly overlapping; leaf blades
ascending, usually less than 3 cm long, and
with sharp tips (on sand bars)
Eragrostis hypnoides
Creeping Love Grass; Sn-n

II Plants sometimes low, but not matted and
creeping; spikelets with 6-20 florets,
these not tightly overlapping; leaf blades
usually spreading outward, and usually more
than 3 cm long
J   Leaf blades usually with minute
glandular bumps along the edges (the
bumps are barely visible with a hand
lens); spikelets with 8-20 florets
(usually 9-12) (spikelets less than 2
mm wide; inflorescence dark, usually
less than 15 cm long)
Eragrostis minor [*E. poaeoides*]
Little Love Grass; eu; Sn

JJ Leaf blades without glands; spikelets
usually with fewer than 11 florets
K   With a tuft of white hairs where the
branches of the inflorescence attach
to the stem; leaf blades often less
than 1 mm wide, the margin rolled up
(inflorescence up to 30 cm long,
spreading, gray-green, the spikelets
grouped towards the end of the
branches, these up to 10 cm long;
plants forming dense clumps)
Eragrostis curvula
Weeping Love Grass; af; CC, Sl

KK Without white hairs where the
branches of the inflorescence attach
to the stem; leaf blades 1-7 mm
wide, the margin flat
L   Spikelets on pedicels about 1-2
mm long, pressed against the
branchlets; leaf blades 1-3 mm
wide (inflorescence generally
less than 20 cm long, sometimes
greenish purple)
Eragrostis pectinacea
[*E. diffusa*]
Spreading Love Grass; Ma, Sn, SF
LL Spikelets on pedicels 3-15 mm
long, not pressed against the
branchlets; leaf blades 2-7 mm
wide
M   Inflorescence generally less
than 20 cm long and often
about half as wide; some
spikelets on pedicels more
than 15 mm long, few on each
branch   *Eragrostis mexicana*
ssp. *mexicana*
Mexican Love Grass
MM Inflorescence up to 30 cm
long and much less than half
as wide; spikelets on
pedicels about 3-8 mm long,
usually numerous on each
branch (inflorescence brownish green; usually in moist
areas)   *Eragrostis mexicana*
ssp. *virescens*
[*E. orcuttiana*]
Nonsticky Mexican Love Grass
HH Without a tuft of white hairs where the leaf
blade meets the sheath; spikelets usually with
fewer than 12 florets
I   Spikelets about 1 mm long, each with not
more than 2 florets; inflorescence often as
wide (5-15 cm) as long; glumes and lemma
without prominent veins (inflorescence
delicate, with many threadlike branchlets
bearing numerous, very small spikelets;
leaves crowded, with overlapping sheaths;
in moist, often alkaline areas)
Muhlenbergia asperifolia
Scratch Grass
II Spikelets at least 3 mm long (often
longer), each usually with more than 2
florets; inflorescence generally not as
wide as long; glumes, and usually also the
lemma, with prominent veins (look at
several)
J   Branches of the inflorescence not
concentrated on one side of the stem
(glumes [without awns] somewhat to very
unequal, both [at least the lower one]
usually shorter than the whole
spikelet)
Poaceae, Division IV, Subkey 4 (p. 296)
JJ Branches of the inflorescence all on one
side of the stem (entire length of the
stem is visible on the opposite side)
K   Inflorescence 2-18 cm wide; stem
flattened; spikelets bunched at the
tip of branches, keeled, and awn-
tipped; glumes 4-7 mm long, hairy on
the margin; leaf sheaths (and blades
to some extent) keeled, the edges
fused; sometimes forming dense
clumps *Dactylis glomerata* (*Plate* 63)
Orchard Grass; eua

       KK Inflorescence about 1.5 cm wide;
        stem cylindrical; spikelets (each
        with 4-10 florets) not bunched on
        the branches, not keeled, and
        without awns; glumes about 2 mm
        long, not hairy; leaf sheaths not
        keeled, the edges not fused; plants
        not forming dense clumps
        Desmazeria rigida [*Scleropoa rigida*]
               Stiff Grass; me

Poaceae, Division IV, Subkey 1: Inflorescence with obvious branches, these
divided into branchlets; spikelets with 1 obvious fertile floret (but there
may also be inconspicuous sterile florets; in *Sporobolus airoides*,
furthermore, when the fertile floret splits open at maturity, it may seem
to consist of 2 florets)
A Apparent floret a dark purple bulblet enclosed in 2 papery sheaths,
 these with awnlike projections up to 2 cm long (there may also be bulbs
 at the bases of the stems)   *Poa bulbosa* (Plate 63)
          Bulbous Blue Grass; eu; SLO-n
AA Single fertile floret normal in appearance
 B Awns more than 1 mm long, sometimes on the lemmas as well as the
  glumes (in *Piptatherum miliaceum*, however, they readily fall off)
  C Florets falling with the glumes at maturity; mature florets
   shiny, hard, and nearly round
   D Inflorescence not especially dense, the main stem and its
    branches visible; glumes sharp-tipped, less than half as long
    as the 2 florets (one of them sterile); lemma awn stiff, up
    to 25 mm long (spikelets 2-3 mm long, crowded into 3-4 rows
    on one side of the many branches, these 2-4 cm long; glumes,
    branches, and branchlets with bristly hairs)
      Echinochloa crus-galli [*E. crusgalli*] (Plate 64)
          Barnyard Grass; eua
   DD Inflorescence dense, the spikelets so crowded that the stem
    and branches usually not visible except where the
    inflorescence is lobed or interrupted; glumes (flattened)
    with awns 2-7 mm long, and much longer than the single
    floret; lemma awn delicate, 1-5 mm long (glumes separating at
    maturity to form a U or V; inflorescence greenish yellow,
    becoming straw-colored or occasionally purplish)
    E Glumes 3 mm long, their tips gradually tapering into awns
     2-3 mm long; lemma awn 1-2 mm long (in salt marshes)
             Polypogon elongatus
       Longspike Beard Grass; sa; CC, SLO
    EE Glumes less than 2 mm long, their tips ending abruptly,
     but with awns 5-8 mm long; lemma awn 2-5 mm long
     (spikelets often in grapelike clusters; in moist areas)
             Polypogon australis
         Chilean Beard Grass; sa
  CC Florets falling from the glumes at maturity; mature florets
   shiny, hard, and nearly round (except in *Piptatherum miliaceum*)
   D Awn of lemma up to 11 cm long, attached at an obvious
    junction at the tapered tip; awns typically bent once or
    twice when mature (lemma more than 5 mm long; glumes without
    awns; upper stems and spikelets often purplish; plants
    forming clumps)
    E Florets about 1 mm wide, spindle-shaped (widest at the
     middle); awn (up to 8 cm long) hairy at the base, stiff,
     the portion near the tip straight (lower glume up to 25
     mm long; inflorescence nodding at maturity; the
     California State Grass!) *Nassella pulchra* [*Stipa pulchra*]
          Purple Needle Grass
    EE Florets less than 1 mm wide, cylindrical; awn either not
     hairy or only sparsely hairy at the base, flexible, often
     wavy towards the tip
     F Lower glumes usually less than 10 mm long; awn less
      than 4 cm long   *Nassella lepida* [*Stipa lepida*]
          Foothill Needle Grass
     FF Lower glumes usually more than 12 mm long; awn 2.5-11
      cm long (usually more than 6 cm) *Nassella cernua*
      [includes *Stipa cernua* and *S. lepida* var. *andersonii*]
          Nodding Needle Grass
   DD Awn of lemma less than 2 cm long, not attached at an obvious
    junction; awns either straight, wavy, or only slightly bent

E Florets with a tuft of whitish hairs at the base (pry out
   a floret and examine it with a hand lens); glumes
   sometimes sharp-tipped, but without awns (*Agrostis*, keyed
   under choice EE, may have inconspicuous hairs at the base
   of the florets)
   F Inflorescence open, the spikelets most densely crowded
      near the tip of the branchlets; spikelets 3-4 mm long;
      plants not forming dense clumps (leaf blades flat, up
      to 8 mm wide; often in wet areas)
                                     Calamagrostis bolanderi
                                 Bolander Reed Grass; 4; Sn-n
   FF Inflorescence rather compact (and somewhat
      cylindrical), densely crowded with spikelets
      throughout; spikelets more than 4 mm long; plants
      forming dense clumps
      G Awn protruding at least 2 mm beyond the spikelet
         (tip of lemma with 4 minute teeth)
         H Spikelets 10 mm long, with awns protruding 8-10
            mm (leaf blades 1-3 mm wide, the margin rolled
            up; coastal)   *Calamagrostis foliosa*
                                    Leafy Reed Grass; 4; Sn-n
         HH Spikelets 5-6 mm long, with awns protruding
            about 2 mm (look at several) (inflorescence
            straw-colored or purple-tinged; on serpentine
            soils)   *Calamagrostis ophitidis*
                         Serpentine Reed Grass; 4; La, Ma, Sn
      GG Awn hidden within the spikelet, or protruding not
         more than 1 mm (inflorescence sometimes purplish)
         H Awn about 2 mm long, straight, protruding above
            the tip of the spikelet; hairs at the base of
            the floret about two-thirds as long as the
            floret (inflorescence usually 4-15 cm long;
            lemma roughened by small, stiff hairs;
            spikelets 3-4 mm long; in wet areas)
                         Calamagrostis stricta ssp. *inexpansa*
                                          [*C. crassiglumis*]
                                 Dense Reed Grass; Ma, Me-n
         HH Awn 4 mm long, bent (pry out a floret to see),
            usually protruding from the side of the
            spikelet (but sometimes completely hidden
            within it); hairs at the base of the floret
            much less than one-third as long as the floret
            I Inflorescence loose, up to 3 cm wide and
               15-30 cm long; leaf blades 6-12 mm wide
               (spikelets about 6 mm long; leaves without
               white hairs where the blade joins the
               sheath; up to 1.5 m tall; in moist areas)
                                     Calamagrostis nutkaensis
                                  Pacific Reed Grass; SLO-n
            II Inflorescence often compact, 1-1.5 cm wide
               and 5-15 cm long; leaf blades 2-5 mm wide
               J Inflorescence usually less than 1 cm
                  wide; spikelets 4 mm long; leaf blades
                  less than 4 mm wide, the margin usually
                  rolled up, with delicate white hairs
                  where the blade joins the sheath
                                     Calamagrostis rubescens
                                      Pine Grass; SLO-n
               JJ Inflorescence usually 1-1.5 cm wide;
                  spikelets 6 mm long; leaf blades up to
                  6 mm wide, the margin flat, without
                  white hairs where the blade joins the
                  sheath   *Calamagrostis koelerioides*
                                   Tufted Pine Grass; SLO-n
EE Florets without a tuft of whitish hairs at the base;
   glumes sometimes with awns or awn-tipped
   F Awn attached to the tip of the lemma (florets shiny and
     hard, these and/or the awns readily falling off;
     inflorescence spreading; lemma awn 3-4 mm long; glumes
     with tapering, pointed, awnlike tips; forming dense
     clumps)   *Piptatherum miliaceum* [*Oryzopsis miliacea*]
                                          Smilo Grass; eua
   FF Awn attached to the back of the lemma

G Awn 4-8 mm long (florets with 2-4 very small teeth
   at the tip [use hand lens]; not more than 45 cm
   tall, not forming clumps)
   H Inflorescence dense, ascending, up to 12 cm
      long; glumes 2.5-6 mm long, sometimes with
      short awns; leaf blades up to 3 mm wide; lemma
      with 2-4 teeth (usually in moist areas)
              Agrostis microphylla [includes *A. microphylla*
                      var. *intermedia* and *A. aristiglumis*]
                              Small-leaf Bent Grass
   HH Inflorescence open, spreading, up to 20 cm long
      (the branches threadlike); glumes up to 2 mm
      long, without awns; leaf blades about 1 mm
      wide (wirelike); lemma with 2 teeth
                      Agrostis elliottiana [*A. exigua*]
                              Annual Tickle Grass; Na-n
GG Awn up to 3 mm long (glumes at least 2 mm long)
   H Branches of the inflorescence up to more than
      10 cm long, ascending or spreading outward,
      with branchlets originating near their tips;
      spikelets few, near the end of each branchlet
      (inflorescence often purplish; awn less than 2
      mm long, straight; leaf blades 1-3 mm wide;
      plants forming dense clumps, the stems
      wirelike; in Coast Ranges)   *Agrostis scabra*
                              Tickle Grass
   HH Branches of the inflorescence not more than 5
      cm long, usually ascending (but sometimes
      spreading outward at maturity), with
      branchlets all along their length; spikelets
      numerous, present on at least half the length
      of each branchlet
      I Inflorescence dense and cylindrical (up to
         10 cm long, sometimes partly enclosed by
         the leaf sheath); plants not forming clumps
         (awns up to 3 mm long; leaves 3-11 mm wide;
         ligules up to 2 mm long; coastal)
                 Agrostis densiflora [includes *A. clivicola*
                      vars. *clivicola* and *punta-reyesensis*]
                              California Bent Grass; SCr-n
      II Inflorescence generally open; plants
         forming dense clumps
         J Awn less than 1 mm long, straight;
            inflorescence up to 20 cm long, with
            spikelets usually along the entire
            length of each branchlet; ligules 2-3
            mm long (often in sandy areas)
                 Agrostis pallens [includes *A. diegoensis*]
                              Dune Bent Grass
         JJ Awn 2.5-3 mm long, bent or straight;
            inflorescence up to 30 cm long, with
            spikelets along only half the length of
            each branchlet; ligules 3-6 mm long (in
            moist areas)   *Agrostis exarata*
                              [includes *A. ampla* and
                      A. longiligula var. *australis*]
                              Spike Bent Grass; SCr-n
BB Awns lacking or not more than 1 mm long (except in *Agrostis scabra*,
   in which they may be nearly 2 mm long)
   C Spikelets conspicuously flattened (in wet habitats)
      D Inflorescence 1-3 cm wide, the spikelets crowded on one side
         of each branchlet and arranged in 2 main rows, sometimes with
         additional spikelets between the rows; spikelets roundish,
         but with a sharp tip, without stiff spines on the margin (but
         may have fine hairs on the surface)   *Beckmannia syzigachne*
                              Slough Grass; SFBR-n
      DD Inflorescence 5-10 cm wide, the spikelets not concentrated in
         2 rows; spikelets (without glumes) elliptical, without a
         sharp tip, but with small, stiff spines on the margin
                              Leersia oryzoides
                              Rice Cut Grass; Na, Me-n
   CC Spikelets not conspicuously flattened
      D Inflorescence, at maturity, loosely spreading, usually at
         least 5 cm wide (often more than 10 cm wide)

E   Inflorescence often as long as wide; spikelets about 1 mm
    long; ligule not more than 1 mm long (inflorescence
    delicate, with many threadlike branchlets bearing
    numerous small spikelets; leaves crowded, with
    overlapping sheaths; in moist areas)
                                    Muhlenbergia asperifolia
                                          Scratch Grass
EE Inflorescence generally longer than wide; spikelets 2-3 mm
    long; ligule 3-5 mm long
    F   Inflorescence often partly hidden in the leaf sheath;
        one or both glumes shorter than the whole spikelet,
        the shorter glume not more than half as long as the
        other one; leaves, at least where the blade joins the
        sheath, with long hairs (mature inflorescence up to 20
        cm wide; grains hard)
        G   Leaf blades up to 4 mm wide, the margin usually
            rolled up, scarcely hairy, except where the blade
            joins the sheath; florets falling from the glumes
            at maturity, not prominently veined; grains dark
            brown, rough; plants forming dense clumps (when
            the fertile floret splits open at maturity, it may
            seem to consist of 2 florets; in moist, often
            alkaline areas)   *Sporobolus airoides*
                                          Alkali Dropseed
        GG Leaf blades 5-15 mm wide, the margin not rolled up,
            hairy; florets (one sterile and one fertile)
            falling with the glumes at maturity, prominently
            veined; grains ivory-colored, shiny;   plants not
            forming dense clumps (the mature inflorescence may
            separate from the stem and roll like a tumbleweed)
            H   Inflorescence up to 25 cm long; leaf blades 5-
                12 mm wide; grain with a prominent scar at its
                base   *Panicum hillmanii*
                                     Hillman Witch Grass; na; Sl
            HH Inflorescence up to 40 cm long; leaf blades 6-
                22 mm wide; grain without a scar at its base
                                            Panicum capillare
                                          Witch Grass; SF
    FF Inflorescence, at maturity, not hidden in the leaf
        sheath; glumes longer than the aggregate of florets in
        the spikelet, the glumes nearly equal in length;
        leaves without long hairs
        G   Branches of the inflorescence (these up to 12 cm
            long, upright to spreading) with branchlets
            towards their tips; spikelets few, near the end of
            each branchlet; leaf blades 1-3 mm wide; stems
            wirelike (not more than 1 mm wide) and usually
            crowded together; inflorescence often purplish (up
            to 10 cm wide and more than 20 cm long) (lemma awn
            up to 2 mm long; up to 75 cm tall, forming dense
            clumps; moist areas in the Coast Ranges)
                                            Agrostis scabra
                                          Tickle Grass
        GG Branches of the inflorescence with branchlets all
            along their length; spikelets many, often crowded
            all along the branchlets; leaf blades 2-8 mm wide;
            stems not wirelike and not crowded together;
            inflorescence often reddish purple to bronze (in
            wet areas) (Note: the following two species are
            difficult to separate.)
            H   Inflorescence usually not more than 5 cm wide
                or 15 cm long, dense, the branches ascending;
                leaf blades 2-5 mm wide; up to 60 cm tall,
                with some stems prostrate and often rooting at
                the nodes, forming a mat
                        Agrostis stolonifera [*A. alba* vars. *alba*
                            (in part) and *palustris*] (*Plate* 104)
                                          Creeping Bent Grass; eu
            HH Inflorescence usually up to 10 cm wide and 25
                cm long, the branches spreading; leaf blades
                up to 8 mm wide; up to 100 cm tall
                Agrostis gigantea [*A. alba* var. *alba* (in part)]
                                          Giant Bent Grass; eu

DD Inflorescence, at maturity, not loosely spreading, usually
less than 5 cm wide (sometimes up to 6 cm wide in *Melica
imperfecta*, however)
  E   Branches of the inflorescence ascending, usually
      completely covered with lobelike clusters of spikelets
      (inflorescence up to 15 cm long; glumes about 2 mm long,
      longer than the florets; florets falling with the glumes
      at maturity; stems sometimes prostrate and rooting at the
      nodes; in moist areas)
                            Agrostis viridis [*A. semiverticillata*]
                                          Whorled Bent Grass; eu
  EE Branches of the inflorescence not covered with lobelike
      clusters of spikelets
    F   Spikelets up to 6 mm long, with a club-shaped pedestal
        arising from the base of the fertile floret (pry open
        the spikelet); edges of leaf sheaths fused together
        (inflorescence usually less than 1 cm wide; glumes
        somewhat transparent, with conspicuous purple
        markings; in dry habitats)
      G   Florets without fine hairs; glumes rounded at the
          tip and not prominently veined; pedestal without a
          knob at the tip (inflorescence sometimes up to 6
          cm wide)   *Melica imperfecta*
                                          Coast Range Melic; SCl-s
      GG Florets with fine hairs (use hand lens); glumes
          with a sharp point at the tip and prominently
          veined; pedestal with a knob (this slightly wider
          than the stalk) at the tip (in shaded areas)
                                               Melica torreyana
                                          Torrey Melic; SLO-n
    FF Spikelets usually less than 3 mm long (but sometimes
        up to 5 mm long), without a pedestal; edges of leaf
        sheath not fused
      G   Glumes unequal, the shorter one much smaller than
          the whole spikelet (a sterile floret, as long as
          the fertile floret, also present); spikelets
          (about 2 mm long) plump and hard, flattened in
          such a way that only one glume is visible when a
          spikelet is viewed with a broad surface facing the
          observer; florets falling with the glumes at
          maturity (ligule consisting of upright white
          hairs; plants vary seasonally, forming a basal
          cluster, clump, or mat; in moist, sometimes
          alkaline areas)
            Panicum acuminatum [includes *P. occidentale*,
                             P. pacificum, and *P. thermale*]
                                               Marsh Panicum
      GG Glumes nearly equal, longer than the aggregate of
          florets in the spikelet; spikelets delicate,
          flattened in such a way that both glumes are
          visible when a spikelet is viewed with a broad
          surface facing the observer; florets falling from
          the glumes at maturity
        H   Florets with an evident basal tuft of white
            hairs half as long as the floret (inflo-
            rescence ascending to spreading, with many
            branches at each node; spikelets 3-4 mm long;
            ligules 4-7 mm long; up to 1 m tall; shaded
            areas in the Coast Ranges)   *Agrostis hallii*
                                               Hall Bent Grass
        HH Florets without an evident basal tuft of white
            hairs
          I   Leaves usually less than 1 mm wide,
              wirelike; ligules less than 1.5 mm long
              (inflorescence dense, less than 8 cm long,
              sometimes partly hidden in the leaf sheath;
              up to 30 cm tall, but usually less than 15
              cm; often forming dense clumps; on coastal
              dunes)   *Agrostis blasdalei*
                                          Blasdale Bent Grass; Me-Ma; 1b
          II Leaves up to 11 mm wide, some of them flat;
              ligules 1.5-6 mm long (Note: the following
              species of *Agrostis* are difficult to
              separate.)

J   Some stems prostrate and often rooting
    at the nodes, forming a mat; inflor-
    escence usually 2-5 cm wide (less than
    15 cm long) (ligules 2-5 mm long; leaf
    blades less than 5 mm wide; in wet
    habitats)   *Agrostis stolonifera*
       [includes *A. alba* vars. *alba* (in part)
                     and *palustris*] (*Plate* 104)
                       Creeping Bent Grass; eu
JJ All stems upright; inflorescence less
    than 2 cm wide
    K   Inflorescence up to 30 cm long;
        ligules 3-6 mm long (leaf blades 2-7
        mm wide; plants forming clumps; in
        moist areas)   *Agrostis exarata*
                   [includes *A. ampla* and
                   A. longiligula var. *australis*]
                              Spike Bent Grass
    KK Inflorescence not more than 20 cm
        long; ligules 1.5-3 mm long
        L   Inflorescence up to 20 cm long,
            narrow but open, its branches
            not obscured; ligules 2-3 mm
            long; stem leaves with blades up
            to 6 mm wide; not forming clumps
            (coastal dunes) *Agrostis pallens*
                     [includes *A. diegoensis*]
                              Dune Bent Grass
        LL Inflorescence not more than 10 cm
            long, dense, its branches
            sometimes obscured; ligules up
            to 2 mm long; stem leaves with
            blades up to 11 mm wide; plants
            forming clumps (coastal)
            Agrostis densiflora [includes
            A. clivicola vars. *clivicola*
                   and *punta-reyesensis*]
                       California Bent Grass; SCr-n

Poaceae, Division IV, Subkey 2: Inflorescence with obvious branches, these
divided into branchlets; spikelets (excluding the awns) usually more than
1.5 cm long, with more than 1 obvious floret; awn at least 1 mm long,
attached at or very near the tip of the lemma; glumes (without awns)
unequal and usually shorter than the whole spikelet; leaves up to more than
10 mm wide, the margin usually flat, the ligule prominent (sometimes torn
and jagged); edges of the leaf sheaths (at least the lower portion) fused
together   *Bromus*
A   Awns sharply bent and twisted below (look at several) (glumes less than
    1 mm wide, pointed, the upper one almost as long as the lowest floret;
    teeth at tip of lemma bristlelike, usually 2-3 mm long; awns 10-17 mm
    long)   *Bromus trinii*
                                       Chilean Chess; sa?

AA Awns straight or curved
    B   Awns 3.5-7 cm long (spikelets few, up to 30 cm long [excluding the
        awns]; lower branches usually spreading and drooping; ligule
        whitish, with jagged teeth; teeth at tip of lemma membranous, 3-6 mm
        long; widespread)   *Bromus diandrus*
                                           Ripgut Brome; eu
    BB Awns less than 3 cm long
        C   Spikelets markedly flattened (less than 2 mm wide); lemma with a
            definite keel (thus V-shaped in cross-section) (teeth, if
            present at tip of lemma, very small)
            D   Awns, if present, less than 4 mm long; lower glume with 5 or
                more prominent veins; lemma usually with 9 or more veins
                                                Bromus catharticus
                                              Rescue Grass; sa
            DD Awns 3-12 mm long (sometimes up to 15 mm); lower glume
                usually with 3 prominent veins; lemma with fewer than 9 veins
                (Note: the following 2 species are difficult to separate.)
                E   Inflorescence ascending (1.5-2 cm wide), its shortest
                    branches less than 1 cm long; stems markedly bent near
                    the base; leaf blades up to 12 mm wide, the sheath
                    generally not hairy near the base (on coastal dunes)
                    Bromus carinatus var. *maritimus* [*B. maritimus*]
                                               Marine Brome

EE Inflorescence ascending to spreading, its shortest
   branches often more than 3 cm long; stems not markedly
   bent near the base; leaf blades not more than 7 mm wide,
   the sheath densely hairy near the base (widespread in dry
   habitats)   *Bromus carinatus* var. *carinatus*
   [includes *B. breviaristatus* and *B. marginatus*] (*Plate* 105)
                             California Brome
CC Spikelets not markedly flattened; florets without a keel except
near the tip (thus mostly U-shaped in cross-section)
  D Teeth at tip of lemma 2-5 mm long (inflorescence often
    purplish; up to 60 cm tall, annual)
    E Inflorescence not compact, up to 8 cm wide and 15 cm long,
      its branches spreading and drooping; spikelets (usually
      hairy, on delicate branchlets) drooping and not crowded;
      upper glume usually less than 10 mm long; lemma about 10
      mm long (awns 12-16 mm long; annual, very easily pulled
      up)   *Bromus tectorum* (*Plate* 105)
                            Cheat Grass; eu
    EE Inflorescence compact, usually not more than 5 cm wide and
      8-10 cm long, its branches ascending; spikelets upright
      and crowded; upper glume 10-12 mm long; lemma 11-17 mm
      long
      F Inflorescence very compact (resembling the head of a
        broom), the branches obscured; teeth at tip of lemma
        3-5 mm long; lower glume pointed, but not needlelike,
        6-9 mm long; awns up to 22 mm long
        Bromus madritensis ssp. *rubens* [*B. rubens*] (*Plate* 63)
                        Foxtail Chess; me
      FF Inflorescence slightly spreading, not so compact that
        the branches are obscured; teeth at tip of lemma 2-3
        mm long; lower glume needlelike, 9-11 mm long; awns up
        to 30 mm long   *Bromus madritensis* ssp. *madritensis*
                        Spanish Brome; eu
  DD Teeth either absent at tip of lemma or not more than 2 mm
    long
    E Lower glume usually with 1 prominent vein (sometimes 2 or
      3 in *Bromus orcuttianus*) (up to more than 1 m tall)
      F Awn 18-30 mm long; teeth at tip of lemma 2 mm long;
        annual (upper glume 10-18 mm long; widespread)
                              Bromus sterilis
                      Poverty Brome; eua; Mo-n
      FF Awn 4-13 mm long; teeth at tip of lemma less than 1 mm
        long; perennial
        G Awn 4-8 mm long; upper glume 6-11 mm long; branches
          of the inflorescence mostly less than 2 cm long,
          ascending (at elevations of at least 3000')
                        Bromus orcuttianus
                        Orcutt Brome
        GG Awn 7-13 mm long; upper glume 10-15 m long;
          branches of the inflorescence mostly more than 2
          cm long (in shaded areas in Coast Ranges)
                          Bromus vulgaris
                    Narrowflower Brome; Mo-n
    EE Lower glume with 3-5 prominent veins
      F Awn 10-17 mm long (dark); branches of the inflo-
        rescence very wavy (and spreading outward); up to 40
        cm tall (upper glume 8-12 mm long with 5-7 veins;
        annual)   *Bromus arenarius*
                      Australian Brome; au
      FF Awn, if present, usually less than 10 mm long;
        branches of the inflorescence more or less straight;
        up to more than 80 cm tall
        G Upper glume (7-9 mm long) usually with 3 distinct
          veins (sometimes 4 or 5) (awns 3-6 mm long; leaf
          blades up to 12 mm wide, densely hairy;
          perennial)   *Bromus grandis*
                         Grand Brome
        GG Upper glume with 5-7 distinct veins
          H Leaf blades 4-17 mm wide; perennial
             (inflorescence up to 27 cm long; upper glume
             6-11 mm long; mostly in shaded areas)
             Bromus laevipes [includes *B. pseudolaevipes*]
                        Chinook Brome
        HH Leaf blades 1-7 mm wide; annual

I   Inflorescence up to 13 cm long, dense, the
    spikelets obscuring the short branches;
    upper glume 6-9 mm long (awns 4-10 mm long;
    widespread)   *Bromus hordeaceus*
                                    Soft Cheat Grass; eu
II Inflorescence up to more than 17 cm long,
    open, the spikelets not obscuring the
    branches; upper glume 5-8 mm long
    J   Inflorescence up to 26 cm long;
        spikelets only slightly, if at all,
        flattened; awns 5-11 mm long
                        Bromus japonicus [*B. commutatus*]
                                    Japanese Brome; eua
    JJ Inflorescence up to 17 cm long;
        spikelets flattened; awn 3-8 mm long
                                    Bromus secalinus
                                        Chess; eua

Poaceae, Division IV, Subkey 3: Inflorescence with obvious branches, these
divided into branchlets; spikelets (excluding the awns) usually less than
1.5 cm long, with more than 1 obvious floret; awn at least 1 mm long,
attached at or very near the tip of the lemma; glumes without awns, of
unequal length and usually shorter than the whole spikelet; leaves usually
less than 2 mm wide, often wirelike, the ligule absent or inconspicuous;
edges of the leaf sheaths not fused   *Festuca* (in part), *Vulpia*
A   Glumes and/or lemma with very small hairs (use lens)
    B   Both glumes and florets with hairs (awn of lemma 3-16 mm long; up to
        40 cm tall, annual)   *Vulpia microstachys* var. *ciliata*
                        [includes *Festuca eastwoodae* and *F. grayi*]
    BB Either only the glumes, or only the lemma with hairs
        C   Glumes hairy, lemma not hairy; spikelets usually with 2-3 florets
            (sometimes 4) (awns 4-14 mm long; up to 40 cm tall, annual)
            Vulpia microstachys var. *confusa* [*Festuca confusa* and *F. tracyi*]
                                            Hairyleaf Fescue
        CC Glumes not hairy, lemma hairy (look at several spikelets);
            spikelets with 3-12 florets
            D   Spikelets with 5-12 florets (these densely crowded); awns 1-5
                mm long (less than 20 cm tall, annual)
                                    Vulpia octoflora var. *hirtella*
                                    [*Festuca octoflora* ssp. *hirtella*]
                                            Slender Fescue
            DD Spikelets with 3-6 florets; awns 2-15 mm long
                E   Awns usually 3-5 mm long, attached between 2 very small
                    teeth at the tip of the lemma; branches of the
                    inflorescence up to more than 6 cm long, sometimes
                    drooping; leaf blades up to 4 mm wide, flat; up to 100 cm
                    tall, perennial (florets 3-4, the lemma with 5 prominent
                    veins and very small hairs; in shaded areas)
                        Festuca elmeri [includes *F. elmeri* ssp. *luxurians*]
                                            Elmer Fescue; Mo-n
                EE Awns at least 8 mm long, attached at the tip of the lemma,
                    but not between 2 teeth; branches of the inflorescence
                    not more than 4 cm long (usually much less), erect to
                    spreading; leaf blades less than 2 mm wide, wirelike; up
                    to 45 cm tall, annual (spikelets nearly sessile)
                    F   Inflorescence less than 1 cm wide, all of the branches
                        upright; leaf sheaths not hairy
                            Vulpia myuros var. *hirsuta* [*Festuca megalura*]
                                            Foxtail Fescue; eu
                    FF Inflorescence 1-6 cm wide, at least the lower branches
                        spreading; leaf sheaths hairy
                                    Vulpia microstachys var. *ciliata*
                                    [includes *Festuca eastwoodae* and *F. grayi*]
                                            Hairy Fescue
AA Glumes and florets without hairs
    B   Awns (usually 3-5 mm long) attached between 2 very small teeth at the
        tip of the lemma (branches of the inflorescence up to more than 6 cm
        long, sometimes drooping; leaf blades up to 4 mm wide, flat; florets
        3-4, the lemma with 5 prominent veins; up to 1 m tall, perennial; in
        shaded areas)   *Festuca elmeri* [includes *F. elmeri* ssp. *luxurians*]
                                            Elmer Fescue; Mo-n
    BB Awns attached at the tip of the lemma, but not between 2 teeth
        C   Awns not more than 4 mm long (up to 5 mm long in *Festuca
            idahoensis*) (plants forming dense clumps)

D Leaf sheaths hairy, at least where the blade joins the sheath
   (lower branches of the inflorescence spreading, usually
   paired, often reaching a length of more than 4 cm before
   giving rise to the first branchlets; spikelets numerous, each
   with at least 4 florets; up to 120 cm tall, perennial; in
   shaded areas)   *Festuca californica*
                                        California Fescue; Mo-n
DD Leaf sheaths not hairy
   E Lemma (excluding the awn) 3-5 mm long; spikelets with 6-13
      florets; up to 60 cm tall, annual
                                    Vulpia octoflora var. *octoflora*
                             [*Festuca octoflora* ssp. *octoflora*]
                                                Sixweeks Fescue
   EE Lemma (excluding the awn) 5-10 mm long; spikelets with not
      more than 7 florets; up to more than 80 cm tall,
      perennial
      F Base of the leaf sheaths disintegrating into fibers as
         the plant ages; leaf sheaths and spikelets usually
         dark reddish purple; stems bending outward, before
         turning upward; awns 1-3 mm long; plants forming loose
         clumps (branchlets appearing on the lower branches
         within 2 cm of the stem; in moist areas)
                                        Festuca rubra (*Plate* 107)
                                              Red Fescue; SLO-n
      FF Base of the leaf sheaths not disintegrating; leaf
         sheaths and spikelets purplish; stems not bending
         outward; awns 3-4 mm long; plants forming dense clumps
         (Coast Ranges)   *Festuca idahoensis*
                                              Idaho Fescue; SM-n
CC Some awns at least 5 mm long
   D Spikelets not sessile (pedicels may be less than 2 mm long,
      however); inflorescence up to 12 cm wide, the branches
      spreading or drooping; leaves mostly basal; up to 100 cm
      tall, perennial (awns 5-12 mm long; florets 3-5; in shaded
      areas)   *Festuca occidentalis*
                                            Western Fescue; Mo-n
   DD Many spikelets sessile; inflorescence not more than 3 cm
      wide, the branches erect or spreading; leaves usually not
      mostly basal; not more than 60 cm tall, annual (Note: the
      following species are difficult to separate.)
      E Awns not more than 6 mm long; spikelets with 6-13 florets
         (upper glume usually less than 6 mm long; branches
         upright)   *Vulpia octoflora* var. *octoflora*
                          [*Festuca octoflora* ssp. *octoflora*] (*Plate* 109)
                                                Sixweeks Fescue
      EE Awns up to more than 10 mm long; spikelets usually with
         fewer than 8 florets
         F Lower glume 1-2 mm long, much less than one-half as
            long as the upper glume; awns up to 19 mm long
                              Vulpia myuros var. *myuros* [*Festuca myuros*]
                                             Rattail Fescue; eu
         FF Lower glume 2-8 mm long, from one-half to fully as
            long as the upper glume; awns less than 13 mm long
            G Spikelets usually with 2-4 florets; branches
               spreading outward or drooping
                                 Vulpia microstachys var. *pauciflora*
                                  [*Festuca pacifica* and *F. reflexa*]
                                        Common Hairyleaf Fescue
            GG Spikelets with 4-7 florets; branches upright
               (sometimes very dense)
                        Vulpia bromoides [*Festuca dertonensis*] (*Plate* 109)
                                              Brome Fescue; eu

Poaceae, Division IV, Subkey 4: Inflorescence with obvious branches, these
divided into branchlets; spikelets with 2 or more similar florets; awns, if
present, less than 1 mm long; glumes (without awns) somewhat to very
unequal, both (at least the lower one) usually shorter than the whole
spikelet
A Leaf blades abruptly curved on one side at the tip, forming (especially
   when folded) what looks like the bow of a boat, less than 6 mm wide;
   glumes and lemma somewhat keeled (lemma usually less than 5 mm long;
   lower glume usually shorter than the lemma of the lowest floret;
   spikelets without true awns; florets often with cottony webbing at the
   base and sometimes with hairs elsewhere; leaf sheaths open above but
   often fused below; spikelets often not open at maturity)

B   Spikelets with purple-black, teardrop-shaped bulblets (these 2-3 mm
    long, with papery, awnlike projections 1-2 cm long) (there may also
    be bulblets at the base of the stem, and sometimes true florets are
    present in the spikelets)   *Poa bulbosa* (*Plate* 63)
                                         Bulbous Blue Grass; eu; SLO-n
BB Spikelets without bulblets
    C   Florets without cottony webbing at the base (there may be other
        hairs, especially on the veins of the floret, but these not
        intertwined to form a web) (plants clumped at the base)
        D   Spikelets 3-6 mm long; branches spreading (sometimes at right
            angles); lemma up to 3 mm long, obviously hairy, with 5
            prominent veins; usually less than 20 cm tall, annual (leaves
            1-3 mm wide, flat; upper glume generally widest at or above
            the middle; plants often forming mats) *Poa annua* (*Plate* 109)
                                                    Annual Blue Grass; eu
        DD Spikelets 5-10 mm long; branches upright; florets up to 4 mm
            long, only slightly hairy, if at all, and the veins scarcely
            visible; up to 100 cm tall, perennial
            E   Spikelets mostly 7-10 mm long, not distinctly flattened
                and not bunched; leaves up to 2 mm wide; ligules 2-7 mm
                long; lemma slightly hairy towards the base; glumes and
                lemma rounded on the back, without an evident keel (thus
                U-shaped in cross-section) (upper glume widest just above
                the base; plants forming dense clumps)
                                            Poa secunda [*P. scabrella*]
                                                       Oneside Blue Grass
            EE Spikelets 5-7 mm long, somewhat flattened and often in
                bunches; leaves less than 1 mm wide (wirelike); ligules
                less than 1 mm long; lemma not hairy; glumes and lemma
                usually with an obvious keel (thus V-shaped in cross
                section) (in alkaline soils near hot springs)
                                                          Poa napensis
                                                 Napa Blue Grass; Na; 1b
    CC Florets with cottony webbing at the base
        D   Inflorescence usually with 1-2 branches at each node (the
            lowest up to 8 cm long), each branch bearing only a few
            spikelets; lemma usually 3-5 mm long (lemma without straight
            hairs, but with 5 prominent veins; leaf blades mostly flat,
            up to 4 mm wide; in moist, shaded areas)   *Poa kelloggii*
                                                        Kellogg Blue Grass
        DD Inflorescence usually with more than 2 branches at each node,
            each branch usually bearing numerous spikelets; lemma not
            more than 3 mm long
            E   Ligules less than 2 mm long
                F   Lower branches of inflorescence usually 3 at a node;
                    leaf blades up to 2 mm wide, the margin flat or rolled
                    up; ligules less than 1 mm long; spikelets with 2-4
                    florets, their lemmas with indistinct veins, sparse
                    cottony webbing at the base, and inconspicuous hairs
                    on the keel; glumes not strongly keeled (plants
                    forming dense clumps; in moist, shaded areas)
                                                        Poa nemoralis
                                                    Wood Blue Grass; eu
                FF Lower branches of inflorescence 3 or more at a node;
                    leaf blades up to 5 mm wide, the margin flat or
                    folded; ligules 1-2 mm long; spikelets with 3-5
                    florets, their lemmas with prominent veins, dense
                    cottony webbing at the base, and obvious hairs on the
                    keel; glumes strongly keeled *Poa pratensis* (*Plate* 109)
                                                 Kentucky Blue Grass; eu
            EE Ligules 2-8 mm long (leaf blades 2-5 mm wide, usually
                flat; glumes and lemma not strongly keeled; lemma with
                prominent veins and either with indistinct hairs or no
                hairs)
                F   Lower branches of the inflorescence usually 4 or more
                    at a node; ligules 2-4 mm long; lower glume not narrow
                    or crescent-shaped; spikelets with 2-5 florets, the
                    lemmas with sparse cottony webbing at the base and
                    sometimes with fine hairs on the back; annual (leaf
                    blades sometimes folded; in the Coast Ranges)
                                    Poa howellii [*P. bolanderi* ssp. *howellii*]
                                                        Howell Blue Grass

FF Lower branches of inflorescence generally 5 or more at
 a node; ligules 4-8 mm long; lower glume narrow and
 crescent-shaped; spikelets with 2-3 florets, the
 lemmas with dense cottony webbing at the base but
 without hairs elsewhere; perennial (in moist areas)
             Poa trivialis
           Rough Blue Grass; eu; SF-n
AA Leaf blades tapering gradually to a needlelike tip, up to 15 mm wide
 (but mostly narrower, and the margin often rolled up); glumes and lemma
 not keeled (except sometimes in *Festuca*)
 B Spikelets with purple splotches and with either a club-shaped
  pedestal hidden among a few fertile florets, or with several greatly
  reduced florets above large fertile florets (glumes papery, the
  lower one with at least 3 veins, the upper one with at least 5;
  lemma rounded on the back; edges of leaf sheaths fused)
  C Glumes at least three-quarters as long as the spikelets, these
   with 2-4 (rarely 1) fertile florets 4-9 mm long (spikelets with
   a club-shaped pedestal hidden among the florets; inflorescence
   usually less than 2 cm wide)
   D Spikelets 7-12 mm long, with 2-4 florets, the lemmas not
    hairy, 5-9 mm long; pedestal knob 3-4 mm long, definitely
    wider than the stalk; stems sometimes with bulbs at the base
    (Coast Ranges)   *Melica californica*
              California Melic
   DD Spikelets 5-7 mm long, with 2 florets (rarely 1), the lemmas
    hairy, 4-5 mm long; pedestal knob 1 mm long, scarcely wider
    than the stalk; stems without bulbs at the base (purple
    splotches on spikelets prominent) (in shaded areas)
             Melica torreyana
           Torrey Melic; SLO-n
  CC Glumes (at least the lower one of each pair) not more than about
   half as long as the spikelets, these with 2-6 fertile florets 8-
   13 mm long (spikelets 12-20 mm long, with reduced florets at the
   tip)
   D Lemma not hairy (except for tiny spines along the edge);
    inflorescence up to 8 cm wide (lower glume 4-5 mm long, two-
    thirds as long as the upper glume; lemma 8-10 mm long; stems
    with bulbs at the base; usually in shaded areas)
    Melica geyeri [includes *M. geyeri* var. *aristulata*]
           Geyer Onion Grass; Mo-n
   DD Lemma hairy; inflorescence usually less than 4 cm wide
    E Lemma 10 mm long, the tip somewhat rounded and sometimes
     with awns up to 3 mm long; branches upright; lower glume
     6-8 mm long; stems without bulbs at the base
             Melica harfordii
          Harford Onion Grass; Mo-n
    EE Lemma 9-15 mm long, the tip tapered, needlelike, but
     without awns; branches upright to slightly spreading;
     lower glume 6 mm long; stems with bulbs at the base
     (shaded areas in Coast Ranges)
           Melica subulata (*Plate* 109)
          Alaska Onion Grass; Ma-n
 BB Spikelets without purple splotches (there may be some small purple
  markings, however), and without a club-shaped pedestal or reduced
  upper florets
  C Lemma narrow, spindle-shaped, the tip tapered to a sharp point,
   without prominent parallel veins; lower glume 3.5-6 mm long,
   upper glume 5-8 mm long (glumes and lemma sometimes with a keel;
   lower glume with 1 vein, the upper one with 3; edges of leaf
   sheaths not fused)
   D Leaf sheath hairy where the leaf blade joins it;
    inflorescence up to 10 cm wide (and 30 cm long); lower glume
    5-6 mm long; lemma (8-10 mm long) sometimes with a short awn
    (leaf blades up to 5 mm wide, the margin flat or rolled up;
    lower branches of the inflorescence spreading, usually not
    branching for the first 4-6 cm of their length; in shaded
    areas)   *Festuca californica*
          California Fescue; Mo-n
   DD Leaf sheath not hairy where the leaf blade joins it;
    inflorescence up to 4 cm wide; lower glume 3.5-5 mm long;
    lemma without an awn

E   Leaf blades less than 2 mm wide, the margin mostly rolled
    up; lemma 4-7 mm long, without an awn; inflorescence up
    to 15 cm long; plants usually forming clumps (Coast
    Ranges)   *Festuca viridula*
                                            Green Fescue; Sn-n
EE Leaf blades 4-10 mm wide, the margin usually flat; lemma
    6-9 mm long, sometimes short-awned; inflorescence up to
    30 cm long; plants not forming clumps *Festuca arundinacea*
                                            Tall Fescue; eu
CC Lemma rounded on the back, oval to almond-shaped, the tip not
    sharply pointed, with 5-7 often prominent, parallel veins; lower
    glume up to 3 mm long; upper glume up to 5 mm long (in wet
    areas)
    D   Lemma with 7 prominent veins; both glumes with 1 vein; edges
        of leaf sheaths fused; usually in freshwater habitats (glumes
        papery; up to 1.5 m tall, not forming clumps)
        E   Spikelets 3-6 mm long, with 4-7 florets (the lemmas
            without spines at the tip); inflorescence up to 20 cm
            wide (and nodding); lower glume 1 mm long, the upper one
            1.5 mm long; lemma 2 mm long   *Glyceria elata*
                                            Fowl Manna Grass; Ma, Sn
        EE Spikelets 10-22 mm long, with 6-13 florets; inflorescence
            usually not more than 5 cm wide; lower glume up to 3 mm
            long, the upper one not more than 5 mm long; lemma 3-6 mm
            long
            F   Lemma 4-6 mm long, usually with sharp parallel spines
                extending beyond the tip (the spines are continuations
                of the veins; lower glume 2.5-3 mm long, the upper 4-5
                mm long   *Glyceria occidentalis*
                                            Western Manna Grass; SM-n
            FF Lemma 3-4 mm long, without spines at the tip; lower
                glume up to 2 mm long, the upper not more than 3 mm
                long   *Glyceria leptostachya*
                                            Water Manna Grass; Ma, Sn
    DD Lemma with 5 veins (these not extending beyond the tip, and
        sometimes difficult to see) (*Torreyochloa* sometimes has 7
        veins); lower glume with 1 vein, the upper with 3; edges of
        leaf sheaths overlapping or partly fused; generally in
        alkaline or saline habitats (glumes shorter than the lowest
        floret)
        E   Inflorescence less than 0.5 cm wide, the branches upright,
            each usually with 1 spikelet; lemma with fine hairs on
            the back (difficult to see even with a hand lens); less
            than 20 cm tall, annual (upper glume with 3 prominent
            veins; lemma tapered to a sharp tip; plants forming
            clumps; inland)   *Puccinellia simplex*
                                            California Alkali Grass
        EE Inflorescence up to 15 cm wide, the branches spreading,
            each with numerous spikelets; lemma not hairy on the back
            (but it may be hairy at the base); more than 60 cm tall,
            perennial
            F   Leaf blades 5-15 mm wide; lemma about 2 mm long,
                rounded at the tip, with prominent veins
                (inflorescence spreading and drooping, up to 12 cm
                wide; edges of leaf sheath not fused; plants not
                forming clumps)   *Torreyochloa pallida* var. *pauciflora*
                                            [*Puccinellia pauciflora*]
                                            Weak Manna Grass; SM-n
            FF Leaf blades 1-4 mm wide; lemma 2-4 mm long, abruptly
                narrowed towards the tip, with indistinct veins (in
                salt marshes or alkaline areas)
                G   Inflorescence up to 6 cm wide, the branches
                    upright; spikelets 8-15 mm long, with 5-12
                    florets; lemma sparsely hairy at the base, 3-4 mm
                    long; plants forming dense clumps
                        Puccinellia nutkaensis [includes *P. grandis*]
                                            Alaska Alkali Grass; SM-n
                GG Inflorescence up to 15 cm wide, the branches
                    spreading; spikelets 4-7 mm long, with 3-6 florets
                    3-6; lemma not hairy at the base, less than 3 mm
                    long; plants forming clumps, but these not dense
                        Puccinellia nuttalliana [*P. airoides*]
                                            Nuttall Alkali Grass

Pontederiaceae--Pickerel-weed Family

Our only native representative of the Pontederiaceae is *Heteranthera dubia* (*Plate* 110), called Water Stargrass. It is a weak-stemmed aquatic that grows in ponds, ditches, and sluggish streams. The leaves, with prominent stipules at the base, are 7-15 cm long, but not more than 0.5 cm wide. They are nearly translucent and do not have a distinct midrib. The flowers open at the water surface. Each one originates within a rolled up bract. The perianth, with a long, slender tube and 6 lobes about 0.5 cm long, is pale yellow. There are 3 stamens and a pistil that becomes a many-seeded fruit.

Eichhornia crassipes, the Water Hyacinth is also a member of this family. This floating plant, native to South America and Africa, has inflated leaf petioles and an elongated inflorescence of white to pale blue flowers. It is widely cultivated in garden pools, and occasionally becomes established in ponds and lakes in frost-free areas of California. It is a nuisance in Florida and in some other places to which it has been introduced.

Potamogetonaceae (includes Ruppiaceae)--Pondweed Family

The Pondweed Family, consisting of aquatic plants, is represented in our region by 2 genera. In *Potamogeton*, the leaves are mostly alternate (only the ones near the tip of the stem are likely to be opposite) and have a conspicuous scale that resembles a stipule. The base of this scale often forms a sheath around the stem. In certain species, there are broad floating leaves that are very different from the narrow submerged leaves. The small flowers, usually in rather dense terminal inflorescences, typically have 4 pistils and 4 stamens. Each stamen arises from the base of a structure that resembles a sepal, but which may in fact be part of the stamen. The pistils ripen into 1-seeded fruits.

In *Ruppia* (ditchgrasses, sometimes put into a separate family, Ruppiaceae), the leaves are always slender, although they have a rather broad, sheathing base. The inflorescence originates within the sheath of the uppermost leaves. Each flower has 4 pistils, as in *Potamogeton*, but there are only 2 stamens. These have extremely short stalks and do not have sepal-like structures at the base.

A   Inflorescence loose, with only a few flowers; leaves all narrow; in
    brackish water of streams entering salt marshes and bays
    B   Peduncles of inflorescence not more than 3 cm long, not spiraling;
        leaves pointed at the tip   *Ruppia maritima*
                                                        Straight Ditchgrass
    BB Peduncles up to 30 cm long, often spiraling or at least showing a
        tendency to spiral; leaves blunt at the tip
                                       Ruppia cirrhosa [*R. spiralis*] (*Plate* 110)
                                                          Coiled Ditchgrass
AA Inflorescence dense, cylindrical; leaves either all narrow or of 2
    types, the floating ones much broader than the submerged ones; in fresh
    water
    B   All leaves submerged and similar (not more than 1 mm wide, pointed at
        the tip) (fruit oblong, 2.5-4 mm long)   *Potamogeton pectinatus*
                                                        Fennel-leaf Pondweed
    BB Leaves of 2 types: firm, floating leaves at least 5 mm wide and
        submerged leaves 1-5 mm wide
        C   Floating leaves 4-11 cm long and up to 6 cm wide; submerged
            leaves 10-30 cm long   *Potamogeton natans* (*Plate* 64)
                                                      Floating Pondweed; SM-n
        CC Floating leaves mostly 1-3 cm long and up to about 1 cm wide;
            submerged leaves up to 6 cm long   *Potamogeton diversifolius*
                                                      Diverseleaf Pondweed; Sn

Typhaceae (includes Sparganiaceae)--Cattail Family

There are two genera in this small family. In *Typha* (cattails), the leaves are somewhat similar to those of grasses, and the pithy, upright stems arise from creeping rhizomes rooted in mud in swamps and at the edges of lakes, ponds, and ditches. The flowers are small and packed tightly into a cylindrical inflorescence. The staminate flowers, intermixed with slender hairs, have 2-5 stamens, and are above the pistillate flowers, each of which produces a small, 1-seeded dry fruit. There are neither petals nor sepals.

In *Sparganium* (burreeds), the stems and leaves may be mostly submerged or floating, but even when this is the case, the inflorescence is raised above the surface of the water. Globular clusters of pistillate flowers, each of which produces a 1-seeded dry fruit, are below the clusters of staminate flowers, which have 3-5 stamens.

A   Inflorescence with several separate globular clusters of staminate
    flowers in its upper portion, and clusters of pistillate flowers in its
    lower portion; fruit 2-20 mm long
    B   Inflorescence branched; pistils with 2 stigmas; leaves, if
        straightened, not reaching beyond the top of the inflorescence; up
        to 2.5 m tall   *Sparganium eurycarpum* (*Plate* 110)
                                                             Giant Burreed
    BB Inflorescence not branched; pistils with 1 stigma; leaves, if
        straightened, often reaching beyond the top of the inflorescence; up
        to 1 m tall   *Sparganium emersum* [*S. multipedunculatum*]
                                                             Burreed; Sn-n
AA Inflorescence cylindrical, a continuous column of staminate flowers
    above a column of pistillate flowers; fruit less than 1 mm long
    B   Staminate and pistillate portions of the inflorescence usually not
        separated by a gap; pistillate flowers green when fresh; leaves 10-
        25 mm wide   *Typha latifolia* (*Plate* 64)
                                                          Broadleaf Cattail
    BB Staminate and pistillate portions of the inflorescence separated by a
        gap of at least 5 mm; pistillate flowers yellow or brown when fresh;
        leaves 4-18 mm wide
        C   Inflorescence usually shorter than the leaves; pistillate flowers
            usually brown when fresh; leaves 4-12 mm wide *Typha angustifolia*
                                                          Narrowleaf Cattail
        CC Inflorescence usually at least as long as the leaves; pistillate
            flowers yellow to orange-brown when fresh; leaves 6-18 mm wide
                                                         Typha domingensis
                                                          Southern Cattail

Zannichelliaceae--Horned-pondweed Family

The only member of Zannichelliaceae in our region is *Zannichellia palustris* (*Plate* 110), Horned-pondweed, found in ponds and sluggish streams. It resembles some pondweeds (Potamogetonaceae) in having extremely slender stems, comparably narrow leaves, and tiny flowers. The leaves, however, are opposite, and the flowers are produced in the leaf axils, rather than in terminal inflorescences. There is usually a staminate flower that consists of a single stamen and a pistillate flower with 3 or more pistils, these being located above a broad, nearly cup-shaped bract.

Zosteraceae--Eel-grass Family

Eel-grasses are mostly submerged marine plants, exposed only at low tide. They have creeping rhizomes, and narrow alternate leaves that arise in 2 rows from the stems. The flowers, without petals or sepals, are concentrated on one side of a flattened spike that is at first enclosed within the leaf sheath. Staminate flowers have a single anther; pistillate flowers produce a 1-seeded fruit.

A   Leaves generally more than 4 mm wide, and sometimes more than 10 mm
    wide; each inflorescence with both pistillate and staminate flowers;
    mostly in bays, rooted in mud or sand
                            Zostera marina [includes *Z. marina* var. *latifolia*]
                                                                    Eel-grass
AA Leaves not more than 4 mm wide; pistillate and staminate flowers on
    separate plants; on rocky shores where there is considerable wave action
    B   Leaves 1-2 mm wide, often at least half as thick as wide; flowering
        stems 30-100 cm long   *Phyllospadix torreyi*
                                                            Torrey Surfgrass
    BB Leaves 2-4 mm wide, and not half as thick as wide; flowering stems up
        to 20 cm long   *Phyllospadix scouleri* (*Plate* 110)
                                                            Scouler Surfgrass

ABBREVIATIONS AND GLOSSARY

Abbreviations

af	Africa	mx	Mexico
Al	Alameda County	na	North America (excluding
as	Asia		Mexico)
au	Australia/New Zealand	Na	Napa County
CC	Contra Costa County	PR	Point Reyes
cm	centimeter (about 2/5	sa	South and Central America
	inch)	SB	San Benito County
eu	Europe	SCl	Santa Clara County
eua	Eurasia	SCM	Santa Cruz Mountains
La	Lake County	SCr	Santa Cruz County
m	meter (about 39 inches)	SF	San Francisco County
Ma	Marin County	SFBR	San Francisco Bay Region
MD	Mount Diablo	Sl	Solano County
me	Mediterranean region	SLO	San Luis Obispo County
Me	Mendocino County	SM	San Mateo County
MH	Mount Hamilton	Sn	Sonoma County
mm	millimeter=0.1 centimeter	x	a plant considered to be
MM	Montara Mountain		a hybrid, but to
Mo	Monterey County		which a Latin species
MT	Mount Tamalpais		name has been applied

Abbreviations below, used to designate rare or endangered status, are from Skinner, M. W., and B. M. Pavlik. (editors), 1994, Inventory of Rare and Endangered Vascular Plants of California, fifth edition. Special Publication no. 1, California Native Plant Society, Sacramento.

1a presumed extinct in California (last recorded sighting shown in parentheses)
1b plants rare, threatened, or endangered in California and elsewhere
2 plants rare, threatened, or endangered in California, but more common elsewhere
4 plants of limited distribution--a watch list

Glossary

Achene. A dry (and usually hard) 1-seeded fruit that does not split open. Achenes are especially characteristic of the Sunflower Family, but are also found in some other families.
Acorn. A hard, 1-seeded nut whose base is enclosed by a scaly cup
Alternate. Referring to leaves or branches that originate singly at the nodes, rather than in pairs or whorls
Annual. A plant that produces flowers and fruit in its first year, then dies
Anther. The sac or sacs in which a stamen produces pollen
Awn. A bristle. The term is commonly applied to bristles on glumes and florets of grasses, and on florets of sedges.
Banner. The uppermost petal (usually broad) in flowers of the Pea Family; sometimes applied to a similar petal in other irregular flowers
Basal leaves. The leaves that originate at the very base of a plant. They often form a compact cluster, and may be the only substantial leaves on the plant.
Biennial. A plant that does not produce flowers and fruit until the second year, then dies.
Bipinnately compound, bipinnately lobed. *See* bipinnate
Bipinnate. Referring to leaves that are divided twice in a pinnate manner
Bract. A modified (and usually much reduced) leaf below a flower or an inflorescence
Branchlet. A small branch; in some grasses, equivalent to a pedicel
Bur. A dry fruit covered with spines, scales, or hooks
Calyx cup, calyx tube. The cup or tube formed by union of the sepals
Calyx lobes. The free portions of sepals that are united for part of their length
Calyx. The collective term for the sepals of a flower, whether these are separate or united
Capsule. A type of fruit that consists of more than one chamber and that cracks open when dry
Catkin. A condensed inflorescence of flowers that lack petals, typical of willows, birches, alders, and some other trees and shrubs

303

petiole

alternate,
petioled

opposite,
sessile

whorled,
sessile

Leaf Arrangements

Chlorophyll. The green pigment that enables plants to absorb the light
    energy needed for synthesis of organic compounds
Composite head. In the Sunflower Family, an aggregation of several to many
    sessile flowers attached to a disk-shaped, conical, or concave
    receptacle. The receptacle is usually surrounded by bracts called
    phyllaries.
Compound leaf. A leaf with 2 or more completely separate leaflets, as in a
    clover, rose, or pea
Corolla lobes. The free portions of petals that are united for part of
    their length
Corolla. The collective term for the petals of a flower, whether these are
    separate or united
Corymb. A more or less flat-topped inflorescence in which the pedicels of
    the outer flowers, which generally open first, are longer than those
    of flowers closer to the center
Crown. The persistent base of a perennial herbaceous plant; also applied to
    the top of a tree
Cyme. A flat-topped or convex inflorescence in which the central or
    uppermost flowers open first, and the outermost or lowermost flowers
    open last
Deciduous. Falling after its function has been performed; most often
    applied to leaves that drop off, and to trees or shrubs that lose
    their leaves in autumn
Dichotomous. Referring to stems that branch into two equal or nearly equal
    divisions
Disk flowers. In the Sunflower Family, flowers that have tubular corollas,
    as distinct from ray flowers, in which the corollas are flattened
    out. Some members of the family have a central mass of disk flowers
    surrounded by ray flowers; in others, all the flowers are of the
    disk-flower or ray-flower type.
Evergreen. Retaining leaves from one year to the next (as opposed to
    deciduous)
Filament. The slender stalk of a stamen; also applicable to any threadlike
    structure
Floret. A small flower, especially of a grass or sedge
Flower head. In the Sunflower and other families, an aggregation of several
    to many flowers
Fruit. The ripened, seed-containing structure into which a pistil of a
    flower develops. (In some plants, such as apples and pears, the
    fruit consists partly of a pistil, partly of a fleshy receptacle
    that envelops the pistil. Some so-called fruits, such as
    blackberries, are actually aggregates of fruits.)
Glandular. Referring to a hair, bump, or pit that secretes a sticky
    substance
Glumes. In the Grass Family, the two bracts below the florets of each
    spikelet
Grain. In the Grass Family, the 1-seeded fruit enclosed in the floret
Head. An especially compact inflorescence. In this book, primarily used to
    describe inflorescences in the Grass Family. *See also* Flower head

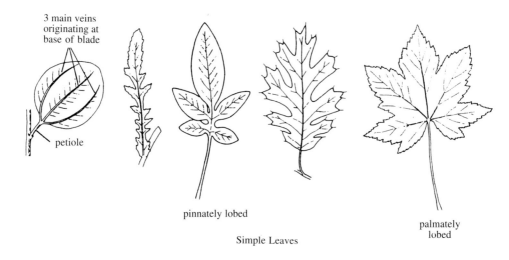

3 main veins
originating at
base of blade

petiole

pinnately lobed

palmately
lobed

Simple Leaves

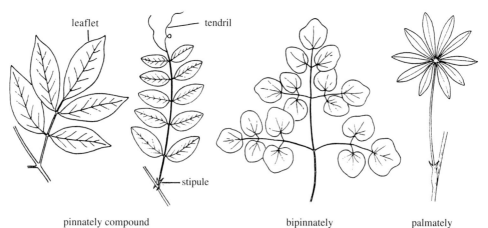

leaflet

tendril

stipule

pinnately compound

bipinnately
compound

palmately
compound

Compound Leaves

Herb. A plant that does not have woody stems, at least above ground

Indusium. In ferns, a fold or shieldlike structure that covers a sorus, at least when the sporangia are young

Inflorescence. A cluster of flowers, or several clusters, on one plant

Internode. The portion of a stem between two nodes. *See* node.

Involucre. In the Sunflower Family, one or more circles of phyllaries (bracts) that partly enclose the flower head; in some other families, a circle of bracts below a cluster of flowers

Irregular. Applied to a flower and a corolla or calyx in which the petals and sepals (or corolla lobes and calyx lobes) are of unequal size and shape, as the petals are in a snapdragon or sweet pea

Keel. In the Pea Family, a structure formed by the union of the two lower petals (the keel partly encloses the pistil); in other cases, a ridge, such as is formed when a bract or the glume of a grass is folded, or when the midrib of a bract is pronounced

Leaf. Generally the main food-producing structure of plants, usually composed of a stalk (petiole) and expanded surface (the blade)

Leaf axil. The upper side of the junction between a stem and a leaf, where a branch, flower, or inflorescence may originate

Leaf blade. The flat, expanded portion of a leaf (excluding the petiole)

Leaf sheath. The base of a leaf blade or a broad petiole that wraps around the stem. The character of leaf sheaths is especially important in identification of grasses and sedges.

Leaflet. Each division of a compound leaf

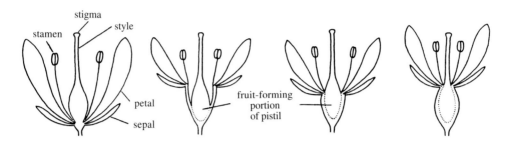

fruit-forming portion
of pistil free of
calyx

fruit-forming portion of pistil fused, to varying extents,
with the calyx

Structure of a Flower

Ligule. In the Grass and Arrow-grass Families, a collarlike outgrowth that
    originates at the base of the blade and partly encircles the stem
Lip. The upper or lower portion of an irregular, united corolla, as in the
    Mint, Orchid, and Snapdragon families
Lobe. One of the deeply separated divisions of a leaf, such as that of a
    maple or sycamore. (The separations are not so deep as to make the
    leaf compound.) See also Calyx lobes, Corolla lobes
Needle. A narrow, stiff leaf of a pine, fir, or other cone-bearing tree
Nerves. In the Grass Family, riblike lengthwise thickenings on florets or
    glumes
Node. The "joint" of a stem, where one or more leaves are attached and
    where a branch, flower, or inflorescence may develop; in the Grass
    Family where spikelets attach to the rachis of the inflorescence
Nutlet. A small, dry, 1-seeded fruit, or comparable structure into which a
    fruit separates early in its development, as is typical of the
    Borage Family
Opposite. Referring to leaves or branches that originate in pairs or whorls
    at the nodes
Palmate. Divided in such a way as to resemble a hand, with fingers spread
    (used in describing the way lobes or principal veins of some leaves
    are arranged)
Palmately compound. Referring to compound leaves in which the leaflets are
    arranged in a palmate pattern, as in a lupine
Panicle. A corymb, cyme, or raceme type of inflorescence in which the
    pedicels branch (thus compound corymb, cyme, or raceme)
Pappus. In the Sunflower Family, modified sepals, appearing as scalelike,
    bristlelike, hairlike, or plumose structures at the top of the
    achene
Parasitic. Drawing nourishment from another plant, as members of the
    Mistletoe and Dodder families do
Pedicel. The stalk of each individual flower in an inflorescence; in the
    Grass Family the stalk of a spikelet
Peduncle. The stalk of a flower that is born singly, or the stalk of an
    inflorescence that consists of several to many flowers
Perennial. Applied to plants that live indefinitely, as opposed to those
    that live only one or two years
Perianth segments. The divisions of a corolla or calyx
Perianth tube. Tube formed by the union of the petals and sepals
Perianth. The complex formed by the corolla and calyx. The term is
    especially useful in connection with flowers in which the divisions
    of the corolla and those of the calyx are similar, as they are in
    some members of the Lily Family.
Petals. The separate segments of a corolla
Petiole. The stalk of a leaf
Phyllaries. In the Sunflower Family, the bracts that surround the
    receptacle of a flower head
Pinnate. Referring to compound or lobed leaves in which either the leaflets
    or the lobes are arranged on both sides of the rachis (as in a pea
    or rose)
Pinnately compound, pinnately lobed. See Pinnate
Pistil. The portion of a flower in which a seed or seeds are eventually
    produced. (A flower may have more than one pistil.)

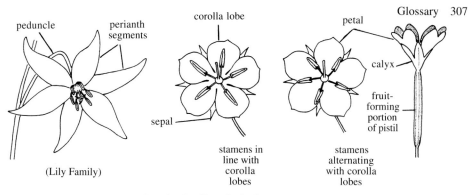

Regular Corolla or Perianth

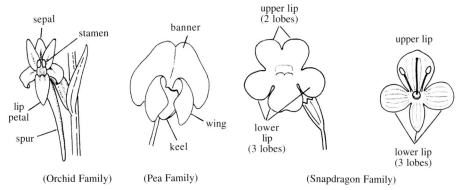

Irregular Corolla

Flower Types

Pistillate. Referring to flowers that have one or more pistils, but no
      stamens; also applied to plants that produce only pistillate flowers
Pollen. Microscopic reproductive structures produced by stamens of
      flowering plants and by certain short-lived conelike structures of
      gymnosperms. Each pollen grain is at first a 1-celled spore. Before
      it is released, however, the nucleus of the cell divides, initiating
      the formation of a microscopic male plant. Development is not
      completed until the pollen grain reaches the stigma of a flower or
      the scale of a cone that is destined to produce seeds.
Prostrate. Applied to stems that lie on the soil
Quadripinnate. Referring to a structure, usually a leaf, that is divided
      (lobed or compound) four times in a pinnate manner
Raceme. An inflorescence in which the flowers are borne all along the
      peduncle, on short pedicels of more or less equal length
Rachis. The main axis of a compound or nearly compound leaf; in the Grass
      and Sedge families, the main axis of a spikelet
Ray flowers. In the Sunflower Family, the marginal flowers, with flattened
      corolla, that surround the disk flowers in the central portion of
      the flower head. The distinction between ray flowers and disk
      flowers is clearly seen in daisies, asters, and sunflowers. Some
      members of the family (such as thistles) lack ray flowers; in
      others, all flowers are of this type.
Ray. In the Sunflower Family, the flattened corolla of a ray flower; in the
      Parsley Family, the primary stalks of a compound inflorescence
Receptacle. The portion of a flower peduncle or pedicel on which the flower
      parts are mounted. In the Sunflower Family, several to many sessile
      flowers are attached to each receptacle.
Regular flower. A flower that has perfect or nearly perfect radial
      symmetry, as in a lily, poppy, wild rose, or morning-glory
Resin pit. A microscopic glandular pit on the scalelike leaves of some
      conifers of the Cypress Family
Rhizome. A horizontal underground stem from which leaves or other stems
      arise
Saprophyte. A plant that lacks chlorophyll (the leaves therefore not
      green), subsisting on decaying organic matter, usually with the aid
      of fungi that penetrate its roots

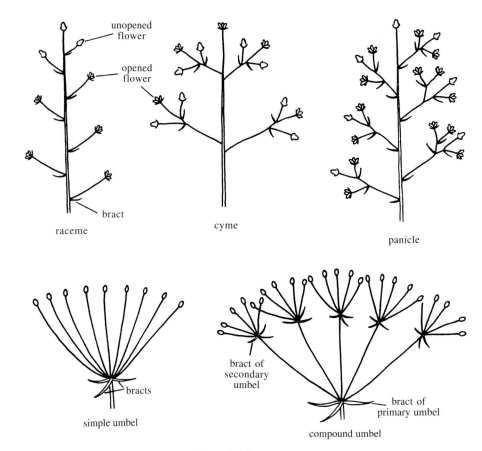

Types of Inflorescences

Sepals. The segments of a calyx
Sessile. Attached directly to a stem or some other structure (applied
        mostly to leaves that lack a petiole and to flowers that lack
        pedicels)
Shrub. A perennial plant whose stems are woody, but which usually does not
        have a distinct trunk and is not often more than 3 m tall. In this
        book, shrubs are defined as being at least 1 m tall. Some large
        shrubs may have the form of a small tree, so the distinction between
        the two categories is not absolute.
Sorus. In ferns, a cluster of sporangia. See also Sporangium
Spikelet. In grasses, each unit of an inflorescence, consisting of a pair
        of glumes and 1 to several florets; in sedges, a similar group of
        florets, but without glumes
Sporangium. In ferns, fern allies, and lower plants, a structure within
        which spores are formed
Spur. A hollow, saclike, or tubular extension of a petal or sepal, as in a
        larkspur, columbine, or violet
Stamen. The pollen-producing part of a flower, usually consisting of an
        anther and filament. Most flowers have at least 2 stamens, and some
        have many.
Staminate. Referring to flowers that have stamens, but no pistils; also
        applied to plants whose flowers are staminate
Sterile. Referring to modified stamens that do not produce pollen, and to
        flowers that do not function in reproduction
Stigma. The sticky tip of a pistil, which traps pollen
Stipules. Paired appendages at the base of a petiole. These may be so large
        as to resemble leaves.
Stomate. A microscopic pore on the surface of a leaf or herbaceous stem.
        Stomates permit escape of water vapor and exchange of oxygen and
        carbon dioxide.

Style. The usually slender upper portion of a pistil, at the tip of which
the stigma is located. In many plants, the pistil has more than 1
style.

Succulent. Referring to leaves and stems that are thick and juicy

Tendril. A slender, twining structure, usually part of a leaf, by which a
climbing plant, such as a pea or grape vine, clings to its support

Tripinnate. Referring to a structure, usually a leaf, that is divided
(lobed or compound) three times in a pinnate manner

Tripinnately compound, tripinnately lobed. *See* tripinnate

Two-lipped. Referring to flowers in which the corolla, consisting of united
petals, and often also the calyx, have distinctly different upper
and lower portions, as in a snapdragon or mint

Umbel. A nearly flat-topped inflorescence in which the pedicels of the
flowers originate at the top of the peduncle; in compound umbels,
the peduncle branches into rays which support the pedicels

Urn-shaped. Referring to a corolla or calyx that is more or less cup-
shaped, but also slightly constricted near the opening

Vegetative. Non-reproductive

Vein. In a leaf or petal, a branching structure consisting of tissues that
distribute water and nutrients. In the case of petals, the term is
often applied to colored lines.

Vernal. Referring to spring. The term is most commonly used, however, in
connection with ponds that become filled with water in winter or
spring, and that dry out in summer.

Whorl. A type of leaf or branch arrangement in which 3 or more of these
structures originate from a single node

Wing. A thin expansion of a dry fruit, as in that of a maple; also applied
to the 2 side petals of some flowers in the Pea Family

INDEX

Common Names of Plant Families

In the text, Families are arranged alphabetically according to their Latin names under Dicotyledons and Monocotyledons. They are not listed in the Index unless they have Subkeys or are synonyms of currently accepted names.

General Index

In addition to species names that are in use at the present time, this Index lists some synonyms.